Treasures Hidden
Within the Empire

Martin Concoyle

 www.trafford.com

North America & international
toll-free: 1 888 232 4444 (USA & Canada)
fax: 812 355 4082

These are new intellectual treasures which must be excluded from the main-stream by the materialistic, militarist master-investor-class of the empire

Or

A "Mathematical Map" Which Gives a Guide…

{which contains the descriptive context of physics as a proper subset, and which leads into higher-dimensions by means of a context of both microscopic and macroscopic stable shapes (it has nothing to do with string theory, see below)}

…. Toward Our Spirit's Natural Context of Both Experience and Creativity

(it describes the higher-dimensional context of the mystical proclamation "to see the world as it really is," and the description is a mathematical model, which leads to an amazing treasure of possible human experience, and a new context for human intent)

Or

Stable Patterns and Quantitative Consistency Can Be Used in a Valid and Logically Consistent Descriptive Context so as to Form an Invitation to New Creative Contexts, so as to Re-interpret the World in an Abstract but Measurable Context of Many-Dimensions, and to Conjecture a New Set of Stable System Components which are Stable Shapes, within a Descriptive Construct which Can Possess Actual Meaningful Content

[It is a context which solves the problem of providing an accurate and sufficiently precise quantitative description of the stable many-(but-few)-body systems which possess stable spectral-orbital properties, where the solutions are the stable math shapes of discrete hyperbolic shapes, which model the spectral, angular-momentum, and spin-rotational properties associated to the substructures of these types of very stable systems, where these stable many-(but-few)-body spectral-orbital systems exist at all size-scales from: nuclei, to general-atoms, to molecules, to the lattice structures for liquid-condensed-material systems, to crystals (condensed solid material), to a new global model for system-containment in regard to living-systems, where chemistry exists in a new context as an array of lattices, which are, in turn, controllable by a global living-system construct, which generates its own energy, to an actual model for a mind, to the solar-system, to the orbits of stars in galaxies (dark matter), to sets of group conjugations acting on SU(11,1) where these conjugation group actions define the expansion of the, so called, dark energy, and to create new models of elementary particles, where there are new models for elementary-particles, which are also stable shapes]

These are new intellectual treasures which must be excluded from the main-stream by the materialistic, militarist master-investor-class of the empire, where these are (new) intellectual treasures are denigrated by the publicists and agents of the empire, where this is done by means of using the deep illusions, which the propaganda-education-indoctrination system is capable of weaving into the minds of the people of the empire, and the subsequent intellectual inertia of the society, which results from the repetitive and circularly re-enforcing the claims of the media, and it results in a refusal (by the people) to make obvious decisions about the invalidity of modern physics, and because modern physics has failed, then to try to develop new sets of assumptions, contexts, interpretations…, and new ways in which to organize the "patterns and language" of a quantitatively based descriptive language…, so as to develop a more accurate and wide ranging descriptive-language based on a general set of ideas which result in both a new set of practically useful descriptions and a new set of practically creative contexts

Note that an interesting thing about

"what are considered to be scholarly and valid intellectual efforts by a society" can deeply affect that same society's vulnerability to the onslaughts of the western empire.

That is, one can be sure that a country (or a culture) will not be able to resist the west if the culture adopts

either A notion of economics "as being a valid descriptive context which explains the structure of a society," such as Marx's theories based upon "the collectivist nature of capitalism," but economics is not an objective, quantitatively based description about how money functions in society, rather it is an arbitrary descriptive context which identifies an arbitrary way in which society is organized, which is upheld through both propaganda and violent institutions.

or The society takes seriously the ideas of modern physics, and its associated set of math constructs (whose sole purpose, which seems to be hidden from its actual practitioners, is to detonate a mythological gravitational singularity)

By adopting these viewpoints this will result in the society…, which has not been within the western culture…, becoming a part of the western empire.

A "Mathematical Map" Which Gives a Guide…

{which contains the descriptive context of physics as a proper subset, and which leads into higher-dimensions by means of a context of both microscopic and macroscopic stable shapes (not string theory, see below)}

... Toward Our Spirit's Natural Context of Both Experience and Creativity

(it describes the context of the mystical proclamation "to see the world as it really is," and the description is a mathematical model, which leads to an amazing treasure of a wide ranging set of possible human experience, and a new context for human intent and creativity)

Or

Stable Patterns and Quantitative Consistency Can Be Used in a Valid and Logically Consistent Descriptive Context so as to Form an Invitation into New Creative Contexts, so as to Re-interpret the World in an Abstract but Measurable Context of Many-Dimensions, and to Conjecture a New Set of Stable System Components, which are Stable Shapes, within a Descriptive Construct which Can Possess Actual Meaningful Content [It is a context which solves the problem of providing an accurate and sufficiently precise quantitative description of the stable many-(but-few)-body systems which possess stable spectral-orbital properties, where the solutions are the stable math shapes of discrete hyperbolic shapes, which model the spectral, angular-momentum, and spin-rotational properties associated to the substructures of these types of very stable systems, where these stable many-(but-few)-body spectral-orbital systems exist at all size-scales from: nuclei, to general-atoms, to molecules, to the lattice structures for liquid-condensed-material systems, to crystals (condensed solid material), to a new global model for system-containment in regard to living-systems, where chemistry exists in a new context as an array of lattices, which are, in turn, controllable by a global living-system construct, which generates its own energy, to an actual model for a mind, to the solar-system, to the orbits of stars in galaxies (dark matter), to sets of group conjugations acting on SU(11,1) where these conjugation group actions define the expansion of the, so called, dark energy, and to create new models of elementary particles, where there are new models for elementary-particles, which are also stable shapes]

This book is dedicated to my wife M. B.

The hidden treasure, to which the title refers, is the set of: ideas, constructs, new ways in which to organize language around stable shapes, the new central idea through which existence is to have its description based (different from the idea of materialism, reduction, and partial differential equations), and methods for providing a wide ranging, accurate, and practically useful set of precise descriptions of physical systems, which are identified in this book, and which can be used to precisely describe the observed stable properties of the world. This new descriptive context contains the valid (or true) parts of the current context, along with a number of corrections, as a proper subset of the new description, so that the new set of assumptions can account for, in a rational, and logically consistent, and in a quantitatively consistent context the observed set of stable spectral-orbital properties of a general many-(but-few)-body physical system, as well as providing a new model, in regard to the "context of set-containment," for a living-system, while modern (2014) overly authoritative physical and biological science is based-on materialism, reductionism, and that all measurable descriptions depend on (finding) solutions to partial differential equations, which, in turn, are identified by physical law, but, in general, the pde are: non-linear, non-commutative and indefinably random so there is no content in modern science.

These new ideas depend on some of the simplest and most elementary math patterns based on some of the simplest stable shapes defined within metric-space which possess natural coordinates which have the simplest sets of continuously local independent coordinates, natural coordinates which continuously diagonalize a metric-invariant metric-function's local matrix-representation, and whereas, in the current descriptive context, the basic shapes upon which the new description is centered can be related to pde where the pde's associated to the solution functions of these shapes are linear shapes which are identified by differential-forms.

Not only can this new descriptive context be applied to: elementary-particles, the properties of nuclei, general atoms, molecular properties and their shapes, crystals, chemistry, chemistry in liquids, and it provides: a global and many-dimensional model of life, and new ways to view living systems, but they are ideas which also apply to: the stability of the solar-system, the effect of, so called, dark matter, on the motions of stars in galaxies, the mind, an alternative interpretation of, so called, dark energy, but furthermore it expands the context of creativity of life beyond all of these primary categories which might be associated to both physical and living existence.

Contents

Part II
(social commentary)

Note: The symbol pde stands for the words "partial differential equation".

Preface

A mathematical model or "map" of measurable properties which are stable and which lead to our spirit's natural creative context, in regard to both experiencing many new connections within the world, eg higher-dimensional experiences, many-worlds, and creating existence itself, where the "map" is a description of a many-dimensional descriptive context ie describing an over-view of a new containment set's properties, wherein a metric-space and its associated (isometry) fiber group exist in a slightly complicated construct of metric-space sets and an associated complex of set-containment, and set-separation (ie the sets possess a property of being independent), and which is defined over many-dimensional levels, wherein the shapes of "the metric-space building-blocks" (of this new descriptive context) for any dimension can be either large or small, though needing to satisfy the properties of set-containment, and where this new contexts takes over the models of "measurable descriptions of physical properties" from the partial differential equation (or pde) which, along with materialism, is now used as the basis for describing material-interactions, which have been assumed to be the cause for the formation of stable systems, but this has not been demonstrated within the context of the pde, and its associated assumption of reduction of material to a random structure of elementary-particles (as the basis for the (or of the) organization and order of material-components)

That is, the metric-spaces and their associated fiber groups..., which are defined over a many-dimensional context..., are the main attributes which are needed for the descriptions of observed physical properties of the world, and not pde's, where in the context of a pde's the metric-space defined by the idea of materialism is seen as the containment sets for the pde's model of local measuring.

A similar set of pde based descriptions of material-interactions also exist in the new descriptive context, though it is altered to be consistent with the discreteness of the time-intervals defined by the periods of the spin-rotations of (what are now, in the new description) metric-space states, but in the new context... in which the new "discretely dependent pde based description of material-interactions" ..."the pde described material-interactions," themselves, are not the main cause of (or for) the formation of stable physical system components. For example, the elliptic 2nd-order pde is the usual pde which is associated with the descriptions of bounded systems, but the context of an elliptic 2nd-order pde (and/or its solution-functions) does not identify a cause for a stable system to become bounded (that is, the solutions do not describe a formation of a new stable system), rather (in the new descriptive context) the elliptic 2nd-order pde describes the "perturbations" of the various sets of 2-body orbits (in a stable many-(but-few)-body system), where the perturbations are made in regard to the geodesic orbits of condensed material, which exists within a stable metric-space shape (which, in turn, defines the set of stable geodesic orbits for the various many-(but-few)-bodies which compose the system), where the way the stable metric-space shapes dominates the interaction structure, so as to allow new organizations of bounded-ness for the components in a newly formed system, where the newly formed stable shape identifies a stable material-component, which is itself occupied by other lower-dimension components, and where the new stable shape's formation is allowed within the newly defined discrete context of the pde, where in the discrete context (of a pde) one can deal with separate interacting components, but which come to be "in resonance" with an allowed set of spectral values, which, in turn, are defined outside of the context of the pde, except when (or unless) the pde is modeled to be discrete, in its (the pde's) new fundamental definition a new combination of discrete geometric properties of an interacting-dynamic system comes to be in resonance with the finite spectral set (ie a finite spectral-set which defines the entire set of stable component shapes) and when the interaction is within a set of the proper energy bounds, in the interaction-structure, then this allows for a new stable material-component to form, due to it being in resonance with a finite spectral set, where this property of being "in resonance" is defined between discrete time intervals (which, in turn, is a small time interval defined by the time period of the rotation between [or around] the pairs of opposite metric-space states)

The context which the metric-spaces define is a context of the generation of a countable quantitative set within which the measuring of physical properties is defined (and/or contained), ie thus one does not need a continuum to model measurable properties of physical systems, ie where these are the measured properties which distinguish (or identify) a system by its identifying "spectral (or orbital) set." That is, hyperbolic metric-spaces, modeled as stable discrete shapes, in turn, are related to Coxeter's classification of the entire set of discrete hyperbolic shapes for all the dimensions for which the discrete hyperbolic shapes exist.

Note: The last discrete hyperbolic shape exists in a hyperbolic metric-space of dimension-10, and this shape is formed (or identifiable) in a hyperbolic metric-space of dimension-11, eg the shapes of 2-surfaces are identified in 3-space. Metric-spaces which are discrete hyperbolic shapes, as identified by Coxeter, are used to "partition" an 11-dimensional hyperbolic metric-space in order to define a countable quantitative-set which can be used, in turn, to define measuring for a precise descriptive language, which is quantitatively consistent and it is associated to a set of stable patterns…* (*this is the central idea)

Such a new logically consistent, valid, and meaningful math construct based on both quantitative consistency and stable patterns is excluded from the academic research halls of the empire.

This is because, though the math community in the empire deals with a fairly wide range of patterns, the empire itself is totalitarian and arbitrary, and held together by violence, so it (believes it) must focus on military instruments and the control of all communications channels.

The research of the empire is always controlled by through funding and administrators and spies and agents of the police-state, where E Teller is a prime example of an agent for the empire, of secret-police, determining the way in which academic institutions would have their knowledge structured to serve the empire's greatest interests, ie expanding their interests through violence

The empire, is the western culture, and it was first defined by Rome, where Rome emerged as a community regimented militarily, and its law was based on property rights and minority rule, ie which emerged from the process of property owners fighting one another for control of the property, so that the community was a collectivist regimented community, which was run by the aristocracy, where each aristocrat had their own personal militia, and the rule (or law) was determined by the aristocracy either fighting or meeting in a Senate. Since their lives were based on lying, stealing, and murdering, and they had "had enough murdering between themselves," so they eventually based their collectivist society on murderous expansions, and looting, but also being based on a brick-laying technology, wherein they would build: town squares, water systems, sewers, and roads, so the conquered-people liked the new civic structures, which had the effect of causing the community to become built around a fixed set of categories, related to their fixed life-styles, in turn, built around a narrowly defined technology, ie limited categories based on limited knowledge and limited contexts for creativity.

This regimentation of life, into a small set of fixed categories, somewhat loosely organized around technical capacities, along with the propaganda-education-indoctrination system of (Paul's and) Constantine's church, was what was left when "Rome went east (to Constantinople, where Constantine wrote the Bible)," leaving the western empire to the church, so that the bankers and their debt enforcers emerged from this limited set of categories, which defined a regimented life-style for the communities of what was now Europe.

The bankers thrived during the process of "mechanization of the means of production" in regard to the limited set of categories, about which the life-style of the European society was based, and, subsequently, the Anglo-US society was also based on the investor-class's investments in the mechanization of the narrow set of fixed categories of technical development, upon which the society was organized in a very regimented fashion. This defines the US society. Furthermore, economics is more about fixing a small set of categories which define a society's life-style, and being able to force the people into these life-style categories, than economics, actually, being about invention, production, and distribution, ie modern markets are all about nearly total control of people's thoughts and actions through the propaganda-police-state institutions.

But consider the difference between the investor-class, who get someone to slightly manipulate an off-the-shelf technology, so as to make communication systems and to subsequently, become powerful and rich, compared to the people, such as Faraday and Tesla who did the most to allow these communication systems to come into being, but who died poor, if not paupers. One sees that the main idea of economics is not knowledge and creativity but rather it is still about the power related to lying, stealing, and murdering, codified by property rights and minority rule, in a collectivist and (militarily) regimented society whose collective actions are to support the interests of the investor-class, with the prime mover still being violence.

Police-state control, as characterized by military funding and administrative agents, such as E Teller, along with the formation of axiomatic-formalization, by D Hilbert, are the social constructs which lead researchers to, essentially, only consider quantitatively inconsistent constructs, and thus, meaningless descriptive contexts, as being their focus of research. Nonetheless, this formalization and associated coercion leads to nearly complete control of the set of descriptions, in regard to certain categories as being identified, "as knowledge," and the set of creative contexts (mostly about military and oil) to which the "properly identified set of knowledge" is to be used.

Where now (2014) "knowledge" implies absolute authoritative truths, which is a very anti-scientific construct. Science is really about equal free-inquiry, while dogma is about fixed limited contexts of existence.

But the treasures identified in this book are the treasures of knowledge, associated to equal free-inquiry, and its relation to new practically creative contexts (ie not lyrical creativity), ie one wants knowledge to be tied to experience and "practical" creativity, where, now, "practical creativity" might mean in higher-dimensions.

The geometric properties associated to the real-numbers (lines) and the complex-numbers (a 2-plane or a circle which defines a ray [or half-line]), together, imply that the geometric patterns of the lines, line-segments, and circles, can be quantitatively consistent, and, subsequently, only shapes composed of these simple shapes.... (lines and circles) which always possess the local property of orthogonality (or independence of the variables, or continuous local commutativity) between any two pairs of the natural coordinates of these simple shapes (coordinates within the shape) (except possibly at a single point)...., can be quantitatively consistent shapes, where the quantitative sets of coordinates are both independent and parallelizable (ie the parallel transport of any local coordinate direction around a local closed coordinate curve is commutative, ie the beginning vector is commutative with the parallel-ly transported vector) and these simple shapes are also the main set of stable shapes as defined in the Thurston-Perlman geometrization theorem, namely, the set of discrete hyperbolic shapes

The set of natural stable shapes, which are defined over many-dimensions, can be organized to "partition" both the different dimensional levels and the different subspaces of same dimensional level, where these different dimensional levels and the various subspaces for each particular dimensional level are all a part of an 11-dimensional hyperbolic metric-space.

A finite set of these stable metric-space shapes can be used to define:

1. (The generation [or building] of a quantitative set) A finite spectral-set, from which a number-field..., which is similar to the rational-numbers..., can be generated, where this generated quantitative-set is the basis for the quantitative-set within which all measurable properties of the physical systems...., which are contained in the various independent sub-metric-spaces (in the 11-dimensional space)... can be made where the measuring can be thought of as being made either in a metric-space (shaped) subspace or in an 11-dimensional hyperbolic metric-space, but different topological properties will be "seen," dependent on the containment set of which one assumes to be a part when measuring, where the quantitative-set is applicable to all of these measuring processes, ie the given finitely generated quantitative-set with which any of the existing-system's properties are to be measured, is based on a finite spectral-set defined by the "partition" by a finite number of stable shapes defined within the 11-dimensional space (all measurements can be based on this newly generated quantitative set).

2. (Partition the 11-dimensional space into independent metric-spaces, by using stable metric-space shapes) The set of independent containment-sets..., which are defined for the subspaces of any particular dimensional level..., are stable metric-space shapes, within which the properties of the set's sub-component shapes can be described (where sub-component shapes are, usually, the adjacent one-lower-dimension stable shapes).

3. (The rule of resonance) The existence of a stable shape, or stable material component, or stable metric-space sub-containment-set..., within any particular one of the subspaces of any particular dimensional level which is within the 11-dimensional over-all containment-set hyperbolic metric-space..., depends on the shape being in resonance with the finite spectral-set defined by the "partition" of the 11-dimensional over-all containment-set hyperbolic metric-space.

It needs to be noted that this new descriptive context, when modeled (or chosen) to be inside a 3-space, is consistent with

1. Classical geometrically based physical description

2. It is also consistent with the quantum-randomness of quantum physics, where this is due to the Brownian motions, which the new descriptive language describes for the (local) interactions

of the micro-components, in turn, the Brownian motion is due to the sudden changes in charge distributions which occur at the micro-level of material-interactions, and where Brownian motions for micro-components is equivalent to the randomness assumed by quantum physics (shown by E Nelson, Princeton, 1967)

Note: If one uses the indefinable randomness of quantum physics as a basis for the types of physical descriptions of the properties of one's interest, then it would be recommended to use a set of stable discrete hyperbolic shapes as the set of functions which compose the harmonic function-space one is using to model the physical system (in a quantum physics context). That is, this does not eliminate quantum physics, it simply states that there is a geometric context which will (most likely) be a better fit to a physical system's properties, and this geometric context is, perhaps, best identified as the context which defines a finite spectral-set for the 11-dimensional over-all containment-set hyperbolic metric-space.

3. That is, a locally measured pair of single material-component interactions are geometric, either fields or component-collisions, but now the collisions are not about the small components being reduced to elementary-particles, or simply being elementary-particles, which can only be related to particle-scattering models of particle-cross-sections, in turn, only related to probabilities of particle-collisions,

but rather material-interactions, such as component-collisions, are (in a discrete process) related to sets of spectral resonances, which might exist, and these resonances can enter the description when the (geometric) interaction is within the energy-ranges needed for a new system to emerge based-on the new system being "in resonance" with the finite spectral-set, in any particular interaction.

Furthermore,

There are also new models of the lattice structures, particularly in liquid condensed-matter systems, of some complexity, where these new sets of lattice structures offer new ways to understand the organization of a system's components and in relation to a greater containing (higher-dimension) construct, associated to the system (see chapter 24).

5. It can also be noted that a particle-collision scattering model, based on the events defined by unstable elementary-particle events, is organized around SU(2) x SU(3) fiber group, but where quarks, which are associated to SU(3), are not observable, ie they are outside of science, but SU(2) and SU(3) are naturally fiber groups which can be associated with stable discrete shapes which possess a distinguished point, about which an apparent particle-path can be formed, ie these shapes may be in a brief transition which occurs after a collision, but (one can assume that) these

transitioning shapes are not in resonance with the over-all finite spectral set, but they are brief shapes which possess a distinguished point about which the so called particle-trail is identified, ie they are 2-shapes or 3-shapes associated to the associated discrete subgroups of either SU(2) or SU(3) respectively, and which are a part of the disintegration process caused by a particle-collision, they are always present, since these 2-dimensional and 3-dimensional (complex-coordinate) shapes would also always be present.

Furthermore, the material-particle types (called elementary-particles), which do possess some properties of being stable are the proton, the neutron (how stable?), the electron, and the neutrino, so one should model these components as stable shapes, where the neutrino is seemingly an unbounded shape, but it is a component shape which would (might) be bounded by the metric-space within which the system… "to which the neutrino is a component"… is contained.

This would be a much better interpretation of the observed patterns of particle-collisions, since the construct of particle-physics has <u>not</u> been able to describe the relatively stable properties of the general nuclei based on the, so called, laws of particle-physics. Thus, it is a failed construct, and other constructs should be considered. [but particle-physics is not considered a failed model by the empire, since particle-physics is related to particle-cross-sections, and thus it is directly related to the properties of nuclear weaponry.]

The treasures can also be considered the work done by: Thurston-Perlman, D Coxeter, L Eisenhardt, Einstein, Newton, Gibbs, Faraday, Tesla, Boyle to Carnot, Godel, Poincare (did some good work, but mostly he was too abstract), Hodge, etc etc …… but they are really built on the idea that only the stable, solvable, and controllable math constructs, in regard to pde, have any valid relation to describing the observed stable properties of physical systems, and these pde describe the stable shapes of Thurston-Perlman and Coxeter, ie they are the stable discrete hyperbolic metric-space shapes.

Part I
(more math)

Introduction and Review

(for Review go six pages ahead, also see Chapter 12)

A review of the rather (or comparatively) simple math, and geometrically-dependent construct of a descriptive language, which is to be used to describe the observed stable physical properties of the world (or of existence), where for descriptions of stable properties, especially in a geometric context, ie a context which requires descriptions of spatial positions, it is required that the description be based on the set of stable geometric shapes: cubes, cylinders, tori, and the very stable shapes (which exist in energy-space) composed of toral-components, ie the set of very stable discrete hyperbolic shapes, and where these shapes (of energy) exist over multiple-dimensional levels, ie the description is not based on the idea of materialism.

Such a new descriptive context is required by the laws of both science, and the logic of Godel's incompleteness theorem, (namely) where if there are patterns which a precise language cannot describe then the language needs to be changed, and the change needs to be at the most elementary levels at which languages can be built, ie at the level of assumption, context, interpretation, ways to organize descriptive patterns etc etc.

The set of observed physical patterns which cannot be described, in any valid manner (based on the, so called, laws of physics), are the very prevalent and very stable many-(but-few)-body systems which possess stable spectral-orbital properties, where these stable systems are the most prevalent, and the most fundamental, structures of our experience. Where these stable systems exist at all size scales, and they include the: nuclei, the general atom, the molecule, the crystal, living-systems (both stable and highly controllable, well beyond "what the idea of materialism" is capable of describing), the solar-system, and, apparently, the orbits of stars in galaxies (dark matter) etc, etc, ie these systems should be the most natural focus for physics.

Whereas, now (2014) attempts to re-define mass for particle-physics by a mass-field-particle which is claimed to be seen by interpreting vague relations (of questionable data) to the theoretical descriptions of particle-scattering models the more highly defined images (due to higher-energy collisions) of unstable events (of what measurably descriptive value is an unstable event?), and to then to use such information in order to try to detonate a mythological gravitational singularity, based on the quantitative descriptions of cross-sections of particle-collisions, where these are the types of constructs about which the "technical staff" (the mythological intellectually-superior types of people) is most heavily involved, in their (misguided) endeavors (and their obedient and uncritical viewpoints).

But the new alternative viewpoint, that is, a new stable-geometric and many-dimensional alternative descriptive language is a knowledge, ie is a descriptive language (a new form of knowledge), which is at odds with the desires of the investor-class's investment structures, where the knowledge, which the investor-class wants to use, is based on descriptions of a material existence, where the narrowly defined knowledge of material is used for military and communication instruments, and the expected gradual development of these instruments to which all institutional knowledge of the society are focused so as to serve the interests of the investor-class, the over-lord pay-masters ruling over their police-state domain.

So truly alternative viewpoints are at odds with the interests of the investor-class

Where the investor-class has isolated, the functioning technical personnel (who work at, and control, the technical institutions), by means of the highly constrained cultural constructs of both axiomatic formalization, and peer-review (two ideas [or social constructs] which are authoritarian, and thus, anti-scientific)

And

Whereas, the associated propagandists (associated to the technical intellectual-class, where these propagandists play a role as "the guardians of the absolute truths of the Empire")…, whom legitimize and promote the actions of the technical staff (ie society's, so called, superior-intellects)…, are ignorant, and they are wage-slaves, and the main part of propaganda (the main part of the job of a propagandist) is the development of an illusion of high-social-value…, (they express their belief in personality-cult, which is masked as an expression of [or as a fable about] a very difficult to attain absolute-truth, which only superior intellects can acquire), eg high-value, in regard to an absolute truth, about which the technical staff (the superior-intellects) deals,

And

Thus, they (the associated propagandists) also decide things based on their "faith in the proclamations made by the (also) indoctrinated technical staff," where the technical staff have been isolated, intellectually, by a wall of language based on both axiomatic formalization and peer-review, ie the propagandists express ideas which are mostly an uninformed expression of personality-cult.

Thus, one can see that there will be all sorts of institutional inertia involved in any such a challenge to authority, ie providing an alternative vision about "what is true," where the authority of the intellectual institutions is all about using the knowledge of a culture to both (only) serve the interests of the investor-class, and where those interests are about the relation of military and communication instruments to a scientific model of a material-world, and where this, so called, scientific knowledge is used and controlled by the investor-class.

The result of all this highly controlled and manipulated baloney, is the claim that "this is the only way things can be" but there is the further proclamation that the stable many-(but-few)-body spectral-orbital system "is too complicated to be able to use the current (2014), so called, laws of physics, so as to be able describe these properties," ie the stable many-(but-few)-body spectral systems are too complicated to describe. (ie the claim being made is both that the technical staff is following the absolute truths of science, "the laws of physics," and people in general are violent and selfish [the personality traits which best describe the ruling-class]), where it (ie the proclaimed scientific absolute truth) is really a manipulated illusion, built by the investor-class for their selfish interests, and it (ie building such an institutional structure based on illusions) involves a secret network of: spies, agents, managers, administrators (all of whom are wage-slaves), and they (these institutions of social-control and control-of-technology) are defined in the usual US cultural context of terror, and criminality, emanating from the justice-military-police-state institutions of the US (eg the Bundy-ranch had two groups of the same set of secret agents of the police-state facing one another)

Inequality, social hierarchy, personality-cults, and wage-slavery, will always bring-out the worst traits of the people which compose such a society.

In contemporary society (2014) the "two sides," ie the intellectual-elites and the "top-dog people-of-action," are sort-of represented by "Amy Goodman's Democracy Now" and "Rush Limbaugh," respectively, and both represent arbitrary authority and arbitrary high-value, where, apparently, one side wants "democracy guided by the intellectual-elites" (where the intellectual-elites would continue to serve the same types of interests [the same intellectual constructs] as they did for the bankers-oilmen and military), and the other side, apparently, wants a "Calvinist Theocracy," but neither side seems to stand for anything different than from the M Thatcher phrase, "Things can only be this way," ie the Roman debate between the Roman Emperors: Caesar Augustus and Constantine, which seems to best characterize both Western thought and Western culture, a material-based culture based on both violence and propaganda, which requires some form of expansion to maintain its power.

The one side (the left) is supposedly materialistic and follows the authority defined by the investor-class (ie the only way in which knowledge can be truthfully represented), yet, the intellectual-class wants themselves, ie the intellectual-class, to run-things (they claim to be better technically capable of running the society which is ruled by the investor-class), while the other side claims to not be materialistic, but Calvinism is both material-worship and a collectivist form of personality-cult, and its seems to revere the investor-class, the differences are too slight, so as to not really be able to "distinguish between the two," ie the two, so called, opposite sides of the propaganda-system.

Both sides are composed of deluded hypocrites. One side (the right) is reverent, and protecting traditions, but active, while the other side (the intellectual-class) is authoritarian, protecting the investor-class's vision of very narrow progress, and rational and reflective within the context of established institutions of rationality.

Both sides support a form of a collectivist society: either the God-given rich-few, or the authoritative rational basis of (for) the investor-class's interests.

Both sides are far too narrow, and extremely arrogant, but, effectively, both sides are also extremely timid of the unknown.

A many-dimensional math model, based on both macro-shapes and micro-shapes, which is used to describe the properties of existence, and where the material-world is a proper subset of the new context of description, extends past the idea of materialism, into a new context for both existence and creativity (ie the "science vs. religion" or "materialism vs. idealism" "contest," or debate, loses its context).

Note that it is foolish for the scientific intellectual-class to engage in such a debate (science vs. religion), since there is so much failure of the scientific community in regard to its inability to describe [based on the, so called, laws of physics], the most fundamental stable physical systems, while claiming to be in possession of the basic tenants from which "life will eventually be explained," is simply an example of disgusting arrogance, and a major sign of corruption and dis-honesty on their part.

On the other hand, religion is all about defining an arbitrary hierarchy, and though some of their criticisms of science are valid, particularly, in regard to the parts of science which are based on a descriptive language which, in turn, is built from (or built upon the idea of) indefinable randomness, ie a totally useless technical language, nonetheless if religion is going to criticize science, then religion needs to provide an alternative precise language, which gives the idealist's a valid math model of existence,

But

A truly religious person, especially, a person of action, or a curious intellectual, would (should) want to have an opportunity to explore a world of the spirit, a world which is many-dimensional, and which is a type of mysterious world, beyond the material-world, where the religious seers and prophets seek (sought) to have experiences.

This is what is being offered by this new model.

But

Apparently, everyone is a totalitarian, and each person seems to be dependent upon someone, other than themselves, to lead them. Someone who has the social rank of an authority, so that it is socially acceptable for the authority (ie an institutional authority, as proclaimed by any of the failing institution) to tell them "what to believe," and about "what the actual possibilities of the spirit are," or which "actually exist, and are there for the taking," and subsequently, the arrogant materialist authorities tell a material-based story, to the people, that the authoritative absolute truth about existence is the story of arbitrariness and violence in the context of materialism. Apparently, this is the only world which "is there for the taking" so as to take it by violence and to dominate others by violence and (intellectual) arbitrariness,

but the religious side refuses to try to either put-forth alternative ideas about existence, which possess any practically useful meaning or they are too timid to consider a new context through which religion, like the religion of the ancient prophets, might be experienced.

But

The spiritual-world beyond the material-world has, actually, already been carefully described in a measurable context, in the language of mathematics, in the math structures reviewed in this paper (Chapter), where the material-world is a proper subset of a high-dimensional model of existence (perhaps a model of an ideal existence).

Did the old (ancient) seers, actually, deal with many giant, and powerful beings, in a realm, which might be called, the spirit, but which is really in "higher-dimensional contexts of perception" ?

Other aspects of the propaganda-education-indoctrination system

In today's collectivist capitalist system, the propaganda-education-indoctrination system is populated by the, so called, intellectual-elites, who are well indoctrinated in the, so called, knowledge, which is defined by the interests of the investor-class. Based on its context of high-social-value, the knowledge, held by the intellectual-class, and which defines the intellectual-class, is considered to be an absolute knowledge, ie things which can be described by science cannot be described in any other manner (ie the, so called, laws of physics have been chiseled into stone). But this is far from being true, in fact, it is quite false.

This means that all published material..., except self-published material..., which is allowed to be considered by the intellectual-class, either in the media or in the academic surroundings, is all fundamentally flawed.

It is based on an authoritative form of knowledge which is failing to be able to describe, and, subsequently, use in new practical ways, the observed properties of the world.

This failure is mostly due to the fact that the discussions within intellectual institutions are not based on assumptions and criticisms of existing expressions, especially, if the existing technical expressions are not capable of describing the observed patterns, eg of stable many-(but-few)-body spectral-orbital systems, where no expressions in modern physics or modern math are capable of describing these fundamental systems,

All participants in the intellectual-class believe in both inequality (especially, the inequality of intellectual capability, ie those in the intellectual-class are the superior intellects of society), and subsequently, they believe in a social hierarchy, where this is (circularly) related (in logic) to the social construct of an absolute authority, which, in turn, is associated to the intellectual-class, ie they believe in the particular form of formalized axiomatic which defines an absolute authoritative context, upon which is based the (academic) competition through which they came to be determined to be through examinations to be, "the superior intellects of society," and, thus, validating their belief in their own intellectual superiority over the public,

But

This is really a belief in arbitrariness (as can be demonstrated by means of the actual implications of Godel's incompleteness theorem). Thus, it can be stated that holding onto a belief in an arbitrary authority is exactly equivalent to the arbitrary system of beliefs related to racism. Apparently, this is the (or a) central social construct of an Empire.

Without equal free-inquiry, there will be a belief that an authoritarian culture is an exceptional culture, thus, justifying imperial expansion

Instead (of equal free-inquiry, and elementary discussions about assumptions contexts etc etc) then academic intellectualism is about developing, in a very narrow context, namely, slowly developing the instruments which already exist, and which the indoctrinated intellectual-class maintain, (eg the nuclear bomb) and

It (this intellectual narrowness) is (has come to be) based on formalized axiomatic, whose, apparent, intent seems to be to be able to tailor-make knowledge to fit any purpose which might be of interest to the investor-class, (but according to Godel's incompleteness theorem, this is an impossibility for a language...., which is based on a monolithic [and absolute] set of: assumptions, contexts, interpretations, and a single way in which to organize the descriptive patterns for the authoritative and..., which is assumed to be an...., all encompassing language, where there is no room in the authoritarian discussion for criticisms of the dogmatic doctrine, which is implied in the fixed context of formalized axiomatics)

The review of the new math constructs of the introduction begins

But rather than (axiomatic formalizations of math and physics) making the language versatile and widely applicable, instead it makes the language dysfunctional, where this has come about by trying to extend the descriptive context defined by both the idea of materialism and a partial differential equation (or pde)...., which is used to make locally measurable models of material-interactions...., by considering a mathematically wider pde context (but maintaining the idea of materialism), than the original context of the classical physics language, which was a context of being: linear, metric-invariant, and solvable where in this wider context of pde it is assumed that

(1) the pde is non-linear, based on some general metric-function, and in

(2) the contexts where the sets of operators, which model physical properties, and related to the definition of systems of pde which are non-commutative...., this is usually in the context where shape is not the focus of the descriptive properties but rather the containment-set of the (usually indefinably random) pattern one is trying to describe is a function-space where, in turn, this is quite often related to issues of probability..., and

(3) the descriptive context of indefinable randomness, where the, supposedly, discernable patterns, are, in fact, patterns which are unstable, and/or where the elementary-events of a probability space are unstable, this is an improperly defined probabilistic context (so the probabilities which such an improperly defined context tries to define are not valid) and where the set of observed patterns are not capable of being identified, through an invalid (or quantitatively inconsistent) set of formulations of "system material-interaction models" based on pde, and subsequent calculations, in regard to solving for these pde's descriptive contexts, are contained in a quantitatively inconsistent math structure, such as the contexts when there is an assumption about modeling a particular type of system as a function-space, ie where it is assumed that the physical system is

not a shape which is contained in a coordinate-set, but some type of shapeless physical system, which nonetheless determine stable sets of spectral-values, but these shapeless models have not been shown to possess any reliable and stable properties (such as the very stable properties associated with a stable many-(but-few)-body spectral-orbital system, just mentioned), but in the, quantitatively inconsistent, yet, nonetheless, still assumed to be describable, contexts (of invalid pde [physical] system-models), the quantitative structures, ie the system containing quantitative sets, upon which the description depends..., eg either in coordinate spaces (with coordinate functions) or in function-values or for function in function-spaces, etc, these quantitative sets (ie coordinates and/or function-values)...., are quantitatively inconsistent, where this is because they are, either non-linear, or non-commutative, and/or indefinably random operators (or operator-contexts)

Which are used in such a quantitatively inconsistent description, where the operators (or maps) possess the above mentioned properties (of being non-linear, non-commutative, indefinably random) which actually, cause the resulting quantitative structures of containment..., (or description) such as the base-space coordinate structures or the set of function-values..., to no longer have:

(1) reliable measuring-scales, and

(2) the measuring-scales become dis-continuous, and

(3) the (function-dependent) quantities become many-valued,

Thus, the number-systems upon which the description depend are not reliable number-systems, so that, the description cannot carry..., within "such a failed and unreliable quantitative structure"..., any reliable (or stable) patterns, ie the description cannot contain any meaningful content.

That is, the context upon which formalized axiomatic was formulated, namely, using pde to model the fundamental (material-interactive) aspects of a (physical) system which one is trying to describe in a quantitative context, is a failed descriptive context.

Note: pde's are used to, supposedly, determine (by math-solution procedures) the measurable properties of a (stable) physical system, where pde's (for physical systems) are, supposedly, determined by physical law. That is, the pde's used to model the material-interaction properties of material systems (or of what are supposed to be models of material-interactions, which, are supposed to, turn into stable systems) are models based upon the idea of local measuring (or sets of operators which represent a system's measurable set of physical properties), which, supposedly, identify (in the set-containment context assumed by the pde) the actual set of inter-relationships of the particular quantitative types, which, in turn, cause the system to possess the measurable properties which, in turn, are observed.

Unfortunately it is a descriptive context which only works for a few cases, eg the 2-body system.

Other ideas

Thus, one needs to conjecture other new ideas, eg it is not "the pde model" which is valid,

rather [in the descriptive context of the pde] it (instead) is the principle fiber bundle of the system's isometry (or unitary) fiber group and its subsequent base-space of a (resulting) stable metric-space, and their (the metric-space and isometry group) associated set of stable shapes (resulting from the: covering group, discrete isometry subgroups, and conjugation-class, properties of the fiber group)…,

That is, discrete time-intervals, stable shapes, and folding by Weyl-angles are derived from a metric-space's isometry group properties.

{which can naturally be associated to a metric-space of (what need to be the stable) non-positive constant curvature metric-spaces, whose metric-functions have constant coefficients, and (the metric-function) is continuously commutative (so as to defined the shape's natural coordinates), or whose local (natural) coordinate parallel transport structure around a closed-curve is locally commutative [and so that (in such a shape, which possesses holes in its shape) where the holes in these metric-space shapes are related to

either different energy-levels (in which a set of lower-dimension material shapes, which occupies the spectral-faces, fit perfectly into the spectral-orbits of the stable shape), or to different stable orbits for the condensed material which occupies both the (large) metric-space and its shape and its geodesics which induces the orbital structures (and where the condensed material can be related to intricate lattice patterns in both fluid and crystalline materials [which is condensed material])]}

Stable shapes with holes identify a stable spectral property for each hole.

And furthermore, the context is many-dimensional.

That is,

There are alternatives (but the typical person has been brow-beat so badly by the violence and intellectual bullying, which are all defined in an arbitrary context, that a person is afraid to challenge intellectual authority, but it (that narrow, and uncurious, authority) is a narrow authority, which only serves the (creative) interests of the investor-class) (thus, it is an intellectual structure which really should be contested, and it should be contested by everyone at an elementary level of language ie at the level of assumption, context, interpretation, and the way in which the descriptive language is organized etc)

There are alternatives (things do not "have to be done this way")

For example:

In regard to math and physics, why not, try using E Noether's symmetries, so as to associate physical properties directly to (metric-invariant) metric-spaces, of non-positive constant curvatures,

and the discrete isometry (and/or discrete unitary) subgroups… of the metric-invariant (or Hermitian-invariant) fiber group of local coordinate transformations, which is associated to the metric-space, can be used to define the stable shapes, upon which both stable patterns (shapes) of (physical) systems (as well as the metric-spaces themselves) can be modeled, so that various different metric-space sets, which are a part of a set of properties associated to a stable physical system, all together, identify properties (for the system) based on stable shapes, ie stable shapes which identify the stable patterns which are associated to a stable system's measurable physical properties, in turn, associated to the different metric-space types, there are the main different metric-space types of:

(1) the stable energy-shapes of a hyperbolic metric-space, and there are

(2) the continuous mathematically consistent contexts of inertia and spatial positions of Euclidean metric-space shapes,

Physical properties are associated to sets of metric-spaces.

And then also, the new description is:

Based on new ways of defining the local processes of measuring change (which can occur), and where the changes are discrete, based on two (opposite) metric-space states

(Which are locally the inverses of one another, ie the local inverse structure identifies an equivalent opposite metric-space structure),

and

The local descriptions of (system) changes are also

Associated to stable shapes (or stable patterns) through both non-local constructs (based on relatively stable discrete Euclidean shapes, which can define a brief action-at-a-distance geometric relationship),

And

Through the discrete processes of interactions, which relate the changing-components to resonances… of the systems which are interacting… to a finite spectral-set,

Where, in turn, this finite spectral set (which is defined for an over-all high-dimension containment hyperbolic metric-space)

is defined on a high-dimensional metric-space containment set,

The new model of material-interactions defined in a discrete context, where a system's shape can resonate with a finite spectral-set so as to define a system, which emerges between discrete time-intervals.

Where the high-dimensional metric-space containment set is, in turn, "partitioned" into stable shapes, ie the set of stable patterns upon which the description is based, ie both metric-spaces and material-components are both modeled to be stable metric-space shapes (but defined on different dimensional levels, or for the same dimensional level they are defined for the different subspaces), furthermore, the shapes (used in the "partition") cause the higher-dimensional containing metric-spaces to "not be related to their lower-dimensional subspaces through the math property of continuity,"

rather there are discrete differences between dimensional levels (and between the set of subspace-shapes contained within the same dimension) which topologically separate the different subspaces (of the high-dimensional containment set) [ie topologically separate the different subspaces by means of various topological properties, defined between the shapes of different dimensions, topological properties of being open-closed and/or being closed, depending on the dimensional level within which the topology (ie properties related to continuity) is defined], ie a high-dimensional metric-space containment set, which is "partitioned" into stable shapes, is "partitioned" by a method of "partition" which is based on the properties of: dimension, size of a shape, and a set-containment structure-tree, and a rule about stable discrete isometry (hyperbolic) shapes, namely, for each subspace the rule is: only one biggest stable shape for each subspace, which requires that a finite spectral-set be defined for the entire high-dimensional containment set, but where resonance can occur between any shape (contained in any subspace) and the entire finite spectral-set.

Defining the finite spectral-set for a containment-set and the topological properties which make each of the different metric-space shapes independent of the others.

Furthermore, transitions between two-different stable systems, which is considered to be a part of the properties of physical systems, can occur (1) in a context of many different (and stable) sets of lattices (ie stable lattices related to condensed material), (2) during some special interactive contexts, and (3) during collisions between different stable material-components, in particular, when the energy-of-the-collision is within the correct range.

That is, systems do not enter into new stable system-types based on detailed locally measurable relations, described through pde, which…, it has been assumed…, cause system formation during material-interactions, eg as Newton was able to describe the construct of a stable system for two-bodies in regard to gravitational interactions, (but no-one has extended this description, based on pde, to the very stable many-(but-few)-body spectral-orbital systems, which are so very prevalent within our existence.

Instead, the new descriptive context, is more about stable metric-space shapes, and their relation to being in resonance with a finite spectral-set, which, in turn, is associated to a high-dimension containment context and its "partition" into stable hyperbolic metric-space shapes.

It might be noted that discrete structures become central to the descriptive context in this new descriptive context, based on stable shapes and resonances of these stable shapes with a finite spectral-set defined for an 11-dimensional hyperbolic metric-space, which has been "partitioned" into shapes (so as to define the finite spectral-set)

But where

The discrete properties emanate from a fiber group of both each different metric-space and

for the 11-dimensional hyperbolic metric-space,

Where these new types (or new interpretations) of discrete properties (which emanate from a fiber group) are:

1. The double-cover of a simply-connected spin-group, whose spin-rotation period (for hyperbolic spaces) define a discrete time-interval, and material-interactions are to be defined as discrete local operators, which are related to both (1) a local inverse structure and (2) to a toral geometry which is redefined at each different discrete time-interval, by the geometry of a toral-shape connecting (touching in a tangent manner) the two interacting materials, so that the tangent geometry of the discrete Euclidean shape (which defines the local geometry of the interaction) is re-defined by an action-at-a-distance toral shape, ie re-defined for each discrete time-interval.

2. The discrete isometry (hyperbolic and Euclidean) subgroups which define the set of stable shapes…, [and their two-metric-state complex-number coordinate system's relation to the discrete unitary subgroups, namely, one metric-space state is contained in the real subset of the complex coordinates, and the other opposite metric-space state is contained in the pure-imaginary subset of the complex coordinates],

3. The discrete set of conjugation (fiber-group) classes, which, in turn, define the discrete set of Weyl-angle folds, which, in turn, can be defined on (or between) the sets of toral-components, where the toral-components, in turn, define the toral-component shapes of the set of very stable discrete hyperbolic (or unitary) shapes.

4. The shapes define topologically separate sets of metric-spaces, where this definition is between the different shapes of the subspaces of the same dimension, and/or between the different dimensional levels, where in the higher-dimensional metric-spaces the lower-dimensional metric-space shapes appear as closed shapes with boundaries.

And

(outside of the isometry or unitary fiber group)

This set of discrete relations (identified in 1-4 above) are defined at either particular discrete points or boundaries, and this allows each separate shape to have its size be defined by a discrete multiplicative constant, which is (can be) defined between either discrete structures or discrete processes

5. The discrete multiplication by a constant, which can be defined between discrete processes, mainly between dimensional levels and between subspaces of the same dimensional level, (so as to identify a change in relative size between the various discrete structures, which are identified in this model of existence), can be defined between either discrete structures or discrete processes (note, this can be done during the discrete operation of introducing Weyl-angle folds between toral components within discrete hyperbolic shapes).

Note that,

The material-interactions are still defined between material-components, but they are

Either

The usual types of second-order pde's of either the hyperbolic (collisions), or the parabolic (free-projectile) types, while the elliptic-type second-order pde is a part of a perturbing process where the elliptic pde is defined when a stable orbit for condensed material is defined by geodesics (ie for condensed-material contained within a stable metric-space shape, a shape which defines stable orbits for the condensed material which is contained in the metric-space), but where each separate orbit (of the different planets of the solar-system) is then perturbed in the context of an elliptic second-order 2-body pde,

Or

the mostly irrelevant, ie non-linear, relations, which can be related to feedback systems (externally defined patterns imposed onto a system's context),

Or

(as above, for the elliptic context of a second-order pde) they define a perturbing construct, which is attached to the orbital structures of (usually condensed) material, where this condensed material is in the orbit structures, which, in turn, are defined by the geodesics of a stable discrete hyperbolic shape of the metric-space within which the condensed material is contained,

but where the material-interaction structures can perturb these geodesic orbits.

That is, pde play a relatively minor role in the descriptions of stable material properties.

The natural coordinates for a metric-space of non-positive constant curvature are defined in regard to the set of stable shapes, which, in turn, are associated to any of its (the metric-space's shape, which is now to be identified) discrete isometry (or unitary) subgroup lattices, ie a partition of the metric-space, where the partition is similar to a checkerboard pattern; and where a square of this checkerboard lattice (pattern) would be called a fundamental-domain of the lattice, or stable shapes, obtained from moding-out (the process of identifying opposite faces of the cubical-type polyhedron, so as to form a shape) the [cubical-type] shape of a fundamental-domain, (ie a stable shape which is associated to a metric-space's isometry or unitary fiber group)

Natural coordinates of shapes and the topology of shapes and their dimension

Where the natural coordinates of such a shape (or of a coordinate metric-space) are the coordinates which continuously diagonalize a metric-function's local matrix representation… (ie representation of the metric-function, where such a locally determined matrix is always a symmetric-matrix), and on these natural coordinates the process of local parallel transport along these natural coordinate curves (or directions) in a closed curve are always commutative (ie the original vector and the parallel-ly transported vector commute), ie the shape is parallelizable.

These properties define a stable shape, and they emerge from discrete isometry or unitary subgroups of a (metric-space) base-space's fiber group.

Furthermore, depending on the dimension and the set-containment context, the shapes of these natural coordinates define

Either

The natural coordinate's of the shape of a stable material-component, which is contained in a metric-space, where when (if) the (lower-dimensional shape) is viewed within its (adjacent higher-dimension) containing metric-space then the material-component shape is seen to possess a closed boundary, and so that the shape possesses holes within itself,

Or

When viewed within the shape, itself, (then) one sees the natural coordinates of a simply connected shape (ie no holes are seen within the shape) and a topology within the metric-space shape is an open-closed topology.

That is, the set of stable (math) shapes, which are associated to either Euclidean metric-space shapes or hyperbolic metric-space shapes [but the stable shapes are primarily the (discrete) hyperbolic shapes], determine either the structure and the set of stable spectral-orbital properties of stable material systems, or

The (planetary) orbital properties for the condensed material, which is contained in a metric-space's orbit-determining stable metric-space shape.

These, above mentioned, topological properties along with properties of set-containment, which are related to shapes and their boundaries, allow for… all of:

1. the different subspaces of the same dimension, and

2. different sized shapes of the same (dimensional) subspace, as well as

3. shapes associated to different, eg adjacent, dimensional levels

... to all be separate and topologically independent metric-spaces, each with their own independent topologies, which are needed to identify, within themselves, the idea (or property) of continuity

Since the property of "place in space" is a property defined within an Euclidean space, and the Euclidean space, R(4,0), has a fiber group of SO(4) = SO(3) x SO(3), which acts on R(3,0) x R(3,0), where there is a common R(2,0) subspace to both of the R(3,0) subspaces which are partitioning R(4,0), so as to fit into the context of allowed local transformations of local coordinate properties (so as to remain metric-invariant), then the natural discrete isometry shapes of R(4,0), can (will) fit into the subspace patterns which are defined by the group actions of SO(4), which will be partitioned into the given (above) subspace constructs of R(4,0), thus there can be two sets of 3-lattices in R(4,0) which possess a common R(2,0) subset (or 2-face structure). This also seems to mean that R(4,0) is experienced in regard to two different 3-subspaces, which may be filled with two different types of shapes.

These two different types of shapes could be the two sets, where one set is a set of 3-dimensional and even-genus-number shapes (inert material-components), and the other set being a 3-dimensional and odd-genus-number shapes, where the odd-genus-number shapes are models of a set of naturally oscillating, and thus energy-generating shapes (oscillating material-components).

These naturally oscillating and energy generating shapes would be good models for the shapes of living systems, and where living systems could be composed of both inert 3-lattice structures, and oscillating 3-lattice structures, so that energy could circulate and be controlled in the oscillating 3-lattice structures, where some of the 2-faces of the two different material types could be common to both sets of material. Namely, common to both inert and oscillating 3-shapes, ie the 2-faces could be either odd-genus or even-genus, but where it is the 3-shapes of an odd-genus which possess the property of "the shape oscillating," ie (in this context) whether a 2-face (shape) "oscillates or not" is determined by which of the two 3-shape types to which it is a 2-face, if it is common to both 3-shape types then it will (can) oscillate and generate its own energy)

Thus, one can model material types (within a living system) which become active or non-active (ie molecular-components become prominent or not prominent) within the material structure of a living system, where the living-system is composed of many, simultaneous, 3-lattice structures, if energy is flowing in a particular 3-lattice or not

And

Such (closed-and-bounded) stable material systems are <u>not</u> determined by material interactions, which are modeled by (elliptic) partial differential equations (or pde).

The generally formulated (form-invariant, when transformed between general frames) pde (associated to what is called physical law) generally exist in a "quantitatively inconsistent" context, where this means that the, supposed (but non-calculable), solution functions for the non-linear, non-commutative general context of these (general) pde..., (ie the, so called, laws of physics), cannot describe a stable quantitative-geometric pattern, ie the general descriptions of material-system interactions based-on solutions to pde possess very limited (if any) useful information.

Due to the nature of the very stable discrete hyperbolic shapes, up-to and including hyperbolic-dimension-5, being relatable to either bounded or unbounded stable shapes,

And

The distinctly different bounded and unbounded properties of nuclei and neutrons and (or as opposed to) electron-clouds and neutrinos, respectively, One can easily accept the idea that a small (hyperbolic) 3-lattice..., which is contained within a large 3-shape hyperbolic metric-space, (the small 3-lattice) can have its 2-faces (of the cubical lattice structure) formed into 2-shapes (which can exist as 2-shapes in 3-space) which can stay connected (by a neutrino structure) to their defining 3-lattice context (when within a system of condensed material).

Thus, if the system of condensed material (eg a solid or a liquid) is also composed of multiple sets of discrete hyperbolic lattices then since the 2-shape component structure of the system is attached to its separate 3-lattices, then different sets of 2-components can appear and then disappear into the different sets of 3-lattices, which are composing the system. (see chapter 24)

Chapter 1

A new way to interpret the "mass equals energy" formula

The "mass equals energy" relationship of "special relativity" can be interpreted to mean that stable material-components are comprised of the set of stable "discrete hyperbolic shapes," which can exist by the property of their being in a relation of resonance, to a very large finite spectral-set, so that these stable shapes represent the fundamental forms in which stable energetic-systems can exist in space.

These shapes exist over many dimensional-levels. Furthermore, these stable shapes are very simple shapes, namely, they take-on the form of "linear-strings of toral-components," which are folded at the places where the toral-components are connected to one another.

An material-interaction-field which is to be defined ....*

....* between these stable energy-forms is, in fact, defined by means of a discretely identified and instantaneously-formed "discrete Euclidean shape" (or a simple torus shape, or, equivalently, the shape of a doughnut) which spans the distance between the (strongest) interacting material-components, ie it is defined by a discrete action-at-a-distance process, and (this distance spanning torus) changes with discrete time-intervals (where the time-interval is defined by the period of the spin-rotation between metric-space states). This distance-spanning toral interaction-component (or field component) does two main things (1) it relates a force-field to a geometrically determined local measuring process, which, in turn, defines a derivative-connection, which, in turn, determines the geometric structure (and motion-changing action) of the resulting force-field, and (2) when it (the distance-spanning torus) is of the right-size, and when the interacting system is within an allowed range of "energy-bounding values," the entire shape-complex of toral-components can define a new resonance structure, so that a new stable form of energy (or stable new material-component, or new discrete hyperbolic shape, or new stable geometric-form of energy) comes into existence, due to it being resonant with the finite (but very large) spectral-set, {where the original interacting-components either are absorbed into the new structure, or they break-up into different components which either remain, but might be deformed, or the pieces move away}. That is the toral interaction-component blends with the energy-shapes of the discrete hyperbolic shapes (which are in the original part of the interaction) by means of a very consistent toral-shape, ie like a toral-component of the original interacting material-components, so as to be highly relatable to energy-transformations which can be associated to the energy-forms of material-components (a complex of folded toral-components) and toral-interaction shapes, which form during a material-interaction. The formation of stable systems result from an interaction… (which is modeled in a way related to models of local measuring which is quite similar to the (pde) interaction models of regular physics, but)… now the new system which emerges from the interaction forms, because the new system also possesses, both the necessary energy-level and the (new) property of it (the new emerging system) being in resonance with a finite spectral-set, which, in turn, is defined for a new high-dimension over-all containment-set, where this high-dimensional…, or an 11-dimensional hyperbolic…, metric-space is, in turn, "partitioned" into stable shapes of various dimensions, and which possess various qualities of being independent metric-spaces.

That is, physical description is provided with a new context, which is associated with a set of stable shapes (where discrete hyperbolic shapes are the stable forms of energy) and it is a context which is defined over many-dimensions, all related to a finite spectral-orbital set, in turn, defined for (or within) a many-dimensional…, ie an "over-all" 11-dimensional…, hyperbolic metric-space containment-set.

The externally defined, locally measurable models for an interaction-structure is now of secondary concern, in regard to the set of stable shapes, where a particular set of stable shapes define both the metric-spaces (which contain material) and the stable material-components, and both types of shapes (either metric-spaces or material-components), which are of different size scales and of different dimensional-levels, are stable geometric forms-of-energy, and both the orbits (of material in metric-spaces) and the spectrum (of the usually charged-material which exists within a stable material-component, eg of the charged-material occupying a stable system's spectral-levels) of these systems are stable, and (the orbits and the spectra) are determined by the geometric properties of this particular set of stable shapes [the particular-set which defines the finite spectral set for the 11-dimensional hyperbolic metric-space],

especially, for the stable many-(but-few)-body spectral-orbital systems, which exist at all size scales, so that these discrete hyperbolic shapes are in resonance, with the finite spectral-set, which, in turn, is associated to the finite spectral set defined within an 11-dimensional over-all containment set.

That is, the stable (hyperbolic metric-space) shapes, which are formed of linear rows of folded toral components (of various sizes), where these stable shapes are (can be) both the material-components, as well as the metric-space shapes associated with the different levels of the various dimensions of which existence is composed,

So that

on the folded strings of toral-components, which are forming metric-spaces, so that within such a metric-space shape (or material containing metric-space) a set of lower-dimensional material-components are (ie can be) contained, so that on such a folded orbital-metric-space-shape the (condensed) material-components (ie the lower-dimensional material-components which are contained in the metric-space shape) will follow a dynamic structure which is defined by the metric-space-shape's well defined geodesic structure (ie this is the idea of general relativity), so as to define stable spectral-orbital paths on the metric-space shape.

For example, condensed sets of either atomic or charged material-shapes, will follow the geodesics of the metric-space shape (within which they are contained).

That is, in the new descriptive context physical description is based on stable shapes of energy.

Physics is not (primarily) the study of either local models of (linear) measuring of properties (or functions) which are defined within small neighborhoods about the point (in space) where the measuring of the system's properties occur, or the neighborhoods of the point (in the system containment space) where operators identify a (eg quantum) system's eigenvalue properties, so that

Either

I. two forces are defined (1) inertial force (of a material component), and (2) force-fields due to the affect which the distant material-geometries have on the point of measuring,

and set equal to one another, so as to identify the Newtonian equations of systems to be built from material-interactions (but this is possible only for the 2-body system) [where only a few such systems can be defined and subsequently solved]

Or

II. That a massive object has an inertial relation to the geodesics of a (general) metric-space's shape,

where a (tensor) metric-function's second (exterior) derivative (or local Ricci-curvature-tensor's) is set equal to the metric-space's energy-density tensors (associated to material distribution in the metric-space) so as to solve for the general metric-function,

Where this general metric-function is then related to a local (neighborhood's) metric-space shape, &

So as to find the… metric-space shape's… geodesics, which, in turn, define a massive object's inertial paths on (or within) a (the) metric-space-shape [where, in this descriptive context, only the 1-body system with spherical symmetry has been solved]

Note:

Where this same descriptive context (of I. or II.) can be placed in a (classical) context of energy and, subsequently, a new property of "action" can be defined, so as to be based on both the property of energy and the property of time, where energy is a line-integral of force, both (a) inertial-force and (b) force-fields, so as to define both

(a) kinetic-energy and (b) potential-energy,

where, in turn, "action," Edt, can be defined so that the above set of Newtonian equations result from finding the extreme values of the "action"

Where (according to the new descriptive context for physics) the reason that an extreme value for the action-function should be related to the system's defining set of (partial) differential equations, is due to

Both

the discrete separation of the spectral-values defined on stable spectral-orbital shapes, and the average toral shapes, which identify the dynamic-change part of an interaction process, and which (during the interaction) the system's context moves {or changes}… (by means of the changing of the discretely defined "average interaction-torus" shapes for each discretely defined time-interval)… between the "allowed" set of spectral-orbital shapes ("allowed," in regard to the finite spectral set of the over-all containment set), where the material-components (or stable spectral-shapes) are a part of the interaction component-complex along with the changing "average interaction-torus" shapes.

Note: where these allowed set of spectral-orbital shapes {associated to the set of stable discrete hyperbolic metric-space shapes} are folded so as to define toral orbital components, either for the shapes of the stable material-component systems, or for the shapes associated with the stable metric-space which contains the material-components.

Furthermore the folds can be on an orbital-plane or the folds can define 3-dimensional (or higher-dimensional) angular-momentum properties.

That is, this interaction process is dominated by toral-component shapes, whose dynamic actions are based on these toral-components always moving (in a time sequence) either toward… or away from… a spectral-orbital extreme, in regard to a set of discretely distinguishable spectral-orbital properties of the set of allowable systems, an allowable set which is defined by the over-all containment-space's finite spectral-set so that this process is always within the geometric context of a set of toral-components of interacting materials and the spatially spanning (solitary) toral-component.

(end note)

Or

III. It also needs to be noted that the idea of local measuring, which is placed into a context of harmonic functions (ie quantum physics) modeling spectral-systems whose spectral-components are found to be related to probability functions defining random-spectral-particle-events found in the system's containing metric-space, where this idea is modeled with function-spaces, and sets of operators, as well as function-transformation operators eg Fourier transforms, does not expand, in any meaningful way, the range of accurate descriptions (to a sufficient level of precision) of general spectral systems, eg the stable many-(but-few)-body spectral systems, where such quantum descriptions depend on some stable geometric context to be abstractly identified, with sets of operators and subsequently, the quantum-system's shape and related spectral-sets identified, where these are spectral-values which one is trying to uncover by means of the techniques of local measures and function-space operators….

[But this has not been accomplished, except for two or three quantum systems: H-atom, harmonic-oscillator (whose physical reality is questionable), and the box…(and…?).]

Only a two-body system either for a 3-dimensional charged-system, or 2-dimensional inertial-system, can be (has been) modeled in the descriptive context of I. II. and III.

But there exist very-many quite-stable many-(but-few)-body spectral-orbital systems at all size scales, which are quite fundamental systems, in regard to all of physics, but which go without a valid descriptions based on applying the, so called, laws of physics, so that material-interactions define the set of stable physical systems.

Alternative viewpoint

The stable physical systems of the fundamental physical systems, which we experience, are stable shapes which, in turn, are an intrinsic property of the containment set, ie the stable forms of energetic-systems. That is, the containment set is endowed with a set-structure which is organized around stable-shapes and dimension, where this set-structure is needed in regard to being able to describe the observed stable properties of physical systems.

Rather (consider an alternative context for [or consider a new context within which to contain] the descriptions of the observed properties of physical systems)

Consider a context within which the many-(but-few)-body spectral-orbital system can be solved, where it is solved with "simple pictures of discrete hyperbolic shapes," as well as a (new) need to (experimentally) find the finite spectral-set of an (our) 11-dimensional containment-set, (so as to give the exact measurements for the picture solution) in order to determine all of the possible ways in which stable spectral-orbital properties of the many-(but-few)-body spectral-orbital systems can be identified, in regard to the set of all the

(A) different-dimension discrete hyperbolic shapes (both for the orbital-plane and for systems with more general angular-momentum properties, ie discrete Weyl-angle folds-angles which are different from an 180-degree fold), and

(B) different sizes of these (fundamental) discrete hyperbolic shapes, and

(C) different containment-trees for the set of all the discrete hyperbolic shapes which can exist (where their existence is due to their resonating with the finite spectral-set) within the 11-dimensional containment hyperbolic metric-space, and where

(D) the possibilities for these different, both bounded and unbounded, stable discrete hyperbolic shapes was found by D Coxeter, where he identified the different types of discrete hyperbolic shapes, which can exist within a hyperbolic metric-space of any dimension.

Note: The discrete hyperbolic shapes of dimension-6 up to and including dimension-10 are all unbounded discrete hyperbolic shapes

Note that there are no 11-dimensional discrete hyperbolic shapes, and there are no other discrete hyperbolic shapes of higher-dimension.

The properties of stable physical systems do not emerge from

1. local models of measuring, defined on a local neighborhood, nor

2. from function-spaces similarly related to the structure of local measuring methods, rather they emerge from sets of stable shapes defined in a many-dimensional context, where

This new descriptive context can be thought of as a new interpretation concerning the "mass equals energy" statement, which came from special relativity,

Where this formula (mass equals energy) was found by considering that, motion was modeled on space-time coordinates, and then the local coordinates (dx, dy, dz, cdt) were multiplied by mass, and divided by time, ie (mdx/dt, mdy/dt, mdz/dt, mc) so that the time-component, ie mc (but each term also has a factor induced by the Lorentz transformation of space-time coordinates) was (which has to be) interpreted to be an energy term, ie mc times a Lorentz transformation-factor equals energy,

Where this equality is because of both of E Noether's time-energy and position-momentum symmetries, so that space-time, or equivalently a hyperbolic metric-space, is the metric-space in which time is an intrinsic property (as opposed to a Euclidean metric-space) so that (due to Noether's time-energy symmetry) hyperbolic space is related to the physical property of either time or energy, which are to be the primary properties which are defined (or identified) in a hyperbolic metric-space, so that discrete hyperbolic shapes are the stable forms of the energy representations of stable physical systems, ie (new interpretation) special relativity is saying that discrete hyperbolic shapes are the forms in which energy can manifest as stable patterns.

But discrete hyperbolic shapes are defined up to, and including, 10-dimensional discrete hyperbolic shapes (as identified by D Coxeter), and thus, this dimensional-span from a 1-dimensional hyperbolic metric-space to a 10-dimensional hyperbolic metric-space, is the context of stable geometric energy-forms, wherein stable forms (of energy) exist. Whereas all of these stable shapes of energy would be contained in an 11-dimensional hyperbolic metric-space.

Physics is the study of quantitatively consistent stable shapes, which are placed into a higher-dimensional context in regard to determining the containment of stable forms of energy (or stable discrete hyperbolic shapes), where material is a stable hyperbolic metric-space-shape, but is at least 1-dimension lower than its (usually adjacent) higher-dimensional containing hyperbolic metric-space,

And

The different subspaces of the of the various dimensional levels of a higher-dimensional structure is partitioned into different dimension and different size metric-space shapes, or there are defined containment sequences of different dimension, and different size, metric-space shapes, so that an

11-dimensional hyperbolic metric-space is (can be) organized around a set of stable metric-space shapes,

so that it can be estimated that a finite spectral-set can be defined within this context if each subspace of each dimensional level, and for all the different dimensional levels, there is defined both a maximum and a minimum (though the minimum size is automatic, ie all discrete hyperbolic shapes are at least (¼)-distance from zero [size]) pair of sizes for each such subspace up to and including the 5-dimensional hyperbolic metric-spaces, so that this pair maximal and minimal sized stable shapes identifies the allowed stable hyperbolic metric-space shapes for that particular subspace and for that particular dimensional-level (this would be an upper-estimate for the size (or estimate of the exact number) of the finite spectral-set for the 11-dimensional hyperbolic metric-space.

Wherein an above (mentioned) descriptive context..., ie in regard to the stable discrete hyperbolic shapes, which is based on a type of a local measuring methods, this new descriptive construct is a context in which the local measures are: linear, metric-invariant, and continuously commutative, so that the metric-function

Either only has constant coefficients

Or

Is defined on a shape whose (natural) coordinates are related to a continuously diagonal metric-function.

In a shape associated with a metric-space (of non-positive constant curvature) where there exists a diagonal metric-function, the structure of the set of (natural) coordinate curves of the shape each coordinate-curve maintains an orthogonal (or parallel) relation to both the shape's equatorial curves, as well as to the longitudinal curves, so that the parallel transport of a local vector is commutative, when transported around a local closed-curve (where the closed curve is defined by (local) coordinates).

In the three isometric-ally-isomorphic metric-space-types, which possess constant curvature, ie [(+1)-constant-curvature, (0)-constant-curvature, and (-1)-constant-curvature] only the non-positive constant curvature shapes can maintain, in a global, as well as local manner, the parallel relation between the independent coordinates, which define the stable shape, thus they can define the set of stable patterns of shape in metric-spaces. These shapes are the shapes of (1) the torus [genus-1, or 1-hole in the shape] and (2) the shape composed of many toral-components [of various genus values]

The idea that an inertial component will follow a shape's geodesics can only make sense when such a component is contained within (or on) a stable shape

Note that space-time is a metric-space with a constant negative-curvature, so it is isometric-ally-isomorphic with a 3-dimension hyperbolic metric-space.

The relation of "mass equals energy" obtained from special relativity, and which is based on an "energy is associated to time" variable-dependence, which is analogous to the "momentum is associated to spatial-position" variable inter-dependence structure (related to the Noether symmetries), and it also is (clearly) related to the idea of action, ie Edt or pdx, can be interpreted to mean that mass should be interpreted to be the lower-dimensional discrete hyperbolic shape, which, in turn, can be related to a spatial-position (where this ability to identify a position-point is due to the existence of a distinguished point on any discrete hyperbolic shape), in an adjacent higher-dimension, either hyperbolic or Euclidean metric-space, within which the lower-dimension shape is contained, in regard to either energy, or in regard to spatial-positions [or equivalently, inertial-properties] (respectively).

That is, when an (n-1)-dimensional (hyperbolic) material-component is contained in an n-dimensional hyperbolic metric-space then the material-component's distinguished-point has both a spatial-position, as well as a relative position in regard to other (n-1)-dimensional (hyperbolic) material-components which are also contained in the same n-dimensional hyperbolic metric-space

But these properties of spatial-position must be in an n-dimensional Euclidean metric-space because of E Noether's symmetries. Thus, if an (n-1)-dimensional (hyperbolic) material-component is contained in an n-dimensional hyperbolic metric-space, then an (n-1)-dimensional (hyperbolic) material-component is also contained in an n-dimensional Euclidean metric-space, is an equivalent statement to the "mass equals energy" formula (statement)

Thus (the lower-dimensional shape is) to be related to (associated with a shape which has the property of possessing) inertia, because it is contained in the higher-dimension metric-space,

but where the containing higher-dimensional metric-space is either a general hyperbolic metric-space,

Or/and (can be both) a general Euclidean metric-space,

Depending on "what physical property is being identified" in regard to a material-component

For example,

Either

its energy (a hyperbolic space)

Or

it spatial-position (an Euclidean space)

Note that an n-dimensional hyperbolic metric-space is related to generalized space-time ie R(n,1),

And an n-dimensional Euclidean metric-space is related to generalized Euclidean space, ie R(n,0) or a generalized space, wherein a component (with a distinguished point) can have the property of possessing a spatial position, ie a spatial-position is a property within a generalized Euclidean space, ie R(n,0).

An n-dimensional Euclidean space is an "absolute" space in relation to the bounded shape (or structural-shape) of the n-dimensional hyperbolic metric-space shape within which a component's

(or an (n-1)-dimensional material-component's) spatial-position (of its distinguished point) is being determined, and a hyperbolic metric-space is a space which can be thought of as a space within which the physical property of time can be identified, (time can be consistently identified on the shape of a discrete hyperbolic metric-space, where a hyperbolic metric-space possesses time as a physical property {associated to such a metric-space}), ie a metric-space shape based on the physical property of time, so it is also the space of energy due to E Noether's energy-time symmetry.

However, an Euclidean metric-space is a space which can be thought of as identifying spatial-position

(relative spatial-positions can be identified by the shape of a "discrete Euclidean metric-space shape," where an Euclidean metric-space possesses spatial-position as a physical property {associated to such a metric-space}), ie a metric-space shape based on the physical property of spatial-position, so it is also the space of momentum (or inertia), where this is due to E Noether's {momentum-"spatial-position"} symmetry.

That is, the stable discrete hyperbolic shapes are stable hyperbolic metric-space shapes, and they are the shapes upon which patterns of energy are defined, ie stable conserved properties of energy associated to relatively stable material systems.

But, by being a stable shape, which is contained in a metric-space, then it is also a stable pattern with a distinguished point, which also possesses a spatial-position within a general Euclidean metric-space, and thus "the stable form of energy" also has inertia, or mass,

And

Thus, this is also an expression for the "mass equals energy" formula.

And

this is also an expression of (for) the conservation of mass (or conservation of material, eg conservation of charge, where charge would be a stable shape, either of the same dimension as the metric-space. [in which case, its shape is not well defined within the given metric-space, except as the faces of rectangular-faced simplexes], or a stable shape which is less by at least 1-dimension than its containing metric-space)

But a charge is a stable "discrete hyperbolic shape" which is contained in a hyperbolic metric-space, so that within the hyperbolic metric-space (or within generalized space-time space) local geometric-measures (or differential-forms) can be related to other geometric-shapes within the metric-space, by means of their properties of "boundaries of regions" where these properties of bounded regions are associated to "sums of local geometric-measures," ie the charge and its motions (or electric-currents) can be related to differential-forms. In the ideal representation of these ideas about charge and currents the currents are to be pictured as the spectral-flows of the stable discrete hyperbolic shapes, ie 1-dimensional-flows on the 2-dimensional discrete hyperbolic shapes, which, in turn, models a stable electromagnetic system.

The simple geometry of the non-positive constant curvature shapes is built from the quantitatively consistent shapes of line-segments and circles

[this is related to the real-number-line and the complex-plane both identifying the same algebraic-properties as that of a number-field, ie the circle and the line can be made quantitatively consistent]

Where the shapes which can be built fro m the circle and the line-segment are:

Lines, circles, and combinations, rectangular-faced simplexes (or cubes of general dimensional values), orthogonal-sets of circles (or, equivalently, parallel circle-spaces), and cylinders, where the parallel circle-spaces, or sets of orthogonal circles, are the most common of these stable and simple geometries

Where, these stable "discrete hyperbolic shapes" are characterized as being "a set of toral-components which are pieced together along a "linear-string of such toral-components," and then "folded between the places where the toral components connect" (where the folds between the toral-components are due to Wyel-angles)," then this means that a material-interaction, which is modeled in relation to a toral-spatial-bridge between material-components…, and which is defined between distant materials (where the materials are modeled as stable discrete hyperbolic shapes)…, so that the toral-bridge's tangent properties (which also defines a normal subset of local vectors in the [relative] material-containing metric-space) are, in turn, related to the fiber group's geometry, and it is through the relation of the torus's local geometry, and it is the fiber group's geometry, in regard to its local diagonal matrix structure, through which a force-field (or bridging toral-component) is being related, so as to subsequently, cause cause a change in a material-component's motion, in relation to both tangent and a normal directions at points of tangency on both material-components, which are defined by a spatial-bridging torus, which both exists between the two interacting material-components," and which is tangent to both material-components (this is a newly defined idea), so that these geometrically defined directions, ie tangent-directions and normal-directions, define the changes in motion of the material-components in both radial and tangent directions of the interacting material-components (actions are re-actions defined in relation to the [newly modeled] spatially spanning <u>torus</u> where these changes in motions are based on the geometry, where, in turn, the toral tangent geometry determines the fiber-group elements, which in their diagonal representation can be related to the spatial geometry of the interacting material-components (in the metric-space which contains the material-components), (ie in a diagonal representation a relevant "group transformation element (acting on the metric-space's local coordinates)" would possess the same dimension as the local spatial-vector of (or associated to) the local interaction geometry {of both the material-component's shape and the geometry of the spatial-bridging toral-component} [in the material-containing metric-space]), and (or so that) these geometrically determined changes are either normal or tangent to the local tangent vectors of the tangent geometry of the toral-component {where the toral-component forms the spatial-bridge of the material-interaction},

Where it might also be noted that

Where these toral components (which bridge the space separating the interacting material-components), are defining the geometry of a material-interaction in such a way, so that this shape (of interaction), (ie the shape formed by both the interacting material-component-shapes along with the toral shape which bridges (or connects) the interacting components), would be similar to an energy structure (a stable energy-shape), ie a set of (or all the) toral-components which make-up the material-components along with the toral-bridge, within the material-containing metric-space, where this is an "energy-structure" in which the interaction-toral-component expands or contracts in an average manner, over time, so that such a change would be easily relatable to a set of harmonic-functions (where a harmonic-function's value can be determined by averaging its function-values in a symmetric local region about the point of the harmonic-function's evaluation) and thus relatable to a spectral-set, but now the spectral-set is a finite spectral set, and the spectral-values are discretely separate from one another, so this defines an average motion of Edt toward the next upper or a lower (extreme) value of action, ie Edt, within the spectral-set of allowed energy-forms.

{in a metric-space which contains the material-components and the metric-space within which such an interacting shape would be contained (within which the interaction exists [or is contained])}.

This is, essentially, a model of material-interaction, which is similar to the idea of a derivative-connection of a classical material-interactions, but the connection (or fiber-Lie group element) is both more geometrically integrated... and more causally integrated... into the description of local changes of the physical system due to the toral-bridge being used to define the interaction (or to define the cause of a force-field)

{note that the radial, inverse-square, material-interaction geometry defined in 3-space can be derived from this context}.

But the interaction shape-construct is only stable if it:

Either enters into resonance (or is in resonance) with the finite spectral-orbital set defined by the over-all 11-dimension hyperbolic metric-space containment set or if the interaction is perturbing a material-component's stable orbit, which is being defined by the geodesics of the shape of the metric-space within which the material-component(s) are contained.

That is,

This model of a material-interaction is, essentially, a classical viewpoint, though it is also closely related to general relativity (but the geometric context is quite different, ie the geometry is stable),

and this model of material-interactions, in this new descriptive context, allows for:

1. a random force-field interaction context for the small atomic and molecular interactions, thus causing the appearance of quantum randomness {where the distinguished points of any material-component's discrete hyperbolic shape accounts for the point-like properties of (random) quantum, or spectral-point events

(eg consider the context of an apparent point-spectral event actually being a material-component [which is encountered as a spectral-point event] which is within a stable spectral orbit of a larger system, (which is also a stable shape, and which is composed on many-(but-few)-components) then the point-event would occur at the distinguished point of both the stable system and this distinguished-point would also be the distinguished-point of the material-component contained within the larger spectral-system)}

2. The usually classical context, which is mostly about non-linear interaction structures, ie a derivative-connection is usually non-linear, where such an unstable pattern eventually disappears into the background of other unstable patterns.

3. When the size of the toral-bridge of an interaction-geometry approaches the next energy-level of the containing space's finite spectral-set, then if the system-complex of both toral-components and total energy (of the individual interacting components and the energy of the interaction) is within a certain "range of energy" and if the system's total geometry, composed of toral-components, is in resonance with some aspect of the finite spectral-set of the containment 11-dimensional hyperbolic metric-space then a new stable system (a new stable discrete hyperbolic shape) can form from the resonances and the existing geometry, but if not then the description reverts to a classical system (ie #2, just-above).

4. If the material, which is interacting, is already in an orbit, which is caused by the material following the geodesics (or minimal-geometric property of the face-structure) of the shape within which the orbiting-material is contained, then the other (external, but still within the metric-space) interaction structure can perturb the orbital motions of the already orbiting-material, but this is usually done in the context of a 2-body system, since the geodesic structures are stronger than are the interactions of the other (more distant and less important) bodies (of a many-body system) in the spectral-orbital system (more distant and less important bodies, eg the sun's central mass is always important in a planetary 2-body perturbing system).

This is a much different way in which to interpret the "mass equals energy" relationship which was identified in regard to special relativity.

But now general relativity can be defined as a material-component's inertial motion being defined by the stable shape of a metric-space's geodesic curves, but where a discrete hyperbolic shape is a model of a stable (hyperbolic) metric-space, which possesses a geodesic structure in which the geodesics are limited in number and they are discretely separated from one another. That is, the idea of general relativity can be modeled in a context of quantitative consistency so it is (it becomes) an idea which can possess meaningful content.

Summary

Note 1:

1. The proof that these discrete hyperbolic shapes exist and are fundamental to physical description is easily shown by the fact that the solar-system is stable.

Note 1b:

The structures of the: general nucleus, general atoms, and molecules, as well as dark-matter…,

(where dark-matter seems to manifest on the size-scale of galaxies, or perhaps the size of solar-systems [perhaps less-dark-matter]), can only be modeled, {in regard to the types of physical models which have been expressed at this time (2014)}, by the sets of discrete hyperbolic shapes which can be any of the various dimensional shapes, which are both bounded and relatively unbounded shapes which are of hyperbolic-dimension-1 through hyperbolic-dimensions-4, and which are resonant with the finite spectral set of the 11-dimensional hyperbolic metric-space containment-set.

2. That the spectra…., of the 11-dimensional hyperbolic metric-space over-all containment set for existence…, is finite, {and with definitive values which are separated from one another}, and that the set of stable discrete hyperbolic shapes which are related to this finite spectral-set exist as models for the metric-space structure in regard to each subspace of all the different dimensional levels, so that there is a maximum and a minimum size discrete hyperbolic shape for each subspace and for all the dimensional levels (where it makes sense) where this idea is proved by the validity of the extreme-action principle.

3. That the dynamics of material-interactions are mediated by the toral-spatial-bridges, which are defined between "the stable shapes of the interacting energy-forms [either inertial, or charged, or other stable energy-forms]" is proven by (or is within) the same logical context as the "mass equals energy" formula of special relativity, ie E Noether's symmetries.

4. An important issue remaining (beyond (1) trying to understand and use higher-dimensional existence, and (2) using the naturally oscillating and energy-generating sets of relatively stable shapes in a context of spectral inter-relationships, as well as (3) understanding the types of patterns which can be related to the extension of such existing properties), is to determine the finite spectral-set for the entire 11-dimensional hyperbolic metric-space within which our existence is being determined, [but there may be other similar types of 11-dimensional containment sets organized in an analogous manner but with different values for the finite spectral-sets, in fact, a

galaxy may be the actual site of the 11-dimensional hyperbolic metric-space and the other galaxies are apparently expanding since the relation between the galaxies may be one of conjugation within a high-dimensional but also finite dimensional Lie group]...., so that this information, ie the knowledge of the finite spectral-set associated to our 11-dimensional hyperbolic metric-space containment-set, can be used in a new creative context, To find this finite spectral-set consider that the spectra of nuclei and atoms etc can be used to identify a large part of this finite spectra, namely, part of the 1-dimensional through 3-dimensional (or perhaps 4-dimensional) set of spectral-values. The number of subspaces, though a large-number, is finite, Another example of identifying this stable spectral-orbital set would be the periods of the planetary-orbital rotations of planets around their stars, even at distant stars, (at least for planets within the galaxy, ie planets within distant galaxies are not distinguishable) would also be related to this finite spectral set, namely, some of the 3-dimensional and 4-dimensional set of stable shapes.

Note:

The "mass equals energy" relationship can be interpreted to mean that material-components are the set of stable "discrete hyperbolic shapes," which represent the forms of stable energy which can exist in space, by being in a relation of resonance to a very large finite spectral-set, and where these shapes exist over many dimensional-levels, and so that these stable shapes are, initially, linear-rows of toral components, which are (can be) folded at the places where the toral-components are connected to one another, and can form into condensed material.

Condensed material, will eventually follow (or is following) the geodesics of the shape within which the condensed material is contained.

The material-components which are (ie can be) contained within such a stable metric-space shape, will, eventually, follow the dynamic structure which is defined by the metric-space-shape's well defined geodesic structure, so as to define stable spectral-orbital paths for that material, for example, condensed sets of atomic-material-shapes, will follow the geodesics of the metric-space shape (within which they are contained), eg the earth is composed of condensed material, which as a planet will follow the geodesics of a 2-surface of the 3-dimensional discrete hyperbolic shape, ie the space-time metric-space within which the planet is contained.

Chapter 2

Extrema and physics

Energy representations and their correct (high-dimension) context of stable shapes

and the relation of this context to the math structure of "the extremes of action principle"

In a new model for the context of physical description the process structures of material-component-interactions focus on toral shapes which bridge separation distances and there are sequences of discrete changes in these toral shapes defined over very small time intervals so that the average motions defined by these toral changes determine an average motion which moves toward extreme values of action (or energy-paths).

There is the definition of inertia (or mass), which depends on (or is defined as) the resistance to the changes of a material-component's motion, but in the context of both....

(1) the material-component's spatial-position, where the material-component's spatial-position is are related to the relative geometry of the material-component's surrounding, but distant, material-geometry, determined in relation to the position of the original material-component's spatial-position, so that the distant material-geometry defines a force-field, which is defined at the original material-component's spatial-position (but this force-field is caused by the distant material-geometry),

and

(2) the material-component's changes of motion energy is (can be) defined, where

Energy is the relation of a line-integral to the two separate representations of a line-integral to the two representations of force (inertial and force-field) so as to define the two types of energy, kinetic-energy (motion related) and potential-energy (related to surrounding geometry of (distant) material by the line integral)

The force of (1) inertia, and (2) the force-field (caused by a distant material geometry, which acts on a material-component),

That is,

(1) resistance to inertial-changes

and

(2) the causes (the force-fields) of a need (by a material-component) for an inertial-change, in regard to the relative positions of the distant material (geometry) to the spatial-position of the original (solitary) material-component,

so as to define (in)

the two types of line-integral representations of energy, which can be considered in relation to inertial-force or force-fields,

(1) the inertial-changes of a material-component's spatial position, namely, the kinetic-energy (motion)

and

(2) the spatial-position changes of a material-component due to a force-field acting on the material-component at each position is space, the potential-energy (position), so that the idea of an energy-value unites the two different descriptive contexts of motion and position, but which are both defined as line-integrals of force, where force can be defined in two equivalent ways, while it is the (partial) differential equation (or law of physics) associated to the inertia of a material-component's local spatial-position and this quantitative relationship's "equality relation" to (material-geometry-caused) force-fields, which unites the focus of a description upon a component's inertia at a (local) spatial-position to a force-field model, so as to be definable by (models of) locally measurable properties (or functions) of a material-component at a point in space

By defining the property of action, ie either pdx or Edt,

one can re-establish the locally defined differential equation...,

"inertial descriptive properties which are equated to force-fields" (ie the (partial) differential equations associated to the laws of physics) through a process of identifying a mathematical extrema of an action-integral.

This is the original model of a unification of "the system defining (partial) differential equations" for different material-types, and the (energy-based) description can, in a more general context, describe different coordinate functions for the system's local representation of spatial measures (ie local coordinates), by using the scalar function of energy and defining a quantity called action

(though it is not clear why one would want to unify the local measuring relations (ie the system's (partial) differential equations), which are defined for the different material-types (where this, so called, unification is, simply, about using the unifying-word, (of) energy)

Thus, for the single-value of energy, represented as a being contained in phase-space, (then) the seemingly distinct values of both "local inertia" and the "local values of a force-field" [defined in relation to distant, material geometry (relative to the particular spatial-point)] become mathematically, seemingly, a much more unified idea, in regard to the representation of physical law, when it is represented as sets of math operations, which act upon the action-function (associated to finding a critical point of the scalar action-function so as to find the action-function's maximum or minimum values), which, in turn, depends upon the definition of a scalar energy-function for a system (locally defined), and, subsequently its (the action-function's) extreme (action) properties, indeed it is the action, defined for a discretely restrained spectral-orbit of an electron-charge, which first defined the H-atom's spectral properties, as identified by Bohr,

ie {[integral-(around a circle)-of (pdx)] = nh}

Note that the elliptical perturbations of this energy-spectra (as done by A Somerfeld) are as good (ie sufficiently precise) a description of the H-atom's energy-spectrum, as any other such description, if not much better (it is also much simpler).

Though this (relation of physical law to an extreme value for (or associated to) a system's action-function) inclines one to believe in the intrinsically unifying property of energy (or the action-function), one must wonder,

What is the reason that a math operation associated to an extreme condition for an action-function would be related to the equations of physical law? (?) eg the local changes of an action-value (for an action-function defined over a time-interval), so that the time-interval can be used so as to identify different line-integrals associated to the energy-function (ie different lines or curves for the energy defining integral), but defined for the same time-interval, over which the action-value is defined. [and where it is assumed that there are no holes in the space over which the action is defined (but it is a locally defined operation which defines the condition of an extreme-value for the action-function.)],

The question which pop's-up is,

Why is an extreme value for an action-function desired (apparently, by the physical system itself) for a physical system?

Or

What is the (physical or descriptive) mechanism through which a physical system is pushed so as to possess an extreme-value for its action-function?

Is the answer related to the Bohr model of the H-atom?

Do holes in a space (or in stable spectral-shape) which is associated to a stable many-(but-few)-body spectral-orbital system, as well as periodicity (or the bounded-ness of a stable system), play a much bigger role in physical description, so that the existence of such holes in a system's-shape are related to the (spectral) resonances a system must have in order to exist, as a stable spectral-orbital (material) structure.

ie (hypothesize that in the context of the new descriptive construct) resonances exist between a relatively stable system and the finite spectral-set of the over-all high-dimension containment set, which, in fact, really defines the containment structure of existence.

That is, the holes in a stable (bounded) system's shape, are immanent (inherent), in the relation it (the stable shape of the system) has to the possible resonances, which are contained-in (or intrinsic to)…,

{a stable system's relation to the spectra of}

… the entire high-dimensional containment set, and this high-dimension containment-set's natural set of a finite spectra (which the high-dimension set possesses).

That is, the bounded system {bounded either by time and its spatial-positions, or by the other systems of which a component can be a part (ie part of a larger stable system)}, requires that the system's motion (or local changes) be related to the set of possible stable shapes (or equivalently to a set of certain-sized holes) to which an energy-structure might be related, where this is modeled as material-interactions which are defined to be determined by an extra-added, toral-component, which is to be added to the new model of interaction between material-components, so that the extra-added toral-component…, which defines the contact of an material-interaction (…contact… like either a force-field or like a particle-collision)…, also defines an energy-shape, which is seeking (so as to be stable) a relation to resonances (or of the entire toral-component complex, associated to an interaction, to be resonant) with the finite spectral-set, in turn, associated to the high-dimension containment set, ie the 11-dimensional hyperbolic metric-space; whose spectra and dimensional levels are organized around the existence of stable shapes, where these stable-shapes compose the different dimensional levels, in a context of defining a sequence of dimension-and-size which, in turn, defines a (shape and dimension) containment sequence, which is defined (for) as the dimensions of the shapes increase (smaller to larger set-containment construct as the dimensions increase).

The toral-complex of an interaction…,

is defined in relation to a discretely defined set of time-intervals, where the time-intervals are the length of a rotation-period of the spin-rotation around the pair of opposite metric-space states, which are, in turn, associated to a hyperbolic metric-space, so that the "toral-component of interaction" changes with each discrete time interval, so that the discretely changing toral-component (which defines the local geometry of the interaction) comes closer to…

(either larger or smaller, depending on the sign of the interaction-field, at least in the radially directed spherically symmetric-model of a (stationary charge) force-field material-interaction, which exists within 3-space [or in 4-space-time] [see note*]) (as quickly as possible)

…. the next size hole, in regard to the set of shapes which, in turn, are related to the finite spectra, so that a stronger resonance is realized between the finite spectra and the size of the toral-interaction-component.

This might be better related to the expression for action, $p\{d(x)\}=p\{r\ d(angle)\}$, ie a model of a math expression for action, where if, p, is thought about as a tangent motion to a radial center, r, so that the changing process, associated to local changes of action (in the math method, or math-process), is related, by $\{r\ d(angle)\}$, to a spectral-value of a stable shape (where the stable shape's hole-defining spectral-value possesses a circle of radius r), where such a stable spectral-shape is a part of (or which is contained in) the "partition" of the 11-dimensional hyperbolic metric-space into stable shapes, where this set of stable shapes in turn, is associated to the finite spectral-set of (or which is defined for) the 11-dimensional space.

[note*] only in inertial 3-space, ie Euclidean 3-space, where the fiber group is SO(3) which is shape-wise isomorphic to the 3-sphere (and the 3-sphere is 3-dimensional), so that only in this

case does there exist a possibility for a spherically symmetric geometry with a radial direction (ie normal direction to the surface of a 2-sphere) force-field for inertia, and this also applies to (for) stationary charges, while there will (at least) be other tangent-forces for material-interactions modeled as derivative-connections in relation to the geometry of an SO(n) fiber-group, where the dimension of the fiber-group is greater than the dimension of the base-space (or coordinate space), and whose geometry (of such a fiber Lie group) is more difficult to identify, though the shape of a maximal torus within SO(n) could be a way to distinguish an interaction-geometry, where this would be done by relating the tangent of the interaction-torus to the local vector-geometry (ie the Lie algebra of the Lie group) of the fiber group, ie a relation to a connection-derivative, where this geometry (of the local geometry of an interaction-torus being related to the local geometry of the fiber Lie-group) so as to determine an inertial direction of the force-field of the interaction-torus acting on the material-component, where the transformed action (due to the force-field) would be done by the (local geometry of the) fiber Lie group and determined by a law "relating the local toral geometry to the local (fiber) Lie group geometry."

New ways of thinking and new ways of using language (new sets of assumptions)

In a quantitative construct, where the quantities are generated by the finite set of spectral-orbital values, and where these spectral-orbital values are the precise values which, also, represent the type of stable shapes which are allowable in regard to both (1) stable material-components, (as well as)/ or (2) the stable (material-containing) metric-space shapes,

Where metric-space-shapes are stable shapes, which exist throughout the high-dimensional context of either material or metric-space containment, where their existence (within an 11-dimensional hyperbolic metric-space) is dependent on these stable shapes being resonant with the finite spectral-orbital set of values (which are defined within the 11-dimensional hyperbolic metric-space, where this is done by organizing the subspaces of the different dimensional levels by using a finite set of stable shapes of various dimensions and various sizes).

One can understand why action seeks an extreme value)....,

Begin with stable (1, 2) material-component, or (3) stable material components, or (4) stable metric-space-shape, there are four types of "changes of system configurations":

1. (The system) stays the same

2. (the components of the original system) being separated (or broken apart, or pulled apart) (collision, or interaction with outside force-field, or with a new externally caused internal resonance relations)

(the separated components are)

(a) stable (with new systems either higher-energy [energy-added] or lower-energy [energy given-off])

(b) unstable (resulting in a new set of interactions, plus the (possibly, other) stable broken-apart component products)

3. The (original) different components interacting, so as to

(a) come together (until they: unite, or until they separate, die-out to other (distant or closer) interactions)

(b) move apart (until the interactions is of no force-field significance)

4. Orbits of stable metric-space shape become unstable (from outside force-field, eg other distant stable metric-space shape, from new resonances, collision between stable metric-space shapes)

In all four cases the system is moving between stable shapes, where the new shapes (or the shape transformations) are transitions (or changes) between either higher-energy shapes or lower-energy shapes (between maximum or minimum; stable energy-structures). That is, the changes will be defined by changes between a finite set of spectral-energy values (or stable shapes), where the stable components (stable shapes which are in resonance with the given finite spectral-set) define the set of shapes to which a transition can occur.

So that, this is defined in a context of a finite-set of separate and discrete energy-levels, so the change (of a component's spectral-energy) will go up or down (in energy), in regard to the spectral-orbital-energy, so that entering into one or another of these different spectral shapes (or discrete spectral path-ways) of arranging the spectral-energy structures, but whatever structure is realized (up or down), that particular realized component-spectral-structure (or state) will be a maximum or a minimum energy-state in regard to the separate discrete changes of spectral-orbital-energy state which can occur.

But in all contexts, the new total energy before and after the interaction (or the instability occurs, eg often caused by a collision, which results in changes and interactions) is in a context of stability, the new system(s) forming into stable shapes which resonate with the high-dimension containment space's finite spectral-set, so that the other energies (different from the newly acquired stable energy-value of the new system) are discretely separated from the stable energy-value of the new system, and thus, these other states will be above or below the energy (or action) value of the energy state, either into which the system finally realizes (and is, thus, a local extrema), or back into the energy state from which the changes began, and thus (in both cases) will move to either to the upper or lower energy-states which are discretely separated from the original state or stay in the original state, but these are effectively the only energy states (or action values) into which the changes can take-place, and thus, again, they define a set of extreme energy-values (or extrema of action), in regard to an energy-comparison with the original state (or the beginning state).

This can also be related to the structure of both stable systems and to the structures of interactions both of which are "complexes" of toral components but

(1) one type of toral-component-complex is both consistent with the math properties needed to be a discrete hyperbolic shapes as well as being consistent with the (needed) property of the finite spectral-set (or resonating with the finite spectral-set) which is associated to the 11-dimensional hyperbolic metric-space over-all containment set (of existence), note: the material-component..., contained in this type of a geometric construct..., follows geodesics (unless perturbed by second-order interaction force-fields) and such paths are consistent with an extreme value for the action function,while

(2) the other type of toral-component-complex is a combination of a set of (or pair of) stable discrete hyperbolic shapes (or stable material-components) along with an extra set of interaction toral-components which join the interacting stable material-components together (so as to define the local material-interaction structure), so that the "interaction toral components" either get smaller with each discrete time interval (associated to the discretely defined interaction process) or larger, so that in either case (smaller or larger) the toral-component-complex is approaching a new stable shape, which defines a local extrema of action.

This means that the context for change of energy (or the value of action), which results from an interaction (or an un-stabilizing event), always defines an extreme-value, relative to the adjacent (higher or lower) allowable energy-levels of stable system properties, in regard to other transitions of the system's energy levels (or energy representations) where these transitions can be defined in regard to the action-values of the transition process, and the transition process is defined by expanding or contracting interaction toral-components so that the interaction toral-component-complex is approaching the new extreme action-value as directly (or as quickly) as is possible (the transition defines a path of minimal action).

In regard to the main principle of general relativity, that "a dynamic system, eg a moving material-component, follows a general metric-space's (or a shape's) geodesics" but this principle only makes sense if the shape (or general metric-space) is a stable shape (or a well defined metric-space [shape], eg a discrete hyperbolic shape) otherwise the quantitative context has no meaning, if a shape is unstable (or a general metric-space's metric-function is quantitatively inconsistent) nothing is definable within such a descriptive structure.

That is, the new descriptive context gives a clear reason as to why the action of a system-interaction leads to an extremum (maximum or minimum),whereas in the current descriptive context it is an empirical law, with no descriptive context within which it is put-forth (other than it being a mathematical process), rather there is only the empirical (or calculation-al) context that "doing this method leads to the (natural) coordinates and related system defining (partial) differential equations"

Description and verification (inter-relationships)

How one organizes the (a) precise descriptive language which one uses to describe measurable patterns, eg the Copernican language vs. the Ptolemaic language,

And

How one organizes a new language, eg its set of assumptions its contexts its interpretations etc, greatly affects what one can describe.

For example (Ptolemy vs. Copernicus),

motions (of planets) based on an arbitrary epicycle structure

vs.

motions deduced from a different but well defined geometry (of concentric circles with the sun in the center).

This is about the set of assumptions, which are a part of the language's structure, and subsequently, these assumptions limit the range of patterns which such a precise language is capable of describing (this is the essence of the Godel's incompleteness theorem).

For example: materialism, reduction, internal particle-states, and (indefinable) randomness, a descriptive structure which is based on an arbitrary quantitative context, and thus it is a descriptive language without any meaningful content, (proof it is a language which cannot be used to describe the fundamental stable properties of general nuclei, general atoms etc) vs. A finite spectral-orbital set, from which a quantitative description's quantitative structure is generated, and the set of stable shapes about which an 11-dimensional hyperbolic metric-space has both

(1) its metric-space-separated dimensional-levels (separated by metric-space shapes) as well as

(2) the set of material-components which can be contained within any of these metric-spaces, where all of these well defined constructs are dependent upon being in resonance with the finite spectral-orbital set upon which the quantitative structures of the descriptive language are being generated

In this new descriptive context, the SU(2)-related particles (of particle-physics) are (or may be interpreted to be), in fact, really 2-dimensional (real-number) shapes, associated to the properties of both (1) the breaking-apart of stable 3-dimensional shapes and (2) the nucleus and (electron) atomic shapes, while the SU(3) are (similarly) related to 3-dimensional shapes associated to both (1) the nucleus and atom, and to (2) the breaking-apart of similar dimension stable shapes, in regard to both the nucleus and atomic systems. (see below)

Continuous spectral sets in the new context

However, in the (new) descriptive context of there, essentially, being only discrete separated energy levels associated to stable shapes, then

In order for a physical system to have a continuous spectral property means two things:

(1) that the system is being defined in an essentially classical context, eg electrons moving freely within conductors

and

(2) there are some discretely defined systems which have well defined orbits to which the material-components are adhering so as to mainly manifest the energy-orbits to which the charged material-components are adhering so as to define the main spectral properties of the system but that either there are many orbital possibilities as in a crystal or as in a nucleus which has a much fewer set of possible orbits but there is not a well defined central charge for the nucleus which is (causing or) organizing the perturbing interaction structures (ie the classical [partial] differential equation dependent perturbations), so that there are many types of interaction directions which can exist with each spin-rotation period (ie very short discrete time-interval) so that the perturbing interactions are, apparently, random (or, seemingly, chaotic) but still adhering mainly to a stable spectral-orbit of the nuclear shape, and its necessarily unusual positive and negative charge distribution with protons, electrons, neutrons, and neutrinos, where neutrons and neutrinos are combinations of electrons and protons in either a bounded shape (neutron) or in an "unbounded" shape (neutrino).

The idea of an action extremum in the current physical language

Though the quantum descriptive context can…, in an implied (but never actually realized by calculation) context…, identify the same structure of discretely separate energy levels---, however, there is no reason to believe that there is a finite set of discrete (quantum) energy levels, which define all energy-values---, and thus, should also be related to a similar discrete-energy based argument in regard to maximum and minimum energies for physical systems, though each system can be related to a fixed set of discrete energy levels, any other quantum system can have discrete energy values which are arbitrary so that the full (or complete) set of quantum energy levels for all quantum systems can define a continuum of spectral-energies,

And

there would still be the issue about the same properties of extrema of action for the solar-system, about which discrete orbital properties (or discrete energy levels) for planetary-orbits are not expected to exist, yet it is a part of the Copernican description of the solar-system, and thus the same discrete energy-levels argument would not be valid (within the current context of physical law and the assumed context of its description).

Other places in our culture where it is claimed that a descriptive context is based on measurable verification (eg particle-physics, economics, psychology, etc)

But it is not true

Since these other descriptive contexts, which are (also) quantitatively inconsistent (usually it is in a context of indefinable randomness) used in a process of fitting data not to well defined (natural) random events, but events which are in a context of language, where the language is being controlled by both the propaganda-education-indoctrination system as well as by social institutions (such as the social condition of being a wage-slave), that is, what is really being measured is the degree to which the propaganda-system is controlling the language and thought of the public, but it is being framed as an objective property of a set of subjects such as the economy and psychology, so these descriptive languages are claimed to be quantitative and measurable but the math… upon which this is being done… is invalid (the quantitative structure is not properly developed or maintained within the descriptive processes).

That is,

It should be noted that supposedly quantitative descriptive languages such as economics and psychology depend on the fixed structures of language, which exist due to both the propaganda system, as well as the economic-and-legal social structures (or the fixed ways in which social institutions are organized), which hold language and thought and many behavioral-patterns fixed within society, and not the so called rules (or laws) of either economics, or psychology, [these criticisms concerning quantitative consistency even Darwinian evolution (ie mutation and the survival of the fittest)], where all three (economics, psychology, and evolutionary biology) of these descriptive contexts are based on indefinable randomness, and thus their, so called, precise descriptive constructs (or vague references to "what are considered valid intellectual viewpoints") possess no meaningful content, other than the content of the institutions and propaganda into which the investor-class invests.

This is very much about the relation which language has to numbers, which is the content of Gödel's incompleteness theorem, whose conclusion is that (math, or precise) descriptive languages have limits to the patterns which the (quantitative) language is capable of describing, where these limits are due to the set of assumptions contexts and interpretations upon which the language is defined and used.

But in the context of indefinable randomness there are no limits to the describable patterns since in the context of indefinable randomness the descriptive language possesses no meaningful content, so the arbitrary descriptions of properties can be made about any arbitrary context.

That is, any fixed category of precise expression has limits to its validity and to its practical usefulness.

How propaganda is used to define arbitrary categories…, whose statistics is based on indefinable randomness…, so that within those arbitrary categories vague distinctions are made and statistics of the arbitrary category developed so that the arbitrary category is endowed with the property of being a measurably verifiable descriptive language.

That is, the categories of intellectual interest to which social-actions (or behaviors) and the value associated to behaviors are assigned… which, in turn, are assigned to (or selected for) society (so as to be the basis for people's motivations) by a society's propaganda-education-indoctrination system, so as to define within a statistical context of a taking many samples-of-averages (eg of a product's value) so that the distribution of these samples of averages form a normal curve, upon which it is claimed intellectual-value (or discernable behaviors) can be measured, are, in fact, false measures of "distinguishable properties which possess value," let alone intellectual-value, and if used will lead to the failure of the society's intellectual efforts, which is what has happened

Intelligence has been effectively defined as an ability to memorize vast strings of meaningless symbols whose meaning is only about the language inter-relationships of one meaningless partial-string to another meaningless partial-string of memorized symbols (ie the symbols are describing a fantasy-world, a world of only symbols)

For example, the only valid definition of intelligence is "ability to discern truth" and within the context of descriptive knowledge this is outside the capacity of a descriptive construct to determine (ie not capable of determining what the truth is, due to Godel's incompleteness theorem), the verification of such a descriptive construct will be self-referential, and thus, blind to its own failures, only referencing, how it is used in a practical context, and/but the vast majority of current (2014) math and physics is not used in a practical manner (but all the members of this community will have high IQ's, as measured by the education-indoctrination system).

This is what a totalitarian state is based upon, trying to extend a descriptive context beyond its capability to remain a valid descriptive language, so new creative development comes to an end, and one is left with arbitrary complexity which is meaningless.

Thus, if one interprets data exclusively within a fixed, but arbitrary, descriptive language, then this will mean that the fixed, but arbitrary, descriptive language will be verified by any data which is interpreted within the given fixed, arbitrary, language, which can be adjusted to fit with the data.

The language structures of: economics, psychology, and law are about the arbitrary way in which our society is organized, based primarily on the way in which the propaganda-education-indoctrination system controls language and thought in society, so the public will conform to the selfish interests of the oligarchy (ie the investor-class).

Chapter 3

Curvature

Curvature and quantitative inconsistency

The dominant expressions (in relation to the current models) of physical description are (mostly) based on non-commutative and non-linear locally-measurable patterns which are quantitatively inconsistent math structures, and thus, have no meaningful content in regard to that which they are claiming to be describing, eg quantum physics, particle-physics, general relativity, and non-linear classical physics (ie most of classical physics),

The real purpose for ignoring that this type of meaningless descriptive structures is to make things appear more complicated than it is, and subsequently to define an intellectual-class within the propaganda system.

The intellectual-class, and their invalid intellectual constructs, are used to falsely identifying both an invalid descriptive context, and to identify a, so called, (social) class of "the illuminati intellectual-class," where, in fact, these "illuminati" are mostly selected by the administrators of both academic institutions and related industries, ie businesses which use the academic knowledge in the business's own limited narrow ways, and they are (simply) those people (illuminati) identified (by administrative mangers) as being of an autistic personality-type, who have limited (or rudimentary) language skills, so that complicated but invalid descriptive constructs, such as the non-commutative math models, which this group (of, so called, intellectuals) focuses upon, in turn, defines an arbitrary set of expressions, whose content is about nothing, (a formula [of focusing on "nothing"] which might make TV "situation comedies" successful, eg J Seinfeld show in the 1990's, as well as allowing [or causing] "useless arbitrariness" to "posture as (or pretend to be valid) science and math" in the propaganda-system), so that these content-less set of descriptive categories mainly define categories of propaganda where the main ideas expressed within the propaganda system is that people are not equal and to make practical creativity more difficult to achieve.

The illuminati claim to prove propositions about quantitative patterns, but since these quantitative patterns, ("most" often) depend on non-commutative math patterns, so that this means that the domain space, where these patterns…, which are being proved…, are supposed to exist, is quantitatively inconsistent, so, in turn, this means that such a base-space cannot carry (or sustain) within (or upon) itself these, so called, proven patterns, ie the proofs are meaningless.

Especially, when applied to physical descriptions these proven patterns (based on non-commutative quantitative constructs) fail to be able to solve the most fundamental questions of physical description (the stability of the many-(but-few)-body systems of stable spectral-orbital properties so that these systems exist at all size-scales), ie it cannot be used to either solve fundamental problems (cannot solve problems) or to be able to provide practically useful creative contexts (has no, or very limited, practical uses), but where almost always these, so called, precisely proven patterns depend on non-commutative math properties.

Thus, these descriptive structures cannot describe the physical patterns which are observed, ie the descriptions are not accurate, and the descriptive context has no relation to practical creative development.

The ruling-class (by institutionally managing and manipulating the, so called, illuminati [they, the ruling-class]) have forced science and math into a context of meaninglessness (so the creative capacity of such a culture "has come to falter," and subsequently, this has led to cultural collapse (due to a lack of practical creative development), and the world [society] naturally descends into chaotic and arbitrary violence, eg the rise of the violently dominated European kingdoms after the collapse of Rome (though Constantinople was still around).

This paper is about how a shape's curvature affects the local vector properties defined on the shape

This paper is about how a shape's curvature affects the local vector properties defined on the shape, and subsequently, how curvature affects the shape's coordinate set-containment structure in regard to quantitative, or measuring, consistency, such as examining the consistency of the curved-shape's geometric measures, where the local vector properties are associated to both local coordinate properties and to the "local closed-curve parallel transport" of these vectors, as well as to a local coordinate-vector's relation to the infinitesimal sums of local geometric measures (so as to define geometric measures on the shape), again, based on local coordinate-vector properties.

The main idea is that, except for the non-positive constant curvature spaces (for metric-spaces in which the metric-function's matrix representation only has constant coefficients), the non-commutative nature of the curvature properties on (local) vector-values, where the non-commutative nature is seen to be due to (or is… in regard to, or… its nature is exemplified by the, or causes the) parallel transport of a given vector around a closed curve, defined within a local region about the point of the origin of the transport (and assume that there are no holes in the local region)., (where this process of parallel transport of a vector around a local closed curve) causes the local vector properties of the (or defined on such a) shape to be many-valued, ie the non-commutative nature of the local vector-values on spaces which possess general curvature properties causes the local vectors to have the property of being many-valued math-objects in regard to their (local) closed-curve vector parallel transport quantitatively inconsistent properties, ie they are not quantitatively consistent, and thus, a local vector (ie a vector defined by a derivative process) is not well defined on a space which has the property of possessing general curvature properties.

Note that, if a space has hole in itself, then the line integral around a closed-curve on the shape, which encloses a hole, is not zero, so that such a line-integral (around a hole) would define a many-valued function. Thus, when a physical system's energy is defined (by a line-integral) then it is assumed that the space which contains the energy of the material system has no holes in itself.

However, if the holes define separate and discrete energy-properties (or line-integral properties) then the holes can be a basis for "a set of many different stable energy-levels," which are (can be) defined on a shape with holes in itself, especially, if the shape is a shape of non-positive constant curvature, ie a space isometric to either 0-constant-curvature, or (-1)-constant-curvature, in particular for the (-1)-constant-curvature shapes, whose geometric properties are very stable, and its (the shape's) distinct spectral-orbital pathways consistently separated.

Thus, such a stable shape with discretely separated holes can be the basis for defining a system's stable energy, or stable spectral-orbital, properties.

Nonetheless, the many-vector-valued-ness…, caused by parallel transport of any (given) vector around a (locally defined) closed curve…, is not related to holes, nor is it related to any other (obvious) topologically different set of relations associated to the shape, where, in turn, the vectors defined on the shape are (is) associated to all the different many-vector-values which are (can be) associated to a general parallel transport of the given vector around the general (local) closed curve, and thus, it seems to be a much more fundamental property (or fundamental problem) in which any given vector seems to not be well-defined on the local region of a shape which possesses general properties of curvature.

Curvature is about shapes which have general curved coordinates placed onto their shapes, if the curved coordinates are either not (almost) everywhere continuously commutative (or, equivalently, locally independent) or if the shape's (or the space's) metric-function, though locally (its matrix representation is) diagonal, does not have constant coefficients,then the parallel displacement (or transport) of local vectors around local closed curves will not bring a vector back to itself, ie the parallel transport process for vectors (around local closed curves) will be many-valued, and the quantitative properties of local vectors on such curved shapes are not reliable, so the quantitative descriptions have no meaningful content.

Consider that, curvature is about the parallel-transporting of vectors around a set of general curved coordinates, which are defined on (of for) a shape [in particular, around local closed curves] and where the ideas about (or constructs for) either identifying quantitatively consistent geometric measures on general curved shapes also depend on the local changes of a metric-function when evaluating a parallel-ly transported vector (or pair of vectors) about a locally defined closed curve, and is determined in regard to the exterior-derivative of a metric-function related to a shape's (local) coordinates, where such considerations, if the coordinates are not continuously commutative and the metric-function's representative matrix does not have (only) constant coefficients then the geometric measures are not reliable, and or

The identification of the general types of curves, ie geodesics, which allow for vector-invariant parallel vector transport (defined on curved coordinates), lead to the definition of non-linear partial differential equations, thus the quantitative consistency of such (non-linear) geometric-measures is very questionable.

In fact, the only geometric property which is identified within the context of these considerations of a space with general curvature is that parallel-ly transported vectors along a geodesic-(coordinate)-curve on a curvilinear-shape leave the vector invariant in its length, however, on a shape with general curvature properties, the direction of the parallel-ly transported vector (whose beginning direction is consistent with the local direction of the geodesic) will have a different direction if compared on a global-universal right-rectangular coordinate system for the entire shape.

That is, the parallel-ly transported vector, along a geodesic, would have a different vector-direction after parallel transport than the original vector's direction when compared in a universal coordinate frame (perhaps of higher-dimension than the dimension of the shape) where the universal coordinate frame would be defined at each point on the shape

While within each local frame, along the parallel transport process, the new vector would have the same vector-direction as the original vector direction in each new frame defined along the parallel transport path.

but for a parallel-ly transported vector around local set of intersecting geodesic-(coordinate)-curves, forming a local closed curve on a curvilinear-shape, their directions as well as their lengths are changed during the parallel transport process (unless the original local geodesic, itself, defines a local closed curve).

That is, in general, different geodesics parallel-ly transport in an invariant manner only those vectors which have a particular direction. So a vector which is transported in an invariant manner on one geodesic will not be transported in an invariant manner on a different geodesic-curve, which is the issue when the closed curve is defined by sets of (different) intersecting geodesic curves.

The curvature is the second "exterior derivative" of the metric-function (where a metric-function is a symmetric second-rank (0,2)-tensor product, and where its second-exterior-derivative, in turn, is a fourth-rank (1,3)-curvature-tensor so that the 1 in (1,3)-tensor can be contracted, so as to form the second-rank Ricci (curvature) tensor. Note: The Ricci tensor is symmetric. Note: Einstein equated the Ricci tensor with the energy-density and momentum-flow tensor so as to define a second-order non-linear partial differential equation whose solution is the (local) general metric-function. Then with the metric-function the shape's geodesics can be determined, so as to define material dynamics as the inertia defined by a shape's geodesic paths (or geodesic curves), but this is a quantitatively inconsistent math structure. However, the discrete hyperbolic shapes have separate distinct geodesics defined on their shapes, and their shapes are both continuously commutative and the metric-function has only constant coefficients, so such shapes are quantitatively consistent. That is, such shapes are the only shapes where "inertia defined as geodesic curves upon which material moves" has any meaning. Furthermore, it is only in this context of stable shapes (which define orbital geodesic patterns of material-motion) in which simple two-body material-interactions modeled as partial differential equations (which can be solved) make any sense, and such interactions would perturb the orbital material-motions which would otherwise be defined on the shape's (separate) orbital geodesics.

Natural coordinates

Local regions of metric-spaces have natural-coordinates associated to themselves...,

namely, the local coordinates which diagonal-ize the metric-function's local matrix-representation, where such local coordinates can always be found for a symmetric (local) matrix, where the metric-function is associated to a symmetric (local) matrix, but the diagonal-ized property of the metric-matrix is (in general) only defined for a single-point in the local region.

Thus, the non-commutative properties of the process of parallel transport cause the vector to change its direction when it is transported parallel-ly around a (local) closed curve defined on the shape,

Where the property of a pair of vectors identifying a fixed angle [or being parallel] can be defined in regard to both the given vector and the direction of one of the local coordinate-vector directions...

(defined at a point of origin of the parallel transport)

... (where the property of a pair of vectors identifying a fixed angle [or being parallel] can be defined) by means of the properties of the metric-function as also being used to define an inner-product, through which the given vector-angle (of the two originally identified vectors) which can always be locally identified (if their two lengths are given, eg unit-vectors) as each vector is parallel-ly transported. Note: Two parallel vectors can be thought of as two unit-vectors with the same direction, but defined (or identified) at different points in space.

That is, the parallel transport of the pair of local vectors on a shape (identifying a fixed angle), which possesses the property of general curvature, have their relative vector directions changed when they come back (by way of the local closed curve) to the origin point of the parallel transport process, where this change in the parallel-ly transported vectors is due to the non-commutative nature of the inner-product, which is (can be) used to define a vector-angle property between two vectors...,

(or it can be used to define the length of an [the] initial vector which defines the beginning direction of the parallel transport process, say, along a geodesic), during the parallel transport process about a closed curve.

That is, if (both) the metric-function's matrix representation remains diagonal throughout the parallel transport process and the diagonal components only have constant coefficients, then the vector would return to the origin (of the parallel transport process) so as to have the same relative direction as when it began the local parallel transport process.

But if not, then

This problem leads to another way in which to define natural coordinates on a shape, which possesses the property of general curvature, namely, define the natural coordinates of a shape to be the shape's "geodesics," which are defined to be the curves which determine local coordinate directions at an initial point of a parallel transport process, so that along a geodesic curve (and) where the parallel transport of the given vectors is along the direction of the geodesic-curve's given local vector, so that another local coordinate vector defined at the given point is (can be) parallel-ly transported, along the direction of the given local vector, so that such a pair of vector parallel transportations can have some of their transported (vector inter-relationship) properties determined by parallel transporting their inner-product (which is a scalar, a product of their vector magnitudes and the cosine of the angle defined between the two vectors).

However, it is simplest to consider the parallel transport process only in regard to the local geodesic curve's directions, so as to consider the length of the original vector defining the geodesic curve's (initial) direction in which case the length of such an initial vector which is parallel transported along a geodesic curve would not change.

But if one is considering the geodesic parallel transport of a vector around a local closed-curve, then, in general (and in both cases), the vector will change upon returning to its initial point, ie parallel transport for general geodesics is non-commutative. Thus, the geodesic parallel transport around a local closed curve, in general, is still many-valued, so the quantitative containment set is still quantitatively inconsistent

However, it might (also) be noted that, at some points on a shape either there might not be any curves whose local directions can be used to define a local parallel vector transport based on (such) an initial vector direction, or (there might not be any curves whose local directions can be used to define a local parallel vector transport) based on the local directions of the coordinates (defined at, or) through the given point.

If a local vector is not well-defined in a space, which has general curvature properties, ie its parallel transport around local closed-curves does not return the vector to itself, then the use of such vectors to define geometric properties of geometric-measures of a (general) shape is meaningless, and is without any (valid) quantitative content.

That is, though a geodesic is a possible math property which can be used to define a set of "natural coordinate curves (and, subsequent, locally defined directions) for some shapes

But

In fact, a diagonal matrix representation for a metric-function, which is continuously defined for the shape (except possibly for one-point), and whose matrix representation has only constant coefficients, is (really) the only math property, which can be used to define a (particular type of) shape's natural-coordinates, ie natural coordinate only exist for a few types of geometric-shapes.

However, if the 2-torus is not thought of as a fundamental-domain in a 2-Euclidean-plane lattice rather it is thought of as a 2-shape in 3-Euclidean-space then as in the case of the sphere there are longitudinal and latitudinal angular measures, where the plane that intersects two-concentric circles identifies longitude angular-measures and the perpendicular plane which intersects to separated circles defines (one of) the plane(s) within which latitude is measured, then there are questions about longitudinal measures since in the metric-function's matrix-representation the components associated to the longitudinal arc-length will depend on the cosine of the latitude-angle variable, ie the metric-function for the torus does not have only constant coefficients, yet it still possesses the property of single-valued-ness for parallel transport of vectors around local closed curves on the torus. This is because the longitude arc-length measure will have a local vector direction which can be related to the circle which defines the center of all the pairs of separated circles which are orthogonal to the longitude-circle defined through all of their centers, so this property allows for invariance of direction for parallel transport of vectors around local closed curves defined on a (Euclidean) torus.

Thus, the natural coordinates are only defined for rectangular-faced (convex) simplexes and their associated (distorted rectangular-faced simplexes of the) convex polyhedrons of a hyperbolic metric-space and cylinders, ie circle-spaces (where the circles are orthogonal) and rectangular-faced simplexes. This has to do with the quantitative consistency which can be associated to the shapes of the circle and the line in regard to both the real-numbers and the complex-numbers, where both number-systems satisfy the same algebraic axioms.

Apparently, how the property of general curvature is related to a stable math context is through such a shape being unstable and spontaneously transforming to either

Its total disintegration (eg deforming to a point, or breaking-apart)

or

These general curved shapes transform into one of the stable shapes of (-1)-constant curvature, where this latter type of transformation seems to require a hole structure in the original shape, or developing from the original shape.

Since both the Newton definition of inertia, as well as the definition of force-fields, are second-order (partial) differential equations then, apparently, it was believed (by Einstein) that the second-order "exterior derivative" applied to the second-rank tensor-product structure of the metric-function, so as to give a rank-4 tensor (product), which is then contracted to form a symmetric second-rank, or (0,2)-tensor, and related (equated) to the energy-density tensor (differential-form, or alternatively symmetric two-index (or second-rank) tensors, ie the Ricci tensor is symmetric and rank-2, and of the (0,2)-tensor type) but when energy is placed in a context of generally curved shapes (or spaces) then it is no longer conserved. Note that, this should be the main clue that such a descriptive context is not within a valid quantitative structure, since energy and material conservation are fundamental constructs of physical description, they are needed for a precise description to have any meaningful content (though curved-space does seem to conserve material, but it is not clear that material has a valid definition in a generally curved space since generally curved spaces have no valid quantitatively consistent, base-space (or material-containing metric-space), structure.

The affect of this formulation, that the second exterior derivative of the metric-function whose upper-index is contracted, and is subsequently set equal to an energy density tensor, allows a general metric-function to be solved (as a partial differential equation), at least locally, and then the local geodesics can be determined (in regard to the solutions of the geodesic partial differential equations, which depend on the metric-function for their definition, and which, unfortunately, are non-linear equations, and thus, there is no valid solution-functions to the geodesic equations, in general, geodesics are quantitatively inconsistent and thus they have no (precise) descriptive validity, ie (in general) they are mathematically meaningless constructs), where it is assumed that dynamics is a result of a material-component following the geodesics on a generally-curved shape, ie this is the equivalent of inertia. But this can only have meaning in a quantitatively consistent shape, ie the shapes of Thurston-Perlman's geometrization (basically the Euclidean tori and discrete hyperbolic shapes [and rectangular-faced simplexes and cylinders]).

Furthermore, in this context (of general relativity) it is assumed that force-fields are really the result of a generally-curved space (shape of the system) upon which a material component will follow the inertial paths of geodesics. But only the one-body problem, with the mass of the planet is assumed to be equal to zero, where the one-body, ie the sun (or a star), is assumed to cause the geometry of space to be spherically symmetric, is the only solved "physical" system for general relativity, "if the assumption of the planet's mass being equal to zero is actually a physical problem?"

But this idea about inertia, namely, a material-component following a shape's geodesic paths, can only be placed in a quantitatively consistent context, where the system can have many-(but-few)-bodies (with each body defining independent orbits), when it is applied to the stable discrete hyperbolic shapes, which, in turn, identify the different dimensional levels of a many-dimensional containment space containment space for "all" of existence,

so that within this new context, the partial differential equations related to material interactions…, where these material-interactions are defined within a higher-dimensional space than the dimension of the shape which defines the inertial geodesic paths…, can be considered so as to perturb the inertial orbital-pathway of the material-component, which is contained within the stable shape, but the geodesic structures of the shape isolate the material-interactions so as to only apply to one geodesic-orbital material-component…

(or one material-component along with the central very large material-component in a spectral-orbital structure)

…, at a time (in regard to material-interactions, ie a perturbation of the material-component's dynamic pathway in regard to a two-body model of a material-interaction for material already orbiting on a geodesic path upon a "discrete hyperbolic metric-space shape," so that the geodesic orbit is perturbed due to a further material-interaction which is modeled as a partial differential equation).

The main thing which this new descriptive context (many-dimensions organized around stable shapes of different dimensions) is to provide a new math model for the philosophically identified idea of there being an "ideal world," where one usually contrasts materialism with the philosophical viewpoint of idealism, but idealism is essentially, the viewpoint of institutional religions dedicated to a belief in realizing existence's true nature or a belief in God (or in Gods).

Although the new ideas

1. Solve the many-(but-few)-body problem

2. Identify the correct context in which to apply general relativity (inertia is related to geodesics on discrete hyperbolic shapes where discrete hyperbolic shapes model both material-components and metric-space material-containment spaces)

3. Provide a quantitative structure for measurable descriptions which is finitely generated, so as to "get away from" the logical inconsistencies of the continuum, which is defined for the real numbers.

4. How we are topologically trapped in a particular dimensional space, which deeply affects our belief structures.

5. Explains the apparent random properties of small material-components

6. That physical description depends on a mixture of metric-spaces of (non-positive) constant curvature, a mixture depending on dimension as well as the metric-function's signature, eg where the "signature of a metric-function is about the division of a metric-space into subspaces, such as the spatial subspace dimension, s, and the temporal subspace dimension, t, of (say) R(s,t) [or

C(s,t)], where (s + t) = n, and n is the dimension of the metric-space, ie R(s,t) can be equated with R(n).

7. The (real-number) metric-spaces possess metric-space states and their associated spin-rotations of (or around) metric-spaces states, where these metric-space states can be defined over "the real and pure-imaginary subsets" of the metric-space's associated set of complex coordinates, and thus, there is a natural relation of physical description to unitary Lie fiber groups of the shape-containing (or shape defined) base-space for all the different types of metric-spaces defined in regard to physical description, in turn, defined for some particular dimensional level.

8. The apparent spherical symmetry of the force-fields of matter in 3-space (or in 4-space-time)

Note: The sphere is only a stable shape if it is not perturbed, and if perturbed (as would be true for any realistic model) the spherical shape disintegrates either disintegration or it transforms into shapes with holes. Though the local coordinates of a sphere can be made continuously commutative (ie the longitude and latitude coordinates, except at the two pole-points), nonetheless, the diagonal components of the metric-function's matrix representation are not constant, but rather functions of the other coordinate-variables of the sphere, and it is this property which causes its local (vector) structure to be non-linear, and thus quantitatively inconsistent, furthermore, it is a shape which is so susceptible to shape-disintegration due to perturbations (or deformations) of its shape.

It should also be noted that the 1/r potential function for a spherically symmetric force-field of a material-component (in 3-space) also leads to great mathematical limitations as to its (the potential's, or its associated force-field's) external material-affects.

Thus, a material-interaction model based on the discrete Euclidean shape (or the toral shape) is an improvement for physical description, but/and it leads one to question the idea of a gravitational singularity (though Newton's laws also identify a mass-value which, in turn, identifies a black-hole).

9. Why distant material motions of galaxies appear to be motions in flat space (such motions are Euclidean)

10. Galaxies modeled as high-dimensional regions in space, whose (now interpreted to be) spatially-expansive motions are due to group conjugation of their high-dimensional shapes by high-dimension classical Lie fiber groups related to their shape's dimensional property, in 11-dimensional space this could be Lie-group conjugations between different galaxy regions, defined in 11-space.

11. A new model for van der Waals forces, due to the existence of higher-dimensional shapes and their subsequent forces, which are, nonetheless, very small forces

12. We exist in a mixture of both different dimensional spaces and there are several "different metric-function signature" metric-spaces which form the totality of the mixture of metric-space types.

13. Motion and dynamic (partial) differential equations are to be modeled in regard to their having a definite relationship to "a discrete time-construct which exists during material interactions, where the discreteness is related to the period of spin-rotation of metric-space states defined for each (real) metric-space, thus bringing the description into the context of complex-coordinates (there are other discrete inter-relationships in the new math construct).

14. A new model of life, as a high-dimensional, at least a 3-dimensional, and a unified shape, where such a 3-dimensional shape would naturally fit into a 4-dimensional hyperbolic metric-space, where the high-dimensional discrete hyperbolic shape naturally oscillates, so as to generate the life-form's own energy source. Life can be 7-dimensional or 9-dimensional etc. so that these high-dimensional unified shapes can become important forces in regard to the unified processes which occur in life-forms, eg chemical processes being affected by sets of resonances, defined within the shape, (in turn) being controlled by the unified high-dimensional life-form's shape and the ability of a life-form to control energy-flows between its sub-faces.

15. A new model for a memory, constructed for a life form, where the memory is based on the maximal tori which are a subgroup structure of the metric-space's fiber Lie groups where the maximal tori can carry within themselves spectral properties, and thus carry information about the spectra of discrete hyperbolic shapes.

16. The model of a neutrino might be best represented as an unbounded discrete hyperbolic shape, where such neutrino models could begin at the dimensional level of hyperbolic-dimension-2, where this provides a geometric model of an infinite, but whose quantitative structure is related to the finite set of bounded shapes of different dimensions (which define a finite spectral-set of finite values for the over-all 11-dimensional containment set), and which organize the 11-dimensional containing space.

17. And this all depends on D Coxeter's identification of the properties of the different dimension discrete hyperbolic shapes, which are defined on hyperbolic metric-spaces, where these properties are the properties upon which much of the organization of the higher-dimensional shapes depends, and thus they organize the many-dimensional (11-dimensional) containing hyperbolic metric-space for "all" of existence, but there can be many such 11-dimensional spaces, in fact, a

galaxy may be a region which really identifies a region in space associated to an 11-dimensional hyperbolic metric-space.

18. This new focus on the non-positive constant curvature stable shapes as a basis for physical description and the fact that non-commutative and non-linear quantitative patterns are unstable, is also the basis of the Thurston-Perlman geometrization, so this is a way to use the relatively stable shapes identified by geometrization (as well as using the properties which Coxeter identified for discrete hyperbolic shapes, defined up to hyperbolic-dimension-10) to model a valid, quantitatively consistent model of existence, of which the material-world is a proper subset. That is, this (new idea) is about the use of geometrization to re-organize the language of physical description.

Where physical description has not been able to progress past the 2-body physical system in its ability to describe the observed stable spectral-orbital properties of the physical (and measurable) world

Nonetheless

But it is a model of an existence for what is usually considered to be a "philosophically referenced"

"ideal-world,"

Which is the main contribution which this new math model of existence provides.

But it is a model of an ideal-world (a model of existence) which contains the material-world as a proper subset.

There are the dichotomies:

Materialism and partial differential equations, which are used to determine either geometric or random material relationships between material-components, either material-shapes or (supposedly) random point-material events

vs.

Reliable measuring and stable patterns

This is equivalent to the dichotomy of

Material based science vs. idealism

Which, in turn, is reduced (in the failed culture of the empire's propaganda-education-indoctrination communication system) to:

Materialism vs. religion (where the propaganda-system, in the west, makes religion into a brutal paternalistic material-worshipping form of Calvinism, wherein the rich are identified as "the Gods on earth," and brutality is determined by arbitrary, and, ever invasive, spying and moral-ism)

That is the properties of individual personalities, in regard to the public, are spied-upon and used and manipulated for the benefit of the ruling-class [this is, essentially, the same as the use of personally owned armies and the use of spies by the Roman aristocrats, but now it seems the rich are more collectively integrated, apparently, they all are rich and they all want to keep and enlarge their fortunes so they have become very collective in these efforts].

Tangent about religion and its relation to personality-cult

There is

Materialism as defined by Galileo and Newton…,

And then "incompetently" modified by Einstein,where Einstein's (incompetent) modification was adopted by the propaganda system so as to identify "the intelligencia," or the intellectual-class, which are effectively used (by the propaganda-system) to define "for society" its highest cultural-values [where for this empire, it is about using {the useless} particle-physics in an attempt to detonate a mythological gravitational singularity, so as to create the ultimate doomsday machine, ie the highest cultural-value of the western empire defines a culture of death {an empire of death}]

but as "pointed-out above,"

Einstein's modification has no meaningful content within its descriptive context.

But

Where idealism was also "incompetently" defined by the propaganda-arm of the Roman-empire as arbitrary intellectualism and arbitrary moral-ism (though today the intelligencia are used as a propaganda-ploy, so as to maintain the arbitrariness (of high-cultural value) of the intellectual-class, whom serve the interests of "the empire," in a manner similar to how the church was used, to represent the intellectual-class, by the propaganda-arm of the Roman-empire, ie it was used to justify the arbitrariness of the empire's brutality.

That is, the illuminati (the so called intellectual-class, whom are held so very-high by the propaganda-education-indoctrination system) today, ie the (so called) "Einstein's of society," have been used to turn science into meaningless arbitrary religion, where, effectively, these intellectuals (since they are uncritical of their own intellectual efforts) worship the investment of the ruling-class which has been put into bomb-engineering, which is, effectively, a 19th century model of chemical reactions, ie rates of component-collisions during a chaotic transition process between relatively stable states of material).

The so called illuminati, or today's high-valued intellectual-class, are autistic types who seek social domination (a good example would be E Teller or J von Neumann, and to a lesser extent Einstein ie note that, Tesla did not respect the, so called, great intellectual capacities of Einstein, etc) and they are manipulated by the investor-class, and are mostly used in the propaganda-system, since general relativity never had much to do with socially-important technical instruments, it only has been used to arbitrarily pick the so called intellectual-class.

Whereas the creative efforts of Tesla were substantial, and, subsequently, the ruling-class simply robbed from him.

This is about war waged…. against those who truly advance the culture…. by the rich, so that the war is defined legally in the context of property rights and minority rule.

That is, Roman-law based on property rights and minority rule (in turn, based on violence, which is organized in a collective society which supports the rich) is the context through which the rich wage war against the public

(the justice system is always designed to be a counter-insurgency action which is waging war on the public).

This is Calvinism and the justice system imposes the worship of the material-world on the public (the justice-system makes us all Calvinists, but this deep belief in the material-world is defined as science, but it is (really) the religion of worshipping the material-world {since now (2014) it has been shown that science does not need to be based-on the idea of materialism}, so that the war, by the rich (waged) on the many, centers on property and the violence needed to maintain the {given, or the existing} distribution of property within the (collective) society.

Where the society is a collective since the public is required to worship the material-world

(those who believe that the Bible should be read an interpreted individually (and not how the propaganda-indoctrination-system wants individuals to interpret the Bible), should also be quite aware that Christianity has come to be all about supporting Caesar (this is the propaganda-indoctrination interpretation of the Bible) where the: spy, justice, and propaganda system have all waged war on those who do not interpret the Bible as "the collectivist need by the public to support Caesar,"

where a belief in an ideal world is a belief which opposes the rule of Caesar over the material-world)

Those in the old US (the beginning US) who believed that the Bible should be read an interpreted individually, were called the evangelicals, but "just as Tesla gets robbed" so too the evangelicals were attacked and corralled so as to now espouse (over the propaganda-system) the ideas which support Caesar,

This is how religion within a police-state is to be defined, within a context of wage-slavery and spying and manipulating through spy-information and the management of institutions by essentially, teams who are loyal to the police-state, and who serve the paymasters of the wage-slaves.

And, subsequently, (for those who, supposedly, interpret the Bible on an individual basis) to not worship the "ideal-world," (eg as Jesus commanded to worship the ideal-world, ie the spirit) so as to not worship knowledge, and knowledge's natural purpose, namely, "practical" creativity, through which a person can give in a selfless manner to others in society, and where "practical" can now be considered to be many-dimensional, where there is now a quantitative and geometric description of the many-dimensions to which human-life is most naturally related, in regard to the "natural" creative efforts of the human-being (eg to create existence itself [re-organize the stable patterns] by using their individual intent on a high-dimensional context, so that their intent is led by correct knowledge, so as to make ever larger, the creative context of life's intent).

But the dichotomy (materialism vs. idealism) may be better identified in a very realistic manner by

"material based science" vs. idealism

Galileo's (and Newton's) materialism (leading to (partial) differential equations associated with either material geometry or the random reduced point-particle event model of material (in space), both of which result in non-commutative and non-linear math patterns, which are patterns which are quantitatively inconsistent and thus they are descriptions have no meaningful content)

vs.

The idealism which results form a many-dimensional containment set, within which stable shapes are used to organize the many-dimensional levels

That is, the model of an ideal existence is given a quantitative model of many-dimensions, up-to and including a hyperbolic metric-space of 11-dimensions, and the different dimensional levels, including the different subspaces of the same-dimensions, are separated by means of stable shapes of different dimensions and (often in) different subspaces, in a dimensional-size sequence of set-containment, organized around the stable shapes,

So that

A finite set of stable shapes of different dimensions, as well as being

defined in different subspaces of the same-dimension, a set of stable shapes which possess the property of having an upper and lower bound on the geometric-measures of the singly-circumvented set of spectral-orbital properties of these shapes (where the spectral-values are associated to the hole structures of these stable shapes) which any and all unified systems, which are contained in the 11-dimensional hyperbolic metric-space can possess, due to these shapes (which exist with the 11-dimensional containment set) being in resonance with this finite-set of shapes and their associated finite spectral-orbital set of values, which, define a finite spectral-set, so that this finite spectral-set can be used to finitely generate (by multiplying this finite-set by the integers) a quantitative set within which all measuring of both system spectral-orbital properties and the base-space metric-spaces' local geometric measures are (or can be) based.

Within this new descriptive context (new viewpoint about a containment set for existence, and a new idea [a new mathematical model] about existence a viewpoint which is not based on the idea of materialism) it can also be seen that the model of the measurable properties of the material world can be a proper subset of the new descriptive context.

That is, in this model of an ideal world material-based science and idealistic-religion are unified, and creative intent can be applied to either practical creative efforts in regard to the wide array of shapes and dimensions which one can try to inter-relate, where in all likely-hood life-forms naturally inter-relate, since life-forms are, most likely, high-dimensional shapes or the creative efforts of exploring a many-dimensional existence where models of shape and material-interaction, and size of the different types of systems, which one might expect to encounter, are provided (by the new descriptive context) as a model for a (spiritual) exploration.

The stable shapes are related to the rectangular faced simplexes of (convex) polyhedrons which generate the lattice of both the discrete Euclidean shapes (flat, or 0-curvature, ie tori [or doughnut shapes]) as well as the very stable discrete hyperbolic shapes [(-1)-constant curvature spaces (or shapes), ie strings of toral-components which can be folded]. The rectangular simplex structure allows the local coordinates to be continuously commutative. Thus, allowing for a quantitatively consistent descriptive context for the descriptions of existence.

The metric-functions only have constant coefficients so the second derivative of these commutative math constructs are constants, ie the curvature is a constant, so that these shapes are isometric to either flat or (-1)-constant curvature metric-space-shapes.

The material-component shapes and the metric-space shapes are the same type of shapes [(-1)-constant curvature shapes] but separated by at least 1-dimensional level. The interaction structures are based on the discrete Euclidean shapes.

Consider,

If existence is given a many-dimensional model then one can ask; "what types of 3-dimensional shapes or material-configurations can couple to the higher-dimensional shapes?" and

"What are the natures of the higher-dimensional shapes, which, in turn, are related to the 3-dimensional material world, which, in turn, have a relation to higher-dimensional geometric properties?"

For example,

The oscillating shape, of dimension-3, is related directly to energy generation, but its oscillating geometry is in a 4-dimensional space, but clearly the 3-shapes which we experience have a relation to these 3-shapes and thus also to 4-space.

Some simple examples of oscillating 3-shapes which are contained as geometries in 4-space are:

1. radioactivity, as well as

2. life-forms such as

2a. Plants (eg photosynthesis etc) or

2b. Bacteria

Etc.

That is, the new models require new viewpoints about relationships to shapes and containment spaces.

Another example is that, the life-form of an oscillating 3-shape may be quite large, eg as big as the solar-system (or bigger), but such a shape would have a relationship to the shape of the solar system, while the entire solar-system, including a model of the sun, [where this may also] would (might) require a 5-dimensional shape (as a model for the life-form), which would be contained in a 6-space, and this would require new ways of considering coupling, in a practical manner, to shapes of very-large size, where such a coupling-geometry may, in turn, be relatable to accessing cheap clean energy sources for earth.

That is, such high-dimensional geometry of the solar-system and/or the sun can be related to the geometry of the condensed material of the earth in the earth's solar-system orbit, and this means that the geometry of the earth, eg its mysterious internal structure, might be easily coupled to the oscillating energy-generating properties of the sun,

eg some idea which (say) Tesla might have been trying to realize in his ideas about antennas and the formation of electromagnetic-fields. Such (magnetic) fields would be related to the toral-shapes or the stable circle-space shapes in 4-dimensions.

Consider that,

The old (or current) set of, so called, experts or the old (current) set of people which composed (compose) the, so called, meritocracy-of-intellect, have failed, both at imagination, and they have failed at analyzing the condition of their own descriptive language,

And thus,

It is not such an indoctrinated set of people who should be either the managers or the set of consultants to which the new ways of revolutionizing the way in which creativity and knowledge are both realized, and used in society, eg through information systems which easily couple to descriptive knowledge which might be possessed or be considered by anyone in regard to creative intent, where the set of assumptions upon which the descriptive knowledge is based in made clear.

Chapter 4

Particle-physics verification (also see chapter 23)
Scientific(?) verification of particle-physics

The verification of particle-physics is done by focusing on unstable random events

Modern physics is about modeling reality as an explosion of the material-world, which has been reduced to sets of unstable elementary-particle events, and since the word explosion is a noun, then any unstable, but distinguishable, property is considered to possess great quantitative significance, even when the descriptive context is quantitatively inconsistent, and thus, the language possesses no meaningful content (there only exist relations to particle cross-sections in this empirical construct).

Essentially,

Modern physics is about modeling reality as an explosion of the material-world, and since the word explosion is a noun, then any unstable, but distinguishable property (random event of a, so called, point-particle's geometrically well defined path, for particles which are supposed to be described in a context of randomness), is considered to possess great quantitative significance, even though the descriptive context is quantitatively inconsistent. This is the viewpoint from which particle-physics developed.

The descriptions of the observed <u>stable patterns</u> of the material-world are ignored.

In an empirical construct similar to Ptolemy's arbitrary epicycles, the arbitrary construct of internal-particle-states is also used to fit data.

Modern physics is about modeling reality as an explosion of the material-world, where material is reduced to elementary-particles (or material-components which collide with one another), and since the word explosion is a noun, and it (the explosion), along with particle cross-sections (which are relatable to reaction-rates), are the central focus (of bomb-engineers), then any unstable, but distinguishable property (random event of a, so called, point-particle's geometrically well defined path, for particles which are supposed to be described in a context of randomness), is considered to possess great quantitative significance, in regard to modeling an explosion of material, even though the descriptive context is quantitatively inconsistent (and thus possesses no meaningful content, it cannot be used to describe any, significant, observed physical property (eg the properties of a stable system)).

This is the viewpoint from which particle-physics developed.

The descriptions of the observed stable patterns of the material-world are ignored, but being able to describe the stability of the many-(but-few)-body spectral-orbital system (which exist at all size-scales) should be the main focus of physics and mathematics.

There is no actual "verification of a theory," since such a theory makes no meaningful predictions, rather this is all about determining particle cross-sections and particle-masses, which are being determined at higher-energy levels (higher-collision-energies). This only about bomb-engineering.

Scientific verification of particle-physics

Consider the method by which Ptolemy verified his descriptions of planetary motions based on his model of epicycles and an earth at the center while the so called heavenly bodies rotated around the earth on their epicycle constructs

The motion of a planet would be predicted based on the epicycle model of the planet's motions and then the planets motions would be observed, and if the predictions were wrong then a new epicycle would be added to the descriptive construct now

Consider the method by which the descriptions of particle-physics are verified, a particular type (or set) of particle-types (in regard to particle-path-detections) would be claimed to exist, and then these types of particle-path-types are looked-for in the particle-detection data the interpretation of the data and the design of an experiment are both being determined by the descriptions of particle-physics, ie it is as closed a description-verification structure as was the language-verification structure of Ptolemy and his planetary-epicycles

Similar to Ptolemy, particle-physics emerged from particle-accelerator data, concerning (billiard-ball type) particles with charge mass and spin properties, and it is this very isolated set of data (colliding-particles distinguished by their: mass, charge, spin, matter--anti-matter, and cross-sectional properties) which the language of particle-physics tries to describe, it has tried to do this using the SU(2) x SU(3) patterns in a context of indefinable randomness so as to identify a quite similar epicycle structure for elementary-particle properties as was used by Ptolemy for the planetary-motion properties.

Ptolemy's epicycle-structure could be used to guide ships, while the epicycle-structure of particle-physics can be used to build bombs.

Context of particle-physics

Within the descriptive context of materialism, ie that material-based experience is all that can exist, so that within this context there are only 3-spatial-dimensions, or 4-space-time-dimensions, within which all material-systems exist, or within which all material systems are contained

Alternative interpretation of the SU(2) x SU(3) math patterns

Thus (conjecture)

The algebra of particle-physics is about the way in which all of the 2-dimensional and (possibly) 3-dimensional shapes (of the nucleus or of atoms) break-apart when they collide with high-energy particles

And

It (the process of stable shapes breaking apart) is placed within a unitary context, ie the shapes which break apart (or which are part of the breaking apart process) are related to complex-coordinate structures, where these are the shapes which are always (or can always be made to be) locally diagonal (in local coordinates).

Thus, SU(2) x SU(3) can be related to (T^1) x (T^2) (or (1-Torus) x (2-Torus)) ie the 1-dimensional and the 2-dimensional tori, which would be examples of a maximal tori in SU(2) and in SU(3), respectively (note: a torus in a Lie group is related to a (continuously defined) diagonal matrix representation of a shape in local complex-coordinates).

Apparently, the data of the particle-accelerators is about the breaking-apart process of the 1-dimensional and the 2-dimensional shapes built from toral components, which can be contained in 3-space (or 4-space-time).

This would be the natural set of data which is relatable to the idea of existence being based on materialism, where it needs to be noted that quantum descriptions are based on materialism, namely, the energy-functions of a quantum description give the probabilities of random spectral-particle events in space (or in space-time).

But the idea of materialism cannot be used to describe the stable solar-system or the relatively stable general nuclei etc etc etc. That is, neither the property of the stability of the solar-system nor the property of the stability of the nucleus have been given valid descriptions based on the-so called, laws of physics, etc etc etc.

These shapes (circles and 2-tori) could be interpreted to be all about the relationship of changes which happen to stable shapes when they are broken-apart by collisions, ie where the stable shapes are built out-of both 1-dimensional and 2-dimensional toral components. These shapes [circles and 2-tori] would be physically related to (or interpretively related to): charges, nuclei, atoms, neutrons, and neutrinos, and unstable related shapes which form in a transitory manner when stable shapes are broken-apart, where these unstable shapes (which also possess distinguished-points) are classified based on the math pattern of SU(2) x SU(3) which, in turn, would express the breaking-apart patterns of circles and 2-tori in 3-space, etc

Thus, there are several geometric constructs which could be... both briefly and more-stably... considered to be a part of a decomposition process of the toral-components and circles (where the shapes are decomposed due to these shapes being struck by high-energy small material-components [or bombarded and struck by high-energy particles]).

The shapes (of: circles, and 2-tori, and discs) are naturally a part of the complex-coordinates, since there do exist the main geometries... of both:

(1) the contractible disc, and

(2) the more stable circle shapes

..., which naturally fit into the complex-coordinate planes (in turn, the complex-planes are associated to each different dimension of the complex-coordinates)

Where both of the shapes (discs and circles) naturally fit into the geometry of complex coordinates, but the disc is more unstable, since it is contractible to a point, and the circle is not contractible.

After a few years of focused efforts, one can be sure, that this (above identified, alternative) context could be just as complex, and also made to have deep analogies with the particle-pattern data, concerning the detected-particle scattering-patterns obtained from particle-accelerators, furthermore, it is a geometry which is deeply related to the properties (geometric and algebraic) of the two Lie groups of SU(2) and SU(3).

The current (2014) interpretation

That is, the data from particle-collisions (of particle-accelerators) has been empirically related to...

SU(2) x SU(3).... in a, seemingly, very thorough, and very complicated manner, so as to model the classification of elementary-particles, which, in turn, are supposed to be the basis for material-interactions..., in regard to material being reduced to elementary point-particles..., where this model is developed as a set of derivative-connection partial differential equations, which, in turn, model the transformation of the internal-states of elementary-particles when they collide, where this collision is the material-interaction, so that the collision is divided into internal-particle-states which are attached to a quantum-system's energy-wave and acted upon by an infinite series of Lie algebra matrices where the (diverging) sum acts on these particle-states attached to the quantum system's energy-function (wherein this component model of an energy-function has a diverging character, which is then re-normalized, so as to give a finite value for each of the different particle-state's energy-function representations)

How can any such predicted internal-particle-state's energy-values be verified, in regard to the different particle-states energies in the quantum system of which they are, supposedly, a part?

Does this require that an atom (or nucleus) be placed in a particle-accelerator's detection-chamber, so as to measure one of an elementary-particle-state's re-coil-energy, (so as to validate its, supposedly, predicted energy-value) due to an accelerated-particle... which has been targeted at the atom (nucleus)..., and which collides with the atom (nucleus) so as to strike the elementary-particle, which, supposedly, composes the quantum system?

But

(despite the over-whelming complications of this model, which is quite difficult to understand, especially, in regard to what this means, namely, how are internal-particle states related to a quantum system's energy-function? and how can these energy-values of the internal particle-states... (which may, or may-not, be a part of a physical system's measurable properties)... be measured? or are they ever measured?)

And despite the incomprehensibility of a descriptive language, which seems to have virtually no relation to describing the observed, stable, physical properties of quantum systems, this is (also) a non-commutative and non-linear descriptive context. and Thus, particle-physics must be an arbitrary (quantitatively inconsistent) model, with no valid (or no meaningful) content in its descriptive structures, {this is seen from its inability to describe the stable properties, which are observed in regard to the relatively stable spectral properties of the nuclei, based on the, so called, laws of particle-physics}

Though it is an intricate model of particle-interactions..., which is, apparently, empirically related to

SU(2) x SU(3), it is built from (and/or claimed to be verified by) statistical correlations, which exist in a quantitatively inconsistent math construct.

But the statistics are defined in a "random" context where the elementary-events are unstable, ie nearly all the elementary-particles are unstable (ie they decay rapidly) This is not a valid quantitative structure upon which to build a descriptive language based on randomness.

Alternative interpretations of the particle-collision data

The above ideas (a different context within which to describe and interpret the data of particle-physics, as well as a deep criticism, in regard to some of the very big problems, which "the language of particle-physics has not answered," and which are not brought-up), indicate that particle-physics, and its associated vast set of data, can be interpreted in a much different descriptive context, namely, in regard to the distinguished points... (the apparent, particle-properties upon which particle-physics is based)... of the discrete hyperbolic shapes. That is, the data of particle-physics can be interpreted in terms of both stable shape and unstable shapes which in both cases these shapes can be related to their own distinguished-points, but furthermore, these are shapes which can exist in higher-dimensions, ie they are shapes which can have a various values for their (own) dimensions (not restricted to the idea of materialism)

That is, not only are the descriptive math structures of particle-physics arbitrary and meaningless they are related to data which can be interpreted in different ways than in the context of indefinably random particle collisions and the cross-sections of these assumed to be "particle"-collisions.

Part of the propaganda-education-indoctrination system of the western culture (a cultural defined by domineering authorities)

The usual way in which western culture talks, namely, as if the west has discovered absolute-truths, and the way in which the west uses knowledge is the only way in which it is possible to use (that) knowledge.

But this is very far from being true, since the so called laws of physics, eg particle-physics, cannot be used to describe the observed properties of stable spectral-orbital systems, particle-physics has yet to describe the stable spectral properties of general nuclei. It is a totalitarian-authoritative viewpoint which is in opposition to the content of Godel's incompleteness theorem, if the observed patterns cannot be described by a descriptive language then one must change that language at the level of assumptions and interpretations. Instead western authorities assert their authority, related to either particular instruments, or related to the traditions of intellectual authority.

Though the intricateness of particle-physics is overwhelming, the description's precise descriptive capabilities are non-existent,

If the description cannot describe the observed relatively stable spectral properties of general nuclei, then of what value is such a description, which is claiming to model the nuclear force?

But

The descriptions of particle-collision properties have always been relatable to the "all-important cross-section" properties of "component-collisions," which are needed (to calculate the rate-of-reaction) in a model of a bomb's explosion

That is,

The authority of the instruments, into which the investor-class has invested, has more social importance than developing truth by the institutions who claim to be developing truth

That is, "particle-physics" (as well as related, or derived, theories such as string-theory) is an elaborate deception of the oil-banker-military investor-class, where such a deception can occur because these same people (the investor-class) also control the propaganda-education-indoctrination system.

Limitations of the language of particle-physics

Only a few types of particles are detected, in the detection chambers of particle-accelerators,

the three families of the following different particle-classes

(leptons and the massive field-particles and quarks [or hadrons])

1. Electrons

2. The pair of massive W-field particles of the weak-force (at high-energy collisions, and the particles are highly unstable, such instability means that the event is not, in fact, an actual entity rather a brief instant of an unstable transition process)

(alternatively: the weak-force is really the inertial properties of a perturbing force-field, which is defined between orbiting material-components, which, in turn, are contained within a stable shape)

3. The quarks, but the quarks cannot be detected and are only relatable to the observable hadron particle-like properties in particle-detectors,

And

3a. hadrons (and the subdivisions of hadrons into different hadron particle-classes, baryons, …. etc etc…)

The hadrons are claimed to be different mixes of the internal particle-states, within which elementary-quark-type-particles… (apparently, inside hadrons, within which exclusively the quark-gluon collision field interactions take-place)… are transformed, when they collide with gluons during a material-interaction of the strong-force, ie where material-interactions can be:

particle-particle or

field-field or

particle-field

type collision-interactions

While leptons and quarks (or leptons and hadrons) are supposed to interact with the field (or field-particle-collisions) of the weak-force

And electrons and protons (or charge and spin-type interactions) interact, in regard to the particle-field collision interactions, within the context of electromagnetism ie the photon model of light,

The most disturbing aspect of this elementary-particle model, is that most of the detected elementary-particles are (themselves) unstable, and quickly transition to other, more stable, "particles." (eg the W-particles possess a half-life of "$10^{(-25)}$ seconds," ie they are very unstable events, their decay suggests a relation to some other pattern, which would be more stable {perhaps some form of a stable higher-dimensional shape})

This quantitatively inconsistent, highly unstable, indefinably random, and extremely complicated (and incomprehensible) math construct which is associated to extremely unstable properties, and has no relation to any description of a fundamental property of a general nucleus, but nonetheless it is associated with an interpretation which claims that (or as if) such unstable events are significant indicators of an existing pattern (but, what pattern would that be?), the only pattern to which the reduction to elementary-particles and a subsequent particle-collision model of material-interactions is through an elementary-particle's cross-sectional measurement which is related to the reaction-rates of bomb explosions (it is a model which is only about bomb detonation instruments,

nonetheless,

Its (ie particle-physic's) inability to provide any practically useful descriptions about the nuclei, should mean that such a math construct cannot be taken seriously by any rational person

That is, the propaganda-education-indoctrination system, along with the manipulative use of the population and its various personality-types (manipulated by the administrators of institutions) and the tight control which the spy system has gained over the society has resulted in a deep following of a form of "grand deception" in regard to how the population and its culture is used by the ruling investor-class.

This is a very poor model of elementary-point-particles, since it is indefinably random and non-linear.

Thus, it is quantitatively inconsistent, and thus, the description is quantitatively arbitrary, ie it has no meaningful content.

This (lack of content) is apparent (or seen) since it is a description which cannot give any information about:

Why general nuclei and general atoms are stable,

and

Why these relatively stable nuclei and general atoms possess the relatively stable spectral-properties which they are observed to have.

A better model might be to give (ascribe) the path-detectable properties in particle-detectors "<u>not</u> to a model of point-particles" but rather to a model of "the distinguished-points of relatively stable shapes," or perhaps "to the distinguished-points of unstable shapes," for shapes which are a part of a transition process.

Particle-physics claims to "predict" the types of elementary-particles which exist, and the, so called, particles which are detected in particle-accelerator particle-collision experiments…,

(it predicted the graviton, which seems to not exist, it predicts quarks, but it also predicts that "quarks cannot be detected," [ie it postures as a religion}, but it did not predict the dark-matter particle, nor the dark-energy particle), That is particle-physics does not predict anything, rather it only fit's a vaguely descriptive language to particle-detection data, it is a description which is entirely empirical with no valid descriptive range, in regard to nuclear properties, and it forms such an limited empirical descriptive language in the context of bomb-engineering, but only the particle-types distinguished by charge and spin, are indicated in the language of particle-physicsnot the particle's mass, nor its cross-section, but where the particle's mass and cross-section are needed for the practical knowledge about elementary-particle properties, which are only used (in a practical context) for bomb engineering (wherein, the only particle-property of interest for bomb-engineering is the cross-section property of a particle)? (needed?)

Where the need to know a particle's cross-section is due to a need to (a narrowly directed desire to) know the physical property of an elementary-particle's cross-section, in regard to it (the property of an, "apparent particle's" cross-section) being a part of a description of a nuclear-reaction's explosion, ie where a component's cross-section is associated to the rate of reaction, ie the probability of collision between particle-types,

The descriptive context of particle-physics is that of randomness, yet its main (and apparently only) focus is on a geometric cross-section properties of point-particle-collisions, which (are assumed to) happen between point-particles (where this is a dys-functional model of material-interactions), but within a descriptive context of randomnessmoving point-particles are models which do not (and should not) exist in a descriptive context based on randomness, this is because of (or due to) the uncertainty principle [which emerges from a property and its dual Fourier transform property upon which the uncertainty principle is expressed] where the uncertainty-principle does not allow positions and motions to be, simultaneously, explicitly expressed, but they are exactly (and explicitly) expressed in regard to a (model of a) point-particle collision.

The particle-detectors are also based on

"the geometrically based physics of point-particles (or billiard-balls)" ([this is also] a model which should not exist in a descriptive language based on randomness).

Though there is an elaborate math structure which is used to "justify" (an unjustifiable model) the geometry of point-particles, in regard to particle-collisions and particle-detectors, which is of only the slightest of interest to the physics community (other than such a "justification," actually, being a deception)

(and, in reality, it is of no interest), since the data obtained from the particle-types detected after a particle-collision in a particle-accelerator is used to determine cross-sections and masses of unstable particle-like (supposedly, random) events.

That is, most of particle-physics is an exercise in distraction, it is not related to any valid, practically creative new contexts.

Verification (within a self-referential descriptive construct)

The description and subsequent verification of a model by interpreting particle-detection data within the descriptive language which only has a practical creative relation to building bombs, that is the only interpretation which is of interest in this descriptive language is its data's relation to bomb-building.

Re-iterating the scheme of particle-classification (in particle-physics)

There are two types of charged material particles

I. leptons with spin, the set of negative charged as well as neutral particles, but with spin, leptons particles,

and

II. quark particles which possess positive charged and negative charged and with spin, but since the quarks have properties which can be used to prove that they cannot be seen, or they cannot be physically detected, ie they will not define any particle-paths in particle-detectors of the particle-accelerators, (rather it will be claimed that there will be hadron particle-patterns detected which are related to quark properties) This is an elaborate epicycle structure.

That is, that the quarks are, supposedly, both positive and negative 1/3-multiple-charged properties, as well as their assumed to exist spin-properties, can be ignored in regard to these properties being detected. They are claims that are not a part of science rather they are claims about math patterns (used to describe particle-paths in particle-detectors associated to particle-accelerators) but the math structure of which these math patterns are a part is non-linear, non-commutative, and indefinably random and thus it is a set of patterns which have no meaning.

Nonetheless string-theory is based on particle-physics where it models space-time and 6-other dimensions (or 7-other dimensions) so that the geometry of the string-theory model is essentially the geometry similar (if not exactly the same as) of the quarks, and it is a 6-dimensional geometry (within the models of string-theory) which is excluded from having any of its properties being detectable within space-time, ie not detectable within the material-world (apparently, only to be in the domain of communication between the high-priests of the bomb, which is to be built on the idea of detonating a gravitational-singularity by means of elementary-particle-collisions)

(an aside)

That is, the quarks, apparently, along with the gluons (with which the quarks, as well as the gluons, themselves) interact, in regard to the so called strong-force, in which gluons act on both quarks and other gluons, so as to form the hadrons

Including both the neutrons, as well as the positively charged protons,

Where it is claimed that the proton is unstable, since it is claimed to be built-up of interacting quarks and gluons,

(but the proton is experimentally found to be stable, nonetheless, one can expect that there will be large budget experiments designed to detect the decaying proton, and in this determined context (and within a language which is an elaborate epicycle construct) some type of particle-detection event will come to be interpreted as the evidence that the proton is unstable, similar to R Milikan's fraudulent data, which it was claimed to have proved that the electron is defined to be some smallest discrete negative-charge point-particle, but the (possible) structure of the neutrino-electron lepton family, as well as an inability to isolate and cool an electron so as to get it to hover as a single-point-charge, rather as it is cooled, it forms into an orbital geometric structure [as shown by Dehmelt] leads one to still ask, Is it as yet clear what the charge-properties of the electron actually are?

(Back to the elementary-particle classifications)

where each of these two different types of material, which are both charged particles (ie leptons and quarks) can be placed into 3-different families, where the families of the particles are based on energy-levels of the particle-phenomenon, ie the energy-level of the particle-collisions in the particle-accelerators, and there is a third type of particle, the field-interaction-particles, eg the gluons, where force-fields between material-particles are mediated by field-particle collisions (between the field-particles and the charged particles)

(another an aside)

where mass, it is hypothesized, is realized by the charged-particles colliding with the Higg's-particle, (or a charged elementary-particle has mass because it interactions with [or collides with] a Higg's-particle), where such a collision is supposed to cause the charged-particle to have the property of being massive,

the fields are:

1. the electromagnetic-field caused by photons colliding with leptons (electrons) and hadrons,

2. the weak-field caused by particles colliding with hadrons and leptons, and

3. the gluons for the strong-force colliding with the quarks and other gluons

The Higg's particle is claimed to have been detected, based on the properties of a collision path of W-field particles, where the W-particles are a part of the weak-force, but these so called events occur apparently 34 in one-trillion collision events

Where these 34 events are supposed to generate particle-track data

ie a probability of a data point which is about $[10^{(-11)}]$ probability,. This is a very low probability, so low it is best to question if such events have occurred. That is, this low probability and the short half-life of the W-particles, ie most of the detected elementary-particles are (themselves) unstable, and quickly transition to other, more stable, "particles." (eg the W-particles possess a half-life of "$10^{(-25)}$ seconds," ie they are very unstable events), their decay suggests that these so called particle-events…

(rather particle-properties are to be interpreted as evidence that the "W-particle's related unstable shape" has a distinguished-point, which is a property of its shape), possess a relation to some other pattern, which would be more stable {perhaps some form of a stable higher-dimensional shape}

Another way to interpret such statistical events with such small probabilities

That is, "is this data, in fact, really from cosmic-rays (not from the particle-collisions of the particle-accelerator), and represents something different than identifying a Higg's-particle event?"

But

Where, nonetheless, this data is interpreted in the context of the Higg's particle's cross-section, where the cross-section is the only quantitative structure which is of any concern in regard to these vague models' of elementary-particles relation to their practical use in bomb-engineering, and the cross-section is the only data which would be of any concern to bomb-engineering.

In a probability based descriptive language, if the probability of an event is $[10^{(-11)}]$ then this is an event which cannot be taken seriously.

That is, probabilities are used to determine if one should bet on "an event occurring," but if the event's probability is $[10^{(-11)}]$ then "one should not bet on such an event ever occurring."

That is, the event is non-existent, if one considers the existence of an event which has a probability of $[10^{(-11)}]$ of occurring.

A cross-section is supposed to be a geometric property so its relation to probabilities of particle-collisions….

for particles which have a higher energy than the energies of the particle-collisions in CERN's large hadron collider (LHC) about 10^{13} eV, should remain the same.

(Probabilities of particle-collisions should remain the same for all energies, but high-energy particle-motions focuses a particle-collision event into a smaller area),

It should be noted that the particle-collisions (between quantum-components) and the detected particle-products (other quantum-components) of the collision are analyzed based on geometry, essentially, the geometry of billiard-ball collisions, and within such a (geometric) model of a collision the cross-section of a geometrically-modeled particle should also be a geometric construct.

Nonetheless, in particle-physics (because it is a description of quantum-components and quantum-components are supposed to be described by their probability-wave properties, so) the cross-sections of these quantum-components could be modeled to be "collision-energy dependent."

That is, the descriptive language of particle-physics is not a quantitatively consistent language, so its descriptions (or its descriptive context) can be arbitrary, so as to fit (or conform to) the interests of the bomb-engineers (as long as cross-sections can be estimated),

so the model can change when it is convenient to do so.

Such as "to fit experimental data with a descriptive context related to bomb-engineering."

This is the trouble with a language which has no meaningful content, it resembles the language which the politicians use to talk to the public during elections, where political talk is a good example of a language which has no meaning (as is the language of particle-physics).

It might be noted that the very-high-energy cosmic rays can have energies of 10^21 eV, where there are about 24 such events over a 1000 meter^2 area for 4-month time interval. That is, cosmic rays may be part of the LHC particle-collision data, if one thinks that such small probability numbers for events (as 10^(-11)) have any significance (at all).

(where the main property of these descriptions is that these particles cannot be detected, rather only the products (or particle-patterns) of their internal-symmetry transformations (which occur during a collision-event) can be detected and subsequently interpreted to be related to a unitary Lie group pattern of a particle's internal particle-symmetries, where these symmetries are defined in regard to the different names of the mostly undetectable and unstable particles which make-up the description)

Review

Within such a complicated descriptive context, where there are assumed to be fundamental point-particles but where, virtually, none of these fundamental particles are detectable by looking at particle-paths in particle-detectors (associated to particle-accelerators), rather only the electron and the quark-composite point-particles, or hadrons, can possess detectable-paths (or asymptotically scattered directions of the detectable hadrons or field-particles being detected) in the particle-detectors, so that the types of hadrons and their collision-properties, and their relation to electron-particles and some (massive, weak-force) field-particles in particle-detectors are claimed to all provide evidence that these theoretical particle-descriptions are "always correct," based on the interpretation of this data within this, so called, theoretically correct interpretive context, but the theory is entirely empirical, where it is claimed that the, usually, short-lived hadrons, which are seen in the particle-detectors, can always be related to the SU(3) patterns, ie as many hadrons as the Lie algebra patterns of SU(3), but this is hardly a prediction, rather than an (at best, an empirical) observation.

As well as the observation that the higher-energy families of leptons (apparently also true for hadrons) have the same charge-spin-properties but different masses, where these masses are also energy-values, but the so called theory cannot predict the spectral (or mass) spectra of the elementary-particles, ie it cannot describe anything rather than it always making correct "predictions." This is a farce.

That is, even though this implies that the only valid "prediction" of such a description would be the masses and cross-sections of the elementary-particles, but this is not being presented as the theoretical predictions of the laws of particle-physics.

This seems to be the main point of seeking the Higg's particle, so as to find (measure) its cross-section, and then to, supposedly, predict the masses of the other elementary-particles, but the data which is being considered is in relation to the unstable collision rebound properties of the unstable massive weak-force field particles with particle-half-lives of 10^(-25) seconds.

Thus, this particle is so unstable that it should not be considered to be a particle which exists.

In non-commutative manifold geometries (such non-commutative geometries) are unstable and they decay (or disintegrate), and such unstable manifolds are not a well defined patterns, so such a non-pattern is not useful as a descriptive construct. This is the true status of the Higg's particle.

Should one consider if the whole set of data should be placed in a different descriptive context, for example, model the Higg's particle as a toral material-interaction component which is decaying (and transforming) with each period of spin-rotation of metric-space state (about 10^(-20) seconds), where this is a pattern in a stable quantitatively consistent context so such small numbers may well have meaning in a descriptive context where measuring is reliable and the math patterns are stable.

That is,

this all seems to be a hoax of epicycles and/or hidden patterns, which are the hidden-constituents of the particle-tracks detected in particle-detectors, ie this epicycle structure (of quarks and electrons and unseen [or seen] field-particles etc) causes this type of particle to be detected in these types of particle-collisions, while these other particles (quarks, gluons, etc) which are also part of the model cannot ever be detected etc etc

One can continue this process of description and data correlation with vague math patterns, and then change of descriptive language to satisfy one's own interpretive convenience, until there is a data description correlation which is found, in a similar manner as Ptolemy used language and data to adjust the description and thus verify his claims about the truth of epicycles within its own descriptive context within a very limited context of a physical description, namely that "the physical description is exclusively related to the rates of particle-collisions for particles of certain types which are detected in the particle-detectors in relation to particle-collisions."

Yet there are no valid descriptions of the observed (relatively) stable properties of the nucleus which emerge from this descriptive context, but rather ever more epicycles within a self-fulfilling description-data context of random events

And then dark-matter was detected (or deduced to exist) and it was given a particle identity, as well as dark-energy then being deduced to exist, etc etc, That is, the particle-types are like epicycles of Ptolemy's model of verification when a new force or a new form of material is identified then new particles are added that is this set of particles can hardly be said to have been predicted rather the particles were detected in particle-paths and then organized within the above model of material reduction to elementary-particles but what other things could an apparent particle-path be caused by? could it be caused by a distinguished-point of a stable shape? Yes, it could be, that is some type of a distinguished-point associated with a stable-shape.

When one considers the way in which modern engineers, in turn, consider the problem of finding a physical process by which the energy of fusion…, ie of pushing nuclei together so that the new nucleus is at a lower energy-state, so the total result is a gain in external energy (so can this gain in external energy)…, can (it) be used for practical purposes?

The context is always the same, even though the properties of the relatively stable spectra of a nucleus have never been described by using the, so called, laws of physics.

And

That same context is about using certain types of elements but these elements can only give-off energy due to fusion if they reach a certain very high temperature

And

There are two strategies to accomplish this

Either

By heating and containing (or confining) high-temperature plasma (very hot charged material in a solid-liquid type state-of-matter

Or

By placing (aiming) an extremely high-temperature (or high energy) impulse at a target of the material-types one wants to fuse together in order to get more energy out of the process

This is the same context which the Manhattan-project (the atomic-bomb project) scientists (as well as those who worked for R Oppenheimer or E Teller) would have given to these engineers back in the 1940's and 1950's.

The investor-class will only (almost always) invest in the knowledge of the propaganda-education-indoctrination system, which has always served the investor-class, and this education-indoctrination system is directed by the investor-class, so this is why the investor-class invests according to the education-indoctrination system's dogmas. They believe the story that merit can only be identified by either psychology or by competitions defined within narrow dogmatic categories, and that the western-culture is marching toward the absolute truths which the high-cultural truths of the west define.

But

The nuclei have relatively stable sets of spectral properties associated to a nuclei with a fixed number of its different two types of components (neutrons and protons, identified by the atomic weight and the atomic-number (the atomic-number is the proton-number))

If a nucleus with a given (fixed) number of components can be identified by the spectral-properties of that-type of a nucleus, then the formation of this nucleus has to have been done in a controlled process, though perhaps quite fast or sudden,

Note: the proof of this, (relatively stable nuclei form in a controlled physical process) can be inductive, namely, that

the idea of this relatively stable general nuclei structure "forming in a non-commutative, and/or non-linear, and in an indefinable random context" has been proven wrong, since that is the descriptive context of today's (2014), so called, laws of particle-physics, and subsequently (since the 1930's, or since the 1960's) particle-physics has failed to supply a valid description of the properties of a general nuclei, ie its model of material-interactions based on elementary-particle collisions has failed.

An alternative viewpoint

That a general (but relatively stable) nuclei forms in a controlled process, is deducible from the fact that there are certain values for "atomic weights" and "atomic numbers," which are preferred by the relatively stable nuclei, and that (in the new description) there are only a finite number of allowed spectra from which to build a nucleus, and these spectra are represented as stable shapes, and stable shapes are related to controlled and quantitatively consistent descriptive context within which stable patterns can exist,

And

Thus, various shapes of various dimensions which are put in a relation to one another, and where interacting nuclear-shapes must satisfy certain conditions of being close to one another, (both high-energy and low-energy interaction contexts between various shapes which are a part of the interaction), so that there are extra interaction toral components (defined in regard to the interaction), which can (must) be of significant spectral-values (eg the toral components are of certain sizes) where the shapes, their separations, and the resonances (which exist) all together are the critical factors which allow transitions to occur between the different possible shapes (stable and unstable, in regard to being resonant with the finite spectral-set), thus, at the critical transition points, where size, shape, and resonance are critical properties (for the controlled transition) in a process of nuclear-change then the set of sizes, shapes, and (domineering) resonances (perhaps imposed on the place where the interaction takes place) can be used to define a sequence of nuclear-shape changes which (it is hypothesized) can lead to the desired result,

(to do this one needs the entire spectral set of the various nuclei, and then try to relate the spectra to stable shapes)

Rather than…. (or in an alternative descriptive context, different from)

…, than simply blasting (forcing) the nuclei together (in a high-energy particle-collision) as in "a model of a bomb's explosion."

Chapter 5

Equal creativity

Just as it needs to be known that the oligarchy is fostering both the social division (it is bank-rolling all sides of the world-wide wars) and dysfunctional knowledge ie knowledge channeled only in certain ways (for military and spying, so as to control behavior), where behavior within society is mostly defined by a propaganda-education-indoctrination system, and its fixed viewpoints, eg that "people are not equal," being continually repeated, ie language is being completely controlled….,

(in regard to vaguely definable and fake, but seemingly, statistically-valued distribution properties of, supposedly, "existing" categories of (the only allowed) social behaviors--- [established by the propaganda-education-indoctrination system's nearly absolute-control over the language, the creativity, and the thoughts of, essentially, all of society (in particular, the fraudulent peer-reviewed science-math, so called, research propaganda, where this, so called, research has only the most limited relation to practical creative development, because the research is virtually all nonsense)]--- where these fixed categories of allowed social activity are established by propaganda and the investor-class {where the investor-class imposes certain types of actions and behaviors, which the wage-slaves are allowed to do, where the investor-class's strongest ally, namely, the justice system, has violently imposed wage-slavery on the public}),

…, and it is within this context that, one also needs to see that (descriptive and/or technical) knowledge has become dysfunctional, and it is only being used to express a form of dogmatic and authoritarian society, a society of market-based-monopoly, which is dominated by a few ideas and products, ie only a few types of (mostly oil and military) products are being made, and so that (such a narrowly confined) knowledge has been made ineffective (due to its own self-imposed narrow-ness).

We live in a society of fanaticism (psycho-pathy), where "the many are being dominated by the few," and where the-many are tricked into crushing "the other-so-called-disobedient-many" (where the crack-down is based on incoherent and non-rational use of descriptive language, as well as being implemented within the context of those acting-to-do the crack-down being (important and virtuous) wage-slaves, where the propaganda system is controlling language, thought, and behavior, and in this context of domination and controlled thought, agents, institutional managers, and administrators are used (sent-out by the investor-class's master-spies) to adjust things {based on consistently following a plan of domination of the, supposed, "superior few" natural social position of moral-and-intellectual dominance over the many})

A different way to consider language and thought

The world of human events is (can be) divided into two compartments (two main categories)

the material-world vs. the world which transcends the material-world, ie the world (or philosophical category) of idealism, this would be the world which is often characterized as where a person might (or could) have a belief in God,

but

the "ideal-world" is assumed to play no valid role in the world of human events (rather only a role within the context of propaganda),

Whereas

in a world based-on idealism, if humans are characterized by their individual creative-ness then such a world should have its law based on each individual being an equal creator,

whereas

in regard to the material-world, if people are considered to be selfish in regard to both social position and in regard to their possessing high-valued material-objects, and they expect to be violent, in order to first realize, and then maintain their selfish desires, then the world's law should be based on property-rights and minority-rule, wherein for such a basis of law, it is the violence expressed by those who make and enforce law which mediates who owns the property, and who determines the course of human creativity (especially, in regard to fashioning [or producing] material-creations).

Selfishness and violence leads to the fixed compartmentalization of both society, as well as humans' creative efforts, or humans' methods of doing things, and the way in which the few can organize society so that the rest of the human population, who are usually (or mostly) poor, can live in small, compartmentalized, places, similar to cages.

It is the activity of putting things (and ideas) into fixed categories, which seems to be the main appeal of basing the laws of society on property-rights and minority rule,

and

not basing (the social organization of) human efforts on everyone being an equal creator, who seeks to selflessly give creative gifts to society and to the world where these inventions and gifts sustains the (quality of) life for everyone (or for all life).

Property has come to mean "getting money, based on doing things in ways which are consistent with the fixed categories of material-use," ie getting-money in a way which is consistent with human society's organization.

And

human creativity is being defined in regard to ways of making-money within a fixed way of using material resources in the context of materialism and in a context of a model of monopolistic-markets, where a market is closer to that activity related to public welfare, ie it is an social action which should be a result of using taxes.

That is, central-planning of a materialistic society, based on selfish interest (ie the model of the monopolistic-markets), is claimed to be superior to "promoting the general welfare," something which government should be doing, so as to allow the public to be free to seek new-knowledge and create, in regard to a wide range of new creative contexts, in turn, made into a wider-range by the development of ever new and practically useful knowledge, ie the truly free-market should emerge from an education and productively creative system

The main idea is that,

if knowledge is made clear, and free equal inquiry is the main focus of both knowledge-development, and in regard to the use of information systems (ie communication systems),

then descriptive knowledge can expand and new creative contexts can continue to be opened-up

The, so called, superior intellects…. (an idea created by the propaganda-education-indoctrination system), whom only support monopolistic models of creativity, and whose, so called, intellectual-superiority have been defined by the restricted dogmas of allowed thought…, have failed to show any ability to develop new creative contexts, rather, just to extend by slight variations, the narrow set of instruments within which the investor-class has invested and upon which the investor-class has based its social power

The technical development of society is all about developing the ideas of 19th century physics and the electrical inventions of, mainly, Tesla.

Just as it needs to be known that the oligarchy is fostering, both the social division (it is bank-rolling all sides of the wars), and dysfunctional knowledge ie knowledge channeled only in certain ways (for military and spying, so as to control behavior), where behavior is mostly defined by a propaganda-education-indoctrination system, and its fixed viewpoints, eg that "people are not equal," being continually repeated, ie language is being controlled….,

…, One also needs to see that (descriptive) knowledge has become dysfunctional, and it is only being used to express a form of dogmatic authoritarian society, a society of monopoly, ie only a few types of products are being made, and so that knowledge has been made ineffective, a society of fanaticism

(psycho-pathy) where "the many are being dominated by the few," where the-many are tricked into crushing the other-many (based on incoherent and non-rational use of descriptive language, which the propaganda system is controlling, and in this context of domination and controlled thought, agents are used (sent-out) to adjust things)

The logic and history of science and math

In a world based on either materialism or idealism…, in either case (should?):

1. Should one begin with materialism, a belief in "material containment in 3-space (or in space-time)" where material is defined to be either inertia or charge, and then material systems are reduced to elementary-particles,

2. Then devise a method of local measuring, ie the derivative, which has an inverse operator, and where the derivative acts on functions, in a context where the functions define the measurable properties of a material-system, which is contained within a 3-space or a 4-space-time, and its (the material-system's) geometry is defined within a 3-space or a 4-space-time (and then to identify differential equations whose solution-functions and (natural) coordinates identify the system's measurable properties).

3. Then develop the quantitative properties of the containing coordinate metric-space, related to the idea of a limit, where the limit is needed to define the derivative, where the derivative is a

model of a local linear measuring construct, so that the limit requires a quantitative model of the coordinates, which defines a quantitative-continuum,

ie a continuum is a set which is "too big,"

And

4. Then to construct many different types of quantities, depending on the existence of limits on a continuum, defined by convergences in regard to a material-world contained within a quantitative-continuum all material-systems possess a highly detailed relation to a quantitative continuum, eg the Taylor-series, so that a function's value (eg a measurable property of a material system) can be directly related to a polynomial, which, in turn, models the structure of numbers

So that virtually all the descriptive contexts which began with the idea of a derivative, and the subsequent differential equations (which could be defined if two different representations of physically measurable values could be defined, eg force-fields and inertial-force, or an energy eigenvalue defined on one side of an equation, and an operator [which represents the terms of physical values of a system's energy] on the other-side of the equation) which could be solved, even if only (solved) in a local-neighborhood, it is assumed that these local quantitative relations can be pieced together in a quantitatively consistent manner so as to be able to express all the (material-system) details of the measurable properties of the system extended out to a global (full containment) context (with exact detail)…,

[but, in fact, the solution function of a material-system's properties is only solvable globally in a quantitatively consistent manner if the partial differential equation is: linear, metric-invariant, and the coordinates are continuously commutative {almost everywhere} (also called separable)]

…, because the derivative has an (almost) inverse operator, which acts as an inverse if there are initial conditions or boundary conditions which restrain the context of the differential equation so that the

But such a solution function is unique….. so as to be controllable by the initial conditions which can be controlled by the geometric and controlled placement of the system and the system's conditions within the system's descriptive (geometric, and intrinsic system-property context eg the strength of a spring, etc) context…., only if the partial differential equation is: linear, metric-invariant, and the coordinates are continuously commutative {almost everywhere}.

However, (this construct does not work, and has only been successfully related to a two-body system)

(it is a descriptive construct which does not work because) Most systems are non-linear (or non-commutative)..., ie the differential equation is non-linear

(or non-commutative, eg the set of operators representing the physical system is not commutative)

{where non-linear differential equations occasionally has solution functions if the differential equation can be related to a few special types of manifolds}

And

When the partial differential equation is not: linear, metric-invariant, and the coordinates are continuously commutative {almost everywhere} then the quantitative extensions of a the partial differential equation's local solution function are not quantitatively consistent and there is no valid global model of the system described in such a non-commutative quantitative context

..., However, in a descriptive context which is local and the system depends on (local) feedback then the local system defined by a non-linear (partial) differential equations can be used (if the (partial) differential equations is a reliable local model of the system's local measurable conditions) to navigate through the system's local conditions when the system can adjust the local conditions which are best modeled as limit-cycles and other types of convergence-divergence related (non-linear) system bounding curves related to the critical points of the (partial) differential equations, where (partial) differential equations do (most often) have critical points, and thus the (partial) differential equation's solution-function properties can be related to limit-cycles, or curved boundaries of convergence or divergence, where limit-cycles are usually circles, lines and asymptotes defined on the (system containing) coordinate space.

However, except for a few exceptions, such as a very few exceptional manifolds

(note: Perlman showed that almost all manifolds, which are non-linear....,

{as well as non-commutative}..., are also unstable, and (as shapes they) either dissipate or degenerate to the relatively few sets of stable shapes)

The non-linear and non-commutative quantitative math structures...,

(in regard to either non-linear (partial) differential equations, or in regard to non-commutative properties associated to partial differential equations, when the solution functions or the coordinate-functions of these types of equations)

..., which are associated with the quantitative structures, of both the solution function and its domain space, are (or have the quantitative properties of) (or the result is, that in these quantitative sets):

1. the scale is not defined

2. the quantities are discontinuous and

3. the quantities are many-valued,

For both system-functions and coordinate-functions.

That is there are no quantitative patterns and the description has no meaningful content

(furthermore, all the meaning which determines a feedback system's behavior is put-in from the outside, ie the feedback system is given a purpose from an outside source, ie its local meaning is determined from an outside source).

(in the non-commutative context) What is done in practice is that the mathematician intervenes and adjusts the local approximate structures (of the assumed to be valid "solution-function"), where the intervention is usually done in relation to a converging series, or occasionally (the supposedly) invertible transforms:

Taylor-series, or

Laplace-transforms, or

Fourier series, or Fourier transforms, or

An exponential-series, defined in regard to a Lie group relation between (locally defined) Lie algebras and Lie group elements, where Lie algebras act on local coordinates, or (they) act on function spaces,

But this process, of intervention by mathematicians, (often used to "piece-together a function" defined between a sequence of locally intersecting (local) neighborhoods) is un-reliable, so as to lead to either quantitative inconsistencies,

that is, the small local neighborhoods of the domain space are patched together so as to identify a function (defined on the domain space),

or

to identify divergences and subsequent re-normalization, where re-normalization is an arbitrary method of re-scaling a math pattern which is non-commutative, and thus its quantitative context does not possess a valid model of a uniform-unit of scale in the first-place

(that is, the description can only be thought of as being a description of an arbitrary pattern, not a pattern which an actual, reliable, math method is bringing forth).

Note, that the indefinably random context is just as bad, or perhaps even worse, in regard to its lack of any meaningful content, in regard to its descriptions (or its descriptive context) since it is also a descriptive context which possesses no quantitative consistency.

Nonetheless the context of indefinable randomness is used to model quantum physics, particle-physics, economics, and psychology, etc etc etc…

Criticisms of indefinably random based in supposedly quantitatively based descriptive languages

where for economics and psychology, it is both the propaganda-education-indoctrination system as well as the fixed social institutions, upon which the description depends…,

(ie descriptions about the properties of statistical distributions of, supposedly, distinguishable random events, but the usually unstable-events are only distinguishable if the language or social institutions remain fixed and their stability [expensively] maintained),

…, descriptions about vaguely defined (and supposedly, distinguishable) properties which might vaguely approximate an elementary-event-space upon which a (claimed to be) probabilistic model, supposedly, can be built

{which is what these economic and psychological descriptions are claiming to build,

ie psychological and economic "laws" which are statistical models of noun-verb relations, which are used in their descriptive languages, which, supposedly, can be quantified, if the social structures upon which this way of using language depends for its existence are kept stable by society (or by the investment of the investor-class),

ie models of local linear correlations are identified in (or defined on) a statistical model of a set of random events (associated to counting the frequency of a, supposedly, distinguishable quality, and where that distinguishable quality (or property) needs to define a stable set of elementary-events…

[usually it is this property [of there existing a set of stable events] which the properties [or nouns] used in the language of psychology and economics does not satisfy])

..., which the language describes (or is trying to describe)

[a statistical model in regard to the frequency (or the counting) of the, supposedly, well-defined events which can be "identified" (event-types whose properties the descriptive language is trying to describe in a quantitatively "causal relation," eg local linear correlations ie one type of an event implies (necessitates) another type of an event)]}

In the physical sciences there is also great failure

But (in regard to, how the power of society is organized around its use of knowledge and creativity)

The main problem with the currently followed world-view...,

(of a fixed set of categories for making money within a model of a material-world)

..., is that materialism is not correct.

This can be said with confidence, since,

science has floundered over the last 80 years,

the main problems have not been solved

and

they are not even considered in any significant way,

where the main problems are solving for the stability of the many-(but-few)-body spectral-orbital systems which exist at all size scales and for the various materials, eg the stable nuclei, (general) atoms, molecules, as well as the solar-system,

Rather science and math flounder in a meaningless math context of non-commutativity, non-linearity, and indefinable randomness, so that their so called descriptions are neither accurate not practically useful (in regard to new creativity).

A higher-dimensional mental-construct (or viewpoint within which to place one's perceptive models, or beliefs, or capabilities) for "the map of the world" and the human's creative relation to that world, is a more accurate depiction of the relation of life (as well as material) to existence.

Precise descriptions and building things

In regard to technology, a lot of the problems which are encountered either in regard to solving known problems or in regard to further invention, are placed into the following descriptive and social context:

[And where the main problems considered have to do with military instruments or communication instruments]

1. the motivation for invention, it is based on military-control of both technical instruments and communication systems [and trying to expand this into food production and the control of food, and this is attached to an associated idea about a precise way of extermination of the population through food and biology]

and

2. The idea of basing precise descriptive knowledge, and all related inventions, on the idea of materialism, where the main focus is on the model of chaotic transitions between two relatively stable states of material, within which the rate of the transition (ie the explosion) is based on the probabilities of material particle-component collisions,

and

3. There are severe, or great, limitations in regard to both

putting-forth quantitative models of existence,

and

the subsequent use of these models of existence,

both

to use (control) the described patterns eg the stability of the nucleus is effectively ignored in physical-description,

and

to possess a context within which to think about, and invent within such a context for existence

(or invent within that context)

These limitations have to do with both quantitative consistency and stable math patterns,

The current viewpoint seeks to use (or act within) a descriptive context of: indefinable randomness, non-linearity, and non-commutative math relations between the "property describing" (or property-framing) operators, defined on both functions and the domain coordinate space, where the system is assumed to be contained in the domain space.

Precise descriptions (math and "measurable and usable" properties)

To abstractly identify the structures, and/or sets, as well as the set of ideas (needed to identify a uniquely identifying set of measurable properties) where ideas are usually represented as processes (or operators) acting between particular sets, so as to then identify a deductive process…, or the systematic limiting process…, associated to the symbolic construct of a measurable reality, so as to obtain a description (or an organized set of measured values) of the system, eg its geometry and measurable properties (motions, energies, momentums, conserved properties, etc) or its spectra and the geometry of its random particle-events in space and time (or in space-time). The problem is, that the current construct (which the following is trying to abstractly identify its sets and processes) is not capable of describing the main set in regard to the types of observed systems which are stable, ie the very stable many-(but-few)-body systems which possess, relatively, stable spectral-orbital properties.

There is the outline of the current viewpoint of measurable "scientific" models for describing patterns

1. Contain the system being described, this is mostly (entirely) about the assumption of materialism, where material is contained-in either 3-Euclidean-space and time, or contained-in 4-space-time,

[even string-theory distinguishes between elementary-particle geometries, and the space where material properties of the (assumed to be) indefinably-random point-particle events to which all material systems are assumed to reduce]

2. Measure the physical properties of a system {ie positions, inertia, momentum, conserved values, energy, the force-fields it creates, etc}, where this depends on the construct of an original set of coordinates, ie the containment set defined by the idea of materialism, and a map of that (original) set, to the function-values, which represent the measured values of the system's physical properties, but the functions are related to the coordinate shapes of the system's geometry in its (material-system) containment set (of the material system's geometry [or of the random particle-spectral-events]),

Note: this will (also) be true for any system representation which focuses upon the system's energy-function,

3. Identify a set of fixed measurable patterns (associated to the measurable properties of the system), which cause the system to become a unified construct...,

ie the laws which govern material interactions or the laws of quantum physics which allow the system's energy to be related to a set of random material-particle spectral-events in space,

..., where (in actuality) many material structures (or random spectral-events) occur with a great amount of regularity and stability, eg (1) the motions of the planets (2) the energy-levels of atoms (3) the properties of materials, and (4) the motions of electric-currents in metals etc.

So that these fixed patterns of measured properties can be used to define (partial) differential equations where the coordinates and operators...

(measurable relationships which exist between two sets (1) the containment set {and the reason two representations of the system exist [so that a (partial) differential equation for the system can be defined]} and (2) the system's measured properties)

... model the system's intrinsically necessary properties (for a description and its context to exist), and the solution-function identifies the system's measurable properties (positions, motions, force-fields, energies, [probabilities],...of the system's material-components [or their random events]).

Note: But material interactions, or their (line-integral) relation to energy, which, in turn, relates material interactions to a phase-space containment-set...,

or relating energy to a function-space description of random particle-event properties in space

though these three related (interactions, energies, random spectral-point-particle events in space)---, or similarly motivated,

---, precise descriptive contexts can [theoretically (or in principle)] reduce the assumed containment quantitative structure, so as to form a solution (or a deduced description) for a physical system, nonetheless

..., (but) in reality..., it is a context which cannot realize (describe) the full set of existing stable systems,

..., that is, it is an old context which can (in reality) only deal with the 2-body system, and all the other descriptive contexts of physical systems (which are described in this context) are unstable, so as to be neither accurate nor of any practically useful value (as a map to inventions).

4. Constrain the system's description, describe its geometry (its confining geometry) and the system's conserved physical properties eg momentum energy etc

That is, within a finite descriptive context (defined by materialism) try to limit and reduce the number of unknown measurable properties, eg identify the system's conserved physical properties,

5. Present a quantitative pattern within a context of the system's measurable properties, which either accurately (to sufficient precision) describes the system's properties, and/or provides a context in regard to both geometric relations and processes of control of the system's described properties, eg using initial (or boundary) conditions of the system's set-up (in its geometric context).

But in general, this process, this method, of description fails to provide accurate descriptions.

This is, essentially, because this descriptive method always leads to non-commutative relations in regard to the local measures of its maps, ie maps from coordinates (or coordinate geometric properties) to the properties of the physical system, so that (because of the non-commutative-ness) the coordinates cannot remain quantitatively consistent, so there cannot be any stable quantitative patterns defined on the coordinate containing-space for the physical system,

(one way to understand this failure is related to the next step (in the process))

One of the goals, in this (above classical and quantum) descriptive process, is to determine the natural coordinates of the system's geometry,

where natural coordinates are:

either

1. Those coordinates which diagonalize the matrix representation of the metric-function in a continuous manner ie continuously commutative matrix representation of the metric-function (this turns out to really be the only one which is quantitatively consistent)

or

2. Finding the system's natural geodesic containing coordinate structure. Though there is some relationships between this viewpoint and "number 1" above, this context is most often defined for a (locally) non-commutative context

or

3. The coordinates which best represent the system's: (1) geometry, (2) motions, (3) interaction (or energy) properties, (4) conserved properties (eg conservation of angular momentum), and (5) constraints (eg imposed solid geometries).

But this is mostly a wish-list, as it is both 1 and (to a lesser extent) 2, which best identify a stable system's coordinates so that the measurable properties can be determined in a quantitatively consistent manner

That is, the problem with these classical and quantum methods (in regard to their inability to precisely describe fundamental stable systems) ie the problem with this descriptive context, is that (1) it becomes non-commutative, and non-linear too easily, ie most of its descriptions have no meaningful content, and (2) the idea that either force-fields or a process of engulfing a system (of many-components) in an intrinsic energy (which centrally [or most often] needs a geometrically determined potential energy term defined for each component) apparently, is not the (a) main cause… [or main context (or main mechanism)]…. through which material-components bind together into a unified system

Rather

The ability to define system-binding context which is a natural part of the descriptive context is a way to adjust the descriptive context so that there are stable discretely separate spectral-orbital properties within the quantitative structure

That is, the systems have to be consistent with the descriptive context, eg be in resonance with a finite set of spectral-orbital values, in order to exist,

And

they do <u>not</u> need to seek a minimal quantitative value where the defined-value for which the minimum (or extrema) exists, is an abstract idea...

(defined in a process remote from the quantitative descriptive context, which is related to: place, motion, and stable spectral-orbital properties of stable shapes, associated to shapes whose stability is dependent on the shape possessing holes in itself),

and

... not a quantity intrinsic to the descriptive structure.

That is, though energy can be seen as being intrinsic to an immediate condition of a component, how is it intrinsic to a set of components (even if they are bound together)?

And then,

How would the containment-set keep track of such an ad hoc and abstract value, called "action," so as to have that value become an extreme-value (ie either a maximum or a minimum)?

In fact, it seems that action can become a minimum (or an extreme value) because of the existence of two different (though related) properties of physical systems (identified in the new descriptive context):

(1) a set of independent and separate spectral-orbital values associated to a system's stable shape, which identify extreme-values within the shape,

and

(2) a set of independent and separate interaction processes, which do not become chaotic because the orbital structures exclude a many-body interaction from manifesting (so as to break-apart the [stable] shape)

(or, equivalently, the orbits of the stable shape [within which the orbital-motions of the system's components are determined] exclude a many-body interaction because the many-body interaction does not have a great enough magnitude to break-apart the orbits of the many-bodies contained in the stable shape, where the separate and independent geodesics of the shape determine the system conditions (or the cause) which most defines the orbital structure of a stable many-body system) (though the two-body interaction may possess an interaction-force which can deform the orbital shape of one of the two components in the two-body interaction, for the two bodies being a part of the orbital properties of a stable shape).

A new math context

(where materialism becomes a proper subset of the greater descriptive context)

But instead of using the above 5-stage method to solve for a system, and to find its natural coordinates, (instead) use a new descriptive context, where the math is organized around the stable (math) patterns,

that is,

because of the realization of the very limited number of quantitatively consistent patterns, which such a descriptive process (associated to 1-5, above) actually allows

(due to a requirement that the descriptive relations between domain-spaces and function-values need to be commutative and the commutative property needs to be maintained in a continuously global context),

what needs to be considered is

the questions:

A. "What stable patterns fit into a quantitative descriptive context, so that there exist systems which exhibit stable measurable patterns?" (the answer is the limited number of stable shapes the convex rectangular-faced simplexes and other types of related convex simplexes, the cylinders and the circle-space shapes obtained from defining an equivalence relation on the rectangular-faced related convex polyhedrons (or to mode-out these types of simplexes so as to make circle-spaces of orthogonally oriented circles)

And

B. "How can a new quantitative descriptive context be organized in regard to some of the many known math patterns, which have already been described by mathematicians?" (ie in regard to the very stable, orthogonal circle-space shapes)

Note that, 1. This is a re-statement of (Thurston-Perlman) geometrization (though the new descriptive context, actually, originated in 1998), and 2. This is also a re-statement of Gödel's incompleteness theorem, wherein the conclusion (in regard to the incompleteness theorem) is that quantitative languages have great limitations as to the (stable) patterns, which they can describe (or are capable of describing), ie math language is greatly limited by their set of assumptions,

but one property which must be upheld (in order to have a valid math description) is that of the quantitative consistency within the system and within the metric-space.

Thus instead,

…(the new) the full set of existing stable systems "are systems, whose stability is directly related to the quantitative basis of the new descriptive context,"

so that,

the quantitative structures…. of both the containment space, and its ("contained") systems (or "contained" sub-metric-spaces)

…, are related to (or define) a high-dimension containment set, which is organized around sets of stable shapes,

so that these stable shapes are both

…. to be the basis for a quantitative context,

and

…. to be the set of spectral-orbital values which stable systems can possess (and realize through their resonance with the finite spectral-set), in order to be defined,

where a finite set of stable shapes is defined both in regard to various dimensional-levels,

and

in regard to various different subspaces (of the same dimension),

Where all of the finite number of governing stable shapes, and their associated finite set of stable spectral-orbital properties (together) generate the high-dimensional descriptive-set's quantitative structures (eg by multiplying the finite spectral set by the integers)

Where an 11-dimensional hyperbolic metric-space actually exists, and when organized by means of its properties of (contained) shapes and subspace constructs, can be used for accurate descriptions (to sufficient precision) and for practically useful new creative contexts…, where the main physical systems which get defined by this construct are the relatively stable: nuclei, atoms, molecules, life-forms, solar-systems, and galaxies etc.

History of academic failure (an inability to describe in a valid manner the stable properties of the: nuclei, atoms, molecules, life-forms, solar-systems, etc)

This failure has occurred because the knowledge is defined by an elite group (of, so called, intellectuals), but an elite group, in turn, is defined (within an oligarchic society) by the investor-class, and the oligarchy's arbitrary expressions of high-social-value are what determine high-social-value, and where high-value is defined (by the oligarchs) to be only those things which support and serve the investor-class (thus, knowledge fails to advance and becomes stagnant in a form of social fanaticism [or dogma]).

This willingness of the intellectual-class to deal with non-commutative mathematics (as well as non-linearity and indefinable randomness) as if the following claim is true, ie that,

"'as long as the algebraic properties are assumed to still hold, at least formally' then there is a still a viable quantitative structure,"

(this is not true, but)

This idea came from the intellectual movement of "axiomatic formalization," which was, essentially, begun by D Hilbert, but effectively, debunked by Godel's incompleteness theorem (in the 1930's), but this expression about the great limitations of the descriptions of patterns which a precise (math) language is capable of describing, has been ignored (since the oligarchs have been granted the social power [due to their control over the material-world] to make these decisions), and the expression of axiomatic formalization has been accepted and expressed by "the (French academic math) N Boubaki volumes" (as well as the math and physics departments of the academic universities around the world, ie it defines western thought, or it defines the, so called, "height of western intellectual culture"), where axiomatic formalization expresses the very complicated and very formal, so called, "proven" math patterns, but where "for the most part" these, so called, "formally proven math patterns" are related to non-commutative and to non-linear math patterns, so even though it is formally assumed that the quantitative structure is still consistent, this is not true, and as a result these, so called, proven patterns (related to non-commutative and non-linear math constructs) have no valid quantitative structure within their quantitative containment sets (eg the domain-space of functions) in regard to the patterns being described (ie being formally-proved).

That is, the (so called, proven) patterns cannot be re-constructed from the quantitative set within which they are supposed to be contained. That is, the entire descriptive context is meaningless.

Note that this is a similar model of the irrelevant academics (called authorities) housed by the church in the age of Copernicus, while today it is the irrelevant academics (but now called the authorities of science) housed by the universities.

Why was the science and math of electromagnetism and thermodynamics made to seem "more complicated and formal than necessary" by the universities? To make the subjects of science and math seem as if they are "too complicated" for a person to use, and so as to discourage the use of such knowledge about which a small business could be built, ie to eliminate competition.

This is about restricting and controlling creativity and production, so as to better please the investor-class.

And there are always agents who are willing to help the investor-class (since people have been reduced to wage-slaves) such as D Hilbert.

This arbitrary use of quantitative structures also promotes the idea of an individual's intrinsic "intelligence," (having the patience to wade-through the formalism) so as to again put this "individual property" of an individual possessing "intelligence," ie as being a person's natural talent (a gift of God), though it (intelligence) is defined in a context of narrow restriction and over complication in regard to the consideration of a subject.

Likely, 90% of science and math academics (today, 2014) believe in axiomatic formalization, 9% feel that something is missing, and at best 1% either see the deception or ignore it (or are unaware of it, but), and only use the math which has some reliability

This process of instituting axiomatic formalization within universities, happened around 1910, about the time of the rise of the high-intelligencia personality of Einstein, which was (being) promoted by the propaganda-system,

and

also when axiomatic formalization was being adopted by the universities.

Einstein was actually an "intuitive type of thinker." He most often began his papers talking about simple assumptions, simple, however, about an abstract subject, ie detecting acceleration in regard to frame of reference. Einstein was willing to deal with many different ideas, but after Einstein there is only the limited claims of experts (about a very limited and specific topic), there are not people who have important opinions about the wide-range of subjects (about science) which Einstein expressed.

This seems to be the true strategy of axiomatic formalization, that of populating the, so called, expert technical people (the intellectual-class) with autistic types, who have limited capability to use language, so that the useful knowledge of the culture remains remote from the public and accessible primarily to the investor-class.

The assumptions of physical description

It should be clear that the assumption of materialism and the assumption of describing "the measurable causes of material-interactions" both have great limitations in regard to their range of patterns which they can describe. (eg Most material-interactions are not commutative, thus the fundamental, stable, many-(but-few)-body spectral-orbital systems cannot be solved in the context of materialism).

One way in which to solve this problem is a construct within which the stable systems are a result of the quantitative structure within which the description is contained.

Thus, the containment set must allow for a wide range (but a finite number) of stable spectral-orbital values upon which the contained…, and also resonating (with some subset of the finite spectral-set)…, physical systems must depend, in regard to their coming into being…,

(by means of a new type of material-interaction, which is [has been, see other papers] defined, where resonance of the interacting structure can come to be in resonance with the external finite stable spectral-set which is associated to the containment space)

…., within a metric-space. That is, the basic construct is quite similar to the old description, which is also contained within a metric-space and many aspects of material-interactions are quite similar, but material-interactions are secondary effects in a quantitative structure whose basic principle is that it be based on stable shapes, which possess stable spectral-orbital properties from which both (1) the quantitative structure is generated and (2) the stable sets of material-systems depend on this set of stable shapes for their existence, ie the material-systems (as well as the metric-space of the various dimensions) must be in resonance with a finite spectral-set defined for a many-dimensional quantitative structure.

That is, the main cause of material-system stability is not due to (the current way of defining) material-interactions, but rather it is due to the newly formed stable system's resonance with an externally defined stable spectral-orbital values,

which a high-dimensional hyperbolic metric-space can have attached to its construct.

This attachment is done by means of organizing the high-dimensional space based-on the stable discrete hyperbolic shapes (ie the stable shapes of geometrization) associated to the different subspaces of the different dimensional levels (or different stable shapes associated to the different dimensional levels).

(again see other papers, and other books)

In this context, the stable material-components….

{which are contained in an adjacent 1-higher-dimensional stable metric-space shape (a shape which is big enough to contain the size of the lower-dimension material-component's shape)}

…. are the shapes which are in resonance with the finite spectral-set, which, in turn, is defined on the high-dimensional containment set.

These stable shapes can have the effect of making "the appearance of materialism" seem true.

This is done through the math properties which the shapes possess, that is, by topologically (and semi-metrically) being trapped in a lower-dimensional metric-space shape (lower-dimension than the over-all high-dimensional containment space), where the small stable shapes, which now model material-systems, are (now) inside the metric-space shape, where the small material-components come to be interpreted as the material-components of atoms and molecules within the (an) open-closed topological and metric-space structure which is defined to be a stable metric-space shape, which, in turn, exists within a higher-dimensional space, (which is also a stable shape).

Both metric-spaces and material-systems are stable metric-space shapes, but in adjacent dimensional levels, where the smaller material systems is contained in the larger metric-space (but where the metric-space is also a stable shape)

The high-dimensional hyperbolic metric-space (an 11-dimensional hyperbolic metric-space) is organized in a "dimension and size" sequence of containment in regard to a finite set of stable shapes (where the finite spectral set defined in this sequential dimensional and containment context determine all the stable material systems, based on their dimensional and spectral resonance with this finite spectral-set).

This is related to the following known math facts:

In regard to the set of quantitative models, where quantities are modeled as (say, geometric-measures of) stable shapes, the only shapes which uphold all the algebraic properties which determine a reliable quantitative structure (or set) are the line {or ray, or line-segment} and the circle,

That is, the real-number line

and

the complex-number plane, where the complex-plane is modeled either as a rectangular coordinate space of lines, or as a ray and a circle.

The importance of quantitative consistency is that if there is not quantitative consistency then there cannot exist any stable patterns which can be contained within the quantitative coordinate-space set in regard to the pattern being described and contained in a coordinate metric-space.

The current method of physical description (goes as follows)

The patterns are described by mapping a vague set, where the pattern being described is assumed to be contained, into a set where the system's physically measurable properties are identified (or described) so that these function-values of the system are related to the domain-space's coordinates by means of local models of measuring,

Ie relating the properties being measured (the function values) to local coordinates in the beginning containment-set of the pattern, which one is trying to organize in a quantitative and shape-related manner.

So the set-contained original pattern is mapped into a space of measurable system properties, where this process (supposedly) organizes the quantitative coordinate structure of its original containing space according to the given pattern's shape (or the given system's shape, which need to be related to the system's measurable properties).

Note: That is, if there is not the property of quantitative consistency within the (global) coordinate structures of the assumed set of the system-containing metric-space, then the coordinate structure of the metric-space cannot be determined ie cannot be identified, ie the coordinate metric-space cannot be organized to describe a stable geometric (or any other type of a stable) pattern. That is, patterns will not be stable and measuring is not reliable in such a quantitatively inconsistent model of a system-containing coordinate-metric-space.

History of physics and math

I. Newton was the first to do this (identify a system's stable shape [ie determine its coordinates] based on measurable properties defined on the system's assumed containment coordinate-metric-space), where he devised a quantitative structure used to define inertia (or mass) based on changes of motion (or changes of position) of the (focused-upon) massive component (within 3-Euclidean-space and time), where the mass-component occupies a spatial-position within the system's original containment-set, and then he defined a "force-field," so that the external force acting on the (focused upon) material-component is a result of the spatial positions of the other material, and where the spatial-positions of that other-material are defined in relation to the focused-upon mass-component's position in space

This definition also identifies the idea of material as defining a 3-space within which it (material) is (must be) contained.

Newton defined and solved the 2-body material-interaction problem, but mathematics has not progressed past this mathematical feat of Newton's, until now (with the new descriptive context).

II. These ideas, about physical description, were changed by the use of other operators, eg the line-integral, which are used to define the idea of energy (the path-integral of a material-component), both dynamic energy (kinetic-energy), and energy of position (or potential-energy) are defined, thus (in the energy-based descriptive context) there are two types of variables: dynamic (or momentum) variables and position variables, so as to form a phase-space of $x \wedge p$ defined as an alternating tensor (operation) operator (of [position(x) $\wedge$ momentum(p)]-variables),

[Note: This is the mixing of metric-space types: the mixing of position-space (or Euclidean space), and momentum-space (or hyperbolic space; but hyperbolic space's depiction of motion is transitory (or defined in an incomplete manner) except when a stable discrete hyperbolic shape can form. This is also the basis for the definition of the classic symplectic Lie group (Sp), ie local transformations of phase-space which leave $x \wedge p$ invariant.]

then

energy and time are used (multiplied in a local construct, and then integrated over a time-interval, so as) to define a new quantity, action, so that the extrema of a system's action (where a system is defined in regard to its energy and its action, where the different curves (or time parameterized energy-values, eg different line-integrals used to define energy) which can define the energy-functions and, subsequently, their actions which the system might possess, are (then) locally measurably-compared, in regard to the coordinates upon which the system's action-value is, in turn, defined, ie the different energy line-integrals defined over the time interval) are used (or determined) to identify the system's quantitative relation to (coordinate) shapes (and its other measurable properties, ie this process results in the equations of the system's interaction [math] structures),

where this idea is usually stated as

determining the state of minimal-action for a physical system (which relates the action to the system's correct coordinates),

and

where the whole description is defined in a "new space (not simply space and time)," ie phase-space, where there are both position-coordinates and momentum-coordinates, so that the newly defined natural-coordinates for the system are the coordinates which best represent the geometric properties, and the motion properties, and the energy properties of the system along with the "constraints" to which the system is subjected such as its conserved physical properties and its rigid geometric confinement structures, eg the curves of motion which are in contact with a rigid surface geometry which confines the motion of the system's material components,

so that local directions of momentum-conservation (which are called Killing vectors) can be used to reduce the number of variables..., which inter-relate the dynamic-variables and the position-variables of a system's energy representation...., in which one is interested in "solving" for (or in regard to) the system's (assumed to be) well defined quantitative pattern (or shape) {so that the system's energy is related to the system's containment space}.

This description depends on both energy being well defined (not necessarily conserved) and on energy seeking an extrema (or a minimum [or maximum]).

This minimal-energy method becomes more of an operator-focused descriptive context than is the local measuring context which relates dynamics (or inertia) to force-fields.

Note: General relativity can be formulated as an extrema of action principle.

But general relativity is (also) about defining a second-order partial differential equation of the metric-function and setting it equal to the energy-density so as to solve for a general metric-function which can then be used to determine the general metric-space's geodesics and thus determine the dynamics for material-components which are contained in this general metric-space.

But this entire context is non-linear and non-commutative, and thus, the general metric-space does not possess a quantitative structure which is stable and thus the entire descriptive context has no meaningful content.

Rather

General relativity only has meaning when its principles are applied to stable shapes.

III. Furthermore, quantum physics can also be interpreted to be an extension of this viewpoint, where sets of operators are equivalent to sets of physically measurable properties, which, in turn, define a set of measurable quantities for a quantum system (or within which a quantum system can be defined, or to which a quantum system's measurable properties can be reduced, or contained)

That is,

Quantum physics extends the idea of using sets of operators which are acting upon a system's containment-set, but where the system....,

though its random point-particle events are still contained in the domain space upon which all the functions of the function-space are defined

...., is modeled as a function-space so that the operators, where one of the main operators is the energy-operator, reduce (or diagonalize) the function-space to its (math) spectral-set, (ie all functions in the function-space are (or can be) generated from the function-space's spectral-set [spectral set of functions]) which, supposedly, also defines the system's spectral set.

For real physical systems there seems to be no relation of this descriptive context to be able to actually find a commutative (or diagonal) basis for the function-space, ie (in general) the function-space's spectral-set of functions cannot be found, ie a spectral-set of functions within which all the system's defining operators are all commutative.

All of these methods fit into the descriptive context described above where it was described as having 6-stages of deduction and/or "reduction of unknowns"

All of these currently used math structures [methods I.-III. above]..., used to "solve" the system's stable pattern..., quickly move into the realm of non-commutative sets of local coordinate transformations, or non-commutative sets of system-defining operators, as well as non-linear local contexts, so such descriptions are quantitatively inconsistent,

and thus they have:

(1) no valid determination of measuring-scale (2) they are quantitatively discontinuous and (3) they are many-valued,

so they cannot describe stable patterns, and they cannot be used to define a reliable way of measuring.

And

They all depend on the idea of materialism.

A stable quantitatively based descriptive language is very limited in regard to the valid quantitative patterns which the containment space of a descriptive context can contain and/or uphold.

The big problem in regard to physical description comes from the many-(but-few)-body systems which exist in a context of both materialism, which is the model used for the atoms and molecules, and where the material-components which are assumed to compose the system possess a force-field geometric structure which has spherical symmetry, and an inverse square field-magnitude geometry, thus it has a singular-point (at the sphere's center) in regard to point-particle models of the system's material-components, and in regard to the nucleus, where the spherically symmetric force-field acting on point-particles is supposed to bind the system together,

so that there is supposedly a similar spherically symmetric force-field structure...., but where the force-magnitudes are either exponential or based on the inverse cube of the radius..., type-model for a force-field structure, which is also applied to point-particles, which is supposed to hold the nucleus together

The problems with finding valid descriptions of the many stable systems are many and difficult if one stays within this context, and the definition of the quantitative sets as defining a continuum seems to have allowed the quest within the same descriptive structure to continue in an endless manner, convergences could be defined so as to maintain the fictions about existence being based on the idea of materialism.

Scientific verification

Consider the method by which Ptolemy verified his descriptions of planetary motions based on his model of epicycles and an earth at the center while the so called heavenly bodies rotated around the earth on their epicycle constructs

The motion of a planet would be predicted based on the epicycle model of the planet's motions and then the planets motions would be observed, and if the predictions were wrong then a new epicycle would be added to the descriptive construct now

Consider the method by which the descriptions of particle-physics are verified, a particular type (or set) of particle-types (in regard to particle-path-detections) would be claimed to exist, and then these types of particle-path-types are looked-for in the particle-detection data the interpretation of the data and the design of an experiment are both being determined by the descriptions of particle-physics, ie it is as closed a description-verification structure as was the language-verification structure of Ptolemy and his planetary-epicycles

Similar to Ptolemy, particle-physics emerged from particle-accelerator data and it is that very isolated set of data which the language of particle-physics tries to describe, the SU(2) x SU(3) patterns in a context of indefinable randomness identify a quite similar epicycle structure as was used by Ptolemy.

Ptolemy's epicycle-structure could be used to guide ships, while the epicycle-structure of particle-physics can be used to build bombs.

But only a few types of particles are detected

the three families of the following different particle-classes

(leptons and quarks and the massive field-particles)

Electrons

The pair of massive W-field particles of the weak-force

And hadrons (and the subdivisions of hadrons into different hadron particle-classes, baryons, ... etc etc...)

The hadrons are claimed to be different mixes of the internal particle-states within which elementary-quark-type-particles (apparently (inside hadrons), within which exclusively the quark-gluon collision field interactions take-place) are transformed, when they collide during a material-interaction of the strong-force, ie where material-interactions can be: particle-particle or field-field or particle-field type collision-interactions

While leptons and quarks (or leptons and hadrons) are supposed to interaction for the particle-field collisions of the weak-force

And electrons and protons (or spin-type interactions) interact in regard to the particle-field collision interactions of electromagnetism

The most disturbing aspect of this elementary-particle model, is that most of the detected elementary-particles are (themselves) unstable, and quickly transition to other, more stable, "particles."

This is a very poor model of elementary-point-particles, since it is indefinably random and non-linear. Thus, it is quantitatively inconsistent, and thus the description is quantitatively arbitrary, ie it has no meaningful content. This (lack of content) is apparent (or seen) since it is a description which cannot give any information about:

why general nuclei and general atoms are stable,

and

why these relatively stable nuclei and general atoms possess the relatively stable spectral-properties which they are observed to have.

A better model might be to give (ascribe) the path-detectable properties in particle-detectors not to point-particle but rather to distinguished points of relatively stable shapes or perhaps to the unstable shapes of a transition process

Particle-physics claims to "predict" the types of elementary-particles which exist, and are detected in particle-accelerator particle-collision experiments, but only particle-type, ie charge and spin, are indicated not the particle's mass, nor its cross-section, but where this needed for knowledge about elementary-particle properties (but the only particle-property of interest is the cross-section property of a particle) is due to a need to (a narrowly directed desire to) know the physical property of an elementary-particle's cross-section, in regard to (that property) it being a part, ie associated to the rate of reaction, ie the probability of collision between particle-types, ie the elementary-particles' cross-sections (in regard to), of a nuclear-reaction's explosion

The context of particle-physics is that, it is a description based on the concept of randomness, yet its main (and apparently only) focus is on a geometric cross-section properties of point-particle-collisions between point-particles (which are a dys-functional model of material-interactions), but where moving point-particles do not exist in a descriptive context based on randomness, since the uncertainty principle [which emerges from a property and its dual Fourier transform property upon which the uncertainty principle is expressed] does not allow positions and motions to be explicitly expressed but they are exactly (and explicitly) expressed in regard to a (model of a) point-particle collision.

Chapter 6

Describing the stable properties of material systems

Describing the stable properties of material systems ie general: nuclei, atoms, molecules, molecular shapes, crystals, and the stable solar-system etc. all these physical systems are stable many-(but-few)-body systems which have no valid descriptions today (2014) based on the, so called, laws of physics.

The measuring of the physical properties (of the world, ie the properties of material-systems contained in a coordinate-space) is about the values of functions, or function-values, whereas finding function-values, in regard to coordinates, is about the local measuring properties of functions, and this local measuring process, or differentiation, is most often related to a local non-commutative quantitative structure, which, in turn, causes the precise (or measurable) descriptive context to lose any practical content, either in regard to accurate descriptions or in regard to practical useful applications, ie because the description cannot contain any stable patterns and the quantitative structure, ie the coordinates, are quantitatively inconsistent (ie measuring is not reliable).

Materialism (geometry or randomness) and partial differential equations (determine the broad context of the current set of assumptions about physical descriptions)

vs.

Stable shapes defined over many-dimensional levels, where the stable shapes are used to re-define the idea of materialism, but in doing this the set of spectral-orbital properties for material systems becomes limited in regard to "the existence of these shapes" being dependent upon on these shapes being "in resonance" with "a finite spectral-set," where this finite spectral-set determines the existence of both metric-spaces and material-components,

and

where this finite spectral-set is defined by the spectra of a distinguished set of the stable shapes (discrete hyperbolic shapes) defined for various dimensions, and contained within, as well as (or) over, a many-dimensional containment-set (an 11-dimensional hyperbolic metric-space, or, equivalently, a [general] 12-dimensional space-time space),

(and, subsequently, partial differential equations, defined as the geometry of material-interactions, are mainly used (in the new descriptive context) to perturb stable spectral-orbital properties of both stable material systems and stable metric-space shapes, which are principally defined, "in a fundamental way," [in the new descriptive context] to be stable shapes)

That is, stable discrete hyperbolic shapes, which are shapes of energy-spaces..., and defined over many-dimensions..., trump the ideas of materialism and partial differential equations,

though both ideas (material and partial differential equations) are contained... in the new descriptive context... as proper subsets.

[note that the, apparent, random properties of small material systems can be derived from the properties of stable geometries, particularly, in a many-small-component context, ie in a context where the atomic hypothesis is an important part of the set of assumptions of the description]

Should the context of physical description be materialism (ie which implies a containment set), where the system containment-set of coordinates defines a continuum, within which are defined the measurable properties of local coordinates, upon which models of local linear measuring of a system's measurable properties, ie function-values, are defined and inter-related with the (system-containing set of) coordinates,

and

where the measurements depend on continuity (ie depend on limits, in order to relate the function's values to the system's containment-set, [ultimate] or domain space's, local measuring coordinate structures),

and

the laws of physical-inter-relationships are based on the relationships that locally measured properties have to material, ie both its geometry and its motions (or the quantum alternative of "point-particle random events in space," which, nonetheless, are still related to geometry and motions).

The idea of materialism implies the system…., either "the material and its geometry" or "the random events of a system's components reduced to point-particle events (in space and time, or in space-time)" but "the idea of materialism is" that the model of material…., is "to be are contained in either space and time or space-time."

However, there seems to be an insistence on one or the other, and not both, this is "the unified idea of materialism." That is, "material defines one containment-set" and it must be some set geometrically related to 3-spatial dimensions and time but material, it is assumed, can only be in one type of coordinate containment-set or another, ie material requires one containment set entity.

In this model, material moves continuously in a local coordinate structure, and its properties (or properties related to material) depend on local (continuous) models of local linear measuring processes. Furthermore, the material properties are conserved, or are defined by continuous-functions, ie there is stability to both material and its measuring relationships, within its containing domain space, in regard to the function-properties of the system.

Material, in a coordinate containment-space, imposes inter-relationships, in regard to certain measured properties, namely, the properties of position-in-space and motion, which are inter-related by "distant material geometry," where this inter-relationship is about locally independent and (locally) linear measurable properties of the selected parts of the system's geometry, eg only the geometry related to the 2-body interaction, in regard to the system's measured property's coordinate domain-space (where the system is both a component's motion as well as a set of selected "distant geometry of [other] material") which are measured in an independent manner (the component and the distant material-geometry affecting its properties), but occasionally (the component and the distant material geometry) form a unified-system, but this unified system is either the 1-body or 2-body system, but for a many-(but-few)-body system the math is non-commutative or non-linear (and/or both), but this is a math model which is based on quantitative inconsistency so that the descriptive context has no meaning.

That is, if the material system is assumed to be contained in a coordinate domain-space then a non-commutative model of the independent local measuring of a system's properties…., {ie a system's non-commutative math model}…., cannot account for either system unification or for the observed measured properties of the system.

Independent material entities…,

which nonetheless create a lot of math structure, when represented as a system, where math is the language of quantity and shape

..., where material defines both space and time, and it defines force-fields,

And

when, mysteriously, "material is aggregated together," it identifies (can identify) a system "of orbits"

Or

in the context of electric circuits it identifies (can identify) a system "of consistently determined current-flows" (eg electric circuits),

or

It (hypothetically) identifies (can identify) a system "of consistently determined probabilities" (eg the wave-function solution to the energy-operator), Note: For the stable many-(but-few)-body spectral quantum system the system cannot even be formulated, let alone solved.

but

this is based on

either

local measuring properties acting on material systems or material-positions,

or

material acting on space consistent with local measuring properties, eg force-fields or determining a local and general metric-function,

where

a force-field travels independently of the material, but it is emanating from material, so as to reach spatial positions as a messenger of the existence of distant material

which is actively trying to either neutralize itself and/or reach an equilibrium orbital state, or to occupy one of the system's allowable spectral-orbital states (in the new descriptive context each pair of spectral-states [differing by an up-or-down spin-energy property] is associated to a different hole in the shape)

But

this descriptive context (in regard to the current set of assumptions about physical law) is (in almost all cases) non-commutative and non-linear, and thus, it cannot carry any meaningful content in its descriptions of measurable properties.

And

The most prevalent…, and the most fundamental types of neutral and orbitally-or-spectrally-equilibrated material-system states…., are the many-(but-few)-body systems with stable spectral-orbital properties, which are quite (most) often very stable, but the properties of the local measuring models of these types of system-properties, and the system's material distributions, cannot be used to describe these stable spectral-orbital properties of mostly neutral material systems, since these descriptive techniques (or processes) lead to non-commutative quantitative relationships, associated to non-commutative shapes, as well as leading to non-linear local models of measuring,

and

thus these descriptive contexts are quantitatively inconsistent,

where

quantitative inconsistency means that they are descriptions of measured properties which are (1) without a valid definition of a measuring-scale (2) the quantities have discontinuous relations in regard to local coordinates, within which the system is assumed to be contained, and (3) the functions are many-valued,

that is,

there is no stable pattern which can be related to such a descriptive context (which is quantitatively inconsistent). Thus, the shapes associated to such systems are unstable.

Is it material that does all of these things

1. define space

2. define entities (systems)

3. define the continuum, basis for local measuring

4. define force-fields

5. define motions of force-fields

6. define motions of material?

But is it a descriptive context (based on materialism) which degenerates into non-commutative and non-linear quantitative patterns, which are associated to the local measuring relationships of the description, ie it degenerates into a descriptive context which contains no stable patterns, and which contains not practical content.

Or

Do the natural shapes of metric-spaces (obtained from the discrete Lie fiber subgroups) define:

1. Properties (physically measurable properties, and types of properties)

2. Material (inertia and charge, and other material-types, which exist in higher-dimensions)

3. shapes of systems, ie system coherence, (both stable system-shapes and stable metric-space shapes, where metric-space shapes determine the orbits of the condensed material, which is contained in the metric-space [shape])

Where these metric-space shapes have certain properties which result from the shapes, and which are related to the orbital-organization of the (condensed) material contained within the metric-space shape,

Where the stable shapes of space define the stable spectral properties of some types of material, which fit into the small material-shapes (of nuclei, atoms, molecules, and crystals from which the orbiting condensed material is formed), while the shape of a metric-space, in which material has condensed, defines the stable orbital properties of the condensed material organized into a solar-system?

An internal context of independent locally measurable properties, which, supposedly, cause the system to unify in regard to energy, so how does a passive domain space, which, supposedly, only relates to a measuring structure of local continuously dependent linear measuring processes.....

related to an external set of material-geometries and independent material motions (properties) so as to

.... bind the system within itself

It cannot,

Except for the 1-body and the 2-body systems, where Newton both defined and solved the 2-body system, and Einstein defined the 1-body problem

That is, the properties of position (measured from a point) and motion are independently defined, and either locally or globally... (as in measuring position of distant material-geometry)…. measured properties, where these two properties are arbitrarily put-together, as in phase-space (or symplectic-space), and used to identify a system's energy properties,

where the principle of least action seems to have a wider-range of application than do Newton's laws, in regard to both inertial and electromagnetic-and-inertial properties, though the two sets of laws (ie least action or Newton's laws) are, apparently, equivalent, where energy is derived from both force-fields and inertial properties of material, but this is in regard to the two-types of line-integrals, the Stieltjes-integral transform, applied to Newton's time derivative, in turn, associated to Newton's definition of inertia, which defines kinetic energy, and the "regular" line integral of the force-field, where the well defined aspect of energy (the single-valued-ness of energy) is supposed to depend on the fact that there can be no holes defined on the domain space, where (in turn) the force-field is defined

But

When the stable material system's are modeled as discrete hyperbolic shapes…

(with associated inertial and [primarily, in a hyperbolic metric-space] energy properties)

…, and material interactions are defined as "placing a 'discretely changing' toral-component" together with (or between) the "two sets of toral-components (one set for each interacting material-component) of the discrete hyperbolic shapes" so that the relative position of the two material components will be discretely changed, and subsequently the interaction geometry is (slightly) re-formed, (for each very short time-period of the spin-rotation of metric-space-states)…, ie the new model of a partial differential equation is defined in regard to sets of discrete time intervals…., so that "the two complexes of toral-components" can, either have an independent inertial relation to one another, or when some physical attribute of this discrete interaction process is a part of an interaction-complex, so that then the system comes into resonance…, with the finite spectral-set of the 11-dimensional hyperbolic metric-space containment-set of material existence…, so as to from a new stable discrete hyperbolic shape (where internal perturbations, due to material-interactions, only slightly adjust the spectral-orbital properties of the discrete hyperbolic shape)

The different toral-components… of the new, and folded, stable discrete hyperbolic shape… define the allowed set of stable spectral-orbital properties which can be occupied by material-components which are contained within the new system (these are material-components whose geometry is consistent with the facial geometry of the new system's fundamental domain, so these material-components can be contained in the metric-space shape of the system, but the containment is in a context of a "minimal amount" of material-containment within the shape, ie no condensed material (or only a few components of condensed material)

Such a new type of description is a better descriptive context than is the descriptive context of materialism and partial differential equations, so that everything in the description is not about independent physical properties (eg force, force-fields, motions, and positions etc) which seem to come into being only when a continuous local process…, which is a model of linear measuring…, is used to define relations of change, ie either force-fields related to inertia, between what are originally very stable system attributes, or when energy-operators and other models of physical operators are applied to a function-space model of a quantum system, but these descriptive techniques only apply to a 2-body system.

That is, the local model of measuring and its relation to either geometry or probability seem to not be able to describe the natural condition stability of a wide array of different types of the many-(but-few)-body systems which are very stable, where only occasionally do they provide very useful information about a system, but only when the partial differential equation of the system is solvable, and this is when the geometry of the system (in both geometric and probabilistic contexts) is linear, metric-invariant, and the natural shape of the system has coordinates which are continuously commutative for the entire shape (or perhaps for the entire shape, except for a single point, whose geometric structure can be resolved), and only for the 1-body or the 2-body systems.

That is, for the current descriptive structures for physical description,

How does the mysterious context of materialism, or independent (interacting) material elements, come to be bound into systems, due to either an energy or the force-field construct of motion and relative material positions?

ie How are the two independent properties motion and position consistent with a system being bound together if the system is a stable many-(but-few)-body system?

Ie When does a system bind together? and When does it fly apart?

There are no valid answers to these questions, in regard to the current descriptive context of physical law and physical descriptions, and this is because when the current assumptions about physical law are applied to the stable many-(but-few)-body system the laws, ie the partial differential equations, are locally non-commutative and non-linear, so the descriptive language cannot describe a stable pattern, the descriptive language has no meaningful content (in such a case of non-commutative and non-linear quantitative relationships).

And

It must be noted that, For the very prevalent and most fundamental stable many-(but-few)-body system (which possess the property of having stable spectral-orbital properties) the above "independently measured physical property model" of material interactions, and its relation to energy (and/or its relation to sets of commuting operators, in the probabilistic context), does not provide any answers, in either the geometric context or in the probabilistic context,.

However, the stable many-(but-few)-body system, with stable spectral-orbital properties, and which exist at all size scales, is the most prevalent and most general of the wide-ranging set of systems about which the descriptive context must provide answers, if one wants the descriptive context to be both accurate and to have any "practically useful" significance.

(ie quantum physics, particle-physics, and general relativity, and all derived theorems eg string-theory etc, have no valid relation to answering this most fundamental problem about the stable many-(but-few)-body system, and these descriptive structures are only related to

either

fitting data, as Ptolemy's epicycle structures were a data-fitting model of science [useful only after the data is determined],

or

to modeling a mythological gravitational singularity, to which particle-physics is being applied, so as to try to detonate such a gravitational singularity, ie it is wild useless speculation about an instrument of death and total annihilation,

it is the ultimate wild-card in "the game for psychopaths" the game of "materialism and violence, and the narrow regimentation of all human life")

Furthermore, there are two types of viewpoints about system-defining-energy, either geometric or probabilistic, while only the solvable systems can be described, in regard to both geometry and probabilistic "partial differential equation modeled" systems,

Since

otherwise the partial differential equation of the system becomes non-commutative and/or non-linear so that only feedback systems (which are provided with an external purpose) can fit into such a chaotic context, where there are no stable patterns exist, and quantities have no valid number properties, so that a precise description has no practical meaning, when there are no stable patterns, and measuring is not reliable.

But (or, that is)

In fact, the context of physical description could be based on a different set of assumptions about both containment and measuring, and the continuous nature of the domain set, the set within which the system is to be contained,

where a system is a unified structure, a unified material component, with its own set of independent spectral-orbital properties, which can be measured,

so that, the measurements of a (higher-dimensional) system-component's properties may be different measured-properties from measurement-properties of the usual, spatial and/or energetic based measuring properties of a (usual, lower-dimensional) containment-set.

On both material-components and on the metric-spaces (ie metric-space shapes), where in both contexts measuring is stable and reliable, these structures ie either material-components or metric-spaces, are the same type of math structure, namely, they are both stable metric-space shapes, but where a material-component must be at least one lower-dimension than the dimension of its containing metric-space, and of course the bounded material-component's size-measures must be smaller than the metric-space within which the material-component is contained.

The quantitative sets are not based on a continuum, but rather they are finitely generated, where the quantitative sets are generated from the finite set of spectral-orbital measures, which both the bounded metric-space shapes and the bounded material-component shapes are allowed to possess, where they can (or are allowed to) possess these stable shapes because they are "in resonance" with the finite spectral-set

Of

An over-all 11-dimensional hyperbolic metric-space, and where the main stable shape is the discrete hyperbolic shape (the hyperbolic metric-space is associated with the main physical properties of time and energy, its stable shapes are "shapes of energy").

The stability of these shapes ie the discrete Euclidean shape and the discrete hyperbolic shape is due to these shapes being based on (the moding-out of) rectangular-faced simplexes, and thus these moded-out shapes (moded-out so as to have a hole structure associated to its shape. This hole structure allows for the system to possess several different distinct and separate spectral-orbital properties (or several different spectral values).

When represented as differential-forms, the moded-out shapes have a local (or natural) coordinate structure which is linear, metric-invariant, and continuously commutative, so the differential-forms of the moded-out fundamental domains of the lattice structures of these discrete subgroups (of Lie classical groups) are solvable and controllable and stable.

The finite spectral-set, obtained from a finite number of the stable discrete hyperbolic shapes in the 11-dimensional hyperbolic metric-space, over-all containment-set, can be determined from the properties of the different dimension discrete hyperbolic shapes, which were determined by D Coxeter, and the structure of set-containment, which depends on an (increasing)-dimensional containment sequence, which in turn, depends on the size of the higher-dimensional shapes, which determine the subspace structures,

Now since the 6-dimensional (and higher-dimension) discrete hyperbolic shapes are all unbounded shapes, so the 5-dimensional and 6-dimensional subspaces define sequences of containment, due to size, and dimension, and a need for the original bounded sizes of the 5-dimensional discrete hyperbolic shapes to possess an upper-bound for the size of such shapes, for each 5-dimensional subspace (there are, 11-chose-6 = 462, different 6-dimensional subspaces, in an 11-dimensional containing hyperbolic metric-space), where this is the assumption of a finite spectral set, which "generates the quantitative sets," which are to be used for measuring the properties of metric-spaces and material-components for the over-all containment set.

The material-properties are also associated to the different dimension and different signature metric-functions for all the possible different set of stable shapes which are in the over-all containment set.

That is, the idea of material is not some mysterious fundamental idea, whose origin is unknown but is related to experience, which nonetheless determines the main aspects of measurable descriptions,

rather

its origins are in the metric-spaces themselves,

ie the metric-spaces of different dimensions and different signatures, which go into the composition of the way in which stable shapes are used to organize the over-all 11-dimensional hyperbolic metric-space's subspace structures.

Find a solvable structure for a partial differential equation and then adjust its structure by fitting properties to the solution function's boundary conditions of the partial differential equation,

or

find the set of these solvable shapes, defined over several dimensional levels and for various metric-function signatures, where each relevant metric-function signature is associated to physical properties and the subsequent set of maximal bounded shapes for both the set of material-components and the set of metric-spaces, which together define the over-all containment space's finite spectral set, and then one has the same context of being able to fit boundary conditions, related to the finite spectral-set, so as to affect control over systems

Note:

If one wants (in society) education to become relevant, and knowledge to become relevant to creative efforts, so that the knowledge is quickly learned, and if one wants creativity to become defined over a wider-range of possibilities, then the above type of discussion is the type of discussion one most wants amongst the public, in public educational institutions, rather than memorization of the failed-dogmas of the, supposedly, un-error-ing expert-social-class, who really represent a façade of narrowly defined and dysfunctional authoritative-knowledge which best serves the interests of the business and propaganda systems of the investor-class.

The many-(but-few)-body system needs an valid descriptive structure (and the current (2014) rendition of the, so called, laws of physics is not providing a valid descriptive construct)

Chapter 7

Math descriptions

Does one want to question the idea that numbers… and how they are based…or …which are based on the identification of a "uniform unit of measuring"… do not need to be consistent, when the math professionals insist-on primarily considering math patterns which are characterized by non-commutative-ness, non-linearity, and indefinable randomness ie indefinable randomness is probability being based on elementary-events which are not stable (or not determinable),

Or

Does one want to extend the ideas of quantitative consistency to its greatest extent, and subsequently, to change the context of set-containment, in regard to the descriptive structures of existence, so that the new containment-set can account for the observed stable properties of existence?

Math descriptions

Does one want to question the idea that numbers… and how they are based…or …which are based on the identification of a "uniform unit of measuring"… do not need to be consistent, when the math professionals insist-on primarily considering math patterns which are characterized by non-commutative-ness, non-linearity, and indefinable randomness ie indefinable randomness is probability being based on elementary-events which are not stable (or not determinable),

Or

Does one want to extend the ideas of quantitative consistency to its greatest extent, and subsequently, to change the context of set-containment, in regard to the descriptive structures of existence, so that the new containment-set can account for the observed stable properties of existence?

Does one want to extend the ideas of quantitative consistency to its greatest extent, and subsequently, to change the context of set-containment, in regard to the descriptive structures of existence, so that the new containment-set is many-dimensional, (and) so as to be organized around stable shapes which have different dimensions, both for the different dimensional-levels, and

For each of the different specific-dimensional subspaces, which exist within the over-all containing-space which is a 12-dimensional (general) space-time metric-space, and this is true for each dimensional level, as well as for its set of subspaces, which are best (most intuitively) identified at the 6-dimensional-level, where the subspaces of the different dimensional levels are all stable shapes, which, in turn, define topologically independent metric-spaces, which are separate from..., and rigidly indifferent to..., their motions in the higher-dimensional shapes (within which they are contained, where this is due to the principles of general relativity), and these are shapes which (most certainly) contain within themselves lower-dimensional shapes.

But such a construct which can relate a finite spectra set to stable shapes, so that the finite spectral set can be used to generate a quantitative structure which both works as the basis for measuring, and is not too big of a set, is also "a needed set property," so that the measured properties will stay within a logically consistent set.

Whereas, the current quantitative structure is a continuum where each of the elements..., which makes-up the "measurable" portion of this set..., requires and infinite amount of information in order to be distinguished, so that each quantity becomes and independent quantitative type...,

ie a quantitative set with this very large set structure allows quantitative types to coexist (as seemingly relatable quantities), which are logically inconsistent with one another, thus leading to logically inconsistent quantitative models which appear to be put together by means of rigorous math methods, eg geometry of particle-collisions are put together with an otherwise logically inconsistent basis in randomness of (for) particle-physics, leading to a fantasy world as to what a quantity is, eg renormalization and quantum-foam, where the basis for measuring is arbitrary and determined by "an authority's" interpretation, ie the same science as was practiced by Ptolemy where an authority's model dictates reality but where the model is verified by measuring so as to create a type of shame science

Science needs both accurate descriptions to sufficient precision and the descriptive context needs to be relatable to practical creative development not simply a narrow relation to one type of instrument, this is the standard of scientific truth imposed on science by classical physics: Newton's-and Galileo's gravity and mechanics, faraday's electromagnetism, and thermal and statistical physics, Einstein is interesting, but his models of general relativity are failures due to quantitative inconsistency which emanates from his assumed geometry of spherical symmetry (where the geometry of spherical symmetry seems to, really, be a property of 3-dimensional Euclidean inertial space [ie the space which mediates inertial material-interactions]), that is the (seemingly, expressed as a triumphant) exclusion of Euclidean space (in special relativity) seems to have been premature, since non-local behavior has been demonstrated (inertia and electromagnetism are most likely defined in separate spaces, where the stable patterns we observe seem to best fit into the space which contains electromagnetism).

Some such a way of organizing math patterns is needed for there to exist sets of stable math patterns, which are observed in physical systems, and in order for math-based descriptions to have an actual content, or meaning (and being internally consistent, both quantitative consistency and logical consistency), rather than "the meaning" which is now (2014) being inserted from an outside viewpoint, which is determined within a set of fleeting, briefly distinguishable, features which form the basis for either forming an explosion (nuclear reaction) or making adjustments to a classical and controllable system's (whose system's purpose is determined externally [from the system], eg in relation to global positioning contexts), brief relation to these transitory features, eg drones operating without a pilot (not even a remote pilot) in an unstable context directed by an external purpose (eg spying).

But the properties of the stability of pattern: from nuclei, to general atoms, to molecules, to molecular shape, to crystals, to life-forms, to solar-system stability, to dark-matter (determined from the observed motions of stars in galaxies), etc; have no valid descriptions based on (what is considered to be) physical law.

This fundamental set of stable physical systems, though identifying a set of special categories of physical systems, nonetheless, also defines a wide array of general systems, wherein all such physical systems are observed to possess very stable spectral-orbital properties, but the generality and abstractness, which now defines the (so called) purposeful direction of current (2014) math considerations (by the professionals) is a descriptive context, which does not contain any stable patterns, but rather fits into the non-commutative and non-linear set of very general math patterns, but this is a quantitatively inconsistent set of math patterns.

It needs to be noted, that the, so called, physical laws are, in fact, (defining) the context of the fleeting, briefly distinguishable features of a particular math model of existence, where most (if not all) properties are: unstable, many-valued, and discontinuously chaotic, ie a quantitatively inconsistent context, where this math context results from a descriptive math context of (or based on) non-linearity, non-commutativity, and indefinable randomness.

A context where there is not quantitative consistency, but a context which is related to the chaotic transition, which is (or quite often can be) characterized by component-collisions, and which exist (in a transition between two relatively stable states) between the broken-apart components of two relatively stable states, ie a nuclear reaction. All of the focus of physics and math is on a chaotic and quantitatively inconsistent descriptive context best suited for either the study of the explosions of nuclear weapons or for the fleeting patterns through which a feedback system is to navigate (or be defined) as well as fitting into the main assumption of information theory, ie that the most information is needed for a fleeting chaotically random context for descriptions (ie where there are laws or organizations of state then information can be compressed).

But this way of considering the development of math patterns, based on quantitative inconsistency, completely ignores the stable properties of existence, the stable properties of fundamental material systems, which are so very fundamental to our experience, and to life, and to the context of creating and building practically useful structures, so as to use the internal controllable properties of existence, and so as to explore that full realm of stable existence, and determine its relation to the creativity of life-forms.

Energy does not come from indefinable randomness and quantitative inconsistency, rather it is stored in a relatively stable states, and the internal structure of the systems within which it (energy) is stored need to be understood so as to control the access to… and use of… that energy.

The propaganda-education system

These are very strong and serious criticisms of science and math, but they do not get repeated by others in society, as they should.

This is the result of a society based on terrorism and the, apparently, fearful belief that the intellectual-class possesses deep knowledge, this is quite an illusion, which is coercively imposed on the public by the propaganda-education system, and its (militarized, and spy agent infested and influenced) managers who uphold the elitism of the ruling-class. For example, in a public education system, where egalitarianism should be the guiding principle, ie equal free-inquiry, instead of the support for the bogus belief in "intelligence" and a curriculum based on the public filling the vocational positions of adjusting the instruments of the ruling-class or helping the institutions of oppression and propaganda of the ruling-class, where the coercion comes from the social state of the public being wage-slaves, a social state which is created and enforced by the justice system, and from the authoritarian and dogmatic education system, which is a vocational institution for the "creative" projects into which the banking system, or investor-class, has invested.

This also seems to be the result of the arrogance and false sense of superiority, which the domineering intellectual-class feels they are entitled, where, apparently, this arrogance is due to the main body of work which the propaganda-education system provides to society, in regard to the proclamations by the propaganda-system of inequality, and the intellectually dominating superiority of the intellectual-class…and the role the intellectual-class plays in …. our society's culture, ie it fits into the interests of the ruling investor-class where this ruling class also identifies the illusional property of something possessing "high-social-value" (a notion closely related to the state of wave-slavery), where this is controlled by the propaganda-education system, that is, the propaganda-education system teams-together to present the illusion of intellectual superiority of a few, who are, considered to be, naturally superior to everyone else.

That is, the public has been colonized by the investor-class, with the public accepting the illusion of an intellectually superior class of people.

But these so called superior intellects have been the basis for the great failure of society, and the poisoning and destruction of the earth.

That is, the so called superior-intellects are not smart at all, they also follow a bunch of nonsense and illusions, eg the math-physics professionals follow a bunch of nonsense about math contexts which are non-commutative and non-linear and based on indefinable randomness ie they deeply consider math where the quantitative structures are not consistent with the quantities (the, accepted, uniform units of measuring) upon which the description is, supposed to be, based, where this destruction of intellectual efforts, ie focusing on general abstractness where there are no well defined patterns, is based on the interests of the investor class, where the education system is designed so that the intellectual class is (supposed) to gradually develop the instruments, which are of the greatest interest to the investor-class, but this is the basis for the failure of the knowledge of the intellectual-class. That is, dogmatic authority leads to great complexity, over generalization, and narrow limited viewpoints which lead to such knowledge to fail (and/or to become irrelevant)

That is, these intellectuals have failed in a deep and fundamental manner, as just indicated in the above list of their failures, in regard to describing the fundamental properties, or the fundamental stable physical systems, of existence.

Though they can adjust classically controllable systems, with feedback mechanisms, so as to adjust the classical system to respond to fleeting features of the world, but it is an intellectual-class which cannot describe the basis for intrinsically controlled and very stable physical systems, including living-systems.

Instead

the partial knowledge they have, in regard to the relation of "DNA to life" and the partial knowledge they have in regard to "radioactivity related to nuclear weapons" are examples of partial knowledge (incomplete, and thus failed knowledge) which is leading to the destruction of earth and life-on-earth (by the accumulation of pure radioactive substances whose safe storage will not be possible, and the interference with the molecular structure of DNA in life-forms could, subsequently, destroy the epi-genome properties necessary for life.

That is, though the atomic and molecular models of independent material components may work for thermal systems, which are confined and bounded systems in equilibrium, in regard to the system-measured properties of average-values defined over many-components, where these components are contained in a stable structure, but for a living system, there is clearly a hidden unifying epi-genome stable system structure, which is the basis for the living-system's ability to affect a great amount of control over its material structures, which exists when there is no, apparent, property of system-equilibrium, and which can affect..., through non-local resonances..., the local-cellular molecular-properties of the entire system.

Such a stable, extra-containment structure, eg containment beyond the thermal-physics idea of a confined systems of material-molecular components in equilibrium, could be either of a larger size electromagnetically, but whose inertial properties are small, or it could be a result of an unseen higher-dimensional (electromagnetic) shape whose inertial properties are (also) quite small, or the higher-dimensional inertial properties might be irrelevant, in regard to the lower-dimension molecular component properties of a living-system (eg vaguely related to something like a weak van der Waals force).

These shapes..., through which the living system is controlled, ie the epi-genome..., may be related to lengths along the long-strands of (stable) "cylindrical" DNA molecules, that is, the lengths along the DNA-cylindrical-strands could be related to the characteristic-lengths of the lower-dimensional faces of the higher-dimensional..., hidden, so called, epi-genome..., stable shapes, which can affect the living system in a non-local context.

That is, DNA may have two functions: (1) relate the "marked" lengths defined along its stable cylindrical shape so as to identify the living-system's relation to a, hidden, higher-dimensional controlling shape, and (2) relating the ladder-structures of proteins of the DNA cylindrically-shaped molecule to the cellular local-molecular conditions (around the cell), in response to resonances defined by the system-controlling higher-dimensional shape (of the living-system).

Perhaps, if the two properties of DNA are not taken into account then the DNA alterations could be quite destructive to life-forms.

But the main use of this partial knowledge, in regard to its representation as an absolute truth, is for monopolistic business interests (eg controlling the entire food system), including dominating, by violence, all corners of the earth

(To whom do the elitist spies give their information? Public institutional people are so corrupt, how can they be taken seriously by the spies? Spies work for the arbitrary expressions of control over violent-and-monetary power, they get the "good-life" [both women and things, in a male-controlled society]).

Virtually every subject, about which the intellectual class (or the educated-class, or the schooled public) talks, are contexts which will fit into the interests of the investor-class, unless the fundamental assumptions, ie the simple elementary ideas, upon which the discussion rests are both identified and questioned in a serious manner.

Does one want a society of equal individual knowledge-gatherers and creators, whose creations are expressions of selfless gifts either given away and/or put into an equal free-market, but highly controlled so as to maintain the equality of the market,

Or

Does one want a collective society, which bows-down to an arbitrary high-value of the creativity based on the investments of the investor-class, where the purpose of the creative efforts is to enhance the selfish interests of the investor-class, where the products (or most often weapons) are placed in a highly controlled and propagandized market which serves the selfish interests of the ruling-class, where most of the products made by the investor-class are products of oppression, and all the products are being organized so as to be used to oppress the public, so that the investor-class reaps the market benefits of both intellectual and material production (the investor-class defines both truth and material-sexual life-style)?

Math

A summary of the history of the traditions of math which the professionals follow, but they follow it in a context of memorization of meaningless complications, and in regard to separated and improperly related and very complicated patterns which focus on "too much generalization and abstraction," wherein quantitative consistency is lost, and in focusing on the traditional authoritative things, they miss the central issues of math.

Namely, the central issues of math is about quantitative consistency and stable patterns, where in regard to (partial) differential equations (ie pde's), ie in regard to the laws of physics, this means that the pde needs to satisfy the following pair of properties:

(1) The pde must be: linearity, metric-invariant, and continuously commutative in regard to the "natural coordinates of the underlying set of stable shapes," but it also means the existence of several different "dimensional levels of description," each of which possess various properties (based on the descriptive dimensional context) in regard to

(2) the "existence of,"… or "non-existence of"… holes in the space where the pde's are defined, where he property of "the existence of stable holes" in the space within which the description focuses is related to spectral stability, but where "a stable value for energy" depends on "the non-existence of holes" in the space,

(note) where a quantitatively consistent and single-valued result from an integration of local geometric measures is associated to the definition of a system's energy.

The story of math

Math begins with counting, which is a process associated to both identifying a uniform unit of measurement and then defining a process which is a sequential accumulation of these uniform units, ie defining a number system, and then (subsequently) there are measuring processes associated to different types of measured quantities: length, area, .., time, mass, charge, (dark matter) etc,

But

where measuring might often need to relate two (or more) different measuring scales (of the same type of quantity), so as to be consistent, ie measuring is based on (different) sets which are (can be) based on different uniform units (as well as different number-types, in turn, based on different types of objects within which a uniform unit can be defined), where changes of measuring-scale (of the same type of number) are related to either "grouping-things together" in different ways, or to defining different uniform units of measuring, and subsequently, these changes in "units of measuring" are, in turn, related to the operation of multiplication, ie multiplying by a constant, which identifies changes of measuring scale, or changing the size of the uniform unit of measuring. Note that, this changing-of-scale "by multiplying by a constant" is a linear function.

Note, adding and multiplying have natural inverse structures ie negative values and multiplicative-inverses (or fractions), respectively.

The operations of adding and multiplying satisfy certain math operations properties which are concerned with the issue of "the order of operations" in regard to defined operation processes eg definition of matrix multiplication:

Associative property, and

Commutative property,

(where in many-variables the property of commuting (of matrices) is about (local) sets of orthogonal variable-directions and corresponding (or associated) local transformations, which transform the given variable-directions to other orthogonal variable directions by means of matrices which are applied to the variable's (or coordinate's) local vector-directions, ie locally independent directions for the variables, where inner products can be used to identify both local distances and angles between local vectors, which are related in (certain types of) metric-spaces to both "dual vectors" and metric-functions)

Measuring is related to the simple geometric shapes of line-segments, and circles, and subsequently the measuring of angles between line segments, which intersect at a vertex, in various dimensional contexts, as well as the simple geometric shapes of, conic sections which seem to be of special types of quantitative patterns associated to shapes (of curves, or surfaces, etc) related to (or defined by) geometric measuring relations, or equations, which along with lines and circles can be put into continuously orthogonal relationships: hyperbolas, ellipses, parabolas and their higher-dimensional, eg surface, intersection sets.

A descriptive context is about a description's, assumed, containment-sets, and quantitative, or geometric, (eg measuring) relations, or patterns, which exist between other (different) quantitative or geometric patterns, eg types of motion-changes associated to forces etc, which, in turn, exist in that same containment-set (or set of variables, or set of coordinates).

Within the descriptive context of a quantitative, or measurable, description, there are variables, and other dependent, or particular, values, or formulas, or functions, which are related to the (measurable) patterns of the system, where the measurable properties of the system, represented as functions, are assumed to be contained in the variable-space, and which are assumed to be dependent on the variables, or sets of coordinates, which are associated to the system-containing set, and the relation (of the formula of one of a system's measured property) to other quantitative constraints, or quantitative inter-relationships, eg properties defining an equation or physical laws which define a physical system's measurable properties.

In particular, the property of apparently different ways (processes) of getting a quantity by measuring, where, when it is determined that there are two different processes which both identify the same (quantitative type and) quantity, which exists in the descriptive context, then this can lead to the identification of equations, and, subsequently, it leads to the desire to solve such equations.

There is one main idea in regard to math which should always be the theme in regard to math considerations, and that idea is about quantitative consistency.

Equations of the type which depend on the entire space of (equation containing) quantitative sets are fixed stationary equations (algebra), ie fixed quantitative relations which are represented as being true everywhere, in this fixed algebraic context, and in which case the idea of quantitative consistency is an intrinsic property, ie the uniform unit of measuring, which is the basis for a number system (or for a context of measuring), is always maintained, or if it is changed by an allowed (linear) transformation of the variables, then for each variable it is changed with an explicit "change of scale" for that variable, which is identified by multiplying that particular variable by a constant.

But

Furthermore, the intersection sets of the solution sets of sets of many-variable polynomial equations identify a set reduction…,

eg intersecting surfaces can result in curves, where if the intersection (of the given pair of surfaces) is also a surface (or a subsurface) then the dimension has not been reduced by the intersection process, and the remaining surface intersection-subset has no more useful information (ie failed at the attempt (or desire) to reduce the set so as to compress information) than the original surfaces,, in a similar way, that a sufficient set of equations for a many-variable set of (local) linear maps (or linear equations), the solution method identifies a set reduction process, which can be modeled as the intersection of (linear) shapes (ie line, planes, unbounded 3-space regions (which are subspaces), etc), but where the simplest reduction is that there is one equation and one direction (the function-value is in the same local direction as the variable's local direction) and this property can be continuously maintained at all points on the shape which the coordinates are following.

While (on the other hand)

Local measuring properties often depend for their properties on both

(1) their variable relation of the measured-values (or function-values) to the position in the coordinate space (or pattern containing domain space) where the function-values are being determined by some formula, and

(2) their (possibly) variable (or controllable) relation (of a solution-function to a locally dependent, as well as positional-ly dependent, relations defined in regard to a pde) to boundary conditions of the coordinate region, wherein the pattern (or system) is defined (or initial conditions of a dynamic pattern, also defined in a local context and local properties in regard to the local point's position) (analysis)

But if one mostly interested in equations, and their relations to the dimensions of the coordinate space, this (number) being the number of variables used to represent the measured values (or, equivalently, to provide a formula for the function-values) as well as the dimension of the "function-value space," where this (local vector dimension) is the number of variables which must be solved, if the number of local-coordinate-vector-component-directions and the number of local-function-vector-components-directions (or dimensions) are equal, so that when measuring of a coordinate in a particular direction then one wants this same direction to be related to the same function-vector-component-value-direction (so that this relation is maintained in a continuous manner about the entire shape of the coordinates) (so as to have the correct equation-solution and its associated variable-function-value relationships, ie one equation and one-independent-variable, which are needed to solve the set of equations),

Or

One can represent the set of pde's as sets of operators which act on a function-space, so as to, thus, form a model of the pattern one wishes to solve (or determine), then one needs sets of commuting operators which (since they are commuting operators) effectively do the same "diagonal forming process" for the functions in the function-space, ie finding the spectral-representation of the function-space (for the given spectral system), as the coordinates and function-value-vector-components do (as) in the above process of forming a diagonal matrix relation for coordinate-directions and function-value-component-directions in the above paragraph.

But both the general function of a function-space, as well as a general set of operators, are (in general) not in commutative relationships with one another, so finding the set of commutative operators and, subsequently, finding the set of spectral-functions for a physical system, usually, cannot be done,

So

Both algebraic equations (usually defined in the context of polynomials) and (partial) differential equations, especially pde, both depend on "the structure of the containment-set," or the pattern containing coordinate space,

1. ie measuring in such a space, eg metric-functions,

2. the shape of the coordinate space, where

the shape of space affects the geometric measures of the space's geometric properties,

3. eg geometric measuring constructs, along with

4. an assumption of continuity of the coordinates, and/or functions, so that

5. geometric measures can be integrated over the coordinate's locally-defined structures to identify geometric measures of the coordinate space's shape (using the shape's natural coordinates), as well as the elementary properties of the polynomial functions can be understood, and

6. (to determine) If these "integrated relations of measurable properties (related) to geometric constructs (or structures) are single-valued?"

eg the line-integral of the force-field is used to define energy, and

7. most importantly, the dimension of the containment set, where

there is an implicit assumption that

8. the pattern or the (physical) system is completely contained within the coordinate space of some given (or assumed) dimensional property

In algebra this property of math containment is about if the equations can be solved by sets of confining quantitative relationships which exist, and can be expressed by the variables upon which the equations depend for their definition, so if they are solvable then containment is assured, or in regard to "solving" polynomial equations this depends on the intersection of domain sets with the graphs, or simply the intersection-sets of the graphs of the function's graphs themselves, ie several function graphs being intersected (if the intersections exist)

Whereas

For pde defined in a geometric context (as opposed to being defined in regard to a function-space model of the system's containment-set) the corresponding set of constraining relations are related to the (defined) solution function's relation to the boundary conditions of the system, which are related to geometric measures of the geometry defined in the system-containing region of the coordinate space, but these types of geometric properties require the local and continuous geometric separation of coordinate-values and their associated function-values, ie one-dimension associated to one-(solvable)-equation.

Or (an alternative viewpoint)

Is the main descriptive issue (or property) really about the "set containment structure?"

Where stable patterns exist as a part of the containment space's structure, so that inter-related quantitative patterns, of an incompletely defined quantity, such as mass or charge, in regard to their stable spectral-orbital properties, can exist (as material components) in a context of stability, so that the stable patterns (of material subsets) fit into the "scaffolding" of stability, which is intrinsic to the stable shape structures of the containment-set, in regard to various dimensional-levels and various metric-space types, and their associated stable shapes, which compose the fundamental structures of existence.

That is, the idea of charged-material being related to a quantized value for a charge, assumes that the charge is a point-particle, but if there are many ways in which an amount of charge can occupy stable spectral-orbital geometric structures (or constructs), where the amount of discrete charge would depend on the orbital-size, then (in this case) there may be many various discrete charge quantities.

Note: Most experimenters found that electron-charge could not be related to a fixed quantized value.

That is, the R Milikan fraudulent experiments which, supposedly, established a quantized electron-charge (by Milikan's dry-labbing techniques), in fact, will be related in actual data, to that which indicate there are variations of electron charge, which is dependent on the spectral-orbital shape which the electron charge occupies.

The electron seems to not be a point-particle, but rather it might have an intrinsic geometric structure. This is, perhaps, the best interpretation of the discretely orbiting, very-cooled, solitary electron observed by Dehmelt.

Furthermore, the properties of crystals can be best fitted to observed properties by using a rectangular polyhedron, or moded-out rectangular polyhedron so as to form a toral-shape, in order to model the crystal and to, subsequently, calculate its stable spectral-orbital structure (for its electrons), where, perhaps, a discrete hyperbolic shape would be a better model of a crystal

Is this ultimately about trying to define a function's value when the function is defined on a space which has holes in its structure? (this is the conclusion of the provided alternative viewpoint)

If the holes were discretely separate then the function's value should be related to periodic values, ie n(k(2[pi])), so as to be able to take minimal values for these discrete values, based on both the discrete separation of the pathways, and an integrated minimal length for the separated pathways defined around each of the separated holes.

Modern science is riddled with fraud and deception, so as to serve the interests of the investor-class eg peer-review is a deception.

For example, there has long been a strong correlation between CO_2 in an atmosphere and the greenhouse effect, and this is the basis for using the precautionary principle to stop burning fossil fuels, but peer-review allows the interests of the investor-class to cloud-up this simple story.

Furthermore, because the following properties are a part of the given math model of elementary-particles, then

The decay of the proton will be found in some "deep" deceptive context as gravitons are, supposedly, now being identified, but the, so called, background 3K temperature is not by any means conclusively identified with a mythological big bang and gravitational singularities are the only interpretations of distant spectral signals, and neutrinos are mixing amongst themselves, though this is in opposition to the interpretations of elementary-particle-families (ie one must be in the correct energy range for all the members of a particle-family to exist, otherwise they should not exist) and there is now inflation in regard to the so called big bang, but this discontinuous process nullifies the math proof that there could have been a big bang, which depended on continuity, etc etc, etc. (ever more fraud, and babble, serving the investor interests, but the, so called, intellectual-elite seem to not notice)

Supposedly, the Higg's particle has been found, but there never was a believable prediction of its mass (the, so called, experts must have had 17 alternative scenarios for the Higg's particle's mass, apparently, the only measurable property it has (or is-it that "after the mass is found then its cross-section can be calculated?"), but much energy (cash) will be put into determining its cross-section,

That is, the cross-section of an elementary-particle, ie its probability of collision, is the only elementary-particle-property which is of interest to a bomb engineer, (and) it is, essentially, the only physical property which is calculable in the math structure of particle-physics.

That is, the cross-section (or probability of particle-collision) of the, supposed, "mass-defining" field-particle, will be central to any model for detonation of a [mythological] gravitational singularity, so in a, so called, verification context, which claims 16-significant digits of precision in regard to the two or three, so called, particle-physics predictions, and thus finding the Higg's particle in regard to its dubious estimates for its mass (where these estimates are claimed as predictions) seems to not be a valid verification of any significant prediction (?) but rather a data point which, essentially, only fits into a particle-collision model of a bomb explosion, but it is doubtful it will be a gravitational singularity which will be the object (or basis) of the Higg's-particle based (or modeled) explosion, since it is a useless model, it is wild speculation whose sole focus is realizing an unstable state of a reaction and the general relativity model of a gravitational singularity is more fantasy than science, the math model..., and its single-solitary viewpoint of interpretation of particle-collision data..., is more-or-less a belief in a "type of magic" of a social combination of, so called, "smart and vague math" "stir them together and what does one get," the wishful hope to produce the "magic effects" that the military wants.

The (alternative) interpretation of this set of data...,

(if it is not out-right fraudulent, which is a possibility, ie they seek (vague) data which can only be fit into some very narrow interpretation within a very limited math-model, ie in regard to a failed math model, since there is no predictive value in regard to actual stable physical systems which comes out of the particle-physic's math-model, ie there is no reason to believe its interpretative descriptive contexts)

..., is that "at high-energy" (where dual-variables can be thought of as being in the context of "errors of measurements") there is "a small time interval," which might have resulted in a brief collision-interaction with a discrete Euclidean shape, ie the essential non-local geometric structure, but a discrete Euclidean shape has no relation to stability [however, the discrete Euclidean shape is an inertial structure], but without stability, so even if a cross-section is measured, it will be a model which is based on the phantom-patterns of an unstable event, ie none of the quantitative structure built around material-collisions based on discrete Euclidean shapes has any validity, in regard to stability, where stability is a property of a hyperbolic metric-space (ie where stable physical properties exist in hyperbolic metric-spaces).

Further algebra and analysis

Complex-coordinates fit into both algebraic and analytic contexts, but the solvability of polynomial equations in complex-variables, apparently, make both the algebraic and the local-linear analytic patterns in complex-coordinates simpler to identify, where these patterns are organized around the fact that complex-numbers possess an intrinsic relation to circular-geometry, ie there is a circle-space structure naturally built into complex-coordinates.

In solving pde there is the method of changing variables, so as to diagonalize the local matrix transformation relations, which are defined by the local derivatives of the (solution) functions (or functions which represent the system's measured properties), associated to the, so called, system-defining pde, ie a pde defined by physical law, but for metric-invariant and linear pde's, which is the natural type of structure (ie not using general metric-functions, which are quantitatively inconsistent) used for local measuring of geometric properties, so that measured values of the (physical) system (or the system's measured properties) are consistent with geometric measured values of the coordinate containment-set of the physical system, but where the natural sets of rotation matrices (used to transform the coordinates, so as to be consistent with the metric-function, and to be locally orthogonal, and thus solvable [and thus diagonal] but this is not possible) ie in general a rotation matrix cannot be diagonalized in a neighborhood around the point, where the local measuring (associated to the pde) is being defined.

Thus, diagonal coordinate-function relations only seem to exist at single-isolated-points (ie not over neighborhoods of the point).

However, by placing these coordinates in corresponding complex-coordinates and changing the group's defining-relation from metric-invariance to the metric-function's natural correspondence of Hermitian-form-invariant (or inner product invariant) then the fiber group is the unitary group, and in the unitary group all local coordinate transformations are (can be made) diagonal, by means of a local unitary coordinate transformation, but one also thinks of the dimension of the coordinate space doubling, ie from dimension-n to dimension-2n.

ie the existence and solution of the Cauchy equations is a condition for the function being analytic

But the solvability in the (an) algebraic system... (or sets of polynomial equations defined for several variables) framed in the complex-number systems (or in regard to complex-coordinates, or complex-variables) ie polynomial equations...., leads to branch-points, related to the periodic quantitative structure of circles in complex-number representations, and a lack of single-valued-ness for the algebraic solution set in complex-coordinates, where one can view branch-points as an attempt to change the domain (or variable) space so the branch-point causes the domain space to identify a new hole, which, in turn, possess distinguishable levels of rotations, so as to form a spiral along the new hole, so as to spiral back to the original level (and the original root) associated to the branch-point, where solution points exist along the spirals of the new branch-point defined holes, so that the solution set again becomes single-valued, but the multiple values (for solutions) in this complex-number context possess a periodic structure, due to the circular geometry of the complex-numbers. Where these branch-points lead to new quantitative pathways which, in turn, define holes in the complex-coordinate space. but, in turn,

Can these complex-number structures (sometimes) also be related to real-holes in shapes of real-number coordinate-spaces? That is, can the rotation placed in a complex-coordinate structure lead to a local diagonalization, or lead to a local commutativity which is continuous, but where the real-space model now has holes in the coordinate system space? But then energy defined in such a space would not be single-valued.

To re-iterate

Thus, diagonalizing a rotation matrix can (possibly) be related to new holes in the originally real and now substituted complex-coordinates, and the subsequent diagonal unitary matrix model of the local (real) coordinate system, which can be modeled as forming a branch-point and a new hole providing a spiral model of single-valued-ness in the complex-coordinate space. Can this complex-coordinate model of a hole, also result in a new real-hole, so as to support (or allow) a single-real valued real model? But such real-holes result in a many-valued integration structure.

But many quantitative patterns, or contexts, are about graphs of formulas, where solution sets of equations most often identify a quantitative set which is one lower dimension than the graph, ie the equation defines an intersection-set, eg for a 1-variable polynomial there is the intersection of a horizontal-line and the graph of the formula where this intersection-set defines a set of intersection points which identify solutions to an equation, related to the 1-variable polynomial,

Thus, one considers the intersections of graphs of composed of various numbers of variables

The properties of graphs of functions, which are defined over the domain space (or, equivalently, the pattern-containing coordinate space), possess properties, which are best described in the context of limits and continuity, where topology is about continuity, in regard to maps between sets of numbers, where the domain is mapped into the set of function-values, which may be vector values, and the issue of continuity is about open and closed sets in regard to function images, ie moving between the two sets of the mapping construct.

There is also the considerations in real coordinate-spaces, where topology defined within a metric-space context, about "the regularity of a pattern" ie the quantitative relations are continuous (ie function-values do not change in a discontinuous way as one moves continuously in the coordinate space) and there are stable properties which might be defined as a function with a continuous (stable) value, eg the energy of a system is constant (ie conservation of energy, conservation of mass).

In a context of continuity, the models of (quantitatively consistent) measuring (of a function's local component-values) are linear models of local transformations between local coordinates and the function-value space.

That is, there is a local linear model of measuring which relates a measurable property of the system, ie a function (where the system, whose properties are being measure, is contained in the domain space), to the domain variables.

In the context of continuity, the locally defined geometric-measures can be integrated so as to provide a set of geometric measures defined for the entire shape, but the local coordinate transformations need to be metric-invariant and linear (and one would also expect the local coordinate transformations need to possess the property of being continuously commutative as well as the property of being parallelizable) in order for the geometric-measures (determined by integration) to be comparable in a reliable manner.

Thus, these continuous geometric measures are valid for shapes with metric-invariant fiber-groups.

But one wants even more, in order for the pde to be solvable, where solvability (in regard to coordinates) also means continuously commutative local coordinate properties, and any coordinate vector should be commutative after being parallel-ly transported around a local closed curve which is defined by the shape's natural coordinates, so that in regard to force-fields, or accelerations, single-variable domain local-coordinate-directions should remain consistent with (define the same local direction as) single-variable function-value-local-directions.

Instead of deepening the complications of a math model, why not simplify it?

Or

The limitations of the traditional model should yield to simpler math processes, and to simpler math patterns, but in doing this the math processes may become part of a more diverse containment-set structure (throwing away materialism, so as to get-at stable patterns)

Neither the idea of a fixed set of equations, associated to a particular quantitative relation (which exists between function-values and the system containing coordinate space)…

Nor, the set of (partial) differential equations (ie pde) which relate the local linear models of measuring to further properties, ie pde are defined based on physical law....

.... Can correctly identify the true structure of all the relations..., most particularly the spectral properties..., and still remain consistent with conservation of energy, where these are quantitative relations which can exist between the stable components (or stable shapes), which compose an existence, or a descriptive context, in a many-dimensional context, since they (algebraic equations and pde's) are too dependent on a fixed dimensional set of relationships.

That is, a many-dimensional model composed of..., or organized around..., stable shapes can provide many more stable quantitative relations between various shapes and between adjacent dimensional levels than can the fixed idea of a containment space of a fixed dimension, eg the model imposed by the concept of materialism.

The problem with the idea of constant values, or continuous descriptive contexts, is that there are two types of stable quantitative properties, one "with holes" in a space's shape, and the other "without holes" in the space's shape, where the distinction between "having holes" and "not having holes" in a shape, can be in relation to a lower-dimensional stable shape which is contained in a higher-dimensional metric-space, which is a topological space (since it is a metric-space).

These distinctions, based on the holes in a space, are the true divisions of perspective which exist between quantum and classical, respectively, and this can only be resolved (as to which description to choose) in the context of many-dimensional levels organized around stable shapes, so that a particular dimensional level is chosen for the basis of the description (of system properties).

There seems to be a mystery in regard to the assumed dimensional level at which a system's properties are actually being observed where the distinction between the existence of a stable shape's hole structure and not distinguishing holes in the metric-space is being determined by the size-scale of the system, eg in the size-scale of the earth there are no observed (or distinguishable) holes while crystalline structures the size of big rocks the hole structure of the shape becomes a part of the system's stable quantum-spectral properties.

Each dimensional level of containment can have its own set of equations, algebraic or pde (so as to be associated to stable shapes [and their hole-structure] in particular ways), which can give a partial picture of the observed measurable properties of a system, or a stable pattern ie a stable shape, within the assumed dimension of the metric-space containment set.

The stable shapes defined in the various dimensional levels identify a context where stable quantitative patterns can exist (the containment coordinate-space identifies (or can identify) the dimensional level of a description based on equations).

Note: The system defining pde are mostly defined in relation to geometric measures and thus distinguishing between the holes in the containment set..., or the no-holes in the containment set..., is a central distinguishing feature of the description, in regards to what is stable; spectral properties or the determination of the system's energy in metric-space.

Quantitative consistency in regard to "local models of measuring as in the descriptive context of a pde" depend on the properties of the pde being: linear, metric-invariant, and continuously commutative (or continuously orthogonal, or continuously independent) in regard to the set of local directions for the natural coordinates of the system's stable shape.

This leads to solvability, continuity, non-chaotic, (in the correct context) single-valued for a context of a metric-space where it is assumed that there are no holes in the metric-space, or at least there are no distinguishable holes at the size-scale of the physical system which the pde is being used to model.

Issues of graphs which are related to the idea of limits and continuity, eg the continuity of graph values

A metric-invariant metric-space and a metric-space based topology provides a continuous context for locally stable and reliable geometric-measures; length, area etc. so that the local measures can be integrated (since they are continuous) so as to give geometric measures for shapes.

The categories of math patterns

The basic categories of university math departments are: algebra, topology, analysis, and complex-variables applied to these other three main categories of math inquiry (ie algebra, analysis, topology)

They have been discussed (above and below).

Where

Algebra is about fixed quantitative relations concerning math operations defining the set structure of images and inverse-images resulting from the algebraic relations of: linear maps, identifying isomorphisms (one to oneness) and kernals (ie subsets of the domain space which are mapped to zero), defining equations Identifying the set of math attributes needed for their solutions and finding the zero sets of polynomial functionsetc.

Topology is about the continuous properties of graphs of functions, and about being able to integrate local geometric measures over the regions which contain a system or a shape, and the strong relation between continuity and quantitative geometric patterns for shapes which are bounded

Analysis is about formulating and solving pde, either in a context of the system's geometry in its containing domain space, or in a context of pattern-containment by a function-space, and then finding the function-space's relation to spectral decomposition of a solution function to a pde,

ie this is about both physical law and harmonic (or probabilistic) descriptive contexts.

This is also about local algebra which needs a relation to topological set-properties (or function-image-properties, or constructs).

And

Complex-numbers in relation to either algebra (finding zeros of polynomials) or analysis eg an analytic function is about a function which can be approximated in a neighborhood by a converging power series, ie or (essentially) being approximated by a polynomial (thus, relatable to the algebra of complex-numbers), while harmonic analysis, a similar but stronger relation of a function's values (not to polynomial approximations, but rather) to corresponding function-values of the nearby values (or points) in the domain, where harmonic functions are thought of as trigonometric-oscillating functions, and which fit most naturally into complex-analysis.

The solutions to the equations, defined in these different math contexts, most often deal with constraints and reductions which are related to the dimensional properties of a function's (or equation's) domain space, which is usually assumed to be the system's (equation's), actual, containment set.

These set constrictions can be defined geometrically or algebraically in regard to local linear approximations of local images of (system containment) sets formed by the function, or

The "topologically allowed" Fourier transforms,

ie Fourier integrals, and their subsequent relation to algebraic equations.

The main distinctions of the math categories are algebra and analysis, where algebra is about math relations (or math patterns), in regard to math operations, and in regard to the fixed quantitative sets, ie measuring fits right into the (already) identified measured values, so that both alternative ways of expressing the same quantitative type with different quantitative relations, or other types of quantitative restrictions (or constraints), such as intersecting subsets (or subspaces) which, in turn, can lead to the identification of equations and their solutions (quantitative consistency is intrinsic)

While

In analysis, the measuring is modeled as a local linear relation between the pattern's (or system's) measurable properties, represented as functions, and the domain space (or coordinate space's) set of measurable properties, so that, again (as in algebra), equations are determined by different quantitative relations (or quantitative ways) of identifying the same quantitative type.

In physical description this is also called a physical law.

Thus, solving the equation is not simply identifying a set of numbers (a solution set), but more comprehensively, identifying a formula for the system's defining function-values, so as to be based on formulas represented by the variables of the domain space, so the measurable properties of the model remain quantitatively consistent (for all aspects of the measurable pattern's precise description) so that measuring is reliable and the pattern described is stable.

But functions need to be related to the properties of both (1) continuity (though the solution processes, in algebra, related to intersecting subsets seem to also assume the property of continuity for the graphs of formulas) and (2) quantitative consistency, which is simplest to describe as a coordinate-function structure which maintains, in a continuous manner, the local linear relationships which must exist between the function-values and the domain-values.

But further, since measured values are being defined (only) locally, need to be summed-up (integrated) so as to identify an actual measured value for, say, a geometric measure for a shape, eg its surface area or its energy, where this needs the properties of both (1) metric-invariance and (2) continuity, this leads to a need for the local linear relations…, which are consistent with the coordinate's metric-function…, being continuously commutative almost everywhere.

This rather long list of required quantitative relations for the description to remain quantitatively consistent has a very limiting affect on both (1) how these equations can be defined, and subsequently, (2) solved, and subsequently, on (3) the types of shapes (or patterns) which these local measuring constructs can describe, so as to remain quantitatively consistent, ie so that the patterns are stable and reliable and measuring is a reliable operation (or reliable process).

The solution is nearly always based on "separation of variables" so that either a scalar-function can be factored into function-factors which are each dependent on one-variable, or the vector-function has vector-components which each depend on only one-variable and the coordinate directions and function-vector-directions correspond. This identifies the type of shapes of (physical) systems whose local linear models (of measuring the system's properties) are continuously commutative.

Note: When a quantum description claims agreement between the system's set of operators and its spectral properties then this is almost always associated to a solution based on "separation of variables."

And

Physical law, when placed in a geometric context (ie more geometry than a point-particle), solely considers force-fields with spherical symmetry, so that when the spherical symmetry is perfectly maintained, then spherical coordinates are continuously commutative almost everywhere (except at the poles), but on being perturbed from spherical symmetry, then the quantitative structure becomes quantitatively inconsistent, and (immediately) there is chaos, and the geometry of the sphere deteriorates (or disintegrates) [there is no stability].

One sees that there is a math tradition of following the math patterns described by the previous authorities, thus one sees an emphasis on polynomials, but not much emphasis, in regard to pde, on the context of: linear, metric-invariant, and continuously commutative local quantitative properties (or shapes), and the limited relation these properties have to the simple and stable shapes of circles, cylinders, rectangular polyhedron simplexes, and the relation of these rectangular simplex shapes...., by the process of being moded-out ie making opposite "faces" (of the simplex) topologically equivalent...., to what are continuously commutative circle-space-shapes, eg the tori and the very stable shapes of the discrete hyperbolic shapes,

Where the very stable discrete hyperbolic shapes are, seemingly, built from many-toral components, but so as to still be based on rectangular polyhedron simplexes which are attached to one another at vertexes (but then these vertices are pulled-apart) to form "rectangular" polyhedral simplexes, ie whose moded-out shapes possess continuously commutative coordinates and appear to be composed of toral components, ie one such toral component for each of the "rectangular" polyhedral simplexes, which are first considered to attach to one another at vertices (before the vertices are pulled-apart).

That is, the professional mathematician is not likely to ask certain types of very fundamental questions about the basic properties of math,

Such as,

Should the containment sets be about a many-dimensional structure organized around these (above mentioned) stable shapes, and not so much about:

The artificial confinement to "certain dimension" domain spaces, for either pde, or algebraic geometry, which are (both) "locally" either non-commutative or non-linear (ie for both, pde and algebraic geometry) and, in regard to the professional mathematicians, there is, little questioning of: assumptions, contexts, containment structures, the way math patterns are organized, interpretations, and/or discussion about quantitative consistency, etc

Thus, the identifying the differential equation, and subsequently, determining either what it is capable of describing, or identifying the limits as to what it can describe, and subsequently "defining one's way out-side of those limitations," has not progressed much past the work on differential equations, which was done by Newton.

Though the pde's context for description has gotten ever more complicated, but seemingly more dys-functional, as the complications often deal with non-linear and non-commutative math contexts, and these contexts are quantitatively inconsistent. Thus, leading to multi-valued-ness, discontinuity, chaos and indefinable randomness, in the math descriptions, which mathematicians try to consider in their belief in the high-value of the generality (and abstractness) of math patterns.

The described patterns in this quantitatively inconsistent context are not stable, so the descriptive language has no content, or no meaning, in regard to the practical usefulness of such a, supposedly, measurable description of existence.

This, lack of fundamental questioning, seems to have to do with the selection of the authoritative personnel chosen as professional mathematicians, they are obsessive and border-line autistic, relying on memory and symbolic relations more than challenging and proposing new math contexts

And it is possible to select "narrow obsessive types of people" since they are (often) in the "professional" wage-slave social-class, so that wage-slave people can be preyed upon to serve the investor-class.

The curriculum in math at university level is a memorized sequence of authoritarian traditions, which are made as complicated as possible, and which are poorly inter-related, ie there is not a common theme (such as quantitative consistency) which runs (or should run) through certain types of math categories, as there should be, and these memorized math constructs, which are overly general and abstract, so as to move (in thought) toward complication and formalism, and away from the fundamental issues of valid quantitative description, so as to be: sufficiently accurate, widely applicable, and practically useful (in regard to practical creativity).

That is, the, so called, math experts mostly ignore the issue of "quantitative consistency" in the math patterns which they consider, thus they get embroiled in a lot of delusional speculations about their overly general, and very complicated, math constructs. They rigorously prove patterns which have no relation to the world which we experience, yet they are applied to constructs such as particle-physics with impunity, ie no one questions such patterns which possess such a very limited and very specialized context of applicability, they are applied to contexts which do not possess any stability and no quantitative consistency, eg the random collisions of broken-apart material components which are transitioning between the relatively stable beginning and ending states of the world (it is the stable aspects of the world which go without valid descriptions).

Quantitative consistency comes from both stable patterns, and, subsequently, from stable shapes, eg shapes of the system-containing coordinates, of different dimensions leading to different types of stable properties, eg a system's stable energy properties and the stable spectral-set of a system, where these two different types of stability are distinguished by the relation of the structure of these systems to different dimensional structures, but where it is now (2014) assumed to all be about containment in "the set" which defines the idea of materialism.

Math teachers at the university level spend a lot of energy defining the sets of patterns and properties which are related to a process of solving equations which are non-linear and characterized by non-commutative relationships, and not enough effort in defining new contexts for the containment of these equations (eg pde or algebraic)

A lot of effort, by them, is put into this type of activity ie solving quantitatively inconsistent equations, yet their understanding of the measuring processes and its relation to the quantitative structure of the "pattern containment set" seems to be deficient.

They seem to memorize sequences of thought, while they maintain a shallow understanding of the subject concerning the failure of describing measurable patterns of stable systems, eg how the idea of material has been a failure as a model for physical systems, especially, in the vague descriptive contexts of: quantum physics, particle-physics, and general relativity, and all the other theories derived from these (three) core expressions of physical law (ie quantum systems are function-spaces, and material interactions are to be modeled as point-particle-collisions, and macroscopic geometry is to only be about non-linear spherically symmetric shapes which are to be based on (or related to) singularity-points.

This descriptive context is about detonating a gravitational singularity as a doomsday-bomb, and nothing else.

Teaching

The point of this (above) discussion (about math) is about how mathematicians have not been able to identify the key issues of math descriptions.

They have not understood the idea of the derivative any better than when Newton first formulated the idea as a model of local measuring (and Newton solved the 2-body system, and that is also as far as the professional mathematician has also gotten.

That is, professional mathematicians mostly express their intellectual domination but have very limited intellectual capacities.

The intellectual-illuminati, which are a dominating part of the propaganda-education system (where the main expression of the propaganda-system is that people are not equal, and certain ways of using ideas is better than other ways, namely, technical ideas must only be used for military advantage), might be better described as the "illustrious turds-for-brains" personality types)

Their only additions to the Newton context of (partial) differential equations have been both non-linear connections, ie a useless quantitatively inconsistent idea [though it needs to be expressed, but its useless-ness needs to also be identified], and sets of operators acting on function-spaces, in order to find a function -space's spectral-orbital set, but this is done in "what is, essentially, a non-commutative structure," and hence it is also quantitatively inconsistent (and thus practically useless).

That is, they try to identify spectral values (of [physical] systems), not by considerations about the nature of stable spectral-orbital properties of the observed pattern of the system, and its natural relation to simple geometry as Bohr first expressed, but rather by trying to find sets of operators, which are analogous to physically measurable classical properties, ie they wish is to find enough measured values to reduce (or solve) the system's function-space model, by a set of measurable constraints, so as to identify the set of spectral values, but thus method has not worked (not enough operators can be found).

Its failure is mostly due to the fact that it is a non-commutative context for description. Furthermore, the "complete set of commuting Hermitian operators" is not a sufficient number of measured values needed to reduce the function-space to the (desired, or sought after) relatively large spectral-set, since the property of "non-commutativity of general inter-functional relations" excludes a relation of a function-space to be able to model a quantitatively consistent structure, or domain space (ie the function-space), or operator space. That is, the function types in such a function-space model of a quantum system need to be relatively simple stable shapes, eg tori, whose commutativity needs to be based on a shape which is built from toral components and whose shape's natural coordinates locally commute.

There is also an attempt is to seek solutions-sets by means of finding the intersections of solutions-sets of many-variable polynomial equations, ie algebraic geometry, but this is both non-linear and non-commutative, or either non-linear or non-commutative or both, and thus not quantitatively consistent, and thus not stable.

That is algebra was flawed as a valid quantitative model for stable, measurable patternsboth due to its need for the notion of continuity related to the graphs of its (often) polynomial equations,where this leads to a further flaw in the descriptive context of polynomials, ie it is also flawed, due to its lack of a valid model of measuring, where this would be the context of finding solution sets on (the system containing) domain space, based on graphs, where if these graphs are to have physical meaning then measuring (on the graphs) needs a local linear model, within a context of stable (or reliable) measuring, ie within a metric-space so that the descriptive context is metric-invariant, but this is not done.

They (the professional academics) tend to talk about the wide array of described patterns in various categories of math, but ignoring the fundamental properties of a quantitative description, which are most valid and practically useful.

They try to bring into a useful context; non-linear, non-commutative, as well as indefinably random descriptive contexts, where this involves non-linear pde and non-commutative function-space techniques, as well as algebraic equations ie the intersections of solution sets of many-variable polynomials (a quantitative construct which is locally non-linear and non-commutative),

The only math patterns which are reliable and practically useful are those patterns which are in a context where measuring is reliable and the main focus is about and based on patterns which are stable.

This is about the stable shapes which exist in a many-dimensional context from hyperbolic-dimension-1 (or hyperbolic-dimension-2) to hyperbolic-dimension-10, inclusive. Where this higher-dimensional context (of stable shapes) was found by D Coxeter, and is the core type of shape which lies at the center of the Thurston-Perelman's geometrization, where geometrization is about how non-linear shapes descend (or evolve) to the stable discrete (mostly) hyperbolic shapes.

Chapter 8

The quantitative models and the narrow dogmas required of society because of the interests of the investor-class

The communication channels of the propaganda system controls thought within society, vague statistics and manipulation of society is done within the context of this highly (and artificially) controlled society, where people are both afraid and arrogant in their narrowly defined ways (defined for them by external coercive social forces)

To write about "the nature of US society," Claim: It is a totalitarian society, and about "the nature of its propaganda" Claim: It represents all which is highly-valued, and supposedly, it represents the, so called, objective truth (filtered through the dogmas of the propaganda-education system) a very-particular interpretation of truth, which nonetheless, is (really) represented as an "absolute truth," and this, so called, "absolute truth" is the basis for the claim being made by the western-culture about the, so called, superiority of the western-culture, and, subsequently, there is a need (by the western culture) to form an empire

(this is, a deception, since all the other big nations in the world, eg China India etc, are already trying to adopt, and hold-high, the same absolute truths of the US's propaganda-education-indoctrination system, thus, the empire has already been formed.

That is, the, so called, advanced knowledge which is represented in the institutions of, so called, higher-learning [in the west] is a knowledge which fits into a particular type of social organization, ie a hierarchical collective society (as was the Roman empire), which is best characterized by its arbitrary basis (for truth) and its great violence, where its arbitrary "truths" are fit in such a coercively functional manner into the justice-spy system), but,

In fact, the whole academic scene is colossal failure, which is what is really leading to cultural and world collapse.

The empire… based on a (or based on this narrowly defined and carefully controlled) failed knowledge… already exists, where the empire-"economic" social construct is based on a limited range of instruments which in turn depend on only a few types of natural-resources, but where the social structures basis in violence and associated oppressive institutions, eg the justice-spy system, so this limited construct can be based on 5% of the population (the technical and brutal forces in society) so that 20% are the participants and recipients of the social construct which has been put-together (or dependent upon) a few technical instruments…, where this realization has led to "this type of development, 'defined' over a narrow range of a population" distributed over many regions of the world…, where the fast development of such a society seems to bring the greatest growth, ie the most profits, in a short time, and thereafter (ie after its original growth and development) there is an social-economic-equilibrium, which does not provide as many profits.

So this type of spurts of development and growth, so as to be distributed over various regions (eg nations) is the short-sided vision of the empire of the oligarchs.

Adopting western knowledge is equivalent to adopting the Roman-empire type society, but the emperors are now the investor-class. That is, Mao Tse Tung brought China into the western empire (ie Marxism is capitalism but put into a slightly different relation to the methods of propaganda, but, seemingly, ignoring the fundamentally Machiavellian (or Roman) basis of western culture)

If one expresses ideas which are different from what is expressed in the propaganda-education-indoctrination system, then one must spend a lot of time setting-up the new context, since new ideas require that one also develop a new context for thinking and creating, ie a new vision about existence, but nonetheless, the new ideas will be ignored, (ie there is no time for careful discussion when the "absolute truth" is already in our possession) but established ideas do not need any explanations, rather simply the correct code-words are needed to identify an allowable category of thought, and to enter the world of allowable categories of the, so called experts, but where the established ideas are clearly demarcated and separated from the public, by means of word-strings of code-words, so as to be the sole realm of the, so called, authoritative experts of an empire run by the investor-class.

That is, to review western expressions of ideas, which are published, is to basically get "the truth" as seen by the investor-class, since they control the instruments of publishing, and propaganda can be used to control the politics of education.

The internet is all about opening communication-channels, in order to spy, and then to destroy (or close-off) those communication-channels, so that the instruments remain controlled by the investor-class{in fact, the vaguely realized but not allowed to be expressed, but now spy-documented revelations of Snowdon may be the desired manner (of the oligarchs) by which the public is run-off the internet communication channels)

The ideas which are expressed over the empire's propaganda-system are narrowly defined, dogmatic, and authoritarian ideas, ie the types of ideas which one would expect in regard to such a controlled and authoritarian way of setting-up communication-channels within society, ie through a propaganda-system.

Within an authoritarian society ie the information channels are associated to either propaganda-education or vocations, and the technical channels (of writing) are the exclusive realms of the experts, and the ideas established by the propaganda system are considered to be of extreme high-value, ie essentially absolute truths,

About which academic contests..., which are thought about as the (valid academic) proof of one's 'real" intelligence..., are designed where if one cannot deal-with in the rational cannons of expert technical publishing, so as to compete in regard to these high-truths, (characterized as the highest of absolute truths, which, in fact, are related to a limited number of instruments and the technical experts are dedicated to the gradual development of these instruments, where these few instruments figure deeply into the monopolistic strategies of the businesses, about which these (few) instruments are the basis (for these business monopolies)), then this proves that such a person is of an inferior intellectual capability, so if a person, in this empire of monopolistic businesses, expresses ideas which are different from the, so called, absolute truth (which is really about certain types of instruments), then these expressions will be (must be) ignored by the empire

Consider that C Castaneda expressed some aspects of the cosmology of the ancient knowledge of other cultures of the world, eg the cultures of the native peoples of the Americas, but (1) these cultures are what the western culture seeks to exterminate, and (2) western culture has developed carefully thought-out math structures for physical description, which happen to now (2014) be floundering, but which may provide the correct context to more deeply identify ancient knowledge of other cultures with new practical possibility [for both the adepts of the ancient knowledge as well as for people who are curious and creative, ie everyone]

(3) the new ideas (expressed by m concoyle) about a measurable description of existence [based on many-dimensions, and on stable shapes filling these many-dimensions] seem to bridge this gap of knowledge between two cultures.

The characterizations of the ideas, which are the propaganda-system's absolute truths, are that they are based on materialism, and they are described in a precise descriptive language of partial differential equations, which, in fact, is an inconsistent and/or arbitrary descriptive context, ie a context which cannot carry within itself any meaningful content. This is because local linear models of measuring, ie partial differential equations, cannot become either non-commutative or non-linear and still possess a descriptive content which is quantitatively consistent, and without quantitative consistency the description has no meaningful content. But the partial differential equation models of material systems do exactly become (for a general system) non-commutative and non-linear.

Note: Peer-review journals are owned and controlled by the same people who own and control the propaganda-system, thus the strong connection between "high-valued propaganda" and the expert-class.

This type of tyranny is best upheld by the justice system

where instead of law being based on equality as stated in the US Declaration of Independence, which would lead to the Socratic ideal of equal free-inquiry and the relation of new creative contexts, ie new knowledge, to new creativity

The law has been arbitrarily based, in the US, on THE LAW of the Roman-empire PROPERTY RIGHTS, AND MINORITY RULE.

The main claim of the US intellectual class, is that its, so called, experts are peer-reviewed, so it is considered to be an, absolute truth-value, which can only be possessed by an expert, where this, expert-class, is also a social-class which supports the idea that

"inequality is a scientific truth," based on statistical expressions put into an arbitrary context where vaguely distinguishable patterns are identified ie patterns which are not well defined and patterns within which the interpretations of the, so called, random events of a statistical study are unstable events and thus cannot be counted in a reliable manner, so the statistics based on such vaguely defined qualities (ie vague qualities built up from the constant repetition of the propaganda-system) are meaningless and arbitrary,

But, it is quite clear that

The, so called, "absolute truths" of the expert-class have failed, eg the expert-class cannot explain why the solar-system is stable.

Unfortunately, it is primarily the failure of these, so called, high-valued absolute truths, which is at the core of society's failings, since there has been (subsequently) a false model, which has been followed, and the, so called, "absolute truth" of the empire is either not capable of describing the stable properties of fundamental systems in an accurate manner, eg being able to describe the stable spectral structure of a general (non-radioactive) nuclei, or general atom, or molecule, or the shape of a molecule (where such molecular shapes are being controlled during molecular processes in biological systems) or the stability of the solar-system, ie being able to describe the very stable properties of the many-(but-few)-body spectral-orbital physical systems. and

This colossal failure has caused the creative context, which is supposed to be the realm of the expert, to, instead, become a non-creative context.

This is similar to how the justice system is now best characterized as an "injustice system," or equivalently, the justice-system is the ruling-class's counter-insurgency attack on the public, while the ruling-class breaks the law with impunity, but the public is attacked mercilessly for minor offenses, eg j-walking, or asking for help.

Furthermore, the propaganda system is set-up as a competition between two sides, where the action oriented side (the right) is the voice of the ruling-class, while the intellectual side (the left) represents the intellectual authorities of the high-valued absolute truths, which best serve the technical interests of the ruling-class, but if one reflects on these absolute truths one can easily find that it is a "failed" absolute truth

This has become an issue of the administrative management of an institution's personnel,

That is, the propaganda system is filled with authoritative and competitive types,

And there is no room for those "equal free-inquiry personality types," ie those personality types who question the elementary assumptions of the rational structures of any (authoritative) descriptive language, ie the fundamental issue concerning philosophy, which was identified by Socrates.

The empire only deals with high-valued "absolute truths," which serve the interests of the investor-class.

All aspects of the propaganda-based truth-structure of society (eg the so called news, and the so called, sciences) are promoted by all social institutions, and all of these institutions are failing: economics, war (called defense), law, math, physics, where all aspects of the structure of our culture's authoritative knowledge are designed and organized to serve the ruling-class,

And all aspects of the society are held in place by violence,

The way in which those with social power are trying to wiggle out of (ie to avoid responsibility for) the mess they have created, is to now consider "short term developmental growth" for a wide range of regions around the world

The regimentation of people in a hierarchical social structure is the propaganda mind-set which human-beings do not seem to be able to stop

Furthermore, the investor-class is no longer claiming to protect the US public.

The government protects the rich, so the cruel way in which the propaganda-system uses language, it always lies,

For example, Reagan stated that "the problem was the government," a vague claim, which was said "in code" to the business-class, and it went over the heads of the fairly prosperous public (at the time, 1981), but it was presented as a government which was in opposition to those who sought the US to be a theocracy…

(in fact, a theocracy is quite similar to an Islamist state, as the Islamic state was quite similar to the Roman-empire in 600 AD, and since the "cold war," along with the vile-communists, made the government serve the militarization of society, where such militarization was the real problem of the government (in 1981))

… however, in regard for the drive by the propaganda system for society to become a theocracy, the problem is law, where the government does not uphold the US law which is to be based on equality, ie the government was bent, by the propaganda-system (under Reagan) to serve the investor-class, and to cement the investor-class's relation to the militarization of society, through deeper spy-relationships to the militias (of the right), over government's (true) relation to the public.

The war with the Islamic world by the west, is really about driving the society to become a theocracy, where impunity can more easily be defined by Calvinism, where Calvinism is, essentially, the idea, that "the rich are the messengers and agents of God."

It should be pointed-out, that the "well established modes of technology" (which are derived from 19th century science) are now a part of the action-oriented set of personality-types, not the intellectual-class, (this knowledge is related to established processes), and they form the basis for the monopolistic businesses, which are often organized around some particular types of technical instruments, eg the media is about publishing and broadcasting,

ie control of the communication channels.

Consider that,

If particle-physics can not be used to describe the stability and the stable spectral-structure of the general (non-radioactive) nuclei, then it is meaningless for propaganda to state that "the particle-tracks of a high-energy "particle"-collision validate the existence of the Higg's particle, and subsequently, all of particle-physics is, thus, verified."

Since particle-physics cannot describe the fundamental properties of a general nucleus, which, according to the expert literature, particle-physics was supposed to be designed to actually be able to describe the fundamental properties of a general nucleus, and thus, its failure to be able to describe these observed properties of the general nucleus, means that particle-physics is a failed descriptive context.

How is it known that any of the "tracks" in the detection chambers are a result of a particle's motion? Where some structure is identifying "tracks" in a detection-chamber, and that these "tracks" are, in fact, really a property of a distinguished-point of a discrete Euclidean shape, a better model of the, so called, Higgs mechanism, ie the relation of material-interactions to the property of inertia, and

How does one know if the protons, in the so called "point-particle-collisions," which are formed in the particle-accelerators, are not (really) also, really, the distinguished-points of discrete hyperbolic shapes,

Where it should be noted that an isolated and cooled-electron (of Dehmelt) does not hover as a charged point-charge, but rather forms into dynamic orbits, ie the elementary-particle is better interpreted to be a discrete hyperbolic shape with a distinguished point, than as a point-charge.

The meaning of this announcement (ie that "the particle-tracks of a high-energy "particle"-collision validate the existence of the Higg's particle, and subsequently, all of particle-physics is, thus, verified.") of what is most likely another example of dry-labbing of data, ie scientific fraud or, equivalently, scientific misrepresentation,

(since the descriptive context of particle-physics cannot be used to describe the properties of stable systems, and can only be used in the context of bomb-engineering, the main focus of particle-physics is on particle cross-sections and their relation to rates of reaction in the instrument of the nuclear bomb) and which is an announcement which is, essentially, announcing a heavily funded descriptive math-model, which possesses no content, in regard to the properties of the physical world,

Rather

it is a statement, which is all about stating that, "the research into detonating a gravitational singularity, using the cross-section models of particle-physics, will continue full-tilt."

Data from (particle-detection, or) component-detection chambers of particle-accelerators is interpreted within the framework…, which is the only theoretical language which is allowed by the very meritorious academics…, where that language is non-linear, non-commutative, and indefinably random, so that the implied "quantitative-set of… a material-system's… containment" is quantitatively inconsistent, so that the quantitative elements… of the set which is being used for the fundamental measured properties of the system… (1) do not possess a well defined unit of measuring-scale, and (2) the quantities have a discontinuous relation to one another, and (3) the values of a particular type of measurement are multiple-valued, so that there cannot exist "a describable pattern of any system" within the fundamental set of system-containment which is quantitatively inconsistent, ie the descriptive properties (of the quantities, or of the partial differential equations and their operators) do not allow any descriptive pattern in such a mathematical context to possess any validity [such a descriptive context possesses no meaningful content]. ie particle-collision data is interpreted in a meaningless context, that is, whatever is being claimed is arbitrary and without any practical meaning, but the randomness of particle-collisions automatically fits into the business interests of the military businesses.

Thus, by incorporating the fake-knowledge of particle-physics into the focus of the, so called, scientists, of all the high-valued and uniformly based propaganda totalitarian nations, then the result of this will be that the focus of "the knowledge of physics" will remain on the relation of math and physics to war and destruction, and not on the relation of descriptive measurable knowledge on "the true relation" which "life has to existence."

Thus, the international high-energy particle-accelerator, and its intellectual context (for the global intellectual), is one of the main ways in which other nations are taken-over, and incorporated within the western Calvinist empire, a collective society which is very-much based on the hierarchy of the highly-valued intellectual class, and the subsequent relation of cultural knowledge to the interests of the investor-class.

One sees from the intellectuals (ie those with a voice within the media, and in the meritorious institutions) a group of people supporting a descriptive context, a, supposedly, rational structure, which has been organized and used, so as to no longer have any meaning (incapable of expressing any practically useful content), where this has been true since about 1910 (where this was organized by D Hilbert), and its, so called, pinnacle of intellectual greatness ie particle-physics, is, in fact, an expression of developing the ultimate instrument of destruction, and with no other redeeming creative attribute.

Thus, the intellectual class is really about representing their own arrogance, and their own self-importance, and their individual-superiority, which has been defined for them within the propaganda-education-indoctrination system,

ie they are the social icons of "inequality as a law of nature,"

eg the top .001% of the normal distribution of the random property,

eg of, so called, individual intelligence

(this type of arbitrary mis-representation of math properties, ie the statistically defined property (or quality) of "intelligence," can not be used to identify a stable elementary-event set,

ie this is because such an invalid use of statistics is not quantitatively consistent,

and, subsequently, this way of using knowledge is best described as barbarity, the, so called, "top intellectuals" are those who are best indoctrinated in the knowledge which is failing, thus, it is an arbitrary characterization of high-value, which is being barbarically upheld by violence, especially, since their high-valued truths cannot be used to describe the fundamental stable properties, which are observed for physical systems, ie their (supposedly) superior intellectual interests are, in fact, massive failed intellectual endeavors,

That is, this should not be considered to be the work of those who are claimed to possess a, so called, superior intelligence,

Where a more realistic idea about intelligence might be best defined as "intelligence is an ability to discern truth," but without there being a "known truth," how can the property of "discerning truth" be determinable by others?)

Such a discussion will continue to revolve around arbitrary ways of identifying invalid statistics, a statistics which depends on the control of the population's thought, through the arbitrary constructs of the propaganda-education-indoctrination system.

While the right (the, so called, people-of-action) is also all about arbitrariness, and their, so called, superior ability to "get things done," and subsequently, they express their own type of arrogance, ie their self-righteousness based on arbitrariness, which is used to justify their barbarity (where their barbarity is justified because they have demonstrated their social-worth, by their ability "to get things done," for those things which the investor-class "wants done," mostly through the trained-barbarity of the people-of-action).

This barbarity, based on arrogance, characterizes the justice-military system, as well as monopolistic businesses.

Note that, the Roman-army built the superior Roman-culture with bricks, while the modern business-army builds product, essentially based on the, seemingly, absolute truths of 19th century science (ie the equivalent of the Roman-army brick-laying).

The intellectuals claim to be rational individuals serving a collective (a consensus) on a "rational march to an absolute truth," while

The right claims to be individual people-of-action, but such a "person of, so called, 'individual' action is only successful if they are serving in an army," which, in turn, "collectively supports the few in the investor-class."

Both sides serve arbitrary 'truths" and both sides do it with great barbarous-ness.

The "great sin" of the human is to always mis-represent themselves

The intellectual-class gives no review about violence, which possesses any sense of reality to it,

ie violence is not simply imperialism, rather violence, in a totalitarian state, is constant and unrelenting, it is arbitrary, and it is justified by arrogance, the arrogance of the right, is the arrogance of those serving the ruling-class so as to be granted impunity for their arrogant brutal behaviors, while the arrogance of the intellectual-class is the (an) arrogance based-on the arbitrary intellectual language structures, which serve the same business monopolies, for which the continual violence is needed in order to defend and maintain the social position of the ruling-class and their intellectual helpers,

ie the intellectual-bullies and the action-bullies both collectively serve the same monopolistic businesses, and both arrogantly claim to serve a superior (but arbitrary) belief structure, where these beliefs are given high-social-value by the ruling class, and both of these opposite sides (intellect and action) are controlled by the ruling-class, and both sides are motivated by their own type of arbitrarily (based) arrogance.(and a belief that social inequality is a law of nature, ie circular emotional pathways used by the investor-class to get society to collectively support the arbitrary high-value defined by the interests of the investor-class, supported by the arrogant intellectuals and by the arrogant violence of the, usually arbitrarily moralistic, people-of-action).

Such a philosophical division {(intellect vs. action) for both of these opposite-sides serving the same investor-class, but both side motivated by a, seemingly, opposite set of arbitrary principles} is based on:

Science (materialism) vs. Religion (idealism but moralistic)

But within the constructs of math one finds a polarization of stable patterns vs. arbitrary quantitative inconsistency

Where quantitative inconsistency emanates from arbitrary restrictions on the containment set, ie the idea of materialism, and the need to model a physical system as a partial differential equation, and where the system is contained in a quantitative set structure where the quantitative sets are too-big the real-numbers defines too-big of a set so that this can (does) lead to logical inconsistencies.

That is, Science vs. Religion

Is really Arbitrariness vs. Arbitrariness;

ieArbitrary (material based science) vs. Arbitrary, (arbitrary moralistic beliefs)

(note that, religion is an arbitrary language, which has a deep relation to identifying hierarchical social structures, but it is a language which has no relation to practical useful creative efforts, but science [as it is now practiced, as being based on materialism and partial differential equations] is quantitatively inconsistent, and thus, it <u>also</u> has no relation to practical useful creative efforts [except some parts of 19th century science])

That is, the wonder of science which is more related to religion than one would expect, is about the nature of stable, measurable (geometric) patterns, which can exist in an 11-dimensional context.

Within this mystery resides the relation of creativity to knowledge, and the relation of creativity to existence, and the relation of creativity to life, and the relation of existence to life.

What is an alternative to the intellectual systems of the arrogant intellectual-class?

Consider first

The Contrasting viewpoints about how to organize a containment set used for the descriptions of observed patterns within existence,

or

How can the structure of existence be re-organized to be able to describe the observed properties of material systems which exist?

The problem is that despite there being some development of ideas, and the fact that virtually all the technological development is directly related to a set of relatively new science findings, in particular, the 19th century science-findings about electromagnetism, by Faraday and Tesla,

Nonetheless the ruling-class is at war with any new ideas, since the social power of the ruling-class was developed without any new ideas, and subsequently, any new idea do, in fact, place their high-social position at risk, thus there is all the propaganda baloney about there being so much change and development, but in fact there is virtually no changes in western culture since the Roman empire the Roman-solders built civic monuments with bricks, which cemented in the public, a mental-illusion, of a position of high-social-value for western culture, while the surprise of technical development, despite science and math being oppressed by the ruling-class, has, nonetheless, provided the ruling-class with a new type of building-block, namely, the electronic gadget, as well as instruments based on the thermal properties of systems, through (or about) which the illusion of change and development and what is, in reality, "highly limited innovation" and "highly controlled development," where the development is limited to the development of the instruments about which monopolistic businesses are organized, so that only a particular (very limited) development of exactly one type of instrument are allowed to be expressed in the technical channels of expressions are allowed, while the propaganda system expresses the illusion of there being "fast occurring technical changes" (it is all the development of the TV and the computer, which have been around since the 1930's)

But its is still the same "main idea of Rome" which is being expressed, that of superior hierarchical culture (within a collectivist culture) expressing domination by means of (and held in place by) extreme violence.

The current viewpoint

The current science and math viewpoint is mainly based on the two ideas of:

1. materialism and

2. the partial differential equation, which is being used to identify a physical system's set of "measurable" properties

And

This logical set-up has failed, but the social structure is about the ruling-class being at war with any new ideas, where this is mostly expressed by the social-class of the, so called, experts, who are narrow and dogmatic in their language, and who are only allowed to tinker with the development of a few instruments in a very limited manner, eg physics departments at public universities are almost uniformly only about bomb engineering, or dedicated to the commercial development of some narrow concept or physical property, eg liquid crystals, etc,

And

Thus, the viewpoint of the expert, is that of traditional thought, which follows the authoritarian lines of thought, which was granted-to the previous experts, who were also the favorite scientists of the ruling-class, ie the scientists who study the categories of thought which the ruling-class wants explored,

ie the usual way of the academic institutions

Where this is done in an overly authoritative and oppressive manner, by the, so called, top authorities,

eg S Hawking is one of the, so called, great-authorities, whom leads the way in the attempt to detonate a gravitational singularities through the context of particle-collisions (ie through the route developed by the Manhattan project and greatly extended by E Teller) of particle-physics, but which is really the 19th century model of the chemical reaction.

But the descriptive context of materialism and partial differential equations quickly leads to math patterns which are non-linear and/or non-commutative, where, in turn, these properties (or being non-linear and non-commutative) lead to quantitative inconsistency, eg the measuring-scale is not well defined, the quantitiave containment structure is dis-continuous, and the function representation of a system's measured values, ie the assumed solution to the partial differential equation, becomes many-valued, thus there is no quantitative model which makes any sense, ie there is no meaningful content in such a descriptive structure.

So this leads to the issue of how to understand reliable measuring and stable (math) patterns and logical consistency (where, as a note, the computer is all about operating in a well-timed construct of electric signals which are in a construct of logical consistency,

Where logical consistency is at the core of the argument which states that free-thought by an electronic switching device is not possible, the argument against this statement seems to be that by processing a large amount of data related to a seemingly identifiable subject matter, and setting-up a logical method of making decisions in regard to selecting data which "best" correlates as a "response" to the identified subject, based on the statistical relation of the data to the partially identified subject-material, is an expression of (or a model of) an actual thought process, but the thought is, in fact, the motivation of the circuit switching operation within the logical context of the circuits, and that motivation is being designed, or determined, outside the instrument ie various routes for the set of electric signals to be processed are being selected.)

Physical science

If physical science, as well as the associated biological sciences, are all based on the idea of materialism, ie where the word materialism is about an idea concerning set-containment,

And

There are two basic viewpoints

I quantum theory,

Which is based on an assumption of randomness in regard to the assumed to be point-particle component (random) events of a quantum system where randomness depends on identifying a set of elementary-events upon which probabilities are to be determined, by a counting-of-elementary-events process, or modeled as a function-space and its associated eigenvalue-events, where the events are to be calculated by solving a partial differential equation which, most often is set-up to identify the system's energy values, but for "the most fundamental of the quantum-system types which exist," ie the very stable many-(but-few)-body quantum system, such system operator structures either cannot be formulated or they are non-commutative operators and cannot be solved

Furthermore, the random basis of quantum theory means that the method is best modeled to fit with already existing data and it is a description which has not been able to identify (or predict) through calculation new elementary-events which the method was not already trying to adjust the system's operator structure in order to fit the already known data (or measured properties of the quantum system). That is, a description based on randomness is designed to fit observed data as Ptolemy's epicycles were added-on to fit the observed data

And

When quantum theory tries to model material interactions the description is both based on randomness, ie basically trying to fit the operator structure of the system's model with observed data, but the operator structure becomes non-linear so as to model a particle's internal particle-state structure during a particle-collision, where the particle-collision with an internal particle-state structure is the model of material-interactions in quantum theory

And

II classical physics and general relativity

Which are both based on an assumption of material geometry, but which except for either the 1-body system of general relativity, or the 2-body system of a planet and a sun, in regard to Newton's model of gravity, and which is a model which Schrodinger extended to the H-atom in quantum theory, but the solution to the radial equation in regard to the H-atom diverges due to the point-particle model in regard to a spherically symmetric electromagnetic field which emanates from the two point-particles, there are no other solutions to physical systems composed of many-(but-few)-components

That is, this is the real state of physical science, as well as abstract math, it is either about adjusting a few types of complicated equipment or it is meaningless nonsense which has not progressed much past the original efforts of Newton

And

This is a result of it (science and math) being turned into an overly authoritative bunch of nonsense, an expression of intellectual elitism, more a branch of the propaganda-system than an actual lively science community, and it is noticeably controlled by the investor-class, but where the results of science would cause the formation of very risky competitions for the investor-class, in regard to new knowledge and in regard to new creative contexts, expensive equipment would become obsolete.

Where in both cases...

Except in regard to II where there are many examples of geometric material distributions which are solvable, or linear, metric-invariant, and continuously commutative over the system's shape, and these solvable descriptions have led to a great deal of technical development, eg electric circuits ie linear; second-order; (separable) partial differential equations with constant coefficients, the (partial) differential equations (or operator structures)…, which are being used to model physical systems…, become either non-commutative or non-linear, or both, which also leads into the indefinably random context, and these types of math processes, or math properties, cause the quantitative-set structure… of the system-containing quantitative coordinate (and/or function-domain) space… to become quantitatively inconsistent

And

quantitative inconsistency…. which results from non-commutative-ness, and non-linearity, and indefinable randomness, in turn…., leads to an (1) undefined unit of measuring, (2) discontinuity in value, and (3) many-valued-ness, within both the quantitative containment-set, as well as having these same (failed) quantitative effects on the function-values in regard to (or of) the system's measurable properties., and thus the description becomes meaningless, and without any practically useful content.

I. Either

(physical description…)

is about (a material system's) reduction to point-particle material-components whose particle-event properties can only fit into a random context of event descriptions (or event identifications) in space and time (ie the containing-set) which are, subsequently, related by operator-models "of physically measurable properties" to "a space of harmonic functions" which, in turn, model the random events of a system's particle-components, so that, if one can find a complete set of commuting Hermitian operators (of the physically measuring properties) for the system, so as to identify a set of partial differential equations, whose solutions identify the spectral-energy properties of the quantum system, which is assumed to be composed of point-particle components, then one can identify the quantum system's full spectrum of measurable properties,

But

Such a set of commuting operators, in general, cannot be found, in particular, they cannot be found for the most important, and most fundamental, set of stable physical (quantum) systems, Where such quantum systems would also include, for example, the very stable many-(but-few)-body systems which possess very stable spectral-orbital properties.

Rather

The claim made is that "such systems are too complicated to be able to describe," but virtually no money goes into researching this very fundamental question,

ie "how does one describe such stable properties of many-(but-few)-body systems?"

But

It needs to be pointed-out that the property of being stable, in fact, implies that the system is a simple system,

That is, it will be a system whose existence (or whose description) is to be based on: linearity, metric-invariance (an implicit set of assumptions of the energy-wave-equation), and the solution containment set being continuously commutative, or an almost completely parallelizable system-shape, so that it is both solvable and controllable,

Note that,

if it were not a controllable context in which such stable systems form then the regularity of the physical properties in regard to the existence of this set of equivalent sets of systems…where each such system possesses the same number of components, and where each such system has (nearly) exactly the same spectral properties, which are uniformly identifiable to a very precise level, would not be possible ie these stable systems are forming in a highly controlled (descriptive) context (of existence).

II. Or

(physical description…)

is based on the geometry of material, and sets of (partial) differential equations.

But both descriptions, either I or II, are very limited in regard to the set of systems which they can describe, so that (ie if one desires to have a precise description of the properties of the physical world, so that) the precise description is both accurate and practically useful, both descriptions, of I and II, fail on both of these desired (precise) descriptive properties, in particular in regard to the most prevalent and most fundamental set of material systems which compose our existence, namely, the very stable many-(but-few)-body systems, of general: nuclei, atoms, molecules, molecular-shape, the solar-system, as well as the motions of stars in galaxies (eg dark-matter)

It should be noted that there is a lot of fraud in science where much of this fraud has its main cause in the limited knowledge which is possessed concerning the physical world, eg no explanations for the very stable many-(but-few)-body systems, yet there is a reach far beyond our local region, eg into both the extremely large and the extremely small, and in these regions of which we have very limited experience our models of physical law are very suspect but nonetheless it is the very speculative models of very remote regions through which all interpretations of data are made, and most of these, so called, physical models are based on indefinable randomness and thus they are about fitting data not understanding data, that is these speculative theories, are about detonating a gravitational singularity, and subsequently, form a descriptive structure whose relation to an actual reality should not be trusted at all, by anyone who is a demanding scientist who seeks actual knowledge, but these descriptive contexts which cannot carry within themselves any meaningful content since they are non-commutative and/or non-linear, are being pushed forward by the institution for warfare, where these institutions were created in part by E Teller, and they are heavily pushed by the propaganda system, and the people who participate are either listening to the cheer-leading of the propaganda-system, or they are the personality types which seek dominance and authority

But these people are clearly both not good scientists and they are not very bright where one can use Forrest Gump's definition of smart, ie "smart is as smart does," and these so called authorities are not doing much but spin-their-wheels in one place, the language they use has no content, in regard to solving any practically significant problems in the physical sciences, rather it is only related to indefinable randomness, but where rates of particle-collisions are related to bomb engineering

Try actual rationality (rather than authoritative nonsense, ie Why try to prove things which only exist in an illusionary world?)

That is, what one wants in a precise descriptive language is that "the description is close to the nature of existence," and subsequently, the context of attaching (eg a component) to (an) existence, so as to cause the desired changes and control of the existence, is to be led by the description

But in either of these ways of considering the descriptions of material properties, either in regard to randomness and sets of operators (in particular, seeking, but [in general] not finding, sets of commuting operators) acting on a harmonic-or random-system whose containment is modeled as a function-space but where the events are defined on the domain-space of the functions which compose the function-space, or in regard to material geometry and force-fields (or geodesics of a [general] metric-space's shape, but the context of finding geodesics is still the context of materialism based containment-sets)but (in both descriptive contexts) one finds that when the number of components is more than two, (or, in regard to general relativity, when the number of material-determining energy-density components of the system [or of the general metric-space] is more than one), then there are no (partial) differential equations which are solvable, so as to be able to determine the system's measurable properties by calculations, based on solving these (partial) differential equation models of the physical system, where these (partial) differential equations are based on the, so called, laws of physics, yet there are a large number of very fundamental and very stable many-(but-few)-body systems where this is true for both quantum-description as well as for classical geometric-based descriptions (of either spectrally-based or geometrically-based measurable properties)

That is, in regard to the math patterns and associated math processes (or math-methods), the math-operator structures (or (partial) differential equation structures) of these math processes, (associated to math models of the measurable properties of the physical world), become: non-commutative, non-linear, and indefinably random (the elementary-events are unstable, or not well-defined), and this descriptive context is not capable of describing stable patterns, and hence it is devoid of possessing any meaning,

And

This meaningless math construct has also come to be defined on a containment set-structure of quantitative sets…, ie containing domain space coordinates…, which is "too-big"

so that the math constructs on such a set structure where the set is "too-big of a set" do not have to be logically consistent, since convergence into such a big-set can be used to pull the context of a descriptive structure "out and away from" the descriptive-structure's natural set of logical relationships (with other math constructs, associated to the [too-big] quantitative containing-space).

Alternative

What are some ways in which to organize some "patterns of math" in order to stay away from the quantitatively inconsistent and, subsequently, meaningless set of (math) descriptive processes, eg non-commutative-ness, non-linearity, and indefinably random?

First, the observed patterns which are seen to be stable need to "<u>not</u> be based on a descriptive context which is based on randomness," ie instead the descriptive context must be based-on stable geometric shapes.

And

How can one fashion a containment-set based on quantitative sets, so that the quantitative structures do not need to become too-big?

The quantitative sets need to be quite a bit smaller in regard to the number of elements which are in the set. That is, the real-numbers need to be rational-numbers, which have lower-limits in regard to small sizes, yet the basic idea of an open-set in regard to inverse images of maps, seems to be a math idea of some value. The "type of a limitation of a quantitative set," in regard to sizes of elements within the set, and to have such quantitative sets still be relatable to a system with very-large (idea about the system's) extent, thus requiring both upper and lower bounds, in regard to the sizes which the elements, which a quantitative set can measure.

Such a set of bounds on measurable quantities, in regard to a quantitative containment set, and still be used to identify a wide array of system contexts, and system relationships, seems to require that all the measurable properties of all the systems, which can exist, be a finite set (ie be finitely generated).

To consider these issues concerning quantitative descriptions about stable geometry and relating the quantitative set structure to the set of measurable properties of the full-range of possible existing systems, so that;(or Then) the whole idea of a containment-set needs to be reconsidered, so that the description be based on principle fiber bundles where the base space is a metric-space and the fiber group is about the invariance-group, in regard to the basic function (on the quantitative sets) to be used for measuring geometric-properties, (so as to not be defined in a non-linear context, in regard to determining "the basic function used for geometric-based measuring" (ie in the coordinate domain-space for the functions which define the measurable properties of a physical system), so that the coordinate containing quantitative sets, and their metric-function basis for geometrically invariant measuring requires a limited type of local coordinate transformations which are defined by either an isometry group or a unitary group or a symplectic group (related to the 2n-dimensional base-space of coordinates which measures both spatial-position and relative-motions at geometric positions) where the base spaces need to be quantitative spaces which can fundamentally identify, either (unambiguous) geometric measures or measures of motions, and the systems which are contained in the base-spaces need to be directly related to the set of discrete isometry, unitary, or symplectic fiber subgroups, where these discrete subgroups identify stable shapes or stable periodic motions within the base-space.

Furthermore, there need to be metric-spaces of various metric-function signatures, where different signatures are related to different fundamental physical propertiesbut there seems to also need to be two main metric-function signature types so that

These two-types (or physical properties) are related to the

1. spaces of geometric position and the relation which motion changes (within this geometric space) have to material inertia (ie material with a spatial-position) and

2. the space of relative-motions and the relation which constant (these) motions have to both charge and energy (or to time, ie where time-invariance is related to the property of energy conservation)

These two spaces have metric-functions, which have constant coefficients, so {the second derivative of the metric-function, which is the Ricci curvature, so that}, thus, the metric-space's curvature is a constant, and where the metric-function is a non-degenerate (it is zero only at the origin), symmetric second-order tensor field ie a 2-tensor, and the two constants (of these two main types of metric-spaces) are either zero (Euclidean space) or negative-one (hyperbolic space, or equivalently generalized space-time space).

Thus, in these spaces (or on these two-types of stable shapes) the local coordinate relations to the discrete (hyperbolic and Euclidean) coordinate-functions of the shapes can be continuously commutative,

Thus, there are; linear, metric-invariant, and continuously commutative coordinate relations which exist between the local (rectangular) coordinates and the curved-coordinates of the shapes, and the shapes are parallelizable, (which are the simple math properties which exist) on these two main types of stable metric-space shapes.

These types of stable and quantitatively-consistent shapes are: the circle-spaces, and the cubes, and the cylinders

Note that, the "positive-one" constant-curvature space is based on the Euclidean metric-function being restricted to the sphere, and because of this restriction the metric-function has terms which depend on several variables, and thus it is a non-commutative and non-linear relation and

Thus, it is contained in a quantitatively inconsistent context, and thus it loses any of its meaning as a quantitative description (especially, when deformed for being a perfect sphere) but the stable shapes of these two spaces (ie both zero and negative-one constant curvature spaces) need to be related to conserved energy, so that the set of very stable discrete hyperbolic shapes identify the stable shapes, associated to stable periodic motions, of the stable material systems, while the Euclidean toral shapes are related to the locally measurable properties of material-interaction structures, and can identify non-local, or action-at-a-distance, interactions (spatial) relationships.

Thus, over various dimensional levels these stable discrete hyperbolic shapes define both the stable metric-spaces which contain material, as well as the stable material-components which includes the stable spectral-orbital properties of these shapes,

One wants a mathematical context for physical description where the local measurable context is: linear, metric-invariant, and continuously commutative, so that the math patterns can be stable and reliable and where the locally measurable descriptions (laws) are solvable, stable, and controllable and thus the description possess meaningful practically useful content. This can be realized in a many-dimensional context, ie an 11-dimensional hyperbolic metric-space set containment space which is organized around the set of stable shapes, the very stable discrete hyperbolic shapes and the discrete Euclidean shapes. So that ideas about physical containment need to be modified as well as the models of material interactions where material components and metric-spaces are both shapes which are distinguished primarily by their relative dimensions.

Furthermore,

Thus (or that is) one would think that there would exist a simpler model of physical systems, a model based on linearity, metric-invariance, and continuously commutative (or completely parallelizable shapes), where a completely parallelizable shape is in a class of shapes which can only be relatable isometric-ally, ie a differentiable homeomorphism defined between (just identified) isometric shapes, to either zero or negative-1 constant curvature spaces, for metric-functions (of these isometric-ally isomorphic shapes) with constant coefficients, so that, these shapes have a strong relation to differential-forms, and to the rectangular-faced and convex simplexes which are associated to the torus (for an arbitrary dimension), where the geometric measures of the boundaries of the holes in a torus determine the spectral-values associated to the toral shape at that (or any) particular dimension of a hole's bounding "identified" face structures, obtained from the facial-structure of a rectangular faced simplex, where the hole structures on such simple shapes are related to the set of stable spectral-orbital properties of these stable shapes.

But furthermore, these stable and simple shapes are related to isometry as well as being related to unitary as well as being related to symplectic Lie groups, all of these Lie groups also happen to have sets of discrete isometry, unitary, sympletic subgroups which are, in turn, associated to stable discrete shapes, which are based on the same type of rectangular-faced and convex simplexes, as are all the discrete isometry-unitary-symplectic subgroups, but the shape of a "discrete hyperbolic shape" has (is) the shape of several toral-components strung together, where the number of toral components is the genus of the shape, where these are the "discrete hyperbolic shapes," which were studied extensively by Coxeter.

So that if one places this context of linearity, metric-unitary-symplectic-invariance, and continuously commutative local coordinate properties, which are associated with these stable shapes (which, in turn, are associated to rectangular-faced and convex simplexes), into a many-dimensional and many metric-function-signature context so that the dimensional structure is determined by this set of stable shapes then one has a new containment-set context, where patterns are stable and there can be meaningful content in the descriptive language which can be used in a practical manner, where system control based on control over the finite spectral-set which is defined for all the different dimensional levels and is ultimately contained in the 11-dimensional hyperbolic metric--space, where this control, where the control is related to possessing knowledge of such a finite spectral set, is implied in the context,

And where the properties of general relativity fit into the geometric-geodesic structures which are defined on the stable geometric shapes, where these shapes also possess stable separate and distinct geodesic orbital structures, where these geodesic (or minimal surface spectrally-related geometric-measures) are associated to the geometric measures of the faces, which bound the holes in the shapes, about which stable spectral-orbital properties of the metric-space shape can be defined

This is further adjusted by means of the natural Weyl-angles, which are related to the maximal tori within Lie groups where the Weyl-angles define the conjugation (group) classes, the angular transformation group-conjugations which transform between different maximal tori within the Lie group, so that the stable shapes can be bent at angles consistent with the Weyl-angles

To re-iterate:

That is, the metric-invariant shapes which emerge from the discrete isometry subgroups ie from the metric-spaces and their associated Lie groups...,

Ie the stable shapes which emerge from the discrete isometry-unitary-symplectic subgroups of the Lie groups which define the local coordinate transformations, which maintain the invariance of the geometric measures on the shapes, are the most natural geometric shapes which can be contained within a metric-space.

Again:

The principles of general relativity also apply to many of the properties of the separate stable shapes such as the separate geodesics for each separate hole, in a Weyl-angle-folded shape, a shape with some genus-number, so that the orbit (for some condensed material, defined by the geodesic) can be perturbed by a material-interaction structure, ie perturbed by a (partial) differential equation material-interaction context, which is a math structure which is, exactly, similar to the 2-body problem which was first solved by Newton, ie the orbital-geodesics of the stable shapes allow the 2-body problem to appear to be so significant, in regard to perturbing the spectral-orbital properties of material which is occupying the stable shape's orbital-geodesic structures.

The separate stable shapes also define the metric-space shapes which fit into the different subspaces of an 11-dimensional hyperbolic metric-space, where the 11-dimensional hyperbolic metric-space is the "over-all containing space," so that the shapes…, which are contained in the 11-dimensional hyperbolic metric-space…, also determine a math structure which isolates the shapes

1. topologically, and

2. through material-interactions, and

3. there is isolation between metric-space-shapes due to the relative sizes of the shapes,

where sizes of the shapes can change when there is a discrete change in the containment context, eg between shapes or between dimensional levels, etc, where the change in size is caused by multiplication of the discretely separated contexts which are distinguished by the separate stable shapes by constant (factors), and this context makes the shapes of the metric-space, seem to be quite independent of… and/or closed-off from… any other such shape of the same dimension or of different dimension.

That is, our deep belief in materialism is a result of being mathematically hemmed-in, in regard to the shapes (and their independent topological properties) which organize the different dimensional levels (as well as the way the shapes organize and separate the same dimensional levels)

Note that,

The set of stable and distinct spectra for a discrete hyperbolic shape depends on the number of holes in the shape, in a similar way in which the n-torus can be defined by, n, constant circles, ie n such spectral-lengths, so too each hole of an n-dimensional discrete hyperbolic shape defines, n, separate and (most often) distinct spectral values (or spectral-lengths),

Consider the hole structure of an n-torus

One begins with an n-dimensional rectangular-faced simplex, where the bounding faces of such a simplex are (n-1)-dimensional rectangular shaped faces and there are, n x 2 = 2n, number of faces, which define pairs of opposite faces, where a pair of opposite-faces identify a direction, so that the geometric-measure of the face (for the pair of opposite faces chosen) multiplied by 2(pi)r, where r is measured from the center of the circle which is determined when a pair of opposite faces are identified, based on a circle which is associated with this opposite-face identification process.

That is, there are the (n-1)-dimensional bounding-faces of the n-dimensional holes so there are, n-chose-(n-1) = n, such n-dimensional holes in the n-torus, and, n-chose-(n-2) = [n x (n-1)]/2, (n-1)-dimensional holes in the n-torus, or equal to number of (n-2)-dimensional spectral values (or orbital paths which have a cross-section which is the geometric-measure of the (n-2)-dimensional bounded rectangular-shaped region through which (or normal to which) a circular-path is defined, or (n-1)-dimensional spectral-orbital paths) on the n-torus etc, where the set of combinations of the different dimension subspaces within an n-dimensional space identifies a symmetric sequence of values for the dimensional sequence of the different subspace combinations in the n-dimensional space...,

So

Since a discrete hyperbolic shape can be considered to be a shape composed of separate and different toral-components where the number of these toral components is related to both the holes in the shape, and is called the genus of the discrete hyperbolic shape.

if g is the number of holes, or the genus, in an n-dimensional discrete hyperbolic shape then there are [g x (n-chose-p)] (p+1)-dimensional spectral values (or orbital paths) associated to the discrete hyperbolic shape.

That is, it is the hole structure of a stable shape which allows for separate and (possibly) distinct spectral values to be associated with the shape.

That is, it is the holes in a stable shape, as well as the way in which the shape has been folded by the Weyl-angles, which determine the set of stable spectral-orbital properties of the shape, which is associated to curve-lengths, in turn, related to spectral radii, as well as orbital radii, which, in turn, determine the set of possible interaction-types, or the possible material-component spectral-orbital types,

While (in the current viewpoint about physical description) the curvature, if it is not constant, then it defines a non-commutative (and non-linear) set of quantitatively inconsistent relations, which cannot possess any content or meaning.

The number of spectral-values in the over-all containment set

The number of shapes, for the different dimensional levels, which possess different spectral properties,

Consider that, the (non-positive) constant curvature metric-space shapes are (now, [or can be]) organized so as to all be contained in an 11-dimensional hyperbolic metric-space, so that each of the 5-dimensional subspaces is defined by a set of these 5-dimensional shapes, which fit into 6-dimensional shapes, and which, in turn, identify the same number of 6-dimensional subspaces (in the 11-dimensional hyperbolic metric-space), but the 6-dimensional shapes are all unbounded shapes (note: this property, that 6-dimensional shapes are all unbounded shapes, was demonstrated by D Coxeter) so there are only the same number of such (unbounded) 6-dimensional discrete hyperbolic shapes as there are 6-dimensional subspaces in the 11-dimensional hyperbolic metric-space, and thus, for the 5-dimensional shapes (which fit into the finite number of 6-dimensional subspaces) if the 5-dimensional shapes have both an upper-bound and a lower-bound, in regard to their sizes...

(note for hyperbolic metric-space the lower spectral bound is automatically defined for each subspace, or for each shape), and thus, there will also be a bound in regard to the set of these shape's spectral-orbital properties, and if this is true for all lower-dimensional discrete hyperbolic shapes, where each shape must fit into an adjacent, higher-dimensional shaped metric-spaces, whose shapes are also bounded, (or usually bounded, but if a metric-space shape whose dimension is lower than 6-dimensions is an unbounded shape then this unbounded shape would define a particular subspace of that particular dimension [so that there would still need to be a requirement that there is both an upper bound and a lower bound on the sizes the shapes contained in such an unbounded shape] but this fixed unbounded subspace would cause the finite spectral set to be even smaller in number), then there is also a finite number of spectral values which can fit into this type of organization concerning dimensions and shapes.

Note that this is true because the stable spectra of any discrete hyperbolic shapes is discrete, ie there is no spectral continuum, for a discrete hyperbolic shape, thus the set of spectral values defined between an upper-bound and a lower-bound for spectral-orbital values and confined to a finite number of subspaces will be a finite-set.

Thus, stability of the material systems can be described by using this finite spectral-set associated to a set of (bounded) stable shapes, and

Thus, the set of measurable spectral-orbital values for all the different systems, in this construct, can only be finite.

That is, this requires that any stable shape in the over-all containment-set must be in resonance with the finite spectral set (of an assumed set of 1-dimensional spectral values, or spectral-lengths), defined above,

Thus, the set of stable properties, which material systems can achieve, can be described by stable math patterns, and the quantitative sets needed for quantitative containment can be generated from a finite set of spectral values (associated to the properties of measurable systems) so that the quantitative containing space can be (small enough, so as to be) logically consistent.

Note: Computers can function in a reliable manner because they are designed to remain logically consistent.

The set of different material types, as well as the different physical properties, are associated to the different metric-spaces, which are distinguished by both their dimensions and their metric-function signatures, is the set of properties about which physical descriptions must be built,

And furthermore,the fiber-group and the base-space for a principle fiber bundle (for a particular dimensional level), would be a set product of metric-spaces of different signatures, for the base-space, and a set product of the corresponding fiber Lie-groups as the fiber-group, and so as to be properly related to the particular dimensional level.

The principles of general relativity make a lot of sense when they are applied to the new descriptive context of many stable shapes of different dimensions, defined on an 11-dimensional hyperbolic metric-space, but the principles of general relativity lead to nonsense when applied to a system's, supposedly, defining energy densities in a context defined by materialism, ie where the idea of materialism is a statement about the containment space for existence being 4-dimensional space-time, in a meaningless math process of trying to find a system's shape by means of finding the system's general metric-function in a non-linear, and non-commutative math context.

Whereas in the context where there are many stable metric-space shapes defined for many different dimensional levels, but all of these shapes are contained within an 11-dimensional hyperbolic metric-space, then for condensed material accumulated on a discrete hyperbolic metric-space shape and the material is confined to the geodesic structure of a particular hole in the stable shape, then the condensed material will follow the geodesic paths around the particular hole (on the shape) to which the condensed material is confined, but within this main cause of the orbital properties, caused by geodesics on the shape, there can also be material-interactions defined by partial differential equations whose affect on the condensed material would be to perturb the condensed material's orbit.

Furthermore, the interactions between stable material components which are contained within a metric-space would be determinable by a combination of material-component properties involved in the interaction which would include the properties of the energy of the interaction, eg field strengths and relative dynamic energy between the interacting-components, the size of the interaction region, and the resonances which can exist between the shape of the interaction and the finite spectrum, in regard to the relevant dimensions, where the (outside) spectrum exists in the over-all containing-space.

Thus there is a new descriptive and containment context which possesses the properties of possessing stable patterns and being in a quantitatively consistent context and the quantitative sets are not too-big so the description can be logically consistent, furthermore, nearly all of the physical attributes of the old description are maintained as subsets in the new context

Or

One can try to use materialism and partial differential equations as the basis for describing material properties but in a non-commutative and non-linear context ie a descriptive context in which there is no meaning and this is done on a quantitative set structure which is too-big so that there are logical inconsistencies which can be introduced by a convergence processes which allow convergences to quantitative properties which are logically inconsistent with one another

Quantitative inconsistency which results from non-commutative-ness, and non-linearity, and indefinable randomness, in turn, leads to an undefined unit of measuring, discontinuity in value, and many-valued-ness within both the quantitative containment set, as well as having these same (failed) quantitative effects on the function-values of the system's measurable properties.

Chapter 9

Propaganda, the intellectual-class, science-and-math, and "hidden knowledge"

Those allowed into the intellectual-class, by the investor-class, are required to support the idea of materialism, yet, at the, so called, margins of society (as defined by the propaganda-system) there are some intellectuals allowed to market to a market-niche, but it is really the majority of people, who believe that people can realize "hidden knowledge,"

but those expressing ideas for this market-niche (defined by the mainstream media) are ineffective at providing models for "hidden knowledge."

Yet, there is a carefully considered math-model put-forth by an expert, which is not based on materialism, but the expression of these ideas are opposed by the intellectual-class, ie opposed by all of the media, both alternative and the mainstream media.

Though the alternative media sometime quotes Orwell, who stated that "journalism is all about expressing the ideas which someone wants excluded, all else is propaganda" ie the alternative web-sites are also "simply expressions of the same type of propaganda which serves the investor-class."

Every aspect of the media, ie the communication channels of the US society, is full of expressions of rigid narrow dogmas, which serve the investor-class's interests, and this means that any discussion on the media is due to a propaganda-person who serves the accepted dogmas, so when there are expressions about ideas provided by Ouspensky and B Ehrenreich, to which an author, such as myself, can, in turn, relate new ideas about using math and physics, which are used to describe the measurable properties of existence, then I must always question the motivation of the propagandist (eg Ouspensky or Ehrenreich, etc) who provided an opportunity (by their expressing seldom expressed ideas, or ideas usually expressed in different [occult] contexts [other publishing niches]) for myself to express properties of a new physical theory in the "new" context which is also being expressed by other authors, ie other propagandists who are allowed on the mainstream-media, and this questioning of their motivations is because these propagandists always defer to the, so called, "truth" of the appointed authorities, in the propaganda-education system, which are supported by the ruling-class

The point is made by the media, which is owned and controlled by the investor-class, that the other viewpoints (not expressed by sanctioned propagandists, such as the viewpoints which I put-forth) are not true (nor reliable) and the public should be shielded from such unreliable ideas, they are not accurate expressions when judged in the context of the interests of the ruling-social-class.

But the issue about the efforts needed in determining truth, and what it means to be accurate, in the US society, is determined by the social-class of the person who is expressing the idea. This is certainly true in science, since the science community simply express the same ideas about verification by empirical observation, which were used by Ptolemy, and Ptolemy verified, by careful observation, a fantasy-world (of epicycles).

That is, without relating scientific truth to both (1) wide-ranging applicability so as to get accurate descriptions of the properties of a lot of physical systems which are within the context of the physical law, and (2) to a great amount of creative-development which results from the description and its context, then the description has very little value as an actual "truth."

(but, in regard to physics and math, the top sciences of: quantum physics, particle-physics, and general relativity, and the other derived theories, do not satisfy this criterion, in regard to determining or establishing a physical law's truth, or in regard to the truth of such a descriptive language).

Thus, the media and the peer-reviewed professional, the so called, truth-tellers, state a bunch of arbitrary and quite inaccurate statements, and express viewpoints which are not useful, nor accurate, and whose range of applicability is extremely narrow…., yet since they are claimed to be [….. as expressed by the social-icons of respectability,…..] within the trusted truth-telling-class, then "what they say" is accepted as being true, because the public has been trained to "determine truth" by determining the social-class of the person "doing the expressing," this is about clothing, and if the one doing the talking is on the media (or not),

But those doing the accepting (about what "is to be called" true), ie the mangers and editors of institutions, are expressing their necessary belief in dogmas and their deep belief that those in the higher social-classes are more capable of discerning truth and expressing truth in its "proper manner" than are other people.

That is, the propaganda-education system: mis-leads, lies about… and misrepresents… issues about: governing, politics, law, economics, foreign events, it also proclaims "as being true" a fraudulent truth which is expressed by both science and math, where the validity of the truth of science (which is claimed by the propaganda-education system as being an absolute truth) is at the level of Ptolemy's (also verified) truth about the planetary motions.

Furthermore, math experts rigorously prove math statements, but the math statements are not true in the world which we experience, but only true within the fantasy-world to which their assumptions, contexts, interpretations, etc, are actually a part

That is, the claim to exclude anyone's representation of truth from the media due to it not being accurate is a standard which is sufficient to not allow the media to exist at all. The function of the media is to express the lies and failed dogmas which the ruling-class want expressed to the public over-and-over again.

That is, US (or western) society is arbitrary and is held in place by illusions and by violence.

It is so very difficult to penetrate the fraudulent intellectual constructs, in the US society, which are upheld by the, so called, intellectual communities of the US society, intellectual groups which are bequeathed with high-social-value by the investor-class, where this is done (by the ruling-class) by both controlling, essentially, everything which is produced in society (done through monopolistic business control over society), and everything which is published or expressed within society, ie controlling language and thought, and by controlling the academic institutions whose scholarly departments and their curriculums are all consistent with the interests of the investor-class (even the dissent), and it is done through the use of: investments, institutional mangers, and spies-and-agents-and-their-managers (or controllers, or handlers) who find, by identifying letter-strings and tagging key-words (as the librarians have done for years), ideas being expressed by the public, so that for those ideas being expressed about-which the investor-class opposes "their expression within society," and subsequently, sending agents, or contacting institutional mangers, so as to interfere with the expression of ideas which are "not wanted expressed in society" by the investor-class.

There is but one voice which is allowed to be expressed, and which everyone must obey, in a totalitarian society, built from investments where the investors want no risk and no competition, ie the ruling-class wants only monopolistic social domination.

This is an example of propaganda in regard to the claim by Orwell that, journalism is about publishing something (those ideas) which someone else does not want published, and everything else is propaganda, or one can add that spying is about finding ideas (using word-strings or sound-string searches over all the communication channels) which are being expressed (or being publishing) which someone else (in the investor-class) does not want expressed, and the main purpose of spying is about interfering with the expression of these ideas

One finds very few intellectually-oriented sites on the internet which will publish ideas about which the authorities are not also expressing the same ideas (where the, so called, authorities are in the "stables and barns" of the ruling-class)

And

Where the interference is usually done by manager-editors claiming that someone is expressing ideas which "the authorities do not believe to be true," and the intellectuals, whom administer a publishing-site, then judge the new ideas…. (in the context of the authority which they have been taught, or the authority to which they have been led to believe)… to not be true, the phrase often used in the so called matter-of-fact objective reporting of this, so called, well verified "knowledge" is; What do We know about….(the subject)" where We is the group think…

(or most often called the fraudulent expression of "scientific consensus") of our society (as expressed by the propaganda-system) which is properly expanding as a "victorious Empire," an Empire which only sheds light upon the rest-of the "ignorant" world, and

Thus, the managing editors exclude the expressions (which are opposed to the interests of the investor-class) from being published, based (on this propagandistic mythology, and) on the judgment of the mangers that "the ideas are not sufficiently authoritative," ie "truth" is not being served, where both the mangers and authorities have been taught in the institutions, where these institutions only express the ideas which the bankers believe to have high-value, in particular, high-value to the banker's interests, that is, the mangers and editors and the intellectuals which mange the expression of ideas within society, only believe in "the truth" in which the investor-class wants them to believe (they are wage-slaves in a highly class-conscious society).

But this means they (the managing-class) all believe arbitrary authoritative, narrowly defined, dogmas into which the bankers have invested, ie the knowledge of society must fit with their investment interests.

This shows the way in which a small set of both left and right managers (intellectuals) and agents (spies), whom are intellectual and violent (or taught to interfere), so that both managers and agents are both deluded (in regard to truth) and they are both amoral (in regard to righteousness, or righteous actions), ie in fact both left and right (managers and agents) are deluded and manipulated by their relatively high wage-status (or high intellectual status) in society, and they are exploited by the investor-class, so as to support the ruling-class, where the motivation of the managers is all about group-think and the illusion of high-value, which the propaganda-education system creates for (or around) the ideas which are of greatest interest to the ruling-elite, ie their belief in truth is based on the high-social-value, which institutions…, which serve the interests of the investor-class…, define, and they act on judgments which have nothing to do with either truth or righteousness, ie but these mangers and agents act outside of "true-human-value," deluded by the hierarchical social structures, ie where social structure is based on both (1) narrowly defined (and dogmatic) institutions (both technical and religious), ie propaganda, and (2) wages (where there is always the notion within the US society that those with a great amount of money are favored by God).

By the way, this ability to get institutions and the embodied icons of "authority" and "morality" to serve the investor-class so uniformly throughout society, emanates from the justice system, where law is based on property-rights and minority-rule, and the justice system is organized as a counter-insurgency military action against the public, so as to support the ruling-minority,

eg the Pinkertons whom both spied upon and used extreme violence against the public with (legal) impunity beginning around the 1870's, when they worked as a private army for private corporate and individual interests (Note: Similar to the social-military structure of the Roman society).

Both the authoritative intellectuals, and the "justice-system licensed" agents of coercion and manipulation are expressing their social domination, where one side (the intellectuals) is openly supporting the narrow dogmas of the investor-class, so as to proclaim their great merit (as they are bank-rolled by the ruling-class) and the other side is hidden, since the terror they instill in the public, supposedly, cannot be tracked-down by the "all inclusive, and all-seeing, surveillance system," but the terror is, again, expressing social domination of the ruling-class, but it is being expressed as the violence of a self-righteous absolutism of a totalitarian state expressed (as domination) over the public, who are forced to serve the ruling-class and their narrow interests.

However, the great amount of attention and propaganda used on the right is also about infiltrating and interfering with, or stopping, any unity which might spontaneously form in the lower-classes which might suddenly oppose the interests of the investor-class.

Only those with an authority obtained from an indoctrination process (emanating from the high institutions) "have a say" in regard to practical ideas and creativity (into which investments have been made), and the lower-classes are infiltrated with agents of arbitrary: absolutism, righteousness, and a culture of great violence (and teeming with informants) so as to thwart an uprising of the public, (the non-intellectual public is baffled by (demagoguery) demigods, where the most reactionary are the infiltrators and agents, who are a part of the justice-system).

In fact, what the intellectual-class is implicitly saying is that "one must follow a set of investor-designated intellectuals, whose descriptive interests are narrow, limited, and failing,"

But

According to the propaganda, "the indoctrinated intellectual-class (which serve the investor-interests) must be followed anyway, since (according to the propaganda system) they are the superior intellects of our society," which (the intellectual-class claims) is proven, since the superior-intellectuals with knowledge of particle-physics seem to be talking about things which make no-sense to an ordinary (un-indoctrinated) intellect (but this apparent nonsense seems to be that-way because it is actual nonsense [about particle-physics]).

That is, the intellectual-class is supporting (in a seemingly hidden manner) the idea of superior human-beings, ie racism.

That is, the same type of argument about the, assumed, intellectual superiority of particle-physicists, which can also be applied to the religious fundamentalists (ie arbitrarily calling the religious fundamentalists the superior intellectuals of a society).

That is, as S McElwee from Salon (5-26-14) in "N. dG. Tyson vs. The Right: …" states; the fundamentalists are building an elaborate case for something but which is easily dis-proven by simple (logical) propositions…

But

(this is the central issue of Godel's incompleteness theorem, since this is the nature of precise description, one can change a descriptive language's assumptions and the context [and this is done at an elementary level of language usage]), (eg. similarly, the chaotic context of particle-collisions is not a valid model for understanding the observed stable properties of material systems, where this "refutation" is also a simple statement, which logically dis-mantels both particle-physics and most of the physics-curriculum of the physics departments in the, so called, "public" universities) and/ or (ie back to McElwee's article) they (the religious fundamentalists) form an impressive edifice of logic and evidence upon such a thin foundation of speculation.

Again,

This same type of statement can also be applied to the particle-physicist.

Yet, the intellectual-class (the propagandists and their editors) revere the thinly based speculation of the particle-physicist, whose knowledge is only related to the vaguely empirical and chaotic context of the particle-collisions, and particle cross-sections, which is of so much interest to the investor-class because such a chaotic process is a part of a bomb's explosion (model)

(the particle-physicists claim, incorrectly, that this same chaotic particle-collision context is supposed to be applied to the descriptions of: nuclei and general-atoms etc, so as to understand their observed stable properties (note-that the general nucleus, has a set of "spectral" values, which uniquely identify a nuclear-type), but there are no valid examples of deducing a general nuclei's spectral properties from a description based on particle-collisions. Thus, the claims of particle-physics seem to have very little content, and thus, they are not true [especially, in a context of generality (or wide applicability) which would be needed to make it true]), and the published intellectual-class disparages the elaborate thoughts of the religious-fundamentalist.

The problem is, that both sides, have virtually nothing to say, which is relevant to any practically creative efforts, and

Both groups are effectively expressing an arbitrary measure of human-value (eg determined by one's beliefs),

ie both are expressing a hidden belief in believing-in "arbitrary-high-value," such as the idea of racism,

ie arbitrary traits which identify human superiority of a particular type human.

But the real point of "religious absolutism," being so prominent in the propaganda-system, is that it is "all about" infiltrating and interfering..., by means of using agents and propaganda..., with the people of the lower social-classes, who might, otherwise, quickly unify to oppose the bankers' interests.

However,

The intellectual-class is not about developing knowledge, where descriptive knowledge should be both accurate and practically useful, rather they are fundamentally about arrogance and (intellectual) domination, and the, so called, knowledge which they, supposedly, possess..., and upon which their intellectual domination depends..., they obsess over as if it were the absolute authoritative religion of materialism, so as to make science into an "authoritarian dogmatic religion." (that is, effectively, exactly how it functions within society)

That is, the intellectual-class worships both materialism and indefinable randomness (a language, based on these properties, possesses no actual content). Their, so called, scientific language is a descriptive language which only deals with fleeting, unstable (false-)patterns (or apparent patterns), so as to effectively reduce science to a set of empirical-studies of (observed) physical-system properties, and where, so called, "physical law" is used only in bomb-engineering, in regard to finding elementary-particle cross-sections (related to reaction-rates in bombs), and as the basis for making authoritative, but quite false, statements, such as; "that all the many-(but-few)-body spectral orbital systems, which exist, and at all size-scales, are too complicated to ever be able to describe their properties based on physical law."

That is, this failed physical law, the law which the intellectual-class guards so tenaciously and obsessively, so as to maintain their intellectual dominant position in society, where their dominant social position (as intellectuals) has been granted to them by the ruling-class, essentially, for both their work on the bomb-engineering projects and for their elite arrogance which is needed in the propaganda system.

That is, the intellectual-class does not create an equal scientific community based on equal free-inquiry, rather they form an exclusive authoritarian community, very much like a hierarchical religion, so as to be a community of intellectuals whom obey the orders given to them by their… predacious, amoral, selfish, and ignorant…, rulers, ie the, so called, leaders of society.

The main purpose of the intellectual-class is to uphold an arbitrary and violent society, a (collective) society not unlike the Roman-empire.

Study in propaganda and social control

Consider the, "global-warming vs. science," scenario, but instead of being in a society whose business leaders are the oil-banking industries, but rather a different society where,

(1) human creativity, which is related to building things, should be so robust and capable, so that

(2) all human creative activity should easily be able to adapt to (or change in regard to) all possible problems (about production and resource use) by stopping production or stopping resource use, based simply on a precautionary principle (change production when there are relatively clear (or even any well articulated) indications of a product's destructive effects).

.....

That is, when some designated expert authority, such as, some eminent climate scientist or influence-connected propagandist, whines about "science's authority being silenced by financial interests," then that person needs to re-think "how science is defined within the US (totalitarian) society."

Our society defines science to be materialistic based thinking, which is done by an exclusive (or elite) relatively small set of authorities, put into an institutional hierarchy, where these intellectual institutions have significant roles in the propaganda-system (where the significance of their role in the propaganda-system is more significant than is the subject matter of the intellectual category), where these authorities have been schooled and indoctrinated in materialistic thought, so as to support, or to only be able to think in the contexts, ie in the very narrow contexts (or along the lines-of-thought, or in a narrow category of thought) which best supports the practical creative interests of the investor-class, ie the categories of narrowly defined ideas which best serve the interests of the investor-class.

That is, science is primarily about social-class, and a set of indoctrinated wage-slaves, and it is not about rationality, nor can one express objections to the interpretation of data, where the "authorities" have already made the "correct" interpretations of observed patterns which happen to be consistent with both monopolistic business interests and the propaganda-system,

ie the "correct" interpretations have already been made by the science-institutional managing hierarchy.

That is, science is not based on the much more subtle type of thought of an individual independent thinker who might have a much more correct viewpoint about truth, which is individually developed by means of the methods and processes of equal free-inquiry, and using language in new ways, based on new sets of: assumptions, interpretations, contexts, etc (ie elementary thinking) as Socrates and Godel have proclaimed that science should be developed (but it should be based on this social context of equality),

And, furthermore, science should be organized within society so as to promote many types of "truths" which might be related to many different practically creative contexts.

That is, a robust creative practically creative society can only exist if there is a less authoritative and instead a more wide-ranging viewpoint about "what science is?"

The expert authority is an indoctrinated and intolerant, and most often chosen (by institutional managers) to be a domineering person, but who (now, 2014) is mostly serving a failed dogma, which has limited relationships to the productivity of the investor-class, but this is its primary social function, and a very large relation to destructive affects.

E Teller is the perfect example of a psychopath who seeks to be an intellectually domineering scientist; but when the promise…., by the group of experts he administered…, the promise of their being able to provide a clean, cheap nuclear fusion-energy-source to society, but when this promise was not realized by the end of the 1950's, it was clear that Teller served (administered) a failed viewpoint, but thermonuclear weapon had already established itself in the military-industrial-complex, and, thus, as the thermonuclear weapons chief science administrator, Teller, had already established his intellectually domineering social status (subsequently, the curriculum of physics departments in all "public" universities were determined by the interests of Teller, physics became particle-physics, and derivative theories, such as string-theory, which are really nuclear-bomb-engineering theories).

Whereas global warming is based on an empirical descriptive construct which depends on many factors (or causes) and the description depends on models based on randomness, though many of the correlated claims, eg "temperature rise of the earth and CO_2 in the atmosphere" are so strong that no-one should try to claim that these correlations do not exist (it is like the strong correlation which exists between vitamin-C and scurvy), though the apparent chemical-electromagnetic mechanism causing a high-amount of energy of the earth-reflected electromagnetic spectrum to be re-reflected… back to the earth… by the atmosphere (it) is not "all that clear" as to how it functions…, Is this a re-reflected plane-wave-front model striking an atmosphere modeled as being an effective plane (shape) "at what angle?" and then being reflected by the CO_2 in the atmosphere at "what reflection percentage?"

If one wants a society "which is focused on knowledge and practical creativity" then one wants the law to be based on equality, so that the focus is on knowledge and creativity, and this can be directly related to trade…, but trade cannot be used to dominate "the way in which things are created within society," ie controls need to be placed on those traded objects or processes which are very successful in the market.

The point of knowledge and ideas in our society is that these expressions are to serve the interests of the ruling-class, and that one will not be listened to unless one is high-up in the social hierarchy

That is, the ideas about using stable shapes in a many-dimensional context to describe the observed stable spectral-orbital properties of material systems as well as being able to provide a new model of a living-system as well as a much needed new model of a mind so that the mind is not modeled as a chemical process but is spectrally related to the spectral structure which our senses provide to our mind.

The new math models of existence form a new descriptive context which places physical description beyond the confines of the idea of materialism and the new math structure, which contains the material world as a proper subset, becomes a very general (and abstract) description (or model) of existence

Within this new math-physics model one can hypothesize one life's context is different than the context of the material world, and that life's purpose is to (1) possess correct knowledge (based on equal free-inquiry), and (2) to create a further extension of existence itself through our knowledge and our intent a life-form has a further reach (into existence) than we are led to believe by our continual repetition of the so called necessity of there being only a material-world.

However, this set of valid math models is not allowed to be expressed, in any meaningful way, in a, so called, free country, where free-speech is, supposed to be, the law (of the land).

But instead a Calvinist based idea about social-class (the very rich are at the top of the social classes) determines who can speak in regard to the interests of the ruling-class, ie those who God-favors (by providing these special, favored, people with great material and monetary wealth). This is about license given by society to the ruling-class to lie, steal, and murder, with impunity, and this is essentially about law being based on both property rights and minority-rule.

Practical creativity

In the practical context of building something, one wants a descriptive context... where measurable properties are the basis for the description... wherein

(A) the measuring processes are reliable, and

(B) both,

(1) the patterns used in the description are stable.. either the laws which relate measurable properties so as to identify (solvable) [(partial) differential] equations, or the patterns which identify the proper context for set-containment, and the subsequent descriptions of a system's observed stable measurable properties, eg {where this is, today (2014), assumed to be the context of materialism, but this idea has not worked} such as the new ideas about a containment context of many-dimensions, which are separated into subspaces and dimensional levels (or subspaces

which can be of various dimensional levels) so as to be (or which are to be) organized around the properties of the: dimensions, sizes, and containment contexts (or properties) of stable shapes,

Note: That this last aspect of a valid descriptive context has to do with a new idea about a containment-set for existence, which can act as an alternative to the idea of function-space containment of a (physical) system's description, where function-space containment has not worked, since its needed relation (of the descriptive act, or process) to being able to find "complete sets of Hermitian commutative operators," in general, is not capable of being realized.

Yet, the systems which are being described by this function-space method are stable systems, with well defined stable measurable properties.

Thus, this many-dimensional model of existence is about, amongst other things,

(a), an alternative to the failed function-space methods, where this alternative creates a relation between the functions (of a function-space) and sets of stable shapes, which might be defined over several dimensional levels or defined in regard to several subspaces of the same dimension,

(b), it is also about transcending the idea of materialism (see below), as well as

(c).being about truth,and

(2) the resulting descriptive patterns of the properties of the systems being described are also stable patterns.

In order for this to be possible then one needs both:

(1) quantitative consistency, that is, linearity of local measuring, and metric-invariance, so that the geometric measures used in the description are reliable, and continuous local commutative properties of the descriptions of measurable properties defined-in many-dimensions, or contained-in many-dimensions, ie locally measurable properties (local matrices) can be made diagonal (in a continuous manner) throughout the entire shape of the system's natural coordinate system.

(2) stability of patterns, in math, this property seems to mostly be about stable shapes, and

(3) descriptions of randomness are based on elementary-event-sets whose (random) elementary-events are stable and well defined

Quantitative inconsistency implies: many-valued-ness, chaos, (random) discontinuities, no stable patterns are describable, there are only fleeting, unstable (empirical) contexts, and the indefinable randomness of a system exists as a system's primary measurable property, or equivalently a context of a transitioning between two (or other) vague and unstable sets of patterns which might eventually lead, or degenerate (or decay) to a stable context.

This context of quantitative inconsistency which enters the descriptive language of mathematics in the form of non-commutative and non-linear properties identifies the math properties which are (now, 2014) always present in the math formulations of the descriptive contexts (or such a quantitatively inconsistent form of mathematical language used so prevalently in the contexts of math) of: quantum physics, particle-physics, and general relativity (or more precisely the narrowly defined expert-based context of bomb-engineering, as brought into being, in the US, by Oppenheimer and Teller, and which now (2014) defines the (primary) curriculum of the physics departments of all US universities

How can anyone, actually, take this seriously?) and

It is also the math context of the DNA based biological evolution and the related models of (a molecular based) life-forms, models which are over-whelming-ly complex.

Consider:

Is the model of a life-form in regard to the assumed context of molecules, and molecular interactions…, where molecular interactions are based on molecular-collision-rates…, a valid model of a living system in a system whose chemical-molecular context would be so very complicated when reduced-to molecular-interactions?

That is, the approach of identifying molecular-types (by chemical tests) necessarily reduces the system to this context of the relation between molecules and the living-system,

But

Perhaps the entire system with its complicated but unified over-all context is better related to a crystalline structure, which is realized at both a larger-size-scale and at a multiple-dimensional system-structure.

That is, can a high-dimensional oscillating stable shape spectrally guide a pde defined property of a (living system) which exists at a lower-dimension but whose spectral or limit-cycle sizes is in resonance with the faces of the higher-dimensional oscillating shape?

Or

Can a high-dimensional oscillating stable shape spectrally guide a pde defined property of a (living system) which exists at a lower-dimension but whose spectral or boundary (eg boundary condition) sizes are in resonance with the faces of the higher-dimensional oscillating shape so as to control the boundary conditions of the system-process-defining pde's?

Or

Can DNA molecules, whose cylindrical spiral shape..., whose molecular-shape properties were originally determined by crystallography...., and its internal (protein molecule) sequences along the ladder of the spiral be directly related-to crystal shapes of either higher-dimensional or lower-dimensional stable discrete hyperbolic shapes? Can the lengths of the ladder sequences of the DNA molecule be related to length properties of crystals, and not as much related to the internal-building of molecules by RNA?

That is, could it be that the chemistry of life-forms be better related to large-size-scale crystals which have a relation to various dimensional levels (of the actual living-system structure) rather than local molecular processes?

But such questions cannot be asked in the context of materialism, which requires only a "single dimensional level" which is supposed to contain all aspects of the living-system's properties, where these properties are assumed to be mainly about molecules and molecular-interactions.

That is, science has come to be based (only) on the, so called, expert dogmas, so as to be about the idea of both materialism, and reduction to smaller material-components, and its math models are quantitatively inconsistent mathematical contexts, which cannot possess any meaningful content, but rather only represent the empirical context of chaotic and indefinably-random transition processes, which can (primarily) exist between relatively stable initial and final systems, eg as exist during a nuclear explosion (but also related to the basic ideas of feedback systems).

This (obsession with the idea of the assumption of materialism) also stops "the development of new creative contexts which can be based on useful internally controllable scientific models of physical descriptions."

This, obsession with the idea that Science be based upon materialism, results in science becoming mired in both empiricism, and failed quantitatively inconsistent math representations of physical systems.

This obsession with the assumption of materialism makes investments in science-based projects less risky.

This non-commutative and non-linear quantitative context of a math language is a descriptive context which cannot describe any stable pattern, but stable properties are observed to be intrinsic to physical systems, such as general nuclei etc, and it is these stable properties of the fundamental systems (nuclei, general atoms, molecules, crystals, life-forms, solar system etc) which they (ie the authoritative experts) are supposed to be trying to describe, but the unfortunate truth is that (clearly), they (the expert math and physics professional communities) are not trying to describe the stable properties of observed physical systems, and instead the, indoctrinated, experts use a quantitatively inconsistent mathematical context (which is only relatable to nuclear bomb engineering, or to other empirical contexts, such as feedback systems, where a feedback system is defined by its relation to external empirical properties, the feedback system adjusts to externally measured (distinguished) properties).

It is a descriptive context which is only related to empirical development, [as opposed to the development due to an internal control over systems which the solvable descriptions of classical physics has provided to science].

That is, this type of non-commutative, non-linear mathematical context is a descriptive language which is without meaningful content, and the information, which is a part of these descriptions, is information about transitory and fleeting empirical descriptive contexts, it is not a mathematical context which is about useful descriptions of a system's inner-controlling structures.

When one examines the social forces which exclude the expression of new (technical) ideas from society, one sees that it is due to the belief held by the indoctrinated intellectual-class, or their indoctrinated and trained propaganda-spy gate-keeper managers, that any newly expressed idea cannot be true, since (or if) such a newly expressed idea is not being expressed by the most authoritative of the intellectual-class…., whose belief structure, in turn, results from the narrow authoritative dogmas upon which they have been indoctrinated…., ie it results form the logical-circular confines of the propaganda-education system which is used to determine the set of technical helpers for the ruling-class

But

The brightest and bravest who have climbed into the functional manager social-positions of the social-class structure, apparently, deeply believe-in the failed dogmas, within (or by) which they have been indoctrinated

It is an endless construct of illusion, built (by repetitive propaganda, whose main expression is that "people are not equal") in order to maintain a social hierarchy, which is a value-structure based on arbitrariness, and put in place by violence, lying, and theft.

Math which should be of interest in regard to the development of individual practical creativity

Diagonal local matrix models of physical systems and the continuously diagonal properties of stable physical systems

Diagonal local matrix models of physical systems (related to the system's defining partial differential equations (or pde's)) and the need for a stable system to have a continuously diagonal property which is associated with a stable physical system (both classical-systems and quantum-systems), which are (or now can be) defined as stable shapes.

Also, a feature, which relates energy-levels to the existence of holes in a space, and which clearly distinguishes between quantum physics, and the new ideas about (or new ideas which characterize) the math representation (the math modeling, or the new set-containment structure) of existence, where the single-valued-ness of energy is claimed to be a necessary attribute of physics, but this is a distinction between the dimensional level within which one is defining the property of "holes existing in space," where this distinction between dimensional-levels can only be made in the new descriptive context.

A square matrix, ie the row-number equals the column-number, can be represented as, A = [a(i,j)], where the [a(i,j)] are the matrix's components, which are labeled by the row-number, i, and the column-number, j, so that one can define a diagonal matrix as one in which only the matrix-components which are not zero are the components whose row-number and column-number are equal numbers, ie only the [a(i,i)] components are not zero (or only the [a(i,i)] components can be not zero), ie all the a(i,j) = 0, if i is not equal to j, in a diagonal matrix.

A diagonal local matrix, is a local linear model of a function's (or a locally linear measurable property's) values, means that for each independent (or orthogonal) locally defined direction of measuring of the (system containing) coordinates there are only changes in measured properties which are determined by these properties being defined by the.... (the values of the defined properties) in that.... same quantitatively independent locally defined direction associated to that measured property's local vector component, ie changes in measurable properties are defined both by, and within, the context of the locally defined independent (and locally measured quantity, or coordinate) directions. This depends on the assumption that the system, and all of its measured properties, are all contained in the coordinates and thus they are relatable to the coordinate directions.

Note: Particle-physics changes this, in that, it talks about the inner-particle-states of an elementary-particle modeled as a point-particle, so that these particle-states are not related to the containment space, but rather exist in a different (and separate) dimensional space, different from, and independent of the containment coordinate space.

But this type of a descriptive structure, though it remains consistent with the interests and models of bomb-engineering, concerning the rates of particle-collisions (thus the investor-class is unconcerned about changing this model, unless the new model relates further to a property of bomb engineering, eg detonating a gravitational singularity, eg the point of string-theory in regard to the intent of the investor-class),

But this viewpoint has not led to a useful model of the systems, which these particle-components are supposed to compose.

Note: it is only the shapes of rectangular polyhedrons (often referred-to as cubes) and the 2-dimensional circle which are shapes which can be placed into a continuously commutative quantitative context, or be formed into sets of local independent quantitative directions. That is, it is these simple shapes, and geometric combinations of these shapes, which can be continuously commutative shapes. So, one should try to use these stable shapes to build a descriptive containment context for quantitative descriptions.

Commutativity is an idea which is fairly easy to describe, namely, that the local coordinate relations (the coordinate structure which is assumed to contain the system) which exist in regard to the measurable properties of a system, ie the function-values, in regard to the locally measured properties of the system, which are expressed in a (partial) differential equation (ie pde) model of the physical system, that is, a "physical law" is the basis for defining the equation, and thus the function (or measurable properties of such a system) is the solution function to the (partial) differential equation, form a (local) matrix structure (or matrix representation of these locally measurable relations between functions and coordinates) so that commutativity means that these matrix representations of the system's locally measured properties are diagonal, or can be made diagonal by a suitable choice of local coordinates.

But one wants the property of linear and diagonal matrices to exist (as a property of the local linear matrix) at each point of the system's shape (or of the system's natural coordinate system), where this can happen only if the diagonal property of the local matrix is a continuous property of the system, over the system's entire coordinate system.

This is (also) about the continuous consistency of local geometric-measures defined on the (system-containing) coordinate system, where the system's measurable properties, ie the system defining pde's solution functions, need to be quantitatively consistent, with these local geometric measures, at all points in the coordinate system.

Furthermore, since the metric-function is a symmetric matrix (or can be so represented) this means that local coordinates can always be chosen so that the metric's matrix-representation can always be made diagonal, but such a diagonal representation of a metric-function can only be guaranteed at a single point, where this is due to a particular local coordinate selection, but a "diagonalized" local matrix property (at a single point) for a system's locally measured values (and related to a physical law so that a pde is defined) has little-value, in regard to determining the system's measurable properties, since the system's pde's are then "not" solvable, (though, the property of being locally invertible [or locally diagonalizable], but only at a point, is the basis for the notion of a local fiber group structure defined by "sets of (local) diffeomorphisms," for a math description) and this little-value, in regard to determining a system's description (formulated as a pde), is related to the fact that "it is not enough to diagonalize a description at a point," but rather the diagonalization structure should be a continuous property of the natural coordinates of the system,

ie the shape of a system's natural coordinate-system has a local function relation (where the function-values determine the system's measurable properties) to the local coordinate-structure, which forms a diagonal matrix (continuously throughout the system's coordinate shape),

ie the local linear measurable properties (of a description of a system) need to be continuously commutative and quantitatively consistent, so that the pattern-destroying quantitative patterns of multi-valued-ness and chaos (or discontinuous quantitative changes) do not enter into the quantitative model of the stable patterns of stable systems, so as to make the math model of the system unstable.

It should also be noted that, the ideas of conservation-laws, which are so fundamental to the ideas of physics, are laws, such as the conservation of energy, or the conservation of material, (are laws) which can often be characterized as both continuous scalar-functions and whose time-derivatives are zero.

Note that,

the local matrix representation of a physical system's measurable properties "condenses" the information of the system's coordinate-relatable (or containment) properties down to a small useable set of properties: coordinates (positions in space), geometric-measures within the coordinates, velocities, [or energy (a scalar identification of a stable system, but, apparently, given many temporal pathways) and variation relations used to identify (usually) minimal energy structures within the containment-set, where the ("function") property of action is being identified on the coordinate containment-set] changes of velocity, and force-fields, which, in turn, relate material-geometry to vector-field properties in the coordinate system (or potential-energy terms in the energy-function), where this condensation of a system's set of relevant information…, which is needed in order to use the controllable pattern of the system…, is all based on the idea of a system containment-set and the local measuring of the system's properties being related to the natural coordinate properties of the system's shape, and then identifying the measurable properties of the system which constrain the system so as to conform to some stable controllable pattern (or solvable, or linear metric-invariant, and continuously commutative pattern [and which resonates to the finite (but large) stable spectral-set which the containment space can also determine, or define]).

Note: When the function, which represents physical properties, is a scalar (representing a physically measurable property, eg energy), and its local relations to coordinates is, subsequently, defining a (partial) differential equation (in regard to the physical properties of the system), and if the (partial) differential equation is solvable then the scalar-function will factor into other scalar-functions, where each of the new factor-scalar-functions only depends on one independent variable.

This is possible because the directional physical properties, upon which the scalar function depends for its definition, are all defined in independent (or orthogonal) directions (within the containment-set), so that "that direction" can also define an independent scalar function (of that same physical property) in that one variable.

But this condensation of the information which determines a system's (stable and controllable) properties is only a valid quantitative model if the quantitative relations stay linear and "diagonal" so as to be "diagonal in a continuous manner over the shape associated to the physical system's coordinate containment-set," so that these stable math structures will exclude the math properties of chaos and multi-valued-ness, so that these quantitatively inconsistent math properties do not destroy a stable system's quantitative model, and where these are quantitatively inconsistent math properties which enter into the system's description when the local math properties (of the system's quantitative model) are non-commutative.

However, the attempt to model bounded spectrally stable (quantum) systems …

which are stable, and which are many-(but-few)-body systems, and which are most often composed of charges (or whose components are most often charges), but these, apparently, closed and bounded systems are also, most often charge-neutral, in their relationship to nearby (similar) systems, as function-spaces composed of harmonic-functions where the harmonic functions are global, ie closed (or independently defined) and unbounded functions, so that the system's properties are the spectral-values of a particular spectral-set of these harmonic spectral functions,(has not worked)

And

Furthermore, this set of spectral-functions is to be found by, in turn, finding "a complete set of commuting Hermitian (often differential) operators" so that the property which this set of operators possesses (ie the math property) of being commutative (ie where these are operators which act on the function-space) allows for the idea of "diagonalizing" the function-space into the system's particular set of spectral-Eigen functions (in a similar way in which classical "continuous coordinate diagonalization" of local matrices allows for the solution of the system's equations),

ie and the associated set of stable discrete measurable spectral-values, which are associated to the measurable properties of the system which is being modeled, has not worked.

That is, the attempt to model bounded spectrally stable (quantum) systems by function-spaces of global functions and "diagonalizing" this set of functions by a set of commutative operators, which act on the function-space, has not worked.

That is, this "program" has not worked-out, because, apparently, the algebraic context of sets of operators acting on harmonic function spaces (functions with oscillating periods, and which average-out over circles to the function's value in the circle's center, ie one of the defining properties of a harmonic function) is not a "sufficiently constraining" descriptive context, "which is needed" in order to actually identify the system's spectral properties,

ie the energy-function (or energy-operator) is not sufficient, ie an energy-operator does not constrain the quantum system enough, so as to identify a system's spectral-set.

This is likely a result of the construct of energy, which requires that there exist the property that there be no-holes in the domain-space, wherein the energy-function is defined, so that this "no-holes property" allows the energy-function to be single-valued.

But, in fact, the stable (quantum) systems, which possess stable spectral-values, need to be defined so that there can be (or can have) many different stable discrete values (for the system), and each of these discrete stable spectral properties requires its own "hole in the stable metric-space shape" upon (or around) which the spectral-value is defined. This distinction between the existence of holes in the space within which a system is defined and within the metric-space within which the system is contained cannot be made in regular physics, but it is a part of the new descriptive context where stable shapes which contain holes on their shapes, can exist within a metric-space, but when viewed within a metric-space the holes cannot be detected that is the shape with holes which is contained in a metric-space is a distinction between two different metric-spaces which are defined at different dimensions.

In turn, this is about the set of measurable properties which constrain the system into some stable pattern, and where these constraints are classically measurable properties, and furthermore, energy is defined classically on a space which has no holes, and electromagnetism is also defined on a (star-like) domain shapes which has no holes.

While the stable shapes with holes, which are each (ie each hole, or each set of holes) is associated to stable sets of spectral properties, which, in turn, are defined on convex domain shapes, which can be moded-out, so as to form circle-spaces, ie stable shapes, which define the natural places (natural shapes) where there is both periodicity…,

(defined around each hole in the shape's set of holes, and each hole's spectral-value is discretely separated from the other spectral-values (defined around the other holes) by the property of there being geodesic paths around each hole, which, in turn, is contained within (or defined on) the stable shape), and where the stable shape is based on some fixed set of stable (minimal) geometric-measures, or fixed set of stable (minimal) lengths, in regard to the face-structure (or 1-face structure) of rectangular (convex) simplexes, which when pulled-apart (at the intersecting vertices of these rectangular-based (high-dimension) simplexes so as to) form convex polyhedral simplexes, which when moded-out, in turn, forms a shape upon which a set of continuously orthogonal coordinates of hyperbolas and discretely separated (geodesic) circles (or in higher-dimensions spheres, ie minimal surfaces) can be identified.

Thus, the natural division between a set of measurable properties, which constrain a system into some stable form, and a set of stable shapes which define discrete stable sets of spectral-values for the (stable) shape, is being distinguished by domain spaces without holes vs. domain spaces which are stable shapes, but with many-holes in their shapes.

Dirac's book, "Lectures of Quantum Mechanics" 1964, is about the claim that it is only through the classical models of measurable properties that a set of "(quantum) system constraining values" can be identified, by means of sets of operators acting on a quantum system's function-space model (of system containment, ie the quantum system is contained in a function-space which is a set of unbounded, closed, independent functions), though

Dirac points-out that there are other viewpoints about the definition of "sets of operators" which are to be identified and, subsequently, used to constrain the function-space of a quantum system, so as to "diagonalize the quantum system" (or to find the quantum system's set of spectral-values, and associated set of spectral-functions), but these other viewpoint are all about the fundamental property of a quantum system being "its arbitrary randomness," {whereas the more natural consideration to make, in regard to stable sets of spectral-values, is the opposite viewpoint, ie that these stable spectral properties should be related to stable shapes},

And to not focus-on the "arbitrary randomness" of a quantum system, and its associated sets of operators, where these sets of operators have no geometric relation to spectral-values, that is an algebraic containment context within which to "solve" (or find) a quantum system's (assumed to be) infinite spectral-set, and subsequently, it is a descriptive context which simply leads to vague empiricism, ie the descriptive information about the system cannot be condensed, the information stays in a state of a high amount of probability, in regard to its properties (or in regard to its states) and it is a description which adheres to a language which has no content, ie it is a descriptive context without meaning.

Dirac's conclusion is that, for general quantum systems, the limited sets of operators, which one can use in quantum description based on the analogy of "a quantum operator having a relation to its 'associated' classical measured value" are not going to be sufficient, so as to "constrain the quantum system enough" so as to be able to determine the quantum system's spectral set

(and associated set of spectral-functions, and thus to find the quantum system's correct function-space representation).

That is, Dirac sees quantum physics as a failure.

Note: This descriptive context, ie within which a system's spectral properties cannot be determined from a calculation which is, supposedly, based on the physical-law of quantum physics, is an example of indefinable randomness,

ie an indeterminable elementary-event space, so that such a description "which is based on randomness" has no valid logical basis, and it identifies a descriptive context which has no content or no capacity to compress descriptive information which is either accurate or practically useful.

Thus, it is a descriptive context which pushes science into a more "empirical basis for description," a descriptive structure "not remarkably different from" the epicycle descriptive structures of Ptolemy, though not as effective as Ptolemy's empirical model, since Ptolemy's elementary-events were well defined, ie the motions of planets. Today there are the semi-conductor empirical-models, where holes and electrons occupy energy-levels of a box-model of the crystal, which has been "doped" with atomic impurities, where the "doping" is accomplished by a thermal process, and the bomb-engineering cross-sectional measures of elementary-particles (which can be related to rates of reactions) whose particle-accelerator collision characteristics, are found in in cloud-chambers and seem to fit into a unitary fiber group pattern, but there is no relation of these patterns to anything of importance which one would expect some such relationship, ie an important descriptive relation such as identifying the spectra of a general nuclei by using these particle-collision properties, if it were truly a valid physical model, but instead they are simply patterns which may be related to rates of reactions (in bomb-explosions) and to nothing else, etc etc. That is, the information being gathered by physics labs where this information is assumed to be a part of a descriptive language based on randomness, seems to have very limited empirical value, which is not useful beyond its limited empirical use in specific system contexts, and this limited descriptive value of science's empirical information is due to the non-existence of a systematic descriptive context, which leads to either accurate descriptions (to sufficient precision), or to practically useful descriptive contexts, since the mathematics (of the more careful math models in this descriptive context) is basically non-commutative or non-linear, ie it is quantitatively inconsistent leading to instability in regard to describing patterns, and thus it has no content, it describes only fleeting, unstable patterns, and thus the contexts are also fleeting, and not re-producible.

Thus, there is no descriptive structure based on non-commutativity and/or non-linearity which is useful.

In regard to the idea that the stable spectral properties of quantum systems should be related to stable shapes, in this case (in this alternative descriptive context), one wants the metric-space shapes..., upon which a stable system's shape's holes can be defined..., to be continuously commutative constructs (or continuously commutative shapes), so that this allows for the property of stability.

This property of stable shapes which possess holes in their shapes (where the holes are surrounded by discretely separated sets of geodesics) is the nature (or the natural properties) of the discrete isometry shapes, as well as the nature of the discrete unitary shapes.

The unitary structure enters the description in a natural manner, because metric-spaces have opposite metric-space states defined on them, so that the dynamics of these shapes (or defined on these shapes, and defined in their containing metric-space) are locally invertible, in regard to each of the pair of opposite metric-space states, and this means that these opposite metric-pace states are more naturally contained in the two subsets in the complex-coordinates of: (1) the real subset and (2) the pure-imaginary subset, so that in the complex-coordinates there is a mixture of the metric-space states, but when resolved locally, it resolves into these two "effectively real" subsets, so as to be locally inverse to one another, ie the pair of opposite metric-space states.

The point of much of science, may very well boil-down to (or is about); determining (finding) a new type of "periodic table" or finding the finite (but large) spectral-set of the containment set of an "all engulfing" context for existence, ie both R(11,1) and C(11,1), which are associated to various, but specific, classical Lie groups of both

SO(s,t), for, s, the spatial subspace-dimension, and, t, the temporal subspace-dimension, for an s + t = n-dimensional metric-space,

As well as he set of associated classical Lie groups SU(s,t), so as to also be in relation to R(11,1) so that s<12 and t<s, for s>1 (where when s=1 then t can also equal one).

This is about: subspaces, containment-sets, stable shapes associated to metric-spaces, and the sizes of adjacent dimensional stable metric-space shapes.

The relative sizes of the stable shapes can discontinuously change based on a discrete set of scalar-factors, which can be defined between discontinuous "jumps" in the descriptive context, such as the set of "discontinuous jumps" which can exist between:

1. dimensional levels, between

2. subspaces (or between metric-space shapes of the same or different dimensions), between

3. changes in Weyl-angles (or perhaps only definable between 180-degree Weyl-angles), between

4. discretely defined time intervals (defined by spin-rotation periods defined by spin-rotations between metric-space states), and between

5. the "boundaries" of infinite-extent spaces, which might include

6. "discontinuous jumps" which are made between infinite extent stable shapes, usually in regard to dimension s greater than hyperbolic-dimension-five, and which can also determine a change between different R(11,1) over-all containment sets for existence of material and living systems, etc.

Further math, math patterns which allow information to be compressed so that the information is accurate (to sufficient precision) and it has a practical useful relation to new creative contexts of new creative efforts.

How to organize math properties so as to create a new descriptive context, which is accurate and practically useful.

Math properties which oppose the math patterns, which are used within a descriptive language, from its descriptions being both accurate and practically useful

When one determines an (or a line) integral, that is, the evaluation of the (linear) inverse operator to a (linear) derivative operator, or the inverse integral operator used to solve (partial) differential equations [particularly, partial differential equations for partial differential equations based on solving (or finding) differential-form solutions], then one evaluates an anti-derivative of the (differentiated) function (which is, again, the function which was differentiated) over the boundaries of the spatial and time (based) region, within which the system is contained, or, equivalently, within which the system's set of (partial) differential equation's are defined.

This process of integration assumes both a continuous context for a description, in regard to the quantitative containment of a stable system, as well as a domain space which is free of holes.

Quantitative inconsistency can be characterized by:

1. many-valued-ness,

2. chaos, or

3. arbitrary discontinuity, because

4. the quantitative descriptive context, possesses no stable patterns, so it cannot hold any meaning, ie cannot describe stable patterns in a useful way, the, supposed, "patterns" being described, ie patterns which are supposed to be stable, where, instead all described "patterns" are unstable, fleeting, where the point of instability occurs (or the point at which a discontinuous change might occur) is itself discontinuous in nature, ie out-of-no-where a sudden "big change" occurs, so that, for example, the system's defining partial differential equation changes ie the unstable system (which is being considered) suddenly is to be modeled by a new pde,

5. Loss of quantitative content, or meaninglessness of a descriptive context (which is supposed to be a description of a pattern with measurable properties), results from there being an "externally defined purpose for the system," (the system is defined by both its position and its externally defined purpose) that is the set of fleeting patterns, which has a vague relation to a position in space, is the basis for an externally imposed more stable pattern which is identified by an external purpose identified in regard to the use of the fleeting, unstable pattern (where the unstable pattern

identifies a spatial-position), and which might "suddenly collapse as a pattern," as opposed to both stability of pattern, and an "internal" relation of the description (of a system's internal measurable properties) to the control of a stable system's properties, where this internally-based system control…,

ie the control the sun has over the stable shape of the solar-system, may require that at some point, there existed "the 'correct' beginning conditions" so as to allow the stable system to form (in the "first place").

There are several causes for many-valued functions, most notably (four listed causes):

(1) the existence of non-commutativity in regard to the local (linear) representation of a system's measured properties [in regard to a system's containment within a coordinate "domain space of the system's (partial) differential equation's (or pde's) solution function," which identifies the system's measurable properties….,

ie the system's quantitative model based on the physical laws which, in turn, determine the system's pde], this (or such a) non-commutative operator causes both (1) many-valued-ness and (2) discontinuous chaotic changes of the solution function, in relation to its (the function's) domain values (or in relation to a material-component's position in the domain space), ie some small changes can be (discontinuously) related to large changes in the system, at a latter time, where again the time interval may be big or small, and

Where in such an un-defined quantitative context the system has no stable pattern, and its defining partial differential equation can also change its character in a sudden and discontinuous manner (that is, the system's defining partial differential equation can change its [equation-like] character in a sudden and discontinuous manner), there is also in relation to "the causes for many-valued functions" the related property of

(2) the (partial) differential equation (or pde) being non-linear, ie there are no valid local measuring constructs which allow the system's measured properties to be linear "in relation to the coordinate domain structure," ie the system is quantitatively inconsistent between its measured properties and its system-containing coordinate system, this also causes: multi-valued-ness and chaotic relations (or chaotic behaviors, or patterns), and the system has no stable pattern, and its defining partial differential equation can also change its character in a sudden and discontinuous manner, and

(3) in regard to the spectral-and/or-probability context, where either the events are not stable, or the spectral values cannot be determined from the operator and function-space methods which are

being used in regard to representing harmonic-functions, or wave-functions, as models of spectral (or random) systems,

ie the system is indefinably random (or spectrally undefined), and apparently the system (or material-component) has no stable form, which is either deducible from the math methods (being used), or which can be associated to the system's measurable properties,

ie the description is without any valid content, or it is an incomplete description, or the description (or the discussion about the system) has no meaning,

ie the spectral context of the system is not sufficiently defined, by the method of finding the system's set of operators (where it is desired that the set of operators be commuting operators) to be able to deduce the system's spectral set (where often times this is a stable spectral-set which has been observed for many fundamental stable many-(but-few)-body physical systems)

Consider the H-atom, it is assumed, classically, that the H-atom is a bounded and neutral system,

Yet, in quantum physics, the wave-functions associated to the H-atom's spectral properties are global (or unbounded) functions and the H-atom's angular-momentum states have shapes, which, in turn, imply a multi-pole charge-distribution, ie not a neutral charge-distribution, but atoms are quite often (if not most often) charge-neutral.

Furthermore, the relation of such a wave-function model of the H-atom (or to any physically bounded system) to other physical systems cannot be modeled, other than in the context in which the H-atom collides with some other (bounded) physical system, which is also represented as a wave-function, and the time and spatial-position when and where such a collision might occur is arbitrary and random, but when such systems collide, then these pair of colliding physical wave-function systems will collapse, if the physical state of these two system's position in space and time is identified, ie a quantum-state (of a quantum system-component) is identified, so the wave-function of each of the pair of colliding quantum systems collapses and this collision type interaction seems to be the only model of material-interaction, which can be modeled by systems which are represented as global (and unbounded) wave-functions and as such (ie closed-global wave-functions) these function-representations of the systems are closed to other types of operator relationships, which are different from the system-interaction models of particle-collision interactions,(or a quantum-system's wave-function can collapse spontaneously, as an indefinable random event) but this means that the state of a quantum system's wave-function is always susceptible to discrete and discontinuous wave-function collapse, where such events, such as the event of wave-function-collapse, is only relatable to indefinable randomness,

ie non-local relations of a wave-function are not well defined (since wave-function collapse is always possible)

and models of interaction, which are needed in order to use a quantum system's "superposition of wave-function properties," do not exist, (except, maybe, for a "relatively controlled" collision-beam, which is to be aimed at the (say) H-atom, while the H-atom is otherwise in a vacuum).

That is, the validity of the state of a wave-function model, ie collapsed or not-collapsed, can never be known when the system's description is based on global wave-functions.

That is, neither non-locality not quantum-computing can be provided with valid models (as usable quantum properties) since either the property of coupling (needed for quantum-computing) or the property of wave-collapse (which must be avoided if a quantum-system's non-local properties are going to be used) do not have valid models, ie at best they are indefinably random constructs, ie one cannot know of the existing-state of a quantum wave-function as being either in a collapsed-state or in a non-collapsed-state (of a quantum wave-system).and finally also in relation to "the causes for many-valued functions" there is the issue of

(4) the domain space not having any holes in its shape (if a quantitative model based on calculus is to have any validity).

That is, it is only in a simply-connected domain space, or system-containing coordinate system, that a line-integral can be single-valued for any (arbitrary) curve, in the domain space, which connects two points in the domain space.

This concludes the four listed causes for there being many-valued functions in improperly defined math contexts.

Identifying distinctions between types of systems, eg classical vs. quantum

Identifying distinctions between types of systems, eg classical vs. quantum, based on the properties of holes in a metric-space's apparent shape, the "apparent shape" is influenced by the dimension at which the description is being determined, and the description is being used

Yet, on the other hand, on the very stable discrete-hyperbolic-shapes it is exactly the "holes in the shape" which are associated to the stable spectral-geodesic properties of the (stable) shape, as well as to a finite set of stable "smallest-measures" of finite-spectral-values, for the stable shape, ie a finite set of spectral-values which are defined for each different hole in the stable shape.

These two opposing properties (domain related math processes [eg the value of a path-integral on a domain space] which define a single-valued function, and the stable set of spectral-values defined for a stable shapes which have holes in their shapes) identify a dichotomy:

Other than the rectangular simplex and the cylinder, it is only upon shapes (eg circle-space shapes) with several (one or more) holes (in their shapes) that the natural (local linear properties of the) coordinate systems of the stable shapes can be (made) continuously commutative, where the coordinates on these stable shapes are based-on sets of pairs of orthogonal hyperbolas and (a few defined) circles, which, together, cover the shape, as a set of curved-coordinates, so as to identify a small, and discrete, set of "circular" geodesics, whose smallest lengths define the (finite) set of stable spectral properties of the stable shape.

That is, the math structure has a well defined set, or math pattern, on which to base a spectral description in regard to a spectral-system's stable properties

That is, there is a fundamental paradox (dichotomy), which is central to both

(1) the description of a system's measurable properties, and

(2) if there are laws (pde) which can be used to determine these properties.

The paradox about (1) single-valued-ness and holes in space, and (2) the holes in a space's shape and the stability of the shape.

That is, is this paradox really about the distinction between

(1) stable systems of one size-scale

and

(2) the single valued-ness of a different (larger) size-scale,

in regard to the system, and both its measurable properties, and the shape of (or topological and dimensional properties of) its containment set?

That is, the containment set: its dimension, its independent subsets of various dimensions, the shape of the domain-space, or the shape of the (independent) subspaces of the domain space, their sizes, and their shapes, and their (dimensionally-dependent) topological properties, are the key ideas needed in regard to understanding the observed stable properties of the very-many, general but very similar, ie many-(but-few)-body systems, which are very-stable material systems, eg nuclei, general atoms, molecules, molecular-shape, and the solar-system, where all of these many-(but-few)-body systems now (2014) have no valid description (within peer-reviewed literature) where in peer-reviewed literature the descriptions of these types of stable systems are based on the assumption of materialism, reduction to elementary-components, and the, so called, laws of physics, which are pde's based on the idea of fundamental (indefinable) randomness (as is "to be believed," so as "to be allowed within" peer-reviewed literature, in 2014).

However, the properties of these stable systems can be understood in a many-dimensional context of containment, where the different dimensional levels are defined by stable shapes of each given dimension, and of various sizes, and with different topological properties, where these topological properties are dependent on the viewpoint of the containment-set for a description, where many shapes of many different dimensions are defined on a set of different subspaces, and/or an associated a set of shapes and defined in different dimensional levels (ie the shapes of a lower-dimension are contained in a metric-space of an adjacent higher-dimension, etc)

There can be defined set-containment-trees based on dimension, subspace (or infinite shape), and size, in an 11-dimensional hyperbolic metric-space.

The set of 5-dimensional subspaces equals the set of six-dimensional subspaces, in this metric-space, and that number is equal to 462 such subspaces of dimension-6, so that a 5-dimensional stable and bounded discrete hyperbolic shapes, where several can be defined for each subspace, can each fit into one of these 6-dimensional unbounded (but stable) shapes, so that in each 6-dimensional unbounded shape there is a largest size 5-dimensional shape.

Whereas the spectra of a bounded hyperbolic shape is bounded away from zero, ie there will be a bound to the smallest stable n-dimensional discrete hyperbolic shape, where n is between 1 and 10 (inclusive).

Then the 4-dimensional bounded discrete hyperbolic shapes can fit into the various 5-dimensional shapes, so that in each 5-dimensional bounded shape there is a largest size 4-dimensional shape, and so that there are "6 choose 4" ways of fitting these bounded 4-dimensional shapes into a 6-dimensional space, and so on,

Ie there are various ways in which to "partition" the 11-dimensional hyperbolic space into shapes of different dimensions and of different sizes and different subspaces (or within different unbounded shapes, which may or may-not be associated to subspaces of particular dimensions), so that the stable spectral set, which is associated to this set of stable shapes, is finite (though it could be quite a large quantity of spectral-values)

Shapes and topological properties

There are the issues of linear partial differential equations (ie pde's), metric-invariant (pde's), and continuously commutative coordinates for the domain space [for the shape of the coordinate-domain space upon which the pde is defined], (of the pde's solution function), as well as the property of the simply-connectedness of a space (ie its dimension and shape), where these properties depend on both the dimensional and the topological context within which the description is taking-place, is the description within a particular-dimension so as to have an open-closed topology, so that the system's natural hole structure is only revealed in the next higher-dimensional containment context, which is associated to the topology of the given dimensional shape. That is, the topology in the lower dimension is (can be) open-closed, while the topology of the same structure, but in the next higher adjacent dimension [where the shape has its geometric-form] the topology of the (now) lower-dimensional set (shape) is seen [in the higher-dimension] as being a closed set.

That is, the simple-connectedness topological property in one dimensional level, can be an apparent illusion when viewed from the (adjacent) next higher-dimensional level.

But it is only when the shape…, associated to a stable geometry…, in which its geometry has holes in its structure, that there can be a finite fixed set of stable geometric measures (or a fixed finite set of stable spectral properties) which can be associated to each hole of the stable shape.

That is, for a given dimension metric-space there are the stable set of closed shapes of (adjacent) lower-dimension (contained in the metric-space), as well as the set of stable boundaries of the given metric-space, which when viewed in the next higher adjacent-dimension, can be related to stable sets of spectral-orbital properties.

In the metric-space itself, it is the carefully built linear, stable, metric-invariant, separable (continuously commutative) shapes (or linear pde's with constant coefficients) which can be used in an instrument, so as to possess "regular" stable properties, and be useful and stable, in regard to building an instrument with its own internal mechanism of control

Where these shapes are related to the rectangular shapes and the cylinders and the circle-space shapes. But now this set of stable systems and their geometries (or circuit oscillating structure), which is often periodic in its nature, can be related to other higher-dimensional periodic, stable structures, where some of the stable shapes have periodic-orbital-sizes which are measurable in a clear manner, eg are the periodic structures of the planetary-orbits.

Furthermore, it is this size-scale in which to believe that, we, as living entities, can be related to a higher-dimensional (and well defined) other sets of living-entities, and that our own life-system is more deeply related to (or functions at) this higher-dimensional descriptive context.

There is the stable and continuous descriptive context where it is assumed that the domain-space for functions is simply-connected, ie a simply connected space means that there are no holes which can be surrounded by 1-curves (or surrounded by circles) in the space

And

Then there is the context where a stable discrete hyperbolic shape is stable and spectrally well defined because it is a shape which is not simply-connected where when the shape is continuously commutative (except maybe at one point) the many-holed shape defines a set of well defined spectral-orbital properties on a metric-space shape, and which is discrete, and separated, in its spectral-orbital relationship to its holes.

New viewpoint

That is, physical law is not to be solely based on pde's, but rather the structure of a system's over-all high-dimensional containment set, related to shapes in interesting topological ways, needs to also be considered, so that the pde context of "local linear measures of a stable system's measurable properties" need to be put into a context of these systems (of both metric-space shapes, and material-component shapes) possessing the math properties of being: linear, metric-invariant, and continuously commutative, so that the topology of a system-containing metric-space is open-closed, and, subsequently, the space (within that particular dimension) appears to be simply-connected

(ie simply-connected since the shapes is based on a convex rectangular-polyhedral constructed simplex, which acts as a base-space fundamental-domain of a base-space lattice), or they can also be put into a: linear, metric-invariant, continuously commutative context, but not a simply-connected context, where the stable shapes are related to shapes which derive from rectangular types of simplexes, ie all the shape's lower-dimensional facial-structures (or regional bounding structures) exist down to the point,

ie there are zero-faces, 1-faces, 2-faces etc for all the different dimensional regions of the shapes

ie this means that simply-connectedness is always a relevant topological math construct within the space, but

where one is (becomes) aware of the hole structure of the (stable) shape when the shape is viewed from its adjacent higher-dimensional (containing) metric-space, wherein the shape is seen to be closed and (most often) possessing holes in its shape.

Furthermore

That is, "existence is many-dimensional," as is assumed by both particle-physics and string-theory, but the size of the geometries which these higher-dimensional spaces (namely, metric-spaces which also possess the property of being stable shapes) are to occupy, do not need to be the very, very small (sizes) (as it is assumed to be true, by current (2014) physical science),

But rather the higher-dimensions can "partition" (or divide) a higher-dimensional set into stable shapes of various sizes and dimensions, so that each such independent dimensional level, which is contained within a stable shape, can possess both topological properties and a new type of model for material (a material-component which is also a stable shape, which is an adjacent-dimension-lower than is the dimension of the component's containing metric-space), and a material-interaction structure

which allow for the different dimensional levels to be relatively independent spaces, when viewed within the different dimensional levels of existence.

(note: the higher-dimension spaces, being discussed, are metric-space shapes, which are hyperbolic metric-spaces, where a 3-dimensional hyperbolic metric-space is equivalent to a space-time metric-space)

Furthermore, there can be made a new higher-dimensional model of life, ie naturally oscillating, and subsequently, energy-generating, and relatively-stable, higher-dimensional shapes, ie odd-dimensional and with odd-genus relatively stable shapes, which organize and "govern" the life-form, where the organization is based around a unified spectral-geometric construct (or shape), yet this higher-dimensional meta-structure descends to lower-dimensions, by means of a spectral relationship (eg resonances) to other lower-dimensional stable shapes, so as to be related to atoms and molecules, and "the new model of a living-system" does this through the natural spectral properties of the life-form's (newly modeled) higher-dimensional and unified, stable spectral shape.

Note: the property of radioactivity, where there is material which is relatively stable, if it is not at a critical-mass, but which gives-off energy, could well be an example of a higher-dimensional (odd-dimensional [ie 3-dimensional shape which is contained as a shape in 4-space] and with an odd-number of holes in its shape] which can maintain a relatively stable high-atomic-number nucleus) relatively stable shape, whereas life-forms might begin as stable shapes of 5-dimensional {or higher-dimensional} (so as to form hyperbolic metric-space shapes)

Note: These ideas need to be seriously considered, since there is not any valid model of a general nucleus whose relatively stable observed spectral properties can be derived from what is (humorously) called physical law, ie the observed spectral properties of a (relatively stable) general nucleus derived from the laws of particle-physics.

As well as there being a new model for a

Mind, which is a natural math construct, and which is central to the identification of the set of higher-dimensional stable shapes. Namely, the coordinate space's fiber groups, but which (within the fiber group there) is also a natural structure (in math) through which to carry spectral properties (or spectral-information), as a memory might hold onto such spectral-geometric information, namely, within the maximal tori of "the coordinate base-space's" fiber group. Ie the coordinate-space and its associated classical Lie fiber group seem to possess an equal amount of reality (in regard to their existence).

Thus, there is a dimensional and size-containment context which become inter-related for material existence and living-systems

There can be very many life-forms in this context, of which it is "us" who may be a substructure to some higher life-form, which can exist in a (yet) higher-dimension than the dimension within which we exist. Yet our higher-dimensional structure may be quite different from what we have come to believe to be true.

Thus, our sense experience can (could) be greatly influenced by another entity, in such a new many-dimensional scenario for our existence, ie within higher-dimensions. It is a new higher-dimensional context for existence, which is composed on many size-scales, as well as being based on new types (or new categories) of relatively stable shapes (which can define various materials, including new material-types), and which allows for many-new-possibilities, in regard to life, the knowledge which life can possess, and such information's relation to life's central creative property within existence, the property which human life-forms need to realize, and relate to, through sense-experiences, which is one of the sense-experiences about a further structure to existence,

The "hidden knowledge" which P D Ouspensky seeks

P D Ouspensky in his (old) book 1934 (2ⁿᵈ edition) "A New Model of the Universe," in which he is trying to make "the mind" an important idea in the physical sciences, in the set of (assumed to be) valid descriptions of existence, one sees, as usual, a propaganda-based literary figure, who follows "the authorities who are allowed to be published," so as to slightly permute the expressions of these authorities.

But Ouspensky was more concerned about gaining access to "hidden knowledge," knowledge which transcends both materialism and which transcends our 5-senses (taste, hearing, sight, touch, smell)

But today's science is based on materialism, but this basis has not been capable of describing the observed stable properties of material systems ie systems which are assumed to be contained in either 3-Euclidean space or 4-space-time space.

However,

That is, if a person, such as Ouspensky, is published then the viewpoint expressed "is a viewpoint desired by the ruling-class."

The authoritative dogmas of the intellectual-class are narrowly focused, and subsequently, they are becoming delusional (eg careful proofs of claims which are not relatable to the material world) and this is resulting in the failure of the expert technical capability to expand the creative contexts for society, and subsequently, the main focus of the society's intellectual dogmas deals with their relation to investment interests, and the goals of the investor-class are the only goals which can be realized within society. Thus, leading to continual development of the same types of instruments of oppression, and domination, and instruments which can do great violence and harm, where it is these instruments which now (2014) identify the intellectual context of what is solely being pursued by the intellectual-class.

Furthermore, this is how the economic forces, which govern what society does, and where the subsequent creative actions of society are available to (or are controlled by) the ruling-class, and these forces are being used by the ruling-class, so that both the technical experts are expressing very limited sets of technical ideas, ie they are becoming arbitrary, and subsequently the expert-class is becoming: limited, shallow, and ignorant (a narrow range of instruments are understood, but the physical world is not understood), resulting in the experts becoming incapable, but this leads into a social context which the ruling-class has always depended upon, namely, manipulating arbitrary social contexts and controlling or directing unfounded ideas (or prejudices, most often this is expressed as racism).

What is pursued by the ruling-class is self-interest based on investment, though the economy is not really about markets, but rather the economy is an expression of total domination over the clamoring population of wage-slaves, (where the worst social strategy for the wage-slaves is to plead for jobs, but that is all that their social condition (social-position) allows them to do).

That is, an external intent of a narrow-minded ruling-class is imposed on a material existence, but where the properties of some of the most fundamental material-systems have no explanation within the authoritarian dogmas, but these are dogmas which materially support this arbitrary social-hierarchy, and its violent domination by the few

The only "explanation," about the narrow use and wide ranging failures of the limited dogmas associated to investor-interests, is based on propaganda, that is, the, so called, explanation is that there are a set of elite people who possess arbitrary high-social-value and they know everything which our culture can know. That is, arbitrary high-social-value is placed on a small set of elite people, who, with their failed dogmas, best support the social-intellectual-hierarchy, where this is done in a (dogmatic) context of materialism, the main focus is the dichotomy between the material-world and religion (where religion is a deceptive form of propaganda) and the explanation is that "this is the only way in which these problems (of the failure of the dogma) can be dealt with," since the experts, who are assumed to possess the superior intellects (in all of society), are expressing the only absolute truth which can exist (ie which by-the-way is also the truth which best serves the interests of the investor-class, in regard to their narrowly defined set of products, in turn, associated to their business interests).

Note that, both Constantine and Calvin fundamentally worshiped the material-world, and the relation of their control... of the material-world... to social inequality, in this respect

The idea of "hidden-knowledge" is to be used for the selfish purposes of the ruling-class to control society, where this control is, essentially, based on their brutality and their "somewhat hidden" control over society, where "hidden-knowledge" is to be interpreted in this context, to be about deceptive propaganda and the "hidden from view" expert knowledge, where expert knowledge is the knowledge possessed by the experts, whom, in turn, serve the interests of the ruling-class, so their knowledge is narrowly defined to only be related to the development of the instruments which (best) serve the investor-interests, but because the technical-knowledge is poorly articulated, in fact, the experts themselves do not seem to be able to "place their knowledge in the contexts which make it clear 'what it is that their constructs are trying to do,'"

Particle-physics is supposed to be about describing the observed stable spectral properties of general nuclei, but it fails miserably at doing this, however, the descriptive context of particle-physics can be easily related to the collision properties associated to bomb engineering, so it is knowledge which becomes hidden to the public by a complex description which fails to describe the spectral properties of general nuclei, but Science (which is thus defined to be the expert knowledge upon which the ruling class is based (or depends for its creative projects, ie projects into which the ruling-class invests) is about: materialism, reduction, and indefinable randomness, so as to be consistent with the interests and models regarding both the investor-class and now (2014) nuclear bomb explosions (actually, particle-physic's and general relativity's unified descriptive context, eg string-theory etc, is about detonating a gravitational singularity)

But technical-knowledge is "constructed within society to be hidden (or difficult to understand, since it is poorly articulated)" so that it is used as the basis for a mis-placed reverence by the propagandists, who, in turn, create an illusion (within the propaganda-system) of the great intellectual capabilities of the science and math experts.

Really, the personnel of the expert class are, apparently, on the boundaries of autism, since they seem to obsess on rules which they are required to obey (so as to be peer-reviewed, or in order to join the group-think club of the expert class) and the experts never adequately assess the descriptive context of their dogmas, since assessing their descriptive language's context (describing the properties of a general nuclei [for particle-physics] and describing the properties of the solar-system's stability [for general relativity, but general relativity only focuses on gravitational singularities, (why is this?)]) is hidden behind great complications and elaborate language.

That is, the experts do not really seem to be able to discern truth for themselves (ie their intelligence should be questioned in a fundamental way).

Is intelligence "the ability to discern truth?"

Or

"Is it about obsessing on the complex language, upon which the dominant culture wants the experts to focus?"

That is, the west is a civilization which opposes a true notion about "hidden-knowledge," ie knowledge about existence which lies beyond the idea of material is (beyond the bogus, science vs. religion (baloney) debate)

And instead considering "hidden knowledge" to the point (beyond the limitations of the idea of materialism), where mankind has a purpose of using knowledge to extend both the context and the reach of existence, itself, by controlling the internal processes of existence through its relation to the living-systems.

That is, at the core of the ignorance of mankind, in his civilization, is that his knowledge is based on materialism (and reduction to elementary-particles and its relation to indefinable randomness) so that in the math models of material existence, the distinction between "quantitatively consistent" patterns and "quantitatively inconsistent" math patterns, or math processes, or math operations, is not made.

Where in the descriptive language of the, so called, technical experts, the "quantitatively inconsistent" math properties stem from the use…, by the so called experts…, of the non-commutative and non-linear local properties for their math-models of measured properties, as well as non-commutative properties of function-space relations,

ie the attempt to diagonalize the harmonic function-space model of containment for material systems based on finding a set of commutative Hermitian operators which act on the function space so that the different-operators represent different classically-measured properties, which (all taken) together should limit or constrain the number of variables in the containment-set, so as to solve for the observed spectral properties of the system, but this method (this program for solving for arbitrary spectral systems) has not worked (it can be neither formulated nor calculated ie it cannot be solved).

The spectral-values seem to be independent, while the set of classically measured values (associated to an equivalent set of operators) which should limit the measurable properties of a physical system, seem to be too few in number

In such a descriptive context (in math) the practitioners do not distinguish (in a clear manner) between

(1) internal patterns of existence and control,

And

(2) externally motivated patterns of control, ie based on adjustments made in the system in regard to empirically determined properties outside the system, where this type of non-linear based feedback system-model is both vague and always unreliable (ie several levels of feedback are needed to cut-down the failure-rate of such feedback systems), eg "the drone" is a feedback-system (which is used extensively by the military-industry),

ie the system's purpose is imposed externally, based on the relation of the position of the system to the external patterns (the position of the system is adjusted to the purpose which is determined in regard to the (unstable) external conditions, [ie the properties of the system are not being described, other than in regard to its relation to the external world]).

That is, both "the social structures of the hierarchy of society" are arbitrary, and "the math and science models placed within a material world" are arbitrary and unstable,

ie if they really worked then where are the explanations of the observed internal stable patterns of material systems, ie nuclei; atoms; molecules; life-forms; the solar-system etc.

Yet these arbitrary and failed techniques persist, they keep "getting funded," and more theorems keep getting conjectured, and, subsequently, proved, in this very complicated set of very general and abstract viewpoints about "what a math pattern 'actually is'"

It is in regard to identifying the content of "hidden knowledge" about which the quantitatively inconsistent math constructs should be avoided, in order that the descriptions of higher-dimensions can possess meaningful content, and so as to possess (and to be able to use in practically creative contexts) a precise descriptive language which is based on stable math patterns, which are "quantitatively consistent"

Ouspensky communicates quite clearly in the first chapter of his "A New Model of the Universe" book 1934, the issues involved in man's relation to "hidden knowledge," ie knowledge about existence which goes beyond the three or four dimensions of the material world

He describes the tales of the three-wishes where the people in the story never know "what they want," and often these are stories about "their granted-wishes" in which their "sequence of wishes" leads them back to their original state.

Ouspensky makes the statement (p14) "In our times, theories which deny that mankind is capable of gaining 'hidden knowledge'" and these stories about mankind's inadequacies have become only recently very strongly shouted (through the propaganda-system) by only a few very noisy people.

But, "man, in general, still believes in being able to acquire hidden knowledge, but when it shows itself, {as is provided by the example which B Ehrenreich has given in her book "Living With a Wild God," (talked about below)}, the person is not equipped to deal with the fleeting opportunity."

Ouspensky says that "man is aware of the wall surrounding 'hidden knowledge,' and the unknown, but he also believes that he can get through that wall, and that others have got through that wall, but he cannot imagine what is beyond that wall."

"He (mankind) does not know what he would like to find there, or what it means to possess (hidden) knowledge."

Mankind still cannot articulate what it is that they want.

Do they want, riches and treasures and social domination?

Or

Do they want to have a meaningful relation between their creativity and their knowledge of existence itself?

But now there are math models of existence, which go beyond the idea of materialism, and they maintain the material world as a proper subset, and they are related to the idea that mankind's natural position in existence is to further extend the context of existence itself, eg how to build stable bounded shapes in $R(6,3)$ {contained in $R(7,3)$} so as to (perhaps also) extend the finite spectral set of the containment-set from $R(11,1)$ out to $R(12,1)$ etc

Ouspensky states again (p15) "In this incapacity of man to imagine what exists beyond 'the wall of the known and the possible' lies his chief tragedy, and in this, lies the reason, why so much remains hidden from him, and why there are so many questions to which he can never find the answer."

The new ideas being expressed about existence (by the author of this paper) claim that what lies beyond the material-world are many-dimensions whose subspaces are deeply related to stable shapes, which can be used to construct a subspace structure based on stable shapes.

Furthermore,

Note: The new ideas about math and physics solve the problem of the stability of the many-(but-few)-body systems, where these types of stable systems exist at all size-scales: from the nucleus to the solar system, and apparently beyond this too.

This brings the discussion into the realm defined by B Ehrenreich, the observed context of being in relation to a "new type of" knowledgeable living-entity, where the sensations transcended the sensations of the material-world.

But first review some more of Ouspensky

Ouspensky writes in 1934 that mathematical physics is both too abstract to have any practical value, and none of its math constructs can be solved (though he also a states that the systems are not being sufficiently carefully modeled, "as if this were the problem," but it is not the problem with theoretical mathematical physics (of which general relativity is certainly a part) rather its problem is that it tries to use math structures which are known to "not be quantitatively consistent" such as the math properties of non-commutativity and non-linearity, and the most damaging model… (and now theoretical physics most prevalent physical model)…. the model of an indefinably random based, so called, precise descriptive language, which is used to describe the properties of very stable physical systems, where none these very stable physical systems, ie none of these very stable many-(but-few)-body physical systems, has a valid description of its measurable (observed) physical properties based on descriptions which in turn are based-on non-commutativity, non-linearity, and indefinable randomness.

Namely the sets of stable spectral-orbital properties of physical systems which exist at all size scales, from nuclei to solar-system, are in need of valid descriptions,

And where it needs to be noted that the physical property of stability, as well as the great prevalence of these stable-systems, together implies that these systems are: linear, metric-invariant, and continuously commutative…. in descriptions which are based on.

"Measurable-property-values which are put into local linear, or measurable, relation with the domain set, and which is supposed to exist between the function-values and the domain-set, ie the assumed system-containing set of (locally metric-invariant) coordinates of the domain-set" and this all means that the descriptions of stable systems are solvable and controllable.

That is, the intellectual class of this current social-structure, need to be criticized in the strongest possible way, since "what they are doing is both delusional, and blindly obedient to their dogmas, they quite clearly have no capacity to discern truth on their own," "they seem to obsessively confine themselves to their memories of their, so called, authoritative textbooks" and these are actions which are strongly connected to the most destructive aspects of the society, what is being created with their obsessive intent is military instruments, ie instruments of destruction, ie this is about a process of arbitrary identification of certain people as being superior to all the other-people, where the basis for the identification is a failed dogma, a scientific dogma which has only the narrowest of relations to creative development, ie used to create instruments of the greatest destruction, an instrument of complete annihilation.

That is, the intellectual peer-reviewed bullies of academia (the dominant intellectuals of our society) are much more dangerous, in regard to causing the destruction of the world, than are the very dangerous arbitrarily motivated people-of-action who define the spying and deeply interfering and quite often very violent actions associated to the processes of the justice system and their militias and diplomats and protectors of freedom (ie hired-guns for the rich).

But

In fact the creation of militias and righteous-right (arbitrary) moralists is done by both the propaganda-system and the justice-system so as to divide the set of ordinary citizens, so that the ordinary citizens are thwarted from (or in) their unification (with one another in the public-class [or in the equal citizen-class]) in their opposition to the investor-class.

Ouspensky can sometimes write very clearly, but his book ultimately has been a failure.

But, how to "get done" what the published authorities seek to realize, or "more to the point" how to "get done" what an individual intends to create, eg not basing science on the idea of materialism,… by using different sets of: assumptions, interpretations, and new ways of organizing precise language, so as to build a new accurate, and practically useful, precise language, is what education, and rational-thought, is really all about.

But, to try to answer this question, consider a set of clear messages, or assertions, Ouspensky makes;

In his chapter "A New Model of the Universe," in the above mentioned book, he begins (this chapter) by asking some good questions, which can make the discussion about the descriptions of existence, ie physical description, quite clear, especially, if one can give answers to these questions, where in the descriptive context where existence is: many-dimensional and organized around sets of stable shapes, which possess different properties, in regard to their individual (shape's): dimension, size, (subspace, or) containment-tree sets, as well as genus (number of holes in the shape)

In this context for existence, which is a model of "hidden knowledge," one can give very clear answers to these questions.

Eg 1. Does the world have a form? 2. Is it (the world) chaos or a system? 3. Was it created or an accident?

1. "What form has the world?"

This is a question which is about both containment-set and (stable) patterns---- set-containment seems to be…, in our, apparently, spatially expansive experience…, to be about coordinate metric-space containment-sets, where a metric-space allows for the measuring of coordinates and the geometric measures of shapes which the coordinates might contain. As well as the stable patterns, the material qualities, or consistently measurable material properties, the measurable qualities of material-systems, which are supposed to be related to the system's containment-set, eg

Either to the system's coordinates or to the system's function-space representation,

Where, when the material-system is contained by a function-space then the measurable properties of the system become sets of spectral-values, which, in turn, are related to the spectral-functions of the function-space by means of sets of commuting (Hermitian, differential) operators,

And

The very fundamental idea about the dimensional structure of existence, ie is the description to be based on the idea of materialism or not.

Once within a (metric-invariant) metric-space, where there is consistent and reliable measuring of geometric-measures, and where the metric-space has a particular dimension, then one is also related to the classical Lie groups which are associated to the (base) metric-spaces.

Furthermore, one needs to consider the "signature" of the metric-function, ie a construct which identifies the divide between the temporal subspace and (or from) the spatial subspace of the (locally represented) coordinate space.

The material properties are identified by the dimension and signature of the coordinate metric-space, while the nature of the coordinate spaces needs to be quantitatively consistent, ie general curved coordinate systems are not quantitatively consistent, being either non-commutative or non-linear or both, where the following set of shapes are the shapes whose descriptions can remain quantitatively consistent, ie the shapes of: the rectangular convex polyhedron simplexes, cylinders and relatively stable circle-space shapes (such as the torus).

To re-iterate, one considers only these shapes because they are the only shapes which can be quantitatively consistent, where, again, these are the cubes and the cylinders and the sets of circles, which could include discs (but discs are deformable to a point, so they are little different from the cubes (or the rectangular convex polyhedron simplexes)) (and/or in regard to the sets of circles, these would be the circle-space shapes which are continuously commutative).

These are the natural shapes, which characterize the discrete isomerty (or metric-invariant) subgroups of the real classical orthogonal (or rotation) Lie groups, eg the tori in Euclidean space and the discrete hyperbolic shapes in hyperbolic space) and they are also the shapes which naturally can carry over to the complex, or unitary, classical Lie groups, where circles and discs characterize the shapes found in complex-coordinates or the Hermitian-invariant unitary groups for the classical groups associated to the various signature metric-functions and then introduced into Hermitian-invariant (or inner-product invariant), complex-coordinates, note again, that circles are more interesting than discs.

The physical properties are related to metric-spaces, and thus the model of existence is a mixture of different metric-spaces of various different signatures and different dimensions, but there is a natural progression of dimension and signature which accompanies the different types of dimension and signature metric-spaces which depends on the types of materials and their associated characteristic physical properties, eg $R(n,0)$ is Euclidean space and is about position in space and inertia or mass, while $R(n,1)$ is generalized space-time, or equivalently, it is also a $H(n)$ a hyperbolic metric-space, and is associated to: time, and charge, and energy.

The, "mass equals energy equation," is related to the way in which Euclidean shapes depend on the more stable discrete hyperbolic shapes.

The next new change in signature..., in regard to the dimension of the temporal subspace [different from $R(3,1)$]..., is in $R(4,2)$ since

4-hyperbolic-space can contain a 3-dimensional shape with an odd-number of holes in its shape, where such a shape naturally oscillates, so as to generate its own energy, but it is also a shape which can disintegrate (suddenly). This oscillating material-shape is associated to a new physical property, namely, it is material which generates its own energy. This can be a simple, but higher-dimensional, model for a unified living-system, ie the unified shape and its spectral properties control its internal chemistry rather than local chemistry being the basis for all the living-system's properties of both the system and control over the system, which is given to the living system's consciousness. The memory of a living-system can be modeled within one of the maximal-tori of the fiber group of the living-system's oscillating (but relatively stable) shape.

The ideas of special relativity are simply about metric-invariance and that the time coordinate in the 4-coordinates of space-time in turn related to momentum-energy 4-coordinates so that [mass = energy] is a result of equating the time component, ie m, with energy, where the energy-component of the local coordinates of the 4-momentum is associated to mass,

Note: This energy-time equivalence was determined by E Noether's invariance theorems.

The metric-invariance properties can be used to relate measurable values between two (equivalent) moving frames.

But dynamics is mostly about identifying relative positions in space, ie it is a description which fits into Euclidean space, but most stable material components are built from charged components, which are neutrally-charged, [whose mass is related to both resonances (between hyperbolic and Euclidean shapes) and its inertial affects are related to the changes (relative) spatial positions (as well as relative motion's values) of the material components (the one and the other material components)]

So that, the finite speed of light seems to be the relation of the motion of a "wave-front of a light-wave" to the relative positions (of this wave-front's motion) which can (only) be identified in Euclidean space, but which exist in hyperbolic space.

In regard to the general theory of relativity, the non-linear quantitative (and descriptive) context of general shapes is not measurable within a "consistent quantitative structure," and there are no stable patterns which exist within such a non-linear descriptive context, so it is of no-value to the physical descriptions of useful properties

But

When the principles of the general theory of relativity are applied to the set of stable discrete hyperbolic shapes, either within the shape and/or which exist between different (usually adjacent) dimensional levels where the stable shapes identify geodesic spectral-orbital properties, whereon (the geodesics of these stable shapes) it is Newton's law, which is the better law to use in order to permute a 2-body system from its geodesic orbital properties which the metric-space's shape (within which the material components are contained) determine, so that the application of Newton's law is a good approximation to the material-component's orbit in the metric-space, and it is the types of permutations of orbits, which the stable orbital shapes allow (in regard to the stable shape's geodesic properties), for such a good Newton approximation of perturbed orbit-changes, to be made.

As well as the general theory of relativity being the context, which identifies an effective, rigid-body, in regard to a discrete hyperbolic shape being accelerated in an adjacent higher-dimensional dynamic context, eg so that the acceleration of a 3-shape in 4-space is not felt (not detectable) inside the (rigid) 3-shape.

These few paragraphs deal with what Ouspensky takes several pages to discuss, though this discussion is quite insightful, he misses the main issue, namely, that general, non-linear contexts are not quantitatively consistent, so they have no content, as valid practically useful descriptions,

Furthermore, such a descriptive context degenerates into simply an empirical discussion (of unexplainable measured details).

This is about the main issue: "Are there stable forms about which existence depends?"

The answer given here (in this paper, or equivalently, in regard to a description based on many-dimensions and organized around stable shapes) is yes.

But in quantum physics and particle-physics as well as general relativity, and in regard to non-linear systems (in general), the answer is no

Furthermore, non-locality, or action-at-a-distance, has been confirmed to be a realizable property of material systems in space, but it does seem to be a Euclidean-space property, not a space-time property, though for unbounded discrete hyperbolic shapes, apparently related to both light and neutrinos, there are some other possibilities, in regard to instantaneous inter-connections between material-components and their containing-space.

Ouspensky tries to discuss the issue of defining the measurement of physical properties, but he seems to be mostly interested in quantities like mass, which when placed in a context of "chemical reactions," the form of the mass changes

But (Ouspensky wonders) can the mass of the original system (before the chemical changes) still be measured when after these changes take place, Ie (solution) keeping the chemical changes contained in a closed system.

But

He seems to not be concerned about a much more important issue, as to whether quantities actually can stay well-defined, when function-values (or, equivalently, measurable physical properties) and coordinates are related, by means of non-commutative and non-linear relations, but within such a descriptive context (which is to be determining the quantitative structure of the, so called, measurable description), there is no quantitative consistency and quantities lose their property of being well-defined, eg measured properties become "many-valued and discontinuous."

Ouspensky points out that "mathematical physics" "should be questioned," but he "questions it" based on the grounds that physical systems are not sufficiently precisely formulated (this is a misrepresentation of the problem of quantitative description), and Ouspensky also point out that they (the math formulations of the systems, ie their pde's) most often cannot be solved, and that they are only solvable in the simplest of cases.

Perhaps the physics community should take this criticism much more to heart. The stable systems need to be simple math models where their stability indicates a true descriptive context where the math is very simple and the stable (and uniformly formed) systems do form in a causal manner.

The great value of science was demonstrated by the solvable systems of classical physics where general laws when applied to continuously commutative shapes provide accurate information about the system and the context through which the system can be controlled, but the non-commutative contexts quantum physics, particle-physics and general relativity have not provided this type of practically useful information or control over the systems, which these descriptive contexts "try to describe," rather the scientific context of this type of non-commutative description is a context of both empiricism (ie finding a system's properties by measuring the system) and indefinable randomness (chaos, multi-valued-ness, discontinuity, and/or random events which are not stable) quite similar to the data-fitting descriptive structures of Ptolemy's epicycle (empirically based) descriptions of planetary-motions.

But

this is not the direction Ouspensky takes, rather he follows the authorities, as he pretty-much does (follow the authorities' viewpoints) "all the way through his work," but then if he did not "do this" then he would not get published (and this is back in the 1930's)

In fact, math physics began with Newton who solved the 2-body problem, and the physics community has not been able to progress past this accomplishment,

Especially, in regard to the many very stable spectral-orbital many-(but-few)-body systems, though they (the classical physicists) have described waves and statistical systems composed of many atomic components, and (for these statistical systems) whose measured values are averages over these many components, and classical physics has had a certain amount of success at these descriptions.

Ouspensky points out that Einstein's work is all about math physics, but, as usual, he misses the main point, and focuses on a narrow interpretation of one of his "claimed to be issues" of great importance, namely, about "the form of the world" where he focuses on relativity, both special relativity and general relativity as expressing "the form of the world," though the issue of quantitative formulations in equivalent frames of measuring is important in regard to formulating a physical system's pde's, it is such pde-based descriptions which have missed the main issue of "the stability of pattern of the world, ie the world has a form," where stable forms are needed "in order to identify the stable properties of physical systems."

That is, the main form of the world is that the properties of existence are being determined by metric-invariance and the set of stable discrete isometry shapes which are naturally associated to metric-invariant spaces, and the, subsequent, ambiguity between hyperbolic and Euclidean geometry, but

This ambiguity (ie which containment space is correct?) is best resolved by looking at their sets of discrete isometry shapes, where such shapes in Euclidean space are tori, while such stable shapes in hyperbolic space are strings of toral-components, ie the geometries of the two spaces have a natural relation to one another, so the various coordinate relationships which exist between the two different metric-space types, eg 4-momentum and the invariant metric-function structure of the wave-equation, are difficult to interpret, and to realize.

But special relativity is basically a statement about metric-invariance, so as to fit into a classical Lie isometry fiber group which are related to such space-time (or equivalently, hyperbolic) metric-spaces, which contain energy and charge, while the Euclidean metric-space deals with the measurable properties of spatial-position and inertia.

Whereas time relates..., as an imaginary-number structure (which also identifies the signature of the space-time metric-function)...., the positions of electromagnetic wave-fronts to Euclidean space.

The "Euclidean" time-component of the local position identification coordinate directions is equal to the energy-component of the local momentum identification coordinate directions, but space-time is really the local momentum coordinate properties.

It is exactly this ambiguity about which metric-space property is being identified which is at the heart of the, mass = energy, formula

That is, the world is a mixture of different types of spaces, and these spaces are defined by (or have a definitive relation to) stable shapes

This is the correct answer to "What is the form of the world?"

Note: Ouspensky also makes the mistake of saying that general relativity is about general curved-coordinates as having an isometry fiber group, but rather general curved-coordinates have a diffeomorphism fiber group, which is invertible at a point (diagonalizable at a single point), but the general metric-function (of general relativity) is not related (in general) to continuously commutative local coordinate relations.

That is,

Or Furthermore, if one can deal with many different types of physical quantities, or measurable physical properties, then why must the containment of all of these properties be in one type of metric-space, special relativity triumphantly throws away Euclidean space, but this is an error, since

Euclidean space is the space where the physical properties of position and inertia are contained, while space-time contains other physical properties, but the [mass = energy] property depends on the incorrect formulation of space-time as a space where dynamics can be described as a 4-momentum-vector, but when this is done then E Noether's symmetries (about time and energy) can be used on the (essentially, superfluous) time coordinate (of space-time, but what should be hyperbolic 3-space) to establish the [mass = energy] formula.

Ouspensky does talk around this issue, when he identifies time as an imaginary-number so as to account for the signature of the space-time metric-function, but this is really about the wave-equation in regard to the general Laplace operator defined on space-time, where waves have positions in space, eg positions of the wave-front of a wave's propagation in space (spatial-position is a physical property of an Euclidean space).

2. "Is the 'world' a chaos or a system?"

The answer given, by the alternative set of ideas (which is being endorsed by this paper), is that, existence has many stable and semi-stable (or oscillating) forms (or stable shapes), which exist in higher-dimensions, where D Coxeter has shown that stable discrete hyperbolic shapes, in bounded and unbounded sizes, exist from hyperbolic-dimension-one to dimension-ten inclusive. Where after hyperbolic-dimension-5 all discrete hyperbolic shapes are unbounded. That is, these stable shapes will continue to be a part of any description of a world which is modeled to have higher-dimensions.

That is, for a world composed of mixtures of metric-space-types, and these metric-spaces have stable shapes, the "world" is many-dimensional, and it is a system (metric-spaces are stable shapes), though its regional structure, in regard to its many-dimensional structure, is difficult to discern, due to: 1. topological properties and 2. material-interaction properties and 3. size-and-dimensional properties, in regard to the 4. region's containment structure, where these three (or four) properties of metric-spaces and their material-component properties (where material properties emerge between adjacent dimensional levels of existence, ie existence as stable shapes), can cause one's viewpoint to be trapped within a particular dimensional level of perception. Thus, the higher-dimensional properties around the "true" region of containment are difficult to perceive (some of these properties may have to do with size, as well as our obsessive focus on our (own body's) 2-dimensional material-component composition's (apparent) size, ie our obsession with the idea of materialism.

The equivalent discovery of finding "the periodic table of the elements" would, now, be:

for each R(11,1) containment-space, one (a scientist) wants to identify its existence-defining "finite spectral-set," realized in a context of stable (metric-space) shapes and: size, dimension, and containment-tree set-structures.

That is, "the form of the world," now has a mathematically described pattern, which is about shapes and many-dimensions, and which leads (can lead) to a realization of accessing "hidden knowledge."

That is, (accessing) the knowledge of what might exist beyond the material-model of the world.

Realizing what our own life-form might be in this many-dimensional, and (possibly) many-size-scale, context can be difficult, especially, when, as in 4-Euclidean space, there is a natural division of the SO(4) Lie group into (SO(3) x SO(3)) so that the dynamics and subsequent actions of the oscillating forms and the stable material-forms may be divided between these two spaces, so it is a division in which our "taught to be" "existence as a material entity" can oppose, within ourselves, such a realization about our own structure.

The description of modern physics which claims all stable systems are too complicated to describe

The answer which modern physics has given to this question, is that existence is both chaos and indefinably random, so that science has come to be only about empirical determinations of fleeting, unstable patterns, which somehow (or through many-many boundary conditions, supposedly) "self-organize," so as to form systems, which are only contingently stable.

But self-organizing is about the adjusting of boundary conditions in a (non-linear) chaotic setting, for a description based on non-linear relations (between local coordinate and the system's measurable properties), ie it is completely arbitrary (ie change the boundary conditions of a non-linear system until one gets the type of solution which possesses the set of properties which one wants to obtain, but this type of a system not controllable, except by means of trying many adjustments to the boundary conditions and seeing what happens, again one sees that science of chaos is empirical and not controllable, and non-linear solutions cannot be counted-on to provide accurate information about the system's properties).

"Belief" in self-organization, is a deep belief in the hit-or-miss model of a material-based existence.

That is,

Science is materialism, and reduction to unstable elementary-particles, and indefinable randomness.

But this is clearly not true, since there are so many very stable many-(but-few)-body physical systems and this stability implies: linearity, solvability, and control. That is, stable systems are forming so as to always result in systems which have the same fixed sets of properties, so the interaction which leads to the system's existence is a controllable interaction.

3. "Did the world come into being accidentally or was it created according to a plan?"

This can be answered as follows:

The point of the existence of life, "within the 'world'" is to create existence, itself, but to do this one needs correct knowledge, so one is constrained as to what is possible, and/or the nature of what is created, ie the "plans of creation" must be deeply consistent with the true nature of existence.

This seems to be about the relation between metric-invariant metric-spaces, the stable material, and their (sometimes) oscillating energy-generating material properties

This third question of Ouspensky is also the basis for the bogus eternal debate between science and religion, but both sides are really materialists, but materialism is a failed scientific viewpoint (but it (materialism) is the first loyalty of the society-based religions, which form the basis for the propaganda of an oppressive social hierarchy).

This is similar to the political parties in the US, which are both "for the rich," and their social systems of: money-dominated-markets, violence, and propaganda, and both sides are deep "believers in materialism," except in regard to the bogus debate about science and religion.

Ouspensky claims that neither 3-dimensions nor 4-dimensional space-time are sufficient to contain the observed properties of the "world," but he does not identify the most obvious problem, which now (2014) is made more clear [but it was also clear then (1930's)] that the stable (observed) spectral-orbital properties of the many-(but-few)-body systems are not being described, and the stability of these systems implies that their descriptive structure should be simple, solvable, so that these stable, and prevalent, systems must be formed in a context of a controlled interaction.

Instead he, vaguely, references motion, which he wants to place into a mental context, ie motion's direction points from past to present in our minds

Though there is a locally invertible relation between (immediate) past and present (or immediate future) this is best represented in the context of a particular physical quantity-type, namely, the particular metric-space type, ie dimension and metric-function signature, and to identify the locally invertible property, and the property of local invert-ability means that the local interaction is composed of two sets, ie a pair of "opposite" metric-space-states, eg the advanced and retarded potential 1-forms of electromagnetism.

Ouspensky seeks to find that "mental properties" enter into the descriptive context of physical description.

To do this, Ouspensky defines "the 3-dimensions of time" as the mental properties of:

1. duration (memory),

2. velocity (an instantaneous mental construct in time), and

3. direction from past through present to future, he calls this "the direction of possibilities,"

And Calling these 3-dimensions of time, "the boundary of our senses."

But

It is not clear how these are to be measured (ie how to measure: memory, mental-constructs, and event-possibility), and how they can be used, though in the indefinably random context of quantum physics perhaps the direction of possibilities can be modeled, as a probability wave-structure, (but perhaps a, slightly, better model than the probability of particle-events in space and time model which now (2014) characterizes quantum probability) but again where is the value of a descriptive language which in non-commutative (in regard to its function-space representations for general quantum systems) math structure (?), so as to form a language which has no content, or no meaning, in regard to practical creative development.

Ouspensky gives a spiral geometry to these three new time-dimensions, ie the two-directions defined on a circle, and the line upon which, the cylinder-type circular rotational shape (of a spiral shape) manifests.

This seems to be a pre-cursor to (or the fore-telling of) the useless descriptive context of string-theory, (string-theory is useless, since string-theory (or also particle-physics) is not capable of using its laws, in order to determine (describe, or deduce) the stable spectral properties of general nuclei, but the strong-force is supposed to be about the properties of the nucleus).

Ouspensky says "separate time" manifests as a circle, perhaps he is thinking of the clock, while time associated to the mental construct of motion, ie needing to be able to remember past states of position from current states of position, while the linear-time-dimension is "a way out" of the circle (this is more or less poetry) for mental release from the monotonous periodic time (by itself) structure, so the mind can move on to a (new) mental construct related to the other different from time by itself (again this is more like poetry).

In the new construct the "line of direction" (which Ouspensky associates the possibilities of [local] instantaneous motions) is the locally invertible dynamic construct.

But Ouspensky stays with the notion of materialism, since he states that beyond "the direction of possibilities" which his linear time-dimension represents (of material motions from the instantaneous point of present velocity), there can be nothing (p 378, in his book). {but this assumes the idea of materialism, and the idea of continuity which exists within the limited construct of materialism, ie he remains loyal to the currently published authorities}

Ouspensky wants both material and the mental to be represented in physical descriptions, this fore-telling the "quantum-slit" experiment associated to trying to distinguish between waves or particles, where "the identity of 'what slit,' (of the diffraction grating) which the light passes through, is made" vs. "not making this identification (for the low intensity electromagnetic waves)" in regard to determining diffraction-interference lines for light as only a small light energy (or a few light particles) are incident to the diffraction-grating screen,

Where by doing this experiment it is found that:

"being aware of the slit" which "the particle" went through (thus, insisting on the particle-interpretation) causes the refraction-interference- pattern to be lost whereas without this identification (ie a particle-interpretation of light is not made) the diffraction pattern of light-waves is realized in the experiment.

Thus, the conclusion could be, that "being aware of certain types of information (or being aware of certain types of interpretations) influences the results of experiments."

Or

The conclusion could also be, that the interference at the diffraction-grating-site causes the wave to collapse, so as to then (from the point of wave-collapse) necessitate a particle affect at the distant screen.

Note: A global wave-function can, instantaneously, "collapse to a point," as an expression of the (observed to exist) non-local properties of existence.

But if there is a mental affect then the mechanism of this "mental influence" has not been determined,

But

If there is an influence, it is about a higher-dimensional stable pattern (stable shape), which is capable of affecting such a physical set-up (or causing a physical influence), ie stable shapes with spectral inter-relationships which can allow an apparent hidden control (or influence), so that the higher-dimensional shape will also be of a large (engulfing) size.

Other examples of published authorities who seem to only follow the intellectual trends of the other "allowed to be published" authorities, allowed by the ruling investor-class, where the investor-class feels the need to control thought and language of all society, where this is in opposition to US law, but: spies, agents, spy-managers, and institutional managers, as well as the militias which compose the justice-system are required to terrorizes the public so that the ruling-class can realize thought control over society.

The objective sense experience of an existence, expressed by B Ehrenreich, which has no explanation within the assumption of materialism, reductionism, and indefinable randomness

Though Ouspensky seeks to incorporate the mind into physical description, there is also the more direct experience of the 'spirit." Ouspensky tries to talk about this but there is no such overwhelming experience about which he focuses, but B Ehrenreich (in her book, "Living with a Wild God," 2014) expresses that she did (does) have such a sense-experience, which she interprets as having a mental-relation to some other living entity, but seems to have no way to understand her experience.

Since I have not read her book, what I say about this book is derived from her "book advertising" talks she has given on the media.

Ehrenreich reported (in her new book, "Living With a Wild God") that she had an unusual sense-experience, wherein she experienced (through her senses) the world in a different context than usual, but it is a context (of sense awareness) which our culture identifies (or categorizes) either as the experience of a crazy-person, or the experience of a mystic, or it is a religious, or spiritual experience.

Ehrenreich reports that she had a "non-usual sense-data experience" for a fairly prolonged period of time, which she allowed herself to experience. Furthermore, she accepted the sense-data, which she collected with her senses (apparently, mainly through her eyes (?)), and she deliberated as to "what it was" ie wondering about "that which she had experienced through her senses," and she conjectured about the nature of the experience, claiming it to be about an encounter with another living entity.

In doing this, she is being both a true scientist, and a person with true spiritual interests.

She is accepting her experience over and above the social forces which so greatly influence beliefs and expressions within society, so as to over-come the fears the desires, the authoritative proclamations (ie expressions of high-social-value), and the "group-think" of our heavily laden-ed, propagandistic, and narrow set of allowed set of beliefs

What are (or what might be) explanations of the unusual sense-perceptions which were reported by B Ehrenreich in her book "Living with a Wild God") (2014)?

One way in which to understand such a valid sense-experience depends on an alternative descriptive context

Or an

Alternative descriptive model of existence, a high-dimensional (where all the dimensional levels can be both large and small) model of existence, and which is in a measurable context, where the (math-pattern) models of space and material are stable (note: particle-physics and string-theory do not have these properties) and thus, it is a new model which is better related to extending practical development, and it is relatable to B Ehrenreich's sense-experience and her bold interpretation of that experience (as given in her new book, "Living with a Wild God," 2014)

That is,

Ehrenreich's experience allows one to express an important idea about:

a new math construct based upon a many-dimensional existence where each subspace of each dimensional level is related to a stable shape, is a construct which identifies the (true) spiritual-context of our existence, but this context is also within a scientific descriptive language, the material-world is a proper subset of this new math model, wherein precise (mathematical-geometric, and measurable, and practically useable) descriptions of existence are made, and their relation to creativity, eg experimentation, can be further developed, so as to build practical things and/or use controllable processes in a new context.

For example particle-physics is an expression of the use of a higher-dimensional context, {where the descriptive context of this many-dimensional (particle-physics) context has a very limited purpose} but (in the descriptions of particle-physics) it is required that random events be point-particle events, which define a very small size-scale, (but) where such a small size-scale can have no practical purpose, other than its relation to random point-particle-collisions and their subsequent relation to nuclear bomb engineering, but a different (geometric) context for a higher-dimensional descriptive language associated to various size-scales for all the various dimensions, can have many useful properties, such as the descriptive properties of observed (or measured) stable patterns, as well as practically creative properties,

eg identifying the source from whence come the Van der Waals forces, identifying the cause for the solar-system to be stable,identifying the cause for the nucleus to be stable, to provide a geometric model for dark-matter, a geometric structure which can have practical implications,

And

New models for life,

For example, "what size scales can life, actually, occupy?"and

Do ancient religious myths have many stories about "giant gods," because (for the reason that) our own higher-dimensional structure, also has a relation to a large size-scale?

Ehrenreich's experience seems to suggest that this is true, she described a living entity at least as big as the earth (or at least, it engulfed Ehrenreich's visual horizon).

Or

Perhaps Ehrenreich was experiencing, not another living-entity, but rather she was experiencing the higher-dimensional structure of her "true self," which was larger than one would expect, ie a large higher-dimensional shape

It (the new descriptive math-model for existence) is a useful descriptive context, since it is geometric (not probabilistic, though "the random appearance of events composed of small scale components," can be derived from this new descriptive construct) in regard to her apparent appeal (to the public, or to the authoritarian experts) to describe the perceptive knowledge about which her experience depended, but it was an experience which seems to be outside of science, but where science is assumed to be based on materialism, though its dependence on materialism is not a necessity (ie physics does not need to be based on the idea of materialism, but its current (2014) structure adheres very closely to materialism), so as to place a new type of "perceived experience" into a rational context, which is true [where the truth of a descriptive knowledge should be attached to the idea that "knowledge needs to have a relation to its (ie knowledge's) creative use," ie knowledge and creativity identify the true context of a verifiable (or useable) expression of knowledge], to consider a set of other viewpoints about existence, different from the dogmatic materialistic viewpoints (and to write about these alternative math viewpoints from a person's (eg Ehrenreich's) social position as a published propagandist for society,(materialism, reduction, and (indefinable) randomness, are the basic ideas upon which science is currently (2014) based, ie in current peer-review science literature, whose only real purpose (within the context of peer-review) is to serve military needs, and this type of (military) science is fundamentally based on the idea of materialism), but existence can be modeled mathematically, so as to be based on constructs which are not materialism [though the material-world would need to be a proper subset of any new, more inclusive, math construct which is claimed to be the new basis for science],

ie the peer-reviewed professionals cannot describe "the observed stable properties of the most fundamental physical systems," yet these, so called, professionals ignore this fact about the failures of their authoritative dogmas.

Though all the propaganda ploys are used by the propaganda-system to form (in the public) a (or to be a basis for an individual [but, nonetheless, collective]), personal belief-in the intellectual superiority of the professional science-class, as well as a belief in the intellectual superiority of the science propagandists, and

These propagandists' adopted (or assumed to be equivalent or more accurately the) intellectual-social-class of material based scientists, which the propagandists are upholding, ie upholding the (so called) scientists, where Ehrenreich claims that she herself is connected to material based science, by the fact of her own (careful) academic training..., she has a PhD in cell-biology, but where M Twain's statement, that "he never let his schooling get in the way of his education," is quite relevant, in regard to the dichotomy, or paradox, which Ehrenreich is (apparently) presenting in her new book, "Living With a Wild God" between her perceptions and the exclusion of these perceptions by the scientific viewpoint (which Ehrenreich, nonetheless (or to her credit), is upholding, ie upholding her belief in her own perceptions)

Note: the word "schooling" (to which Twain refers) is the set of narrowly defined dogmas, which best fit into the interests of the investor-class, and the world of most interest to the ruling-class is the material-world.

If one considers the descriptive structure of the derivative and the (physical) system-defining pde's, wherein physical laws are expressed, this context "locks" the description into an assumed containment context of the system-containing coordinates, and an assumption of materialism, while for the operator-structure of quantum physics, the sets of Hermitian operators are defined in relation to their Fourier transformed dual operators, where the position (or coordinate) operators are dual to motion operator, where in quantum physics motion operators are derivatives, so again the dual-pair of derivative and position operators (or operator models of physical properties) "lock" the descriptive context into a containment set of materialism.

Spin was added-on as a needed extra term (in the pde, energy-operator, representation of the quantum system) for some particle-component-configurations.

The Dirac equation made the spin-property of a quantum system's components an intrinsic part of both the operators and the functions which they act-upon.

While particle-physics seems to extend the dimension of the operator dependency of a quantum system's wave-function, but where (in fact,) these extra dimensions represent internal particle-states of (the assumed set of) random point-particle components of a quantum system, but in this context, randomness is both non-linear, ie depends on a derivative-connection operator, and is based on a set of unstable (random) events (in regard to these events being elementary-particles, all of which are unstable particle-events) along with an inability to either formulate and subsequently calculate the spectral sets of general (or the prevalent and fundamental many-(but-few)-body, spectral) quantum systems, but basing randomness on sets of unstable events causes indefinable randomness, which results in a descriptive language without any content.

String-theory geometricizes these extra dimension (of internal particle-states of point-particle random events) in a context of these geometries being, essentially, "undetectable very small shapes" and the dimensions (in this descriptive structure) are divided into a coordinate subspace which contain continuous material-properties, as well as the random point-particle events, and small coordinate shapes for the assumed context of elementary-particle's unobservable internal-states, but in this descriptive context only the cross-sections (or collision probabilities) of elementary-particles enter the material-world coordinate system from their internal-geometry space, and, apparently, the other 6-dimensions (or so) of very small shapes, are used to model a geometric scenario for gravitational singularities (or their field-particles) which seems to be what the military-industries want from "physics models of existence," to be used in regard to detonating a gravitational singularity, but there are more fundamental stable many-(but-few)-body spectral-orbital physical systems of all size scales which go without even an interest in their descriptions, by the dogmatic and overly authoritative professional physics community, where this disregard for more fundamental issues in physics has come about because both particle-physics and general relativity, and "theoretical variations thereof" are not capable of describing the observed physical property of being stable

That is, either the descriptive context of a pde excludes anything beyond an assumed descriptive basis in materialism, of the properties of the apparent point-collisions, which are only attributed to point-particles, but this assumption, of relating apparent point-like-collisions to point-like-particles, is not a logical necessity, and point-collisions can be related to collisions which are centered about the distinguished points of the stable discrete isometry shapes, which can be shapes of any size-scale (though these arbitrary sizes depend on constant multiplicative factors which are defined between both different shapes of the same dimension, and (between) different dimensional levels) and any dimension, eg higher-dimensional stable shapes which can be defined over various size-scales.

The math structures of higher-dimensional shapes can result in independent sub-spaces of either between the same dimension (subspace) shape or between shapes of different (or adjacent) dimensional levels.

In fact, extremely high-energy, so called, cosmic-ray collisions are much better modeled as shapes possessing large size-scales.

Furthermore, the hypothesized, dark-matter, seems to only be detectable in regard to very large-scale systems.

But for life-forms, the knowledge of existence beyond the assumed containment set of materialism, and placing the model of a life-form into higher-dimensions, is about stable shapes (which also exist in higher-dimensions, and all of which have distinguished-points), but the size-scales.... of the shapes of various dimensions which are related to our own higher-dimension life-form structure, as well as the existence which such a life-form-shape will experience.... will be very large,

And... in relation to our 3-spatial-dimensional subspace (of our life-form-shape our higher-dimensional life-form structure, will be very mobile (can move quickly over great distances)

That is, in regard to (1) Ouspensky, (2) Ehrenreich, and (3) C Castaneda the new model of existence "based on many-dimensions and stable-shapes" provides a math-model of an existence

(1) they seek, or

(2) within which they seek explanations of their sense-experience, and

(3) they (should want) or need the new knowledge about in order to categorize their experience and sense-data, and in order to understand (or direct) their intent for deeper creative experiences within existence, respectively.

For those who oppose the society because it is corrupt and rotten certainly in regard to its justice-, governing-, and propaganda-education-systems,

Then consider that

It is pointless to discuss who makes what quantity of money, or how money flows, etc, what matters is who has the controlling stake.

The institutions of economics, justice, military, are all about domination and violence and they represent those who possess the controlling-stake.

Furthermore,

The US justice system has imposed a state-religion of Calvinism (on the US society) where the social condition of wage-slavery is an expression of this imposed religion, and the imposition of Calvinism as the state-religion was done for the stake-holders of society, ie the ruling investor-class.

Calvinism is the worship of the material-world [along with other arbitrary propaganda expressions] and science and math are subsets of Calvinism, ie science and math are controlled by the interests of the ruling-class (those who dominate the material aspect of society). Calvinism worships the material-world and this material world is organized around the social-structures of lying, stealing and murdering, and this is, over-all, what Calvinism stands for, despite its propaganda-system which claims otherwise.

New math models of existence, which are not based on materialism, cause the religion of Calvinism to collapse, so they are expressions which will not be allowed by the dominant culture

The new model of existence based on many-dimensions and stable shapes, is fundamentally about both science and religion (what the religious experience might be) and it is a set of ideas which dismantle Calvinism, so the best efforts of the ruling class will be involved in excluding these new ideas

But

Those intellectuals who claim to oppose the culture's corruption and rotten-ness will (most likely) be the best allies of the ruling-class to oppose the new ideas, about science and religion, since the intellectual-class adheres so faithfully to the idea of material based science, but where the new ideas have the capability of sinking the corruption and social-rot of the society's high-institutions.

Note: that the real nemesis of the bankers' material-world-view, is the [truly] religious people (ie the viewpoint (of true religion) which Ehrenreich tries hard to uphold (by giving great credence to her own sense-experiences) as a result of her objective sense-experience),

ie where the justice system (in accordance with the interests of the investor-class) have required that religion be hollowed-out and embedded in a world-view of a material-world, not a world with both alternative perceptions and alternative interpretations, within which a person can act in a controlled manner, in regard to the example given by Ehrenreich, but the social forces are all aligned against an intellectual residing at the margins, who is expressingeither an objective sense-experience which is outside of a material-world model of reality, or a math model of higher-dimensions which is not (a higher-dimensional scientific model, which is [out of professional necessity]) attached to the failed ideas of particle-physics and string-theory, since these new ideas are espousing a new basis for physical description.

Nonetheless, one can try to appeal to the side of a human that does, within their own person, try to establish knowledge.

However, the knowledge which Ehrenreich is bringing forth is the type of (religious) knowledge, which is most often dealt with directly, (by those who have learned to perceive the world as being different from the material-world, where it is only about a "material-world" in which we are taught to deeply believe), at the sensational and physical level, and acted upon, apparently leading to new ideas, within the (truly spiritually adept) person who is (apparently) experiencing the new type of sense-data, about "what the living-self is," rather than that world being carefully described in a (required) context of materialism, ie the western mind's interest in knowledge, which is a "precisely described and verifiable knowledge," but the interpretation of "verified" is always interpreted (in the western culture) to mean a truth about material patterns, which can supposed to be controlled in a lab, and verified by observing (measurable, or distinguishable) properties, but this is the context in which Ptolemy's model of the solar-system was also verified, and such limited context of verification can result in a fantasy-world being verified, as the example of Ptolemy again demonstrates, and as the particle-physics model also models and subsequently, identifies (by, supposedly, verifying measurements) a fantasy-world though now (for example) science does consider, in a disingenuous way, descriptions in higher-dimensions, eg particle-physics and derived ideas about string-theory, but nonetheless (now, 2014) always with a (required) reduction to the idea of a material world, ie an interpretation of particle-collision data which excludes large-sized shapes, ie the internal geometry of elementary-particles (the realm of string-theory description) is required to be small, so as to preserve the idea of materialism.

However*, there is a (mathematical) descriptive context, based on new ways in which to organize a precise descriptive language (new ways of organizing math patterns) which can contain the material-world as a proper subset, and extend (by means of stable geometric math patterns) into higher-dimensions, in a way which is not restricted to a small size-scale (when beyond the 4-space-time dimensions of a material world), in fact, for dimensions six and above the stable shapes (in the new descriptive context) are all unbounded. Furthermore, the new language allows a geometric model of the infinite, and it provides a finite spectral-set for both the large and small sized geometries of the higher-dimensional context, which has been proposed to model the existence which we experience, and seek to clarify and use the information gained from the clarifications.

That is, (even though) there can be math models where large sized-geometry can be hidden (from our awareness) by other math properties, which depend on (the) new ways of organizing the math patterns (to be used for the description of all-existence), eg topology and the structure of both material and material interactions, ie continuity (ie topology) within a higher-dimensional space is usually the reason large high-dimensions are excluded, but the dimensional-levels can be partitioned (by means of sets of stable-shapes which are associated to the subspaces of a higher-dimensional containment set) so as to be relatively independent of one another, so as to have a topology related to a stable shape, in fact, it seems that the stable properties of material systems can only be described in the new context,

(see above *) as the "over 100 year" failure which has been demonstrated in regard to the ideas of quantum physics and general relativity, namely, the main statement about the stable many-(but-few)-body quantum systems is that they are too complicated to describe, but the stability of these systems, actually implies these systems are linear and solvable ie controllable when they form "out in the world" (where they, actually, have formed). That is, such a statement of being unable to describe the properties of stable many-(but-few)-body physical systems is an admission of deep-failure.

B Ehrenreich has written about her "spiritual" experience, ie apparently, a sense-experience which transcended the "normal sensations," which we usually attribute to our own act of observing the material-world, ie her "spiritual" experience was a small view (or short-lived viewpoint, or sense-experience) where "the world as we think it to be," is found "to be different,"

ie because of a sense-experience (if we believe our own sense experience (or in this case if Ehrenreich believes her own) then the world is not as we think it to be, this is the tenet of Buddhism, and the basis of a Buddhist's, subsequent, quest for enlightenment)

Ehrenreich claims that: [where, apparently, it is assumed that the world is material, and thus,]

The experience of spirit is some undefined "other," which may be some entity.

The intellectual and creative wasteland created by the propaganda-education system

{ie never let one's schooling get in the way of one's education}and the principle actors within this system, ie the published intellectuals, are operatives within the knowledge context which supports the moneyed interests of society.

These intellectual propaganda operatives, operate in a similar fashion (as do) [or in conjunction with] the spies and the spy agencies, which are monitoring (through their spying) the expression of ideas all over society,

So as to find word-strings which identify the expressions of certain types of thoughts, as well as finding people upon whom they can prey-upon so as to manipulate, so as to mostly identify a social mechanism of control, where this controlling mechanism is most often realized by either confusing the population or by terrorizing the population

But such a system, of preying on people in an unsuspected manner through spying, and deceiving them, is made easier due to the fact that the propaganda-education system, which the spies are monitoring causes the thoughts and language of the public to be so very narrow, that techniques of repetition (in the propaganda-education system) can be very effective.

The unfortunate truth is that the propagandists, such as Ehrenreich, are those personality-types which the propaganda-spy-management (militarized) system has preyed-upon, and used (though the social-state of being a wage-slave results in the act by the ruling-class of using indoctrinated people for their own selfish interests (purposes) to be cloaked within society, as the indoctrinated person getting a high-salary, but this is the worst form of preying upon people, in such a hidden manner)

The commandments of the ruling-class;

Thou shalt be:

A materialist

Support authoritative science which is based on the idea of materialism, and it is well related to military interests

The civilization is to be considered technically superior to all other cultures (ie this is a form of racism, or disguised racism)

An anti-communist (whatever that means, but, it means opposing collective, or democratic, or social, efforts which improve a society, it means one must support the collective society defined by "capitalism" which is based on a particular value structure defined by social-class, and defined by the ruling-class)

A self righteous moralist (who really embraces the narrow ideas which oppose life, ie an anti-moralist, such as an imperialist, or supporting colonization, etc etc)

A believer in the intellectual hierarchy, which is, really, "all about" the set of narrow dogmas which best support the interests of the investor-class, in regard to the relation of knowledge to creative efforts of society,

A social-collectivist, who self-righteously proclaims their own (dis-in-genuine) support for individual-freedom, but such freedom is to only be allowed for, or given to... (or exist for) (or defended for, by militias, and by the, so called, libertarians)... the very rich

A believer in the police-state, a system of spying and managing, by an over-whelming number of militarily trained managers, and the resulting social control of all language and thought in society a deep religious belief-in materialism by each individual, but hidden from each individual.

ie A very narrowly focused and obsessive society [unfortunately, this system leads to a destructive narrowness, and intellectual failure, but it regiments the society]

Unfortunately,

This is the set of intellectual values to which Ehrenreich adheres, where this must be true since she is actually published in the, so called, main-stream media, so she is (must be) serving the interests of the investor-class, ie there is no such thing as a free-market "all (so called) market-related things are manipulated by the propaganda-education system to serve the (better) fortunes of the very-rich," and in her book she carefully adheres to the materialist, and its subsequent relation to particle-physics, viewpoint (searching for some acceptable material-based explanations) [perhaps, not realizing that the narrow dogmas of particle-physics and its associated string-theory are not about anything except making a doomsday bomb, which if detonated can apparently blow the solar-system apart, ie using particle-physics properties to try to detonate a gravitational-singularity explosion], ie they are not ideas which deal with the real question of physics, where "a real physics question" would be: "why is the general nuclei stable?"

To re-iterate

Ehrenreich claims that:

She had a sensation (a sense-experience) of a "burning-other set of sensations infused into the background of her regular sensations accompanied by a sense of heightened emotional sensations."

A psychologist wants to reduce this experience to a relationship to "being normal" in regard to psychological categories, eg a fit of epilepsy, in turn, related to chemical activity in the nervous system which is uncontrolled.

This is about a materialistic model in which we are material structures, within which there are other structures, where these other material-structures, in turn, are affected by sense-data, to which a "mental-picture of the world" emerges, within these material structures, ie the deep belief in materialism but which has no model of how the "mental-picture of the world" comes into being, ie it is a logically circularly viewpoint which only upholds the idea of materialism, but it is a picture (or descriptive language) which has no content of its own, rather only "a promise," that the material based viewpoint will provide all answers. It has the same language structure as does organized religion, in which there is an appeal to an obvious truth, namely, that science cannot (actually) describe the nature of the world, [rather the reality within our society is that "science is only about making war-instruments"] so religion describes the large-other which science misses, but like the language of psychology, the language of religion has no content which can be used in a practical manner, other than its (literary or cultural) use within the propaganda-system to manipulate the population.

If one looks at the "The New York Times Book of Physics and Astronomy" one finds the same set of promises, that the materialistic viewpoint will be capable of describing all observed patterns of our experience.

Whereas, in fact, it is a viewpoint which fails to provide any answers. Yet it does provide a context of great complication, but which seems to only reduce to empiricism, since its assumptions, or its physical laws, are not capable of compressing the empirical information into a usable form, rather there are only empirical properties which can be found, so that, if they can be connected to a classical system, then these properties can be used, eg the micro-chip (or micro-processor) where classical thermal and optical properties can be used to form P-material and N-material, or "doped-crystals," and then electric circuit-boards based on light-sensitive chemical techniques ie the small circuit shapes are made optically, ie classical physics, and subsequently these circuit-boards are used in electronic switching circuits.

The intellectual hierarchy promised clean cheap fusion energy (as long ago as the 1950's), ie fusing nuclei together to get energy from mass, but without any valid model of a stable nuclei such an idea is simply a shot in the dark (its only real model being bomb-detonation), and it promises quantum computing, but its quantum model excludes a coupling-mechanism to the quantum wave-function (ie coupling requires the quantum system's wave-collapse), that is, neither of these developments can be achieved within the context of material based physics,

This NYT's book talks at great length about the nuclear bomb and it (ie material based physics) is capable of providing this type of a military instrument (but these bomb-instruments are based on 19th century models of chemical reactions applied to rates of radioactive decay).

They build DNA, but they do not know what they are doing while building it, ie they have only the remotest form of partial-knowledge about life.

The religious person, listening to Ehrenreich, wanted to define Ehrenreich's experience so as to revolve around emotions, where, apparently, it is claimed by the religious person, that there is a certain set of emotions from which the concept of god emerges, and vise-versa, ie from the two undefined sets (emotions and god) of which it is claimed that each one (of the two sets) is to be the essence of the other set,

Whatever this might mean? ie the usual type of circular sentence structures in the material-God dichotomy, which is supposed to be the intellectual basis for the great western efforts, but the main feature of the (Roman-based) western mind is the great violence, upon which a materialistic social structure is based, and the way in which religion is used as a propaganda-mechanism, Constantine's bible is the Roman-manual for the empire's propaganda.

But these intellectual elitists (those published within the propaganda-system) refuse to consider the published set of descriptions of existence, which transcend the material-world......

The propaganda-elitists are in the category of the superior-few, and they refer to the, so called, experts in their intellectual-class, so as to ascertain the truth, then "why does Ehrenreich believe her own sense-data about her "religious" experience?" especially, if she ignores other ideas which are expressed, because of her belief in her own intellectually superior propaganda-class, That is, math models of a high-dimension existence (not string-theory nor particle-physics) based on stable shapes and of which the material-world is a proper subset.

It needs to be noted that the so called elite-intellectuals of the propaganda-education system, ic the scientists, cannot use the ideas about materialism to describe the observed properties of material systems, ie they cannot use their own physical law to describe the observed properties of material systems eg the stable spectral-orbital properties of many of the most fundamental many-(but-few)-body physical systems which exist.

Yet, the new ideas can describe these observed properties, eg why the solar-system is stable and why the nucleus is stable etc

The new set of ideas (about stable shapes in high-dimensions (and not string-theory nor particle-physics)) takes one out on the material-based language which is required to be used by an intellectual who is to be publishable, and/or equivalently, who has any social ranking.

However,

The nature of the human is to seek both knowledge (ie explanations) and its natural relation to creative efforts.

Nonetheless, she (Ehrenreich) is a professional propaganda-person, ie a successful published writer, and thus she is a person who has had her beliefs greatly influenced by the social forces of the propaganda-education system. Yet, in this book she (seems to) stands-up for her inner-being.

As a propagandist she has available a very narrow range of subjects about which she can write (subjects which must be presented in their, "correct," narrow context) and she can only say a limited amount about the subject, ie examining a subject based on equal free inquiry is not allowed, since there are authoritative descriptive contexts (about each of the small set of categories of the subjects which can be discussed) which must be honored by the propaganda-system.

But she is fairly keen on addressing these propaganda issues, where her "market" is the large set of dissenting "slightly educated" types, which are the better part of the population ("slightly educated," since they are reluctant to engage in free inquiry for themselves, where free-inquiry is the true underlying principle of the enlightenment age, eg "should we be a collective society as is a capitalist-state society" or should we be a society of equal individual-creators, encouraged to be free-inquirers, wherein for a society based on equal free-inquirers, there is also an imposed truly free-market, ie popular items (traded in the market) must become collectivized (so that the market cannot become dominated, and subsequently controlled), but nonetheless the production processes can be easily stopped when the "precautionary principle" requires changes in the use of materials (or the use of natural resources))

Furthermore, the reason populations seem to remain loyal to the western viewpoint is the imposed main-state religion of "materialism," which all of the Abrahamic religions wholeheartedly "agree to." (ie where the set of Abrahamic religions can be used to determine a good definition for the category of the western culture, another definition would be that the Abrahamic religions form the hierarchical propaganda structure of a collective society, ruled by a social hierarchy, where the collective purpose of the society is to maintain the social hierarchy, which, in turn, is defined by a, so called, (highly controlled) market-structure presided-over by the investor-class, and their, so called, "expert helpers," where these experts provide (by means of their narrowly dogmas) a set of, so called, intellectually superior technical experts, ie defining the superior intellectual aspect of the social hierarchy (which must be preserved), where these, so called, experts determine the knowledge needed for the instrumentation associated to the productive investment interests of the ruling-class: the masons, the electrical engineers, and the bomb engineers, etc)

But property-rights and minority rule, as a legal basis for social organization (which allows social domination by the few) causes a corporate-state society to adhere to a social structure of high-value whose narrowness (related to the dominant interests of the ruling-class) subsequently causes destruction (related to the [improper] use of resources and the environment), and it excludes the (actual) development (or evolution) of the market-place of trade, since its narrow set of values seek to extend dogmas which are related to the production of instruments, which are used to control markets in a very narrow manner, but perhaps if science goes beyond the notion of "there only being a material-world," then the focus on creativity and "production" might not be so dependent on material resources (and the way in which materials can affect the environment in such a destructive manner, an issue which is ignored, due to the exclusive, or hierarchical, social relation of material-production to market-profits that is:

Property = money = ruler.

Note: The Abrahamic religions seem to greatly oppose the idea of science transcending the idea of "the material world being an absolute truth," but so do the dogmatic and narrowly focused authorities of science, those few who inhabit the intellectually dominant social positions, and are blindly convinced of their dogmas, as well as their intellectual superiority, though they more-or-less occupy the stables of the rich where these rich care-takers of the intellectual-celebrities, seem to milk their herd, and take all the energy from the herd, which the rich feed in their stables.

Though it is not likely that Ehrenreich would actually relinquish her (trained) scientific viewpoint's deep-belief in the material-world as being that which defines science, since, as a wage-slave journalist-author, if she does not continue to adhere to the correct set of dogmas, in regard to particular world-views (as she expresses), she will come to be excluded from being published

(ie one does not earn (or make) money, within society, unless one is performing a service for the investor interests).

So she is treading a narrow line between society's favorite artificial dividing-point, namely, between science and religion, but this dogmatic viewpoint that "materialism is the only valid basis for science," and that religion is a valid "other," is a mis-guided belief.

Thus she is expressing her dilemma of acknowledging her sense-experience yet knowing it has no position in a material-based descriptive language, ie the language from which intellectually allowed explanations must emerge.

But this is not true...., but the alternative also involves acknowledging rational efforts to find math patterns which can super-cede the explanatory language based on materialism, a language which transcends the material-world, but which contains the material-world as a proper subset of its (new) descriptive range.

She courageously proclaims both the validity of her sense-experience, and that her hypothesis of her sense-experience being an example of an experience, of encountering another living entity, which is beyond the material-world's capacity to explain such a belief.

But

The, assumed to be true, and so called, molecular properties of any physical system deal with "the limited way in which molecular types can be identified within systems,"

For example:

one must remember that…, though it is a part of the dogma to believe that "the electron is defined by a fixed discrete value of charge" …, it is not clear that this can actually be determined experimentally where R Millikan got the Nobel prize for proving that "this is true," but his data were the result of dry-lab "experiments"

ie fraud at the highest levels of science,

Thus, the idea of there being (or that there even exists) molecular structure, in very complicated chemical systems, may be a wrong assumption, but nonetheless,

When molecular-types can be identified, then there are theories about correlating "the detection of the, so called, molecular entities" (which are components in very complicated atomic-molecular-system processes) to system processes,

ie it is a description of life's, so called, molecular basis, in turn, based on a statistical shot-in-the-dark,

ie a statistical correlation of molecular presence to a living-system's process.

This is a very weak basis for a molecular description of a living-system.

Furthermore, consider the, so called, theological relation of the so called "big bang" and its, so called, correlated relation to particle-physics, to both God and to the anthropic (anthropomorphic) principle ie that the measured masses (ie inertial properties) of elementary-particles determine if "life can exist" in the universe, or (life) not exist.

This is all a bunch of nonsense, since cosmology and particle-physics cannot explain why either "the nucleus has stable properties" or "why the solar-system is stable?" So using these, so called, principles of the so called expert intellectual-class, to explain an "assumed to exist" "beginning-point (of material-energy and space)," and, subsequently, relating this to a personal sense-experience, is quite absurd.

Furthermore, particle-physics, put into a context of quantitatively inconsistent math models, "predicts" that the proton is unstable, but a proton has never been observed to decay. So there is plenty of money put-into verifying that the proton decays, so there are many viewpoints and models for experiments about observing proton decay, just as it was desired that the electron has a fixed quantized value for its charge, so that by "dry lab-ing" so called experimental data, it was thus found ("experimentally") that such a property was discovered, so too, the "decay of the proton" will be seen, either through dry-labbed data, or

By means of using deceptive interpretational tactics of an indecipherable descriptive language, so as to proclaim the (once again) victorious descriptive structures of particle-physics.

Yet, the stable spectral properties of general nuclei cannot be determined by applying the, so called, laws of particle-physics to the general nuclei, ie the descriptions of particle-physics have no capacity to describe general systems which it is, supposedly, designed to describe, ie the laws of the strong-force (of particle-physics) are supposed to be relatable to determining the observed properties of nuclei, ie particle-physics is an indecipherable (or meaningless) descriptive language.

But one should also note, that the properties of quantum systems are first observed…., that is the properties of quantum systems are not predicted by applying the general laws of quantum physics to the quantum system's structure, and then observing these properties based on the non-existent predictions, rather the properties are observed…, and quantum physics is used, in a manner similar to the way in which the epicycles of Ptolemy were also used, to fit the theory to the observed data.

Chapter 10

The (invalid) mathematics of many-valued functions

(caused by the existence of "holes in space," or quantitative inconsistency caused by: non-linearity, non-commutative-ness, and indefinable randomness, as well as by using quantitative sets which are "too big")

It needs to be noted that the so called elite-intellectuals of the propaganda-education system, ie the scientists, cannot use the ideas about materialism to describe the observed properties of material systems, ie they cannot use their own physical law to describe the observed properties of material systems eg the stable spectral-orbital properties of many of the most fundamental many-(but-few)-body physical systems which exist.

Yet, the new ideas based on stable shape in a many-dimensional context, where shapes of higher-dimensions can be either big or small, can describe these observed properties, eg why the solar-system is stable and why the nucleus is stable etc

The new set of ideas (about stable shapes in high-dimensions (and not string-theory nor particle-physics)) takes one out on the material-based language, which is required to be used by an intellectual who is to be publishable, and/or equivalently, who has any social ranking (note: a social-ranking determined by the investment-class).

However,

The nature of the human is to seek both knowledge (ie explanations) and its natural relation to creative efforts.

Math

Math is about measurability and the patterns which can be measured

Measurability has been related to single-valued-ness when determining the energy-differences, which is a measurable property, where the issue raised is whether the domain space where one is measuring possesses holes in its structure or it does not have any holes in its shape, the point which is made is that if the space has no holes then the energy will be single-valued when it is calculated by an integral

Yet

Quantum systems possess various levels of energy, so if a quantum system's shape (where such a shape can be discussed as long as (or if) particle-physics, is based both on randomness and it is based on point-particle collisions, where a point-particle-collision is a geometric structure which a descriptive context based on randomness cannot allow) was modeled to possess holes in itself, then different stable energies associated to each of the separate different holes, would be a good model for such a system.

When one determines an integral....,

that is, the evaluation of the (linear) inverse operator to a (linear) derivative operator, or the inverse integral operator is used to solve (partial) differential equations [particularly, partial differential equations for differential equations based on solving (or finding) differential-form solutions], then one evaluates an anti-derivative of a differentiated function (which is, again, the function which was differentiated) over the boundaries of the spatial and time (based) region, within which the system is contained, or, equivalently, within which the system's set of (partial) differential equation's are defined, eg the volume integral for a differentiated-function, where the volume integral is "defined over the region where a differentiated-function is defined" is equal to the same function evaluated over the boundary of the region, by integrating the (no longer differentiated) function over the boundary-set of the given region.

This process of integration assumes both a continuous context for a description, in regard to the quantitative containment of a stable system, as well as a domain space which is free of holes.

However,

There are several causes for the existence of many-valued functions, most notably:

(1) the existence of non-commutativity in regard to the local (linear) representation of a system's measured properties [in regard to a system's containment within a coordinate "domain space of the pde's solution function," which identifies the system's measurable properties,

ie the system's quantitative model based on the physical laws which, in turn, determine the system's pde], this (or such a) non-commutative operator causes both (1) many-valued-ness and (2) discontinuous chaotic changes of the solution function, in relation to its (the function's) domain values, or a material-component's position in the domain space, ie some small changes can be (discontinuously) related to large changes in the system, at a latter time, where again the time interval may be big or small, and where the system has no stable pattern, and its defining partial differential equation can also change its character in a sudden and discontinuous manner,

There is also the related property of

(2) the (partial) differential equation (or pde) being non-linear, ie there are no valid local measuring constructs which allow the system's measured properties to be linear in relation to the coordinate domain structure, ie the system is quantitatively inconsistent between its measured properties and its system-containing coordinate system, this also causes multi-valued-ness and chaotic relations (or chaotic behaviors, or patterns), and the system has no stable pattern, and its defining partial differential equation can also change its character in a sudden and discontinuous manner,

and

(3) in regard to the spectral-and/or-probability context, where either the events are not stable, or the spectral values cannot be determined from the operator and function-space methods in regard to representing harmonic-functions, or wave-functions, as models of spectral or random systems, ie the system is indefinably random (or spectrally undefined), and apparently the system (or material-component) has no stable form, which is deducible from the math methods (being used), and which can be associated to the system's measurable properties, ie the description is without any valid content, or it is an incomplete description, or the description (or the discussion about the system) has no meaning, ie the spectral context of the system is not sufficiently defined to be able to deduce the system's spectral set (where often times this is the stable spectral-set observed for many fundamental stable many-(but-few)-body physical systems)

Consider the H-atom, it is assumed classically that the H-atom is a bounded and neutral system,

Yet, in quantum physics, the wave-functions associated to the H-atom's spectral properties are global (or unbounded) functions and the H-atom's angular-momentum states have shapes, which, in turn, imply a multi-pole charge distribution, ie not neutral.

Furthermore, the relation of such a wave-function model of the H-atom (or to any physically bounded system) to other physical systems can not be modeled, other than in the context in which the H-atom collides with some other (bounded) physical system, or when the quantum-system's wave-function spontaneously (and for no [apparent] reason) collapses to one of the system's spectral states

Where the system, to which the H-atom collides, is also represented as a wave-function, and the time and spatial-position when and where such a collision might occur is arbitrary and random, but when such systems collide, then these pair of colliding physical wave-function systems will collapse, since the physical state of these two system's position in space and time will be identified, ie a quantum-state is identified, so the wave-function of each of the pair of quantum systems collapses and this collision type interaction seems to be the only model of material-interaction, which can be modeled by systems which are represented as global (and unbounded) wave-functions (as well as the "self-interaction" associated to the spontaneous wave-function decay of the system's (assumed to exist) harmonic-function) and as such (ie the closed-global wave-functions) these function-representations of the systems are closed to other types of operator relationships, which are different from the system-interaction models of particle-collision interactions, but this means that the state of a quantum system's wave-function is always susceptible to discrete and discontinuous wave-function collapse, where such events, such as the event of wave-function-collapse, is only relatable to indefinable randomness, (either due to spontaneous decay or due to random collisions) ie non-local relations of a wave-function are not well defined and models of interaction which are needed in order to use a quantum system's superposition of wave-function properties do not exist, (except, maybe, for a relatively controlled collision-beam which is to be aimed at the (say) H-atom, while the H-atom is otherwise in a vacuum).

That is, the validity of a wave-function model, ie collapsed or not-collapsed, can never be known when the system's description is based on global wave-functions.

That is, neither non-locality not quantum-computing can be provided with valid models (as usable quantum properties) since either the property of coupling (needed for quantum-computing) or the property of wave-collapse (which must be avoided if a quantum-system's non-local properties are going to be used) do not have valid models, ie at best they are indefinably random constructs, ie within a system which needs to use the quantum properties of a quantum-system (in order for the system to function, so as to be consistent with the system's design) one cannot know about a quantum-system's state in regard to determining either the collapsed-state or the non-collapsed-state of a quantum wave-system is not a controllable property of a quantum-system.

and finally there is the issue of

(4) the domain space not having any holes in its shape (if a quantitative model based on calculus is to have any validity).

That is, it is only in a simply-connected domain space, or system-containing coordinate system, that a line integral can be single-valued for any (arbitrary) curve in the domain space connecting two points in the domain space.

Yet, on the other hand, on the very stable discrete-hyperbolic-shapes it is exactly the holes in the shape which are associated to the stable spectral-geodesic properties of the (stable) shape, as well as to a finite set of stable smallest measures of finite-spectral-values for the stable shape, ie a finite set of values for each different hole in the stable shape.

These two opposing properties identify a dichotomy, which distinguishes current ideas about the laws of physics, from the new alternative ideas about the new type of containment-set in which physical description must be contained, where the currently accepted laws of physics, in regard to quantum systems, it is assumed that the energy of a wave-function is to be determined in a metric-space, which has no holes, while the new ideas require that the energy-structure of a quantum system be related to a metric-space shape which possesses holes.

Note that,

Other than the rectangular simplex and the cylinder, it is only in shapes (eg circle-space shapes) with several (one or more) holes that the natural (local linear properties of the) coordinate systems of the stable shapes can be (made) continuously commutative, where the coordinates on these stable shapes are based-on sets of pairs of orthogonal hyperbolas and (a few defined) circles, which, together, cover the shape, as a set of curved-coordinates, so as to identify a small, and discrete, set of "circular" geodesics, whose smallest lengths define the (finite) set of stable spectral properties of the stable shape.

That is, the math structure has a well defined set, or math pattern, on which to base a spectral description of such a spectral-system's properties

That is, there is a fundamental paradox (dichotomy), which is central to both

(1) the descriptions of a system's measurable properties, and

(2) if there are laws (pde) which can be used to determine these properties.

The paradox about (1) single-valued-ness and holes in space, and (2) the holes in a space's shape and the stability of the shape.

That is, this paradox is really about the distinction between

(1) stable systems of one (smaller) size-scale

and

(2) the single valued-ness of a different (larger) size-scale,

in regard to the system, and both its measurable properties, and the shape of (or topological and dimensional properties of) its containment set.

But

Current quantum physics assumes that (2) is true for quantum systems.

That is, the containment set: its dimension, its independent subsets of various dimensions, their sizes, and their shapes, and their (dimensionally-dependent) topological properties, are the key ideas needed in regard to understanding the observed stable properties of the very-many, general but very similar, ie many-(but-few)-body systems, which are very-stable material systems, eg nuclei, general atoms, molecules, molecular-shape, and the solar-system, where all of these many-(but-few)-body systems now (2014) have no valid description (within peer-reviewed literature) where the description is based on the assumption of materialism and the, so called, laws of physics (as believed to be in peer-reviewed literature in 2014).

However, the properties of these stable systems can be understood in a many-dimensional context of containment, where the different dimensional levels are defined by stable shapes of a given dimension, and of various sizes, and with different topological properties, where these topological properties are dependent on the viewpoint of the containment set for a description, where many shapes of many different dimensions are defined on a set of different subspaces and/or an associated set of shapes and in different dimensional levels (ie the shapes of a lower-dimension are contained in a metric-space of an adjacent higher-dimension)

There can be defined set-containment-trees based on dimension, subspace (or infinite shape), and size, in an 11-dimensional hyperbolic metric-space.

The set of 5-dimensional subspaces equals the set of six-dimensional subspaces in this metric-space and that number is equal to 462 such subspaces of dimension-6, so that the 5-dimensional stable and bounded discrete hyperbolic shapes can each fit into one of these 6-dimensional unbounded (but stable) shapes, so that in each 6-dimensional unbounded shape there is a largest size 5-dimensional shape

Whereas the spectra of a bounded hyperbolic shape is bounded away from zero, ie there will be a bound to the smallest stable n-dimensional discrete hyperbolic shape, where n is between 1 and 10 (inclusive).

Then the 4-dimensional bounded discrete hyperbolic shapes can fit into the various 5-dimensional shapes, so that in each 5-dimensional bounded shape there is a largest size 4-dimensional shape,

and so that there are "6 choose 4" ways of fitting these bounded 4-dimensional shapes into a 6-dimensional space, and so on…,

Ie there are various ways in which to "partition" the 11-dimensional hyperbolic space into shapes of different dimensions and of different sizes and different subspaces, so that the stable spectral set, which is associated to this set of stable shapes, is finite (though quite a large quantity)

There are the issues of linear partial differential equations (ie pde's), metric-invariant (pde's), and continuously commutative coordinates for the domain space [for the shape of the coordinate-domain space upon which the pde is defined], (of the pde's solution function), as well as simply connected, where these properties depend on the topological context within which the description is taking-place, is the description within a particular-dimension so as to have an open-closed topology so that the system's natural hole structure is only revealed in the next higher-dimensional containment context associated to the topology of the given dimensional shape. That is, the topology in the lower dimension is open-closed while the topology of the same structure but in the next higher adjacent dimension, where the shape has its geometric-form, the topology of the (now) lower-dimensional set (shape) is seen in the higher-dimension as being a closed set.

That is, the simple-connectedness topological property in one dimensional level, can be an apparent illusion when viewed from the (adjacent) next higher-dimensional level.

But it is only when the shape associated to a stable geometry that the geometry has holes in its structure, that there can be a finite fixed sets of stable geometric measures (or a fixed finite set of stable spectral properties) which can be associated to each hole of the stable shape

That is, for a given dimension metric-space there are the stable set of closed shapes of (adjacent) lower-dimension as well as the set of stable boundaries of the metric-space in the next higher adjacent-dimension which can be related to stable sets of spectral-orbital properties

In the metric-space itself it is the carefully built linear, stable, metric-invariant, separable (continuously commutative) shapes (or linear pde's with constant coefficients) which can be used in an instrument, so as to possess "regular" stable properties, and be useful and stable, ie the rectangular shapes and the cylinders and the circle-space shapes. But now this set of stable systems and their geometries (or circuit oscillating structure), which is often periodic in its nature, can be related to other higher-dimensional periodic stable structures, where some of the obvious sized stable shapes are the periodic structures of the planetary-orbits.

Furthermore, it is this size scale in which to believe that, we, as living entities, can be related to a higher-dimensional (and well defined) other sets of living-entities, and that our own life-system is more deeply related to (or functions at) this higher-dimensional descriptive context.

There is the stable and continuous descriptive context where it is assumed that the domain space for functions is simply-connected, ie a simply connected space means that there are no holes which can be surrounded by 1-curves (or surrounded by circles) in the space

And

Then there is the context where a stable discrete hyperbolic shape is stable and spectrally well defined because it is a shape which is not simply-connected where when the shape is continuously commutative (except maybe at one point) the many-holed shape defines a set of well defined spectral-orbital properties on a metric-space shape which is discrete in its spectral-orbital relationship to its holes.

That is, physical law is not to be solely based on pde's but rather the structure of a system's over-all high-dimensional containment set, related to shapes in interesting topological ways, needs to also be considered, so that the pde context of local linear measures of a stable system's measurable properties need to be put into a context of these systems possessing the math properties of being: linear, metric-invariant, and continuously commutative, so that the topology of the system-containing metric-space is open-closed, and, subsequently, the space (within that particular dimension) appears to be simply-connected, or they can also be put into a: linear, metric-invariant, continuously commutative context, but not a simply-connected context, where the stable shapes are related to shapes which derive from rectangular types of simplexes, ie all the shape's lower-dimensional facial-structures (or regional bounding structures) exist down to the point, ie there are zero-faces, 1-faces, 2-faces etc for all the different dimensional regions of the shapes ie this means that simply-connectedness is always a relevant topological math construct within the space, but where one is aware of the hole structure of the (stable) shape when the shape is viewed from its adjacent higher-dimensional (containing) metric-space, wherein the shape is seen to be closed and (most often) possessing holes in its shape.

That is, existence is many-dimensional, as is assumed by both particle-physics and string-theory, but the size of the geometries which these higher-dimensional spaces (namely, metric-spaces which also possess the property of being stable shapes) are to occupy do not need to be the very, very small (sizes) (as it is assumed to be true, by current (2014) physical science), but the higher-dimensions can "partition" (or divide) a higher-dimensional set into stable shapes of various sizes and dimensions, so that each such independent dimensional level can possess both topological properties and a new type of model for material, and a material-interaction structure which allow for the different dimensional levels to be relatively independent spaces when viewed within the different dimensional levels of existence. (note: the higher-dimension spaces, being discussed, are metric-space shapes which are hyperbolic metric-spaces, where a 3-dimensional hyperbolic metric-space is equivalent to a space-time metric-space)

Furthermore, there can be made a new higher-dimensional model of life, ie naturally oscillating, and subsequently, energy-generating, and relatively-stable, higher-dimensional shapes, which organize and "govern" the life-form, where the organization is based around a unified spectral-geometric construct (or shape), yet this higher-dimensional meta-structure descends to lower-dimensions so as to be related to atoms and molecules, and it (the new model of a living-system) does this through the natural spectral properties of the life-form's (newly modeled) higher-dimensional and unified, stable spectral shape,

Note: the property of radioactivity, where there is material which is relatively stable, if it is not at a critical-mass, but which gives-off energy, could well be an example of a higher-dimensional (odd-dimensional [ie 3-dimensional shape which is contained as a shape in 4-space] and with an odd-number of holes in its shape] which can maintain a relatively stable high-atomic-number nucleus) relatively stable shape, whereas life-forms might begin as stable shapes of 5-dimensional {or higher-dimensional} (so as to form hyperbolic metric-space shapes)

Note: These ideas need to be seriously considered, since there is not any valid model of a general nucleus whose relatively stable observed spectral properties can be derived from what is (humorously) called physical law, ie the observed spectral properties of a (relatively stable) general nucleus derived from the laws of particle-physics.

Chapter 11

Reviewing P D Ouspensky's 1934 book, "A New Model of the Universe," or the chapter on "A New Model of the Universe"

Ouspensky can sometimes write very clearly, but his book ultimately has been a failure.

But, how to "get done" what the published authorities seek to realize, or "more to the point" how to "get done" what an individual intends to create, eg not basing science on the idea of materialism, ….

Ouspensky can sometimes write very clearly.

But, how to "get done" what the published authorities seek to realize, or "more to the point" how to "get done" what an individual intends to create, eg not basing science on the idea of materialism,… by using different sets of: assumptions, interpretations, and new ways of organizing precise language, so as to build a new accurate, and practically useful, precise language, is what education, and rational-thought, is really all about.

P D Ouspensky in his (old) book 1934 (2nd edition) "A New Model of the Universe," in which he is trying to make "the mind" an important idea in the physical sciences, in the set of (assumed to be) valid descriptions of existence, one sees, as usual, a propaganda-based literary figure, who follows the authorities who are allowed to be published, so as to slightly permute the expressions of the authorities.

That is, if a person, such as Ouspensky, is published then the viewpoint expressed "is a viewpoint desired by the ruling-class."

But

In his chapter "A New Model of the Universe," in the above mentioned book, he begins (this chapter) by asking some good questions, which can make the discussion about the descriptions of existence, ie physical description, quite clear, especially, if one can give answers to these questions

Eg 1. Does the world have a form? 2. Is it chaos or a system? 3. Was it created or an accident?

1. "What form has the world?"

This is a question which is about both containment-set and (stable) patterns---- set-containment seems to be..., in our, apparently, spatially expansive experience..., to be about coordinate metric-space containment-sets, where a metric-space allows for the measuring of coordinates and the geometric measures of shapes which the coordinates might contain.

As well as the stable patterns, the material qualities, or consistently measurable material properties, the measurable qualities of material-systems, which are supposed to be related to the system's containment-set, eg to the system's coordinates or the system's function-space representation,

Where, when the material-system is contained by a function-space then the measurable properties of the system become sets of spectral-values, which, in turn, are related to the spectral-functions of the function-space by means of sets of commuting (Hermitian, differential) operators, and

The very fundamental idea about the dimensional structure of existence,

ie is the description to be based on the idea of materialism or not.

Once within a (metric-invariant) metric-space, where there is consistent and reliable measuring of geometric-measures, and where the metric-space has a particular dimension, then one is also related to the classical Lie groups which are associated to the (base) metric-spaces.

Furthermore, one needs to consider the "signature" of the metric-function, ie a construct which identifies the divide between the temporal subspace and (or from) the spatial subspace of the (locally represented) coordinate space.

The material properties are identified by the dimension and signature of the coordinate metric-space, while the nature of the coordinate spaces needs to be quantitatively consistent, ie general curved coordinate systems are not quantitatively consistent, being either non-commutative or non-linear or both, where the following set of shapes are the shapes whose descriptions can remain quantitatively consistent, ie the shapes of: the rectangular convex polyhedron simplexes, cylinders and relatively stable circle-space shapes (such as the torus).

To re-iterate, one considers only these shapes because they are the only shapes which can be quantitatively consistent, where, again, these are the cubes and the cylinders and the sets of circles, which could include discs (but discs are deformable to a point, so they are little different from the cubes (or the rectangular convex polyhedron simplexes)) (and/or in regard to the sets of circles, these would be the circle-space shapes which are continuously commutative).

These are the natural shapes which characterize the discrete isomerty (or metric-invariant) subgroups of the real classical orthogonal (or rotation) Lie groups, eg the tori in Euclidean space and the discrete hyperbolic shapes in hyperbolic space) and they are also the shapes which naturally can carry over to the complex, or unitary, classical Lie groups, where circles and discs characterize the shapes found in complex-coordinates or the Hermitian-invariant unitary groups for the classical groups associated to the various signature metric-functions and then introduced into Hermitian-invariant, complex-coordinates, note again, that circles are more interesting than discs.

The physical properties are related to metric-spaces, and thus the model of existence is a mixture of different metric-spaces of various different signatures and different dimensions, but there is a natural progression of dimension and signature which accompanies the different types of dimension and signature metric-spaces which depends on the types of materials and their associated characteristic physical properties, eg $R(n,0)$ is Euclidean space and is about position in space and inertia or mass, while $R(n,1)$ is generalized space-time, or equivalently, a $H(n)$ a hyperbolic metric-space, and is associated to time and charge and energy.

The, mass equals energy equation, is related to the way in which Euclidean shapes depend on the more stable discrete hyperbolic shapes.

The next new change in signature in regard to the dimension of the temporal subspace [different from $R(3,1)$] is in $R(4,2)$ since 4-hyperbolic-space can contain a 3-dimensional shape with an odd-number of holes in its shape, where such a shape naturally oscillates, so as to generate its own energy, but it is also a shape which can disintegrate (suddenly). This oscillating material-shape is associated to a new physical property, namely, it is material which generates its own energy. This can be a simple, but higher-dimensional, model for a unified living-system, ie the unified shape and its spectral properties control its internal chemistry rather than local chemistry being the basis for all the living-system's properties of both the system and control over the system, which is given to the living system's consciousness. The memory of a living-system can be modeled within one of the maximal-tori of the fiber group of the living-system's oscillating (but relatively stable) shape.

The ideas of special relativity are simply about metric-invariance and that the time coordinate in the 4-coordinates of space-time in turn related to momentum-energy 4-coordinates so that [mass

= energy] is a result of equating the time component with energy and associated mass-factor from the 4-momentum local coordinates, this energy-time equivalence was determined by E Noether's invariance theorems.

The metric-invariance properties can be used to relate measurable values between two moving frames.

But dynamics is mostly about identifying relative positions in space, ie it is a description which fits into Euclidean space, but most stable material components are built from charged components, which are neutrally-charged, [whose mass is related to both resonances (between hyperbolic and Euclidean shapes) and its inertial affects are related to the changes (relative) spatial positions of the material components (the other material components)]

So that, the finite speed of light seems to be the relation of the motion of a "wave-front of a light-wave" to the relative positions (of this wave-front's motion) in Euclidean space.

also being relations between (relative) positions.

In regard to the general theory of relativity, the non-linear quantitative (and descriptive) context of general shapes is not measurable within a "consistent quantitative structure," and there are no stable patterns which exist within such a non-linear descriptive context, so it is of no-value to the physical descriptions of useful properties

But

When the principles of the general theory of relativity are applied to the set of stable discrete hyperbolic shapes, either within the shape and/or which exist between different (usually adjacent) dimensional levels where the stable shapes identify geodesic spectral-orbital properties, whereon (the geodesics of these stable shapes) it is Newton's law, which is the better law to use in order to permute a 2-body system from its geodesic orbital properties which the metric-space's shape (within which the material components are contained) determine, so that the application of Newton's law is a good approximation to the material-component's orbit in the metric-space, and it is the types of permutations of orbits which the stable orbital shapes allow for such a good Newton approximation of perturbed orbit-changes, to be made.

As well as the general theory of relativity being the context, which identifies an effective, rigid-body, in regard to a discrete hyperbolic shape being accelerated in an adjacent higher-dimensional dynamic context, eg so that the acceleration of a 3-shape in 4-space is not felt (not detectable) inside the (rigid) 3-shape.

These few paragraphs deal with what Ouspensky takes several pages to discuss, though this discussion is quite insightful, he misses the main issue that general, non-linear contexts are not quantitatively consistent so they have no content, as a valid practically useful description, such a descriptive context degenerates into simply an empirical discussion (of unexplainable measured details).

This is about the main issue: "Are there stable forms about which existence depends?"

The answer given here (in this paper) is yes.

Furthermore, non-locality, or action-at-a-distance, has been confirmed to be a realizable property, but it does seem to be a Euclidean-space property, not a space-time property, though for unbounded discrete hyperbolic shapes, apparently related to both light and neutrinos, there are some other possibilities, in regard to instantaneous inter-connections.

Ouspensky tries to discuss the issue of defining the measurement of physical properties, but he seems to be mostly interested in quantities like mass where in "chemical reactions" the form of the mass changes but can the mass of the original system (before the chemical changes) still be measured when after these changes take place,

ie keeping the chemical changes contained in a closed system.

But

He seems to not be concerned about a much more important issue, as to whether quantities actually can stay well-defined, when function-values (or, equivalently, measurable physical properties) and coordinates are related by means of non-commutative and non-linear relations are determining the quantitative structure of the, so called, measurable description, where in such math contexts there is no quantitative consistency and quantities lose their property of being well-defined, eg measured properties become many-valued and discontinuous.

Ouspensky points out that "mathematical physics" "should be questioned," but he "questions it" based on the grounds that physical systems are not sufficiently precisely formulated, and then they most often cannot be solved, and that they are only solvable in the simplest of cases.

Perhaps the physics community should take this criticism much more to heart. The stable systems need to be simple math models where their stability indicates a true descriptive context where the math is very simple and the stable (and uniformly formed) systems do form in a causal manner.

But

This is not the direction Ouspensky takes, rather he follows the authorities, as he pretty-much does "all the way through his work," but then if he did not "do this" then he would not get published (and this is back in the 1930's)

In fact, math physics began with Newton who solved the 2-body problem, and the physics community has not been able to progress past this accomplishment, in regard to the many very stable spectral-orbital many-(but-few)-body systems, though they have described waves and statistical systems composed of many atomic components, and (for these statistical systems) whose measured values are averages over these many components.

Ouspensky points out that Einstein's work is all about math physics, but, as usual, he misses the main point, and focuses on a narrow interpretation of one of his claimed to be issues of great importance, namely, about "the form of the world" where he focuses on relativity, both special relativity and general relativity. But the main form is metric-invariance and ambiguity between hyperbolic and Euclidean geometry, but this is best resolved by looking at their sets of discrete isometry shapes where such shapes in Euclidean space are tori, while such stable shapes in hyperbolic space are strings of toral-components, ie the geometries of the two spaces have a natural relation to one another, so the various coordinate relationships which exist between the two different metric-space types, eg 4-momentum and the invariant metric-function structure of the wave-equation, are difficult to interpret, and to realize.

But special relativity is basically a statement about metric-invariance, so as to fit into a classical Lie isometry fiber group which are related to such space-time (or equivalently, hyperbolic) metric-spaces, which contain energy and charge, while the Euclidean metric-space deals with the measurable properties of spatial-position and inertia. Time relates, as an imaginary structure (which also identifies the signature of the space-time metric-function), the positions of electromagnetic wave-fronts to Euclidean space.

Ouspensky also makes the mistake of saying that general relativity is about general curved-coordinates as having an isometry fiber group, but rather general curved-coordinates have a diffeomorphism fiber group, which is invertible at a point (diagonalizable at a single point), but the general metric-function (of general relativity) is not related to continuously commutative local coordinate relations.

Furthermore, if one can deal with many different types of physical quantities, or measurable physical properties, then why must the containment of all of these properties be in one type of metric-space,

Special relativity triumphantly throws away Euclidean space, but this is an error, since

Euclidean space is the space where the physical properties of position and inertia are contained, while space-time contains other physical properties, but the [mass = energy] property depends on the incorrect formulation of space-time as a space where dynamics can be described as a 4-momentum-vector, but when this is done then E Noether's symmetries (about time and energy) can be used on the (essentially, superfluous) time coordinate (of space-time, but what should be hyperbolic 3-space) to establish the [mass = energy] formula.

Ouspensky does talk around this issue, when he identifies time as an imaginary-number so as to account for the signature of the space-time metric-function, but this is really about the wave-equation in regard to the general Laplace operator defined on space-time, where waves have positions in space, eg the wave-front of a wave's propagation in space.

2. "Is the 'world' a chaos or a system?"

The answer given, by the alternative set of ideas, is that it has many stable and semi-stable (or oscillating) forms (or stable shapes), which exist in higher-dimensions, where D Coxeter has shown that stable discrete hyperbolic shapes, in bounded and unbounded sizes, exist from hyperbolic-dimension-one to dimension-ten inclusive. Where after hyperbolic-dimension-5 all discrete hyperbolic shapes are unbounded. That is, these stable shapes will continue to be a part of any description of a world which is modeled to have higher-dimensions. That is, the "world" is many-dimensional, and it is a system.

Realizing what our own life-form might be in this many-dimensional, and (possibly) many-size-scale, context can be difficult, especially, when, as in 4-Euclidean space, there is a natural division of the SO(4) Lie group into (SO(3) x SO(3)) so that the dynamics and subsequent actions of the oscillating forms and the stable material-forms may be divided between these two spaces so it is a division which our taught to be "existence as a material entity" can oppose such a realization about our own structure.

The answer which modern physics has given to this question, is that existence is both chaos and indefinably random, so that science has come to be only about empirical determinations of fleeting, unstable patterns, which somehow "self-organize," so as to form systems, which are only contingently stable.

But self-organizing is about the adjusting of boundary conditions in a chaotic setting for a description based on non-linear relations (between local coordinate and the system's measurable properties), ie it is completely arbitrary (but not controllable except by means of trying many adjustments to the boundary conditions and seeing what happens). A deep belief in the hit-or-miss model of a material-based existence.

That is,

Science is materialism, reduction to unstable elementary-particles, and indefinable randomness

But this is clearly not true, since there are so many very stable many-(but-few)-body physical systems and this stability implies linearity solvability and control.

3. "Did the world come into being accidentally or was it created according to a plan?"

The point of the existence of life, "within the 'world'" is to create existence, itself, but to do this one needs correct knowledge, so one is constrained as to what is possible, and/or the nature of what is created, ie the "plans of creation" must be deeply consistent with the true nature of existence.

This seems to be about the relation between metric-invariant metric-spaces the stable material and their oscillating energy-generating material properties

This third question of Ouspensky is also the basis for the bogus eternal debate between science and religion, but both sides are really materialists, but materialism is a failed scientific viewpoint.

This is similar to the political parties in the US, which are both "for the rich" and their social systems of: money-dominated-markets, violence, and propaganda, and both sides are deep "believers in materialism," except in regard to the bogus debate about science and religion.

Ouspensky claims that neither 3-dimensions nor 4-dimensional space-time are sufficient to contain the observed properties of the "world," but he does not identify the most obvious problem, which now (2014) is made more clear but it was also clear then (1930's) that the stable (observed) spectral-orbital properties of the many-(but-few)-body systems are not being described and the stability of these systems implies that their descriptive structure should be simple, solvable, so as to be formed in a context of a controlled interaction.

Instead he, vaguely, references motion, which he wants to place into a mental context, ie motion's direction points from past to present in our minds

Though there is a locally invertible relation between past and present this is best represented in the context of the physical quantity-type, namely, the particular metric-space type, ie dimension and metric-function signature, and to identify locally invertible, and thus called, "opposite" metric-space states, eg the advanced and retarded potential 1-forms of electromagnetism.

Ouspensky defines "the 3-dimensions of time" as the mental properties of:

1. duration (memory),

2. velocity (an instantaneous mental construct in time), and

3. direction from past through present to future, he calls this "the direction of possibilities," and

Calling these 3-dimensions of time the boundary of our senses.

But

It is not clear how these are to be measured, and how they can be used, though in the indefinably random context of quantum physics perhaps the direction of possibilities can be modeled, as a probability wave-structure, (but perhaps a better model than the probability of particle-events in space and time model which now (2014) characterizes quantum probability) but again where is the value of a descriptive language which in non-commutative (in regard to its function-space representations for general quantum systems) math structure so as to form a language which has no content, or no meaning, in regard to practical creative development.

Ouspensky gives a spiral geometry to these three new time-dimensions, ie the two-directions defined on a circle, and the line upon which, the cylinder-type circular rotational shape (of a spiral shape) manifests.

This seems to be a pre-cursor to (or the fore-telling of) the useless descriptive context of string-theory, (string-theory is useless, since string-theory (or also particle-physics) is not capable of using its laws, in order to determine (describe, or deduce) the stable spectral properties of general nuclei, but the strong-force is supposed to be about the properties of the nucleus).

Ouspensky says "separate time" manifests as a circle, perhaps he is thinking of the clock, while time associated to the mental construct of motion, ie needing to be able to remember past states of position from current states of position, while the linear-time-dimension is "a way out" of the circle (this is more or less poetry) for mental release from the monotonous periodic time (by itself) structure, so the mind can move on to a (new) mental construct related to the other different from time by itself (again this is more like poetry).

In the new construct the "line of direction" (which Ouspensky associates the possibilities of [local] instantaneous motions) is the locally invertible dynamic construct.

But Ouspensky stays with the notion of materialism, since he states that beyond "the direction of possibilities" which his linear time-dimension represents (of material motions from the instantaneous point of present velocity), there can be nothing (p 378, in his book).

Ouspensky wants both material and the mental to be represented in physical descriptions, this fore-telling the "quantum-slit" experiment associated to trying to distinguish between waves or particles, where

"the identity of 'what slit,' (of the diffraction grating) which the light passes through, is made"

vs.

"not making this identification (for the low intensity electromagnetic waves)" in regard to determining diffraction-interference lines for light as only a small light energy (or a few light particles) are incident to the diffraction-grating screen,

Where by doing this experiment it is found that:

"being aware of the slit" which "the particle" went through (thus, insisting on the particle-interpretation) causes the refraction-interference- pattern to be lost whereas without this identification (ie a particle-interpretation of light is not made) the diffraction pattern of light-waves is realized in the experiment.

Thus, the conclusion could be, that "being aware of certain types of information (or being aware of certain types of interpretations) influences the results of experiments."

But the mechanism of this "mental influence" has not been determined,

But

If there is an influence, it is about a higher-dimensional stable pattern (stable shape), which is capable of affecting such a physical influence, ie stable shapes with spectral inter-relation ships which can allow an apparent hidden control (or influence), so that the higher-dimensional shape will also be of a large size.

Other examples of published authorities who seem to only follow the intellectual trends of the other "allowed to be published" authorities, allowed by the ruling investor-class, who feel the need to control thought and language of all society, where this is in opposition to US law, but: spies, agents, spy-managers, and institutional managers, as well as the militias which compose the justice-system are required to terrorizes the public so that the ruling-class can realize thought control over society.

Chapter 12

Stability and quantitative consistency vs. indefinable randomness and non-linearity

That the world we experience is measurable, and mostly stable, are properties which are best interpreted (or modeled to be) to mean that stability and measurability are the main attributes of existence's structure, where measuring identifies a context which implies the existence of metric-spaces (metric-invariant metric-spaces) and stability implies a context of discrete isometry (or unitary) subgroups of a metric-space's isometry (or unitary) fiber groups, in particular the set of non-positive constant curvature metric-spaces, eg metric-spaces whose metric-functions only have constant coefficients, where the discrete Euclidean metric-space shapes are continuous in the lattice structure's rules for building such a discrete Euclidean lattice, and used in continuous models of material-interactions.

Note that the continuity of material-interactions allows for free-charges to determine a continuous electromagnetic spectral set, for the various ways in which free-charges may be accelerated in the charges containing metric-space, in what is, essentially, a classical-setting for the movement of such free-charges. But stable systems are associated to stable shapes, where these stable shapes include crystals, but now crystals can be associated, in a natural manner, with an energy-lattice (the lattices of the discrete isometry (hyperbolic) subgroups), or there can exist…, within the descriptive context of a crystal…, several simultaneous energy-lattices associated which can be associated to a crystal.

On the other hand, the set of very stable discrete hyperbolic metric-space shapes, in turn, define a discretely separated set of shapes, where the geodesic structures of these hyperbolic shapes are also discretely separated, and are, essentially, related to the (first-level, or the next lower-dimension) face-structures of the fundamental domain, which is associated to the shape, and this allows the definition of a well-defined set of discretely separated spectra.

This also fits in well with the main symmetries which are associated with physical descriptions, as identified by E Noether, where these main symmetries are about spatial positions associated to inertial properties, and time and energy symmetries so as to identify the discrete hyperbolic shapes with these shapes being interpreted to be energy spaces, and these main symmetries can be interpreted to mean that specific physical properties (which can be measured) can be associated with (or are identified with specific types of metric-spaces, where, in turn, the metric-space-types are distinguished by both a metric-function's signature, and its dimension, ie the "signature" is related to the division of a metric-space into subspaces, eg R(s,t) where s is the dimension of the spatial subspace, and t is the dimension of the temporal subspace, and, s + t = n, ie R(s,t) is an n-dimensional metric-space, but where one tends to relate R(s,t) to R(s,0), when one considers a material-component's spatial position.

Where

R(n,0) is associated to the physical property of position in space, and inertia, and mass

R(n,1) is associated to the physical property of time-displacement, energy, and charge

R(n,2) is associated to the physical property of new-time-displacement, new-energy, and new-charge

Etc

note that these properties also define an opposite property eg there are both a positive-time direction and a negative-time direction, and this set of "pairs of opposite metric-space states" can be placed into a complex-coordinate space, ie which is associated to the real metric-space, where, say, positive-time is in a pure-real-space subset of the complex-coordinate system and the other, ie negative-time, is put in the pure-imaginary subset of the complex coordinates, where these pairs of opposite metric-space states, in turn, define the context for the spin-rotation of metric-space states, and placing the two locally inverse but opposite metric-space states into an associated complex-coordinate space relates the description to a unitary fiber group.

Furthermore, this context has a structure which is many-dimensional, since the discrete hyperbolic shapes exist, as well defined shapes, up to and including a 10-dimensional hyperbolic metric-space (as demonstrated by Coxeter), ie the set of energy-shapes are the main ways in which to define the context of existence, and it is a structure, which allows both measurability along with measurability's needed relation to the existence of stable well-defined quantitatively consistent patterns (or stable shapes).

Where the set of stable patterns, ie the set of discrete hyperbolic shapes, can define a finite spectral-set as a result of finitely "partitioning".... the subspaces of the various dimensional levels, where each dimensional-level has a set of subspaces of the same dimension...., within an 11-dimensional hyperbolic metric-space, where the 11-dimensional hyperbolic metric-space contains all the discrete hyperbolic shapes of the various dimensions, which includes the "partitioning set-of-stable-shapes," where the "partition" organizes the shapes into various trees-of-set-containment, where the finite "partition" is constructed by using a set of…. discrete hyperbolic (as well as associated unitary) subgroup defined…. stable shapes, and/but which are physically stable hyperbolic metric-space shapes, which represent both energy spaces and stable material-components, so that each subspace of a given dimension (and for all dimensional levels) has within itself both an upper sized and lower sized pair of stable shapes associated to a particular subspace (note that the upper-sized shape defines the metric-space associated to that particular subspace, in that particular dimensional level), so that this, along with the property of discontinuous discreteness for the set of discrete hyperbolic shapes, implies finiteness for such a "partition."

Note that the finiteness of a "partition-set"… for a "partition" of an 11-dimensional over-all containment-set's set of subspaces… is the simplest viewpoint which is consistent with any particular "partition" (ie where an 11-dimensional over-all containment-set's set is needed for the descriptions of a multi-dimensional existence, and it is consistent with the set of patterns associated with the entire set discrete hyperbolic shapes, where these properties of discrete hyperbolic shapes were described by D Coxeter).

On the other hand

The idea that a consistent quantitative structure, or equivalently, measurability and stable properties, are to be built around (or derived from) the ideas of materialism (which identifies either a 3-space and time, or a space-time metric-space containment-set for all physical systems) and partial differential equations (or pde's), where a pde is based on a model of local measuring of the physical properties in relation to the measurable properties of the containment metric-space's coordinates (or variables), ie a solution function for the system, in regard to a material system's containing coordinates, where physical laws (which relate coordinates to physical properties), supposedly, define a physical system's pde, but this viewpoint has not led to this model being able to quantitatively describe (by the measurable properties of its solution functions) the observed stable properties of existence.

Rather,

it is a descriptive context which leads to quantitative inconsistency, and to an inability to describe any stable patterns, except the 2-body system, which Newton solved, especially, when the fundamental spectra of all of physical description, in regard to a model of systems by pde (or physically motivated sets of operators, which need to be commutative, but which (in general) cannot be found) and the, assumed, reduction of material to elementary-particles., are the, so called, spectra of the, so called, internal particle-states, which, in turn, are associated (related) only to the scattering-patterns of the unstable random events of elementary-particles, which emerge, very briefly, from particle-collisions, are the, assumed to be, mass-spectra of the set of unstable elementary-particles, but it is a spectra associated with unstable particle events.

That is, it is a description based-on indefinable randomness, which is within a random descriptive context, ie the probability wave-function of a quantum system.

That is, in particle-physics, the wave-function of a quantum system is assumed to be based on unstable elementary-particle patterns of scattering (the scattering patterns of unstable elementary-particles are used so as to define particle cross-sections, which are, in turn, are associated with the breaking-apart (of relatively stable-particles during particle-collisions), where a non-measurable (or un-observable) pattern, associated with the fiber group, is associated with an internal mechanism for particle-collision scattering.

This is, supposedly, the basis for the particle-physic's model for material-interactions, but it is a model which only relates to particle-cross-sections, ie probabilities of collisions, which are occurring during a chaotic transition process, ie when systems transition (in a chaotic manner) between relatively stable states, and this is also because it provides <u>no</u> mechanism for a geometric structure within the physical interaction, other than scattering directions determined by an elementary-particle's (hidden) internal state-structure, (only the electromagnetic-part, U(1), includes any geometry, in regard to a geometric based flow of field-particles, and this is possible because U(1) is commutative)

Therefore, there cannot exist any stable quantitative patterns, which such a descriptive context is capable of describing, ie where a description would be a solution function to non-linear field-equations (which either do not exist, or they are quantitatively inconsistent), especially, associated with the observed stable spectral properties of a general nucleus.

Social commentary

One wonders how anyone can take the viewpoint of particle-physics seriously.

Clearly, this is driven by the interest of the investor-class in bomb engineering.

But in turn, it demonstrates that the class of so called superior-intellectuals are, in fact, a bunch of idiots, but it cannot simply be that they are idiots and non-critical thinkers, though apparently, they are, but nonetheless, the entire community of the intellectual-class is a highly manipulated set of certain types of personalities, ie totalitarian personality types.

They are arrogant, extremely competitive, they are obsessive, they are obedient rather than critical, and they possess authoritarian types of personalities, they seek to be domineering.

One could say that these are the totalitarian personalities best suited for a totalitarian state.

They would be called psychopaths, but psychology has little validity, other than when the psychologists make the correlations which they make in a culture which is based on brainwashing the public. That is, indefinable randomness can be related to a context which is sufficiently constrained, this is the subject of feedback systems where the main aspect of the system is the externally determined patterns which are imposed on the indefinably random context of existence.

The, so called, aristocracy-of-intellect is clearly a construct of the propaganda system and the state's spy agencies and it is not a natural property of human populations, where within an un-manipulated human population each person is an equal creator, and according to Godel's incompleteness theorem, both science and math are all about equal free-inquiry, so that the knowledge developed in such a context is to be related by the individuals to their creative intents

But despite the obvious failure of the intellectual-class, whose narrow authoritarian knowledge is mostly associated to the interests of the investor-class,

[note, where this shows that science is not authoritative, rather it is to be based on equal free-inquiry]

The public…, due to the propaganda-education-indoctrination system…, still adhere to such a failed system

The main issue within society is the issue of total violence upon which the society is built, so as to be a society where everything is arbitrary, including science and math, but which has led to failure.

This is because of the arbitrariness of the knowledge which defines the intellectual-class, where this arbitrary idea about truth is being supported by extreme violence associated to maintaining a social hierarchy.

Chapter 13

Diagonal local matrix models of physical systems and the continuously diagonal properties of stable physical systems

Diagonal local matrix models of physical systems (related to the system's defining pde's) and the need for a stable system to be continuously diagonal properties of stable physical systems (classical-systems and quantum-systems), which are (or now can be) defined as stable shapes.

Also, a feature, relating energy to the holes in a space, which clearly distinguishes between quantum physics and the new ideas about (or new ideas which characterize) the math representation (the math modeling, or the new set-containment structure) of existence. There cannot be holes in space in the quantum physics picture, while the new context depends on there being holes in space in order to define stable spectral and orbital properties but when within any particular metric-space for any dimensional level there cannot be any holes detected. In the quantum picture there is only one space, the space defined by materialism, and that one-space cannot have any holes in itself, otherwise energy is not defined.

Commutativity is an idea which is fairly easy to describe, namely, that the local coordinate relations (the coordinate structure which is assumed to contain the system) which exist in regard to the measurable properties of a system,

ie the function-values, in regard to the locally measured properties of the system, which are expressed in a (partial) differential equation (ie pde) model of the physical system, that is, a "physical law" is the basis for defining the equation, and thus the function is the solution function to the (partial) differential equation, form a (local) matrix structure (or matrix representation of these locally measurable relations between functions and coordinates) so that commutativity means that these matrix representations of the system's locally measured properties are diagonal, or can be made diagonal by a suitable choice of local coordinates.

But one wants the property of linear and diagonal matrices to be a property at each point of the system, where this can happen only if the diagonal property of the local matrix is a continuous property of the system, over the system's entire coordinate system.

This is (also) about the continuous consistency of local geometric-measures defined on the (system-containing) coordinate system, where the system's measurable properties, ie the system defining pde's solution functions, need to be quantitatively consistent, with these local geometric measures, at all points in the coordinate system.

Furthermore, since the metric-function is a symmetric matrix (or can be so represented) this means that local coordinates can always be chosen so that the metric's matrix-representation can always be made diagonal, but such a diagonal representation of a metric-function can only be guaranteed at a single point, where this is due to a particular local coordinate selection, but a "diagonalized" local matrix property for a system's locally measured values (and related to a physical law so that a pde is defined) has little-value, in regard to determining the system's measurable properties, since the system's pde's are then not solvable, (though, locally invertible [or locally diagonalizable], but only at a point, is the basis for the notion of a local fiber group structure defined by "sets of (local) diffeomorphisms," for a math description) and this little-value, in regard to determining a system's description, is related to the fact that "it is not enough to diagonalize a description at a point," but rather the diagonalization structure should be a continuous property of the natural coordinates of the system, ie the shape of a system's natural coordinate-system has a local function relation (where the function-values determine the system's measurable properties) to the local coordinate-structure, which forms a diagonal matrix, ie the local linear measurable properties (of a description of a system) need to be continuously commutative and quantitatively consistent, so that the pattern-destroying quantitative patterns of multi-valued-ness and chaos (or discontinuous quantitative changes) do not enter into the quantitative model of the stable patterns of stable systems, so as to make the math model of the system unstable.

It should also be noted that, the ideas of conservation-laws, which are so fundamental to the ideas of physics, are laws such as the conservation of energy, or the conservation of material, are laws which can often be characterized as both continuous scalar-functions and whose time-derivatives are zero.

Note that,

the local matrix representation of a physical system's measurable properties "condenses" the information of the system's coordinate (or containment) relatable properties down to a small useable set of properties: coordinates (positions in space), geometric-measures within the coordinates, velocities, [or energy and variation relations to (usually) minimal energy structures within the containment-set] changes of velocity, and force-fields, which, in turn, relate material-geometry to vector-field properties in the coordinate system (or potential-energy terms in the energy-function), where this condensation of a system's set of relevant information…, which is needed in order to use the controllable pattern of the system…, is all based on the idea of a system containment-set and the local measuring of the system's properties being related to the natural coordinate properties of the system's shape, and then identifying the measurable properties of the system which constrain the system so as to conform to some stable controllable pattern (or solvable, or linear metric-invariant, and continuously commutative pattern).

Note: When the function, which represents physical properties is scalar (representing a physically measurable property, eg energy), and its local relations to coordinates is, subsequently, defining a (partial) differential equation (in regard to the physical properties of the system), and if the (partial) differential equation is solvable then the scalar-function will factor into other scalar-functions, where each of the new factor-scalar-functions only depends on one independent variable.

This is possible because the directional physical properties, upon which the scalar function depends for its definition, are all defined in independent (or orthogonal) directions (within the containment-set), so that "that direction" can also define an independent scalar function (of that same physical property) in that one variable.

But this condensation of the information which determines a system's (stable and controllable) properties is only a valid quantitative model if the quantitative relations stay linear and "diagonal" so as to be "diagonal in a continuous manner over the shape associated to the physical system's coordinate containment-set," so that the math properties of chaos and multi-valued-ness does not destroy a stable system's quantitative model, and where these are math properties which enter into the system's description when the local math properties (of the system's quantitative model) are non-commutative.

However, the attempt to model bounded spectrally stable (quantum) systems which are stable and many-(but-few)-body systems which are most often composed of charges (or whose components are most often charges), but these, apparently, closed and bounded systems are also, most often charge-neutral, in their relationship to nearby (similar) systems, as function-spaces composed of harmonic-functions …

…where the harmonic functions are global, ie closed (or independently defined) and unbounded functions, so that the system's properties are the spectral-values of a particular spectral-set of these functions, (has not worked)

And

Furthermore, this set of spectral-functions is to be found by finding "a complete set of commuting Hermitian (often differential) operators" so that the property which this set of operators possesses (ie the math property) of being commutative (ie where these are operators which act on the function-space) allows for the idea of "diagonalizing" the function-space into the system's particular set of spectral-eigenfunctions, ie and the associated set of stable discrete measurable spectral-values, which are associated to the system which is being modeled, has not worked.

That is, the attempt to model bounded spectrally stable (quantum) systems by function-spaces of global functions and "diagonalizing" this set of functions by a set of commutative operators, which act on the function-space, has not worked.

That is, this "program" has not worked-out, because, apparently, the algebraic context of sets of operators acting on harmonic function spaces (functions with oscillating periods, and which average-out over circles to the function's value in the circle's center, ie one of the defining properties of a harmonic function) is not a "sufficiently constraining" descriptive context, "which is needed" in order to actually identify the system's spectral properties,

ie the energy-function (or energy-operator) is not sufficient, ie an energy-operator does not constrain the quantum system enough, so as to identify a system's spectral-set.

This is likely a result of the construct of energy, which requires that there exist the property that there be no-holes in the domain-space, wherein the energy-function is defined, so that this no-holes property allows the energy-function to be single-valued.

But, in fact, the stable (quantum) systems, which possess stable spectral-values, need to be defined so that there can be (or can have) many different stable discrete values (for the system), and each of these discrete stable spectral properties requires its own "hole in the stable metric-space shape" upon (or around) which the spectral-value is defined.

In turn, this is about the set of measurable properties which constrain the system into some stable pattern, and which are classical properties, and where energy is defined classically on a space which has no holes and electromagnetism is defined on star-like domain shapes which have no holes.

While the stable shapes with holes, which are each (ie each hole, or each set of holes) is associated to stable sets of spectral properties, which, in turn, are defined on convex domain shapes, which can be moded-out, so as to form circle-spaces, ie stable shapes, which define the natural places (natural shapes) where there is both periodicity..., (defined around each hole in the shape's set of holes, and each hole's spectral-value is discretely separated from the other spectral-values (defined around the other holes) by the property of there being geodesic paths around each hole, which, in turn, is contained within (or defined on) the stable shape), and where the shape is based on some fixed set of stable (minimal) geometric-measures, or fixed set of stable (minimal) lengths, in regard to the face-structure (or 1-face structure) of rectangular (convex) simplexes, which when pulled-apart (at the intersecting vertices of these rectangular-based (high-dimension) simplexes so as to) form convex polyhedral simplexes.

Thus, the natural division between a set of measurable properties which constrain a system into some stable form, and a set of stable shapes which define discrete stable sets of spectral-values for the (stable) shape, is being distinguished by domain spaces without holes vs. domain spaces which are stable shapes, but with many-holes in their shapes.

Dirac's book, "Lectures of Quantum Mechanics" 1964, is about the claim that it is only through the classical models of measurable properties that a set of "(quantum) system constraining values" can be identified, by means of sets of operators acting on a quantum system's function-space model (of system containment, ie the quantum system is contained in a function-space which is a set of unbounded, closed, independent functions), though he points-out that there are other viewpoints about the definition of "sets of operators" which are to be used to constrain the function-space of a quantum system, so as to "diagonalize the quantum system" (or to find the system's set of spectral-values, and associated set of spectral-functions), but these other viewpoint are all about the fundamental property of a quantum system being "its arbitrary randomness," {whereas the more natural consideration to make, in regard to stable sets of spectral-values, is the opposite viewpoint, ie that these stable spectral properties should be related to stable shapes}, and not to focus-on the "arbitrary randomness" of a quantum system, and its associated sets of operators, where these sets of operators have no geometric relation to spectral-values, and subsequently, it is a descriptive context which simply leads to vague empiricism, ie the descriptive information about the system cannot be condensed, the information stays in a state of a high amount of probability in regard to its properties (or in regard to its states) and it is a description which adheres to a language which has no content, ie it is a descriptive context without meaning.

Dirac's conclusion is that, for general quantum systems, the limited sets of operators, which one can use in quantum description based on the analogy of "a quantum operator with its classical measured value" are not going to be sufficient so as to "constrain the quantum system enough" so as to determine the quantum system's spectral set (and associated set of spectral-functions, and thus to find the quantum system's correct function-space representation). That is, Dirac sees quantum physics as a failure.

In regard to the idea that the stable spectral properties of quantum systems should be related to stable shapes, one wants the metric-space shapes..., upon which a stable system's holes can be defined..., to be continuously commutative constructs (or continuously commutative shapes), so that this allows for the property of stability.

This property of stable shapes which possess holes in their shapes (where the holes are surrounded by discretely separated sets of geodesics) is the nature of the discrete isometry shapes, as well as the nature of the discrete unitary shapes.

The unitary structure enters the description in a natural manner, because metric-spaces have opposite metric-space states defined on them, so that the dynamics of these shapes (or defined on these shapes, and defined in their containing metric-space) are locally invertible, in regard to each of the pair of opposite metric-space states, and this means that these opposite metric-pace states are more naturally contained in the two subsets in the complex-coordinates of: (1) the real subset and (2) the pure-imaginary subset, so that in the complex-coordinates there is a mixture of the metric-space states, but when resolved locally, it resolves into these two "effectively real" subsets, so as to be locally inverse to one another, ie the pair of opposite metric-space states.

Chapter 14

Math and physics

Math and physics has a meta-question which is important for its development, it is the simple quest about:

"What exists?"

Math and physics is (should be) about the compression of information about a described pattern

But this is not true for today's (2014) peer-reviewed math and physics, today's physics is about big-doomsday-bomb engineering.

Math and physics has a meta-question which is important for its development, it is the simple quest about:

"What exists?" but

Math and physics is (really) about, "what exists," in a context, where measuring is reliable, and the patterns observed are both stable and well defined, and the patterns used in the descriptive language are also stable, ie stable descriptive patterns which uses (or define) "law" so as to compress information, which is sufficient (and needed) to describe the observed stable patterns. Math and physics are about a verifiable "described truth," where the description is useful either as a valid descriptive context (in regard to the observed pattern) or useful in regard to building and controlling a system, which uses the compressed information of the described pattern, and where the "newly built system" is practically useful, ie the information is useful.

There are geometric and spectral (or periodic) patterns, as well as random-event quasi-patterns, or probabilities, which provide information about a fixed (finite) set of stable elementary-events, where sets of random events can be related to relatively stable patterns when:

(1) the events are related to a finite well-defined stable set of elementary-events, and when

(2) the statistical properties of the events depend on very large numbers of {statistical (or thermal) system} components

The question occurs, to what extent are spectral-patterns related to (valid) probabilistic patterns? (eg trying to find a stable, finite set of elementary-events upon which a probabilistic descriptive context depends)

The belief seems to be that spectral-patterns are very much related to random patterns, eg for "finite elementary-event spaces" the random events will repeat in a quasi-periodic manner,

But

It seems that such a relation (of probability to spectra) is only partial, and this partial relation seems to require that the spectral relations be developed based on bounded-ness of the (spectral)-system and/or the (spectral)-system possessing boundary confining conditions

The goal of math and physics is to identify contexts where measuring, or quantities, are stable (or reliable) and the patterns described are (also) stable, so that a description actually compresses information, where the descriptive context of this statement is "the context of information," where random states possess a large (or infinite) amount of information (where quasi-periodic patterns are not periodic patterns), while stable patterns contain less information, and subsequently the measurable, stable patterns are accessible to using their information for building purposes and practical uses, due to a smaller, more manageable, sets of information about which to consider {in regard to the stable patterns which organize the system's properties in relation to a smaller set (of information)}, when building something, (whose properties are described within the descriptive context which reduces the properties so as to be related to a smaller set of measurable categories), eg what type of information is most important in regard to affecting control over what is built (ie internal controls are built into the system).

The issue of physics has been to identify "laws of physics," which have traditionally been about relating locally measurable properties of functions... [defined on domain spaces (where in the domain space, the system {one is describing} is to be contained in its entirety)].... to "global" (or regional) system properties of material geometries, to which a system-component, ie the object of focus for a description, is related by causal force-fields associated to a material-geometry.

The force-fields tend to be about relations between geometric measures of lower dimensions (at the position of a system's material-component)..., than the dimension of the distant geometric properties of material, within the spatial-region within which the distant material-geometry is defined..., and the higher-dimensional regions where the material geometries are defined (or contained), where the relations are resolved at the boundaries of the material's region of definition, and/or a force-field's relation to 1-dimensional motions, ie force-fields related to material-component motions.

But

The math structure naturally ascends in dimension (when applying a local measure to a low-dimensional function), in regard to admitting a new local measuring of geometry, where the lower-dimension constructs are potential-energy functions, whose derivative is the field, but where the field has a relation to the material-geometry of the next higher-dimension (the dimension of the region which the material geometry defines), and thus also a relation to the dimension of the boundaries of the region, wherein the material-geometry is defined, ie fields are integrated over the region's boundaries

In thermal physics, there is this same lower-dimension structure, where again it is the energy function which is acted-on by first dimensional local geometric measure, but the geometric measure is related not to the energy-function but rather to a measure of a volume within which the thermal system is (assumed to be) contained.

However, the concerns of people within society are about sets of arbitrary values, within which they live, and about which they must make decisions about "what will have value for them."

Thus, the condition of life within society is about being within a context of arbitrary randomness (sets of arbitrarily imposed social-values) and being forced to make decisions in such a random context (random in regard to the imposed high-valued-set's relation to one's individual values)

This context is manipulated by propaganda and education, so that there is conveyed the idea, that our culture is marching on a (or an objectively determined) path to truth. This is a "big lie," because of the nature of descriptive truth. Namely, that there are great limitations as to the types of patterns which a fixed, and precisely defined, language can describe. That is, precise descriptive language need to always be re-organized and placed into new contexts depending on the nature of the precise patterns one is trying to describe and use.the condition of "being human" is about gaining knowledge and using that knowledge in creative ways, to build practical things.

Thus, the human in society is about the pull between either seeking truth for independent creative purposes, or determining how to adapt to the arbitrary value-structure of society, in order to survive within society.

The US society has a state imposed religion, which "in a word" is Calvinism, but which is, really, a worship of material, money, and death, where death is associated to the set of arbitrary social conditions within which people are forced to live in their relation, to social institutions, where these social institutions are always organized around a social hierarchy, and which is controlled by a few.

All knowledge and all creative efforts are controlled by those few at the top of the social-hierarchy of society, ie the investor-class, and the money structure is an expression of social dominance (related to the worship of money) which is also a worship by the public of those at the top of the (arbitrary) social hierarchy (ie those with enough wealth [money] to be in the investor-class). The money structure is improperly called a free-market economic structure, since instead it is the structure of social hierarchy which the money-system expresses (or identifies)

The arbitrary nature of social-value is dependent on great violence (and propaganda) which is needed to maintain an arbitrary social-value within society, so the population is required to: both build the social institutions which support the few and to also be a part of, or they must support, the military, which, in turn, also supports the few, where the military is used to ensure that the arbitrary social structure is, and can be, maintained by great violence

This has come to be [is all] about making, in the social institutions:

1. the military instruments, (which keep the public in its place) and

2. Bombs (often referenced in the misleading terms of "clean" nuclear energy), and

3. creating communication channels through which all communication can be sent (done), so that

4. all language and thought is controlled, or corralled,

so as to only be about the interests of the investor-class, where this thought control is done by propaganda and education, and through "militaristic" management, and spying, so as to be able to interfere with any "new expressions of ideas," and to prey upon those susceptible to the manipulations associated to a violent control of all aspects of people's expressions and activities.

It is a capitalist society where everyone (but a few) are excluded from capitalistic endeavors (or excluded from the, so called, free-market) by constrictions on the possession of:

1. "meaningful wealth" or

2. "meaningful new knowledge" eg peer-review dogmas of knowledge, or

by using

3. interfering rules, and

4. By spying

One can join the game only along established, so called, "market" contexts.

Classical vs. quantum

Classical is useful, while quantum is only empirical, apparently the descriptive context of quantum is wrong

One can characterize classical as a descriptive context which is internally focused on local measuring structure which is related to distant material geometric causes

One can characterize quantum physics as a global model of what is, nonetheless, a bounded and stable system, so that sets of Hermitian operators modeling "measurable spectral values" (do) identify invariant system properties, ie the system's identifying discrete spectral properties, but the assumed physical contexts are not commutative, and for the very fundamental and stable many-(but-few)-body systems, the system either cannot be solved, or most usually, ie almost always, the correct operator formulation of the system cannot be found.

This is a fundamental flaw of quantum physics, as was identified by P Dirac, in his book, "Lectures on Quantum Mechanics," 1964, which is about developing a description of a quantum system based on "operators which model physically measurable properties." His conclusion is that, though there may be other ways to formulate quantum descriptions (but these are likely even less satisfactory), there are too many problems when one tries to model quantum systems with "operators which model physically measurable properties."

It should be clear that the obstacles of quantum description are in-surmountable and that new ideas need to be expressed about this effort (eg of quantum descriptions) to go beyond classical description.

The stable many-(but-few)-body quantum systems cannot be formulated, consider the problems of:

when it is assumed these type of many-body systems (also) have spherical symmetric geometrically interacting fields, it also means that the potential-energy term (of the system's energy-operator) possesses a singularity,

Thus, for point-particle models of the system's components there are quantitatively inconsistent structures, due to an inability to find operators, as well as the non-commutativity of the quantum system's set of operators, and along with divergences, due to potential-energy function singularities, and where a general quantum system's, so called, true operator structure is not identifiable, or (if it is, in the two or three such cases) the system's spectral equations cannot be solved (to sufficient precision), so the system's spectra (for the physically measurable properties, which are supposed to determine the system's stable spectral properties) cannot be determined (or cannot be found, or solved)

Apparently there are only three, or so, quantum systems which can be both formulated and solved

1. the potential-wall,

2. the rectangular box, and

3. the harmonic oscillator,

though the exact physical context of a quantum harmonic-oscillator is not described in any type of clear language,

4. It is always claimed that the H-atom has been solved, but this is not true, rather it was adjusted to fit data, since the solution to its radial equation diverges.

The stable many-(but-few)-body system is fundamental; as it is related to: nuclei, general atoms, and molecules, and it is also the basic form of the solar-system

Unfortunately,

These fundamental stable systems have no valid description derived from some form of physical law, ie neither their stable observed properties can be derived to sufficient precision, from some form of physical law, nor can they be placed into the physical context by which their highly controlled interaction structure can be identified, it is controlled since the systems with that same numbers of components all have the same spectral properties, ie they have interacted in the same controlled manner each time they interact (so as to be formed).

That is,

it is (clearly) a highly controlled interaction structure, since interactions, in regard to these types of systems with the same number of components, always result in the same type of system, which possesses the same stable spectral properties, thus their interactions must be highly controlled.

But

What is claimed is that "these types of stable systems are too complicated to describe, using current physical law."

That is, the current formulation of physical law, which is based on indefinable randomness, and non-commutative math, that is, very few, if any (and thus also a general) quantum system's complete sets of commuting Hermitian operators cannot be found for these fundamental stable quantum systems, and thus the descriptive context is based on inconsistent quantitative contexts, and it is a descriptive context which is not capable of compressing information, in regard to large sets of fundamental systems, whose properties are stable.

{this is a major failure in regard to the fundamental ideas concerning the descriptive context of quantum physics}

Particle-physics is supposed to be about adjusting the wave-function of a quantum system, by perturbing it, due to the further particle-physic's model of quantum-interactions, which are modeled within particle-physics, so** as (**that they are supposed) to adjust any given quantum system's wave-function's spectral values based-on an internal particle-state structure of elementary-particles which manifests during collisions where such elementary-particle-collisions define quantum interactions

Thus,

There are no practical reasons to pursue particle-physics, since there do not exist any valid wave-functions which accurately (and to sufficient precision) identify the observed states of the wide range of some of the most fundamental physical systems, eg the fundamental and stable many-(but-few)-body systems, so as to perturb their (non-existent) wave-functions to, supposedly, more precisely identify their spectral-states,

And

Particle-physics is about basing a description of material-interactions on an adjustment to the spectral-values of a quantum description (or of a quantum-system's wave-function) [where these new adjustments are based] on the idea that quantum material-interactions are to be based on random particle-collisions (of a quantum system's (assumed set of) components) with a set of unstable elementary-particles, where the particle-components and the other elementary-particles are (assumed to be) a part of the same particle-family, and that there are changes of particle-states (ie changes between elementary-particle types) during these collisions (which supposedly, define quantum interactions), where it is assumed that these types of particle-collisions exist between the system's components and field-particles and the vacuum state.

Where the vacuum state is defined as a zero-function for a quantum system's function space, but this is not a valid definition, since the vacuum state does not have properties of a function, and since it (the vacuum) is really a "second" model of the domain coordinate-space, where it is a coordinate domain space upon which the functions of a function-space are supposed to be defined, and within which the quantum system is supposed to be contained, so this vacuum-state is an adjustment to the idea of the set-containment of a quantum system, in regard to its descriptive construct.

The commercial context which is really driving this type of failed descriptive construct...

{where the failure is about both

(1) logically inconsistent since there is no longer a valid idea of the system being in a containment-set,

[and it is both a geometric construct, ie the particle-collision geometry, and a description based on randomness associated to random particle-events in space {but where the particle-events are supposed to be identifiable with a specific spectral value of the particle-component in the quantum system} and geometry and randomness are incompatible due to the uncertainty principle which is attached to any description based on randomness, ie the space of functions and its Fourier-transformed set of dual functions together determine the property of the uncertainty principle]

And

(2) quantitatively inconsistent descriptive construct, since the wave-functions of general quantum-systems are not identifiable (findable) and the particle-physic's adjustments (to these quantum system's spectral-values) define diverging (series) adjustments (thus, leading to further descriptive fictions of re-normalization)}, is that particle-collisions are an intrinsic aspect of bomb-engineering

The claim being made (by the particle-physicists, so as to justify their focus on particle-physics) is that since one can quasi-fit the data of particle-collisions, obtained from particle-accelerators, with a

U(1) x SU(2) x SU(3) (fiber) group where this data-fit seems to have no logical origin, ie it is fully open to any type of interpretation, {such as the interpretation about a many-dimensional containment space, whose subspace structure is "partitioned" into stable shapes, which, in turn, identifies and finite spectral set for all existing stable systems in such a high-dimensional model of existence.}

Subsequently (it can be seen to follow from the particular interpretation which the particle-physicists give to particle-collision data), in particle-physics, due to the given fiber group, one can also define a non-linear (derivative) connection associated to internal particle-states (as well as the family (or energy-range) structure of these particle-states) which defines the types of particle-state changes during particle-collisions, so that these particle-states can be attached to the quantum system's wave-function, so as to attach an internal particle-state vector (construct) onto the wave-function, in a manner in which (or so that) the fiber group defines a series, which depends on the connection-matrix acting on the vector particle-state construct, which, in turn, defines adjustments to the wave-function and its spectral-values (but now associated to particle-state vector-components), where these adjustments to a quantum system's spectral values, it is claimed, are associated to the different types of particle-state changes which occur during the particle-collision model of material interactions in the quantum systems.

This imposes (or forms) a new context for measuring, ie a new internal-particle-state spectral context which is arbitrarily imposed on the (claimed to be) vacuum-state origin of the group patterns, where these fiber groups are interpreted to fit particle-collision data (obtained from particle-accelerators).

The so called, vacuum state is an undefined context, it is a mixture of particle-states and space-time relations, ie the non-linear derivative-connection... from which (supposedly) emerge the particle-state corrections (defined as a diverging series) to wave-functions... emerges from an undefined context (or emerges from an undefined set, which is (really) a non-set). {where wave-functions are probability distributions (of the quantum-system's particle-components) defined over space, and which are functions so that each spectral-function is associated to a definite spectral-value defined on space-time (or defined on space and time)}

The entire set of assumptions, which are made by the physics community are interpretations of data, which better fit into the bomb-engineering scenario, than into understanding the properties of stable quantum systems, where these are small, local particles of high-energy (ie small size)

of mysterious origin (ie emerging from the, so called, untapped energy of the vacuum) but the divergence of the perturbation series implies that they are not local high-energy geometric structures, ie point-particle collision geometries, to which the SU(3) fiber group is associated, ie the interpretation upon which is based string theory and its associated small curled-up geometries, {which are derived from both the mass-spectra of the unstable, so called, particle-collision events, and the connection-derivative, which supposedly organizes the particle-collision data}

But instead

The fiber group could be interpreted to be about collisions which exist between the distinguished points of large stable shapes, which fit into a different context about the nature of the stable quantum systems, different from the viewpoint of there being a vacuum-state, ie one does not need to hypothesize a vacuum-state in the new interpretive context.

That is, the fact that particle-physics is about completely ignoring the main issues of physics.

Namely, concerning the structure of the observed stable spectral-orbital systems which exist at all size scales,

This seems to indicate that "the rush" to an interpretation of particle-collisions (of small shapes), which are local, and thus relevant to bomb-engineering, even though the vacuum-state is an undefined, non-quantitative context upon which the so called experts are trying to build math constructs

This is all about data fitting based-on the, so called, expert interpretations of this data, but as R Millikan has shown, this type of data interpretation (ie interpreting data to please investment [or other powerful] interests) is all suspect, and open to fraud, in regard to expressing "dry-lab results" as real experimental results, where the dry-lab results fit into the interests of the investor-class, so as to continue the lines-of-thought which suit the investment interests of the investor-class.

Again, one must note that fixed languages based on formalized sets of assumptions, and interpretations about (or which allow) using vaguely quantitative contexts, in logically inconsistent manner, most often leads to the descriptions of fleeting unstable patterns, which are more likely to be a part of a delusional world, rather than being a description within a real and practically use-able descriptive context

The new interpretation can take a quite unexpected turn, eg

The human life-form can act in a way which transcends the material-world without needing technology.

One needs to examine SU(3) and SU(4) etc and the stable shapes, both microscopic and macroscopic sizes, which are associated to these higher-dimensional (ie spatial-dimension-4) contexts

It is this higher-dimensional, and stable-shaped, type of a descriptive context to which the human efforts should be channeled, rather than trying to make the doomsday bomb which can detonate a gravitational singularity and, subsequently, blow the solar-system apart.

The vacuum state makes no sense, particle-physics has no valid logical basis, quantum theory does no work, general relativity is about one pattern, which cannot exist in the real-world, ie the one-body system with perfect spherical symmetry,

This is not the highest cultural capacity to which mankind can achieve

Particle-physics is a global-local set of unstable, fleeting, quasi-patterns of point-particles, and their, so called, particle-states, defined on a non-mathematical and non-measurable context (the vacuum), so as to be the basis for a model of material-interactions for quantum systems, so that the descriptive context of the interaction only vaguely permutes (or adjusts) the set of spectral-values, supposedly, defined by sets of operators, in turn, defined on a quantum system's function-space but this mythical set of operators, in fact, cannot be found for virtually any observed stable quantum system but

The, so called, adjustments, identified by particle-physics, diverge (anyway, ie they are useless) unless the concept of measuring is again re-formulated so as to fit the data, ie re-normalization, and such a model..., which models interactions between nuclear components..., cannot describe the observed stable spectral properties of any nuclei whose atomic number is more than 2 (or of 2 or more).

Yet this is the descriptive context which identifies peer-reviewed physics and math.

The conclusion is that, this is not a descriptive structure whose purpose is to identify and describe the properties of stable physical systems, ie it is not about physics, nor is it about a valid descriptive context for math

One needs quantitative consistency, and measuring reliability and stable patterns to be able to extend the descriptions of existence into a realm wherein the observed stable patterns can be described, and subsequently, used in a practical context.

Consider:

The movie Platoon, it identifies the two main cultural subcultures in the US the violent thugs (most often people-of-action, who are given a clear choice: join the team which support society's winners ie those who own society, or support one's conscience but in doing this one is choosing to be with society's losers, ie opposed by society's institutions) and the humanitarian types, the normal (intellectually oriented) humans, who have not been successfully conditioned by the propaganda-education system, so as to be brow-beat into a violent viewpoint "about social relations," eg a social-hierarchy of intellectual-value as defined by the filtration process of the competitive and (improperly) authoritative education-indoctrination system.

Only about (a guess) 40% of the population are substantially affected by the violent conditioning, of the propaganda system, which is displayed by the society and its propaganda,

But then the propaganda-education system turns intellectualism into a contest in which the superior intellectuals of society emerge as the winners, where the educational contest is based on narrow dogmas,

But

What the, so called, liberals fail to see, is that peer-review and the attendant claim of, so called, academic excellence is a hoodwink (a lie), and really it is an expression of intellectual violence perpetrated against the people of society, so as to validate the, incorrectly placed, high-social-value associated to the investor-class's narrow totalitarian viewpoints about knowledge and its relation to the "productive" institutions into which the ruling-class has invested, ie the only valid creative efforts are those which serve the interests of the ruling-class, a viewpoint which the justice system violently imposes on the public through the condition of wage-slavery.

Apparently, putting intellectual activity into a competitive context, was seen as an advantage to corporate defined very narrow interests,… where the high intellectual value has come to be defined by the vague property of "intelligence," and intelligence was measured by the corporate and propaganda influenced institutional managers to be (measured) as the rate by which a person can acquire the dominant culture, ie the culture which best satisfies the investment interests of the investor-class,

Measuring the rate of acquiring the dominant culture's values, would also be about both measuring one's obedience to being commanded, and measuring one's desire to possess a socially dominant position.

This is the real (or correct) interpretation of intelligence, when the word is used in such a hierarchical society as, in the US. so that this competition (to acquire the dominant culture and to subsequently have a socially dominant position) is integrated into the propaganda-education system by; propaganda, and by managers of educational institutions, especially since WW II, where the end of WW II was also associated to the great beginning of the militarization of the US society.

Though the Pinkertons and government-police thugs, as well as Masonic Orders, particularly in the western states of the US, had already dominated the social conditions around the process of corporatization of the US, which had begun when the US was originally colonized by Europe: eg Hudson Bay company, the East Indian company, etc

In a society where there are a few who are the winners and the rest of the population scrambles either for a better life (ie to help the [few] winners) or to be true to themselves, but the entire social viewpoint is a deception. That is, in education the curriculum was designed, by its managers, to be about subjects which best serve the interests of the investor-class, so that people have been conditioned to believe that education is about a competition (about (over) narrowly defined dogmas) which are claimed by the propaganda system to be the correct "road to truth," but this is not true. In reality, education is about being trained so as to fit into narrowly defined corporate activities.

But

Equal free-inquiry, and building a new language; based on new sets of assumptions contexts containment-sets, patterns of greatest interest, and new ways in which to organize a precise language, is the correct conclusion of Gödel's incompleteness theorem where these are all intellectual activities which can be done by anyone in an equal manner, where experts are mostly at a disadvantage in this endeavor, since they already possess a fixed viewpoint

But

The first clear example of an intellectual bully in science, (other than the story of how the Church bullied Copernicus and Galileo, and others, etc), and his intellectual bullying is likely the result of his autism, as well as it also being related to the violent orientation of the propaganda system of that age, a society which was still dominated by a very authoritarian church, and the relation of success to being competitive, and the close relation of competition to demonstrating social dominance associated to the life of Isaac Newton, who did not want to publish any of his ideas to be read by any of those people whom he considered to be intellectual inferior people.

Newton was an intellectual bully, which seems to be an expression of his autism, since what he discovered was an intense sequential development of math patterns concerning the ideas of Galileo which Newton put into a set of more refined quantitative structures (than did Galileo) related to local measuring, and Newton also identified the inverse integral structure, which he used to solve the local differential equations which he defined as a physical law, where he followed Galileo very closely. Newton identified the differential equation for the two-body planetary system, and solved it so as to fit with the answers provided by Kepler, about planetary motions.

The descriptions of stable orbital periodic properties of 2-body systems has not progressed past this point, though refinements concerning the descriptions of electromagnetic-fields occurred in regard to differential-forms. The quantum theory has been an expression of a digression in mathematical capability, though its main construct, ie the H-atom model of Bohr re-asserts Newton's idea. That is, the stable spectral-orbital properties of the fundamental many-(but-few)-body material systems remains to this day, in the expert literature, essentially, the same expressions of Newton's 2-body system.

The sense created in the society about (our society) controlling an absolute truth, has to do with the way in which the (actually, a failed) truth is made exclusive, by control over social institutions, both the authoritative people and the instruments upon which their data depends

Corporate interests have used peer-review to narrow the interests of science to a far too narrow set of fixed dogmas, whereas it is difficult to compare today's experts with Newton, yet that is the essence of peer-review ie to protect the "great intellectual winners" of the educational contests (from the onslaught of different sets of ideas, coming from other people) where the contests are based on fixed dogmas (whereas Newton extended a dogma in a surprising manner).

Today's academics, effectively, have contributed very little to development other than adjusting the existing instruments, and the academics are mostly used as props of the, so called, superior-intellectual-ability in the propaganda system

But Newton's math structure has been endowed with "cultural magic," as the differential equation has not been placed in many different interpretive contexts, which are different from Newton's original context, whereas the function-space context, which was used in quantum physics, and which eventually supplanted Newton's viewpoint, was developed based on harmonic-functions in regard to physical descriptions of waves, eg the Fourier series, {originally applied to the heat equation, as well as linear differential equations with constant coefficients associated to periodic signals, eg electrical circuit engineering and (physical) waves}.

However, there have really been two separate viewpoints which have been kept alive (1) classical and (2) quantum viewpoints (where classical is still much as Newton conceived it, but revised a-bit by the energy-formulation viewpoint as well as by differential-forms)

..., ie the result of peer-review (which expresses the idea of intellectual bullying), but which leads to the ability of the bullying-group... those tested (within institutions) to be the superior intellects, or the, so called, experts...,

ie it is believed to be those who quickly absorb the dominant viewpoint, ie not a natural human quality (humans are curious, but also creative), to supply interpretations..., which become the traditional and authoritative thought-patterns for the academic science and math departments, where they are most often quite arbitrary contexts and interpretations,

while it should be clear that other interpretations and contexts should be considered,

ie the main point of Godel's incompleteness theorem is that fixed ways of using language have limitations as to the set of patterns which they are capable of describing, so (therefore) new interpretations and assumptions and new contexts should always be considered by science and math.

Review of particle-physics, and criticisms of its math structure (see chapter 23)

Particle-physics is defined as a set of particle-collision interactions with components of a quantum system, where a quantum system's wave-function, which supposedly, represents the quantum system's probabilities and spectral properties of the quantum system's components, where the particle-collision interactions, supposedly, result in adjustments to the quantum system's wave-function's spectral properties.

This spectral adjustment is defined as an integral, which is, apparently, defined over the interaction process, which apparently takes place at a single point in space (see below), over all possible collision induced particle-state transformations, where the adjustments are about the set of elementary-particles which "form out of a vacuum state," ie the so called zero-state of the "function-space model" of the original quantum system (whose spectral-values are being adjusted by the interactions modeled as elementary-particle-collisions related to both the quantum system's material-components, and the set of field-particles associated to the assumed set of the system's intrinsic set of particles, which become a part of the particle-collision-interaction structure in (or emerging from) the vacuum-state, so as to model many transitions of the colliding particles between the many different particle-state and field-particle-interaction possibilities).

The vacuum-state is a mixture of elementary-particle-states and space-time (or space and time), but it has no valid properties as a function, where the vacuum-state-function is supposed to be a (zero-state) function in a function-space, especially, since the functions of the quantum system's function space are defined on the domain-space, which is the space-time (or space and time) structure, within which both the quantum system and the quantum system's components are supposed to be contained (and the set upon which the quantum system's set of system-defining differential operators are defined, ie the operators which are needed to identify the quantum system's wave-function).

Apparently, the adjustments (to the original quantum system) are determined by integrating "over time" all the various collision-processes, which emerge from the vacuum and provide elementary-particles which collide with the given quantum system's components and field-particles, as the system's components and its field-particles are, supposedly, naturally colliding within the quantum system, and, subsequently, (in the math-constructs of particle-physics) being related to the "family of particles," associated to the quantum system, ie the internal-particle-states of the given quantum system's components and field-particles, which are a natural part of the system, and the system's related energy-range, so that the families of particle are defined in relation to the vacuum-state and the given quantum system's energy-range for its internal interactions.

[Note: However, the neutrinos of the different families can rotate (or oscillate) between the different particle-families (associated to internal particle-states) ir-regardless of the energy-range of the system's components, thus, there is no basis for distinguishing particle-families, yet energy-ranges in regard to the existence of elementary-particles is central to the descriptive context of particle-physics, eg distinguishing between nuclear interactions and electromagnetic interactions, but now the electron-neutrino can oscillate between all the other particle-families, a point which is mostly ignored.] but where the integration (defined on a series, used to determine the spectral adjustments, and the series is defined by the changes in internal particle-states), is defined in regard to its inverse derivative-connection operator (in a Lagrangian density operator), which is part of the model of the quantum material-interaction process.

That is, the descriptive context for the assumed forces which are a part of material-interactions is non-linear.

Thus, the model is: quantitatively inconsistent, it is unstable, it is many-valued, it is chaotic, it is discontinuous, it is indefinably random, and it is defined on an indefinable entity (ie the vacuum state) so it is undefined (thus, it is not mathematical) and it is a description which has no stable patterns associated to itself, except the internal patterns upon which the connection-matrix is defined, subsequently, it (ie the integral of the process) is divergent, which is also readjusted (another re-adjustment of the idea of quantity) by the process called re-normalization, ie a way of re-grouping terms in the diverging series to again get infinity, but, so that infinity is subtracted from infinity to get a finite number (note: this can only be done if there is an empirically determined answer), and the descriptions of these material-interactions cannot be used to describe the stable spectral properties of a nuclei, whose atomic-number is more than two (or, two or more), ie it is without any practical value in regard to the real purpose of physics.

In this descriptive construct the quantum system's (theoretical) wave-function is provided with the structure of internal-particle-state states, ie the wave-function now has particle-state (vector) components.

But

There is no valid reason in which to pursue particle-physics, since the wave-functions of the fundamental set of stable many-(but-few)-body quantum systems, in general, cannot be found.

So particle-physics is only about providing an energy-spectral internal-particle-state context associated to the mass-spectra of elementary-particles found in particle-accelerators, whose assumed set of internal particle-states are interpreted to fit into an, U(1) x SU(2) x SU(3), fiber group, so the internal particle-states are to be "rotated" by this group, in a local manner, and to which a derivative-connection is also fashioned (based on this fiber group), where, subsequently, a manifold structure is (can be) associated to this derivative-connection, and the vacuum-state is given a geometric context, which is (also) the essence of string-theory, this is possible because the derivative-connection is a local construct, so it can be related to a manifold geometry, but when adjusting a wave-function's spectra, the one-point, where the collision process is defined, is a universal global property of the quantum system's spectral properties (but such wave-functions do not exist for general quantum systems), that is, it (particle-physics descriptions) is about information (which is the subject matter of particle-physics, as well as string-theory) which only relates, in a practical manner, to the particle-collision models of bomb engineering projects, where string-theory is, apparently, mainly about a geometric way in which to inter-relate particle-physics with gravitational singularities and the building of a doomsday-bomb, ie to try to detonate a gravitational singularity.

The wave-function of a quantum system is a set of globally defined functions, whose domain space is space-time (or space and time)

The math structure of particle-physics is that of a bounded geometry, which changes in a local context, and is defined on a 6-dimensional space (ie the internal particle-state space), which does not intersect space-time

The hope of the bomb-engineers, apparently, is that the particle-collision model does in fact connect to two separate descriptive contexts, both of which seem to (mostly) be fantasies.

On the other hand

One can interpret the particle-collision data as being related to the collision between the distinguished points of two stable shapes whose dimension (as a shape) could be one, two, three, or more.

Furthermore, there could be many different size-scales associated to these stable shapes.

The shapes also possess both real and complex-number models associated to the metric-space properties of the shapes, and the pairs of opposite metric-space states are also associated to the shapes of metric-spaces. These opposite metric-space states are easiest to mentally-identify in regard to dynamic processes, where in hyperbolic metric-spaces, the opposite metric-space states are "states of time," either forward or backward time-states, which are associated to two pairs of real, but opposite, dynamic processes (of the motions of material-components), where locally the two time-states are inverse to one another. Thus the two opposite metric-space states cause the description to be either in the real-numbers or in the complex-numbers, and this would be related to either SO(n) or SU(n) type fiber groups.

The three pillars of physical science are: materialism, reduction to elementary-particles, and randomness

Materialism is understood in the new model and its new interpretation as the 3-space containment of equal or lower-dimensional material-components (which are also stable metric-space shapes) within an open-closed topology of a rigid 3-space shape, it is in this context where the fundamental assumptions about coordinate-frames and geodesic dynamics of general relativity are best applied, the new model of material-interaction is shape-dependent as well as being dimensionally dependent and it depends on the relative size changes which are fundamental to the description and which occur between dimensional levels, eg 3-shapes interact in four space but the natural sizes of the material components are the size which are comparable to the size of the solar-system

The shape and size of the 3-shape metric-space within which we are contained (so, for us to appear to be material-systems) is the shape and the size of our solar-system, yet light within the subspace which our solar-system metric-space shape defines can extend out past the solar-system to also relate to the similar-dimension shapes which define components in other (external) systems of the galaxy and universe.

In regard to the pillar of reductionism

The claim made by the physics community,

(ie those few who have been hoodwinked--- [deceived by the upper-social-classes appealing to the arrogance and the lure of being dominant {which is an illusion which the propaganda system can provide} which to certain personality-types will be of great interests and appeal, in regard to the, so called, winners of society]--- into doing only doomsday bomb-engineering), is that the only sizes which are of any significance, in all of existence, are the size scales of atoms and nuclei and their, assumed to be, elementary-particle component sizes,

But

The stability of the solar-system makes this claim, obviously, untrue.

In regard to the pillar of randomness

This is absurd, since this assumption (of randomness) has not been able to describe the stable properties of material systems for over 100 years, and it is this set of stable material properties, which is the set of observed patterns of existence, which should be of the most fundamental significance to the physics community____ but the lure of "intellectual domination" has been used to marginalize such an interest in "the natural purpose" of (real) physics.

Chapter 15

Math stability, and how to organize stable math patterns so as to form a new descriptive math language

New contexts for the description of math patterns and the relation of these new patterns to their applicability to measurable descriptions, which are accurate over a wide range of various types of stable system (to sufficient precision) and the stable systems are defined over a wide range of size-scales, so that the context is useful for controlling the pattern, when placed within some greater context, within which a new system can be built and controlled and used in some practical way.

Apparently the experts in the math community are not aware; that if the quantitative sets…. which are the descriptive basis for the description of math patterns…. are not related to one another,

eg in regard to a map or a function defined between two sets, in a "simple" quantitatively consistent manner, then the, so called, patterns{associated to a quantitatively inconsistent, yet, what, apparently, seems to be a general quantitative context, which exists between two sets during a mapping process (or for a mathematical law of mapping),

ie but which does not satisfy the need for quantitatively consistent set of local models for measuring in a changing and interactive descriptive context}, which they are, apparently, trying to describe, in fact, cannot be described (or defined) as stable patterns, and thus, their quantitative descriptions have no content, ie they are not describing a context which has quantitative meaning.

Note: That quantum physics, particle-physics, and general relativity are all descriptive contexts which are framed in what is, apparently, a quantitative context, but nonetheless, they are precise descriptive contexts which have not been able (to be used) to describe the stable patterns, which they were devised to describe. At best they are descriptive contexts which can be fit into an empirical descriptive context, especially, in regard to the quantum, or random based, descriptive contexts, the context of a probability-based description is about identifying the data of the set of random events' which are the outcomes of a random experiment, ie the random system's elementary-event space, so that one can experimentally (or by means of a counting all the possible events) determine event-probabilities, as to identify an event's probability of occurring for an experiment ie information which can be used to bet or to try to make decisions in an uncertain context, but it is not a descriptive context which can predict anything of any significance,

eg the developers of particle-physics did not consider "the structure of the math of particle-physics" and, subsequently, predict the existence of dark-matter particles and dark-matter interactions, yet the dark-matter material-interaction is hypothesized to be based on an set of elementary-particles. Furthermore, the context of particle-physics has led to the idea of "quantum-foam" (also see below) which is an immeasurable set (or state) without possessing any stable patterns of its own, ie from quantum-foam there are supposed to emerge (as random but unstable events) elementary-particles, which are identified as particle-collision events in space, so as to vaguely (or in a small way) adjust the energy-structure of a quantum system, but the unitary math patterns of unstable elementary-particle events, and other sets of operators

(eg the energy-operator for a quantum system, and where a model of material-interactions [within the quantum-system] can be attached to such an energy operator), are defined based on the non-measurable (particle) properties of quantum-foam, so that this descriptive context is "a conceptual-level removed" from (as well as being distinct and separated from) a model of a domain-space, which quantum-foam is trying to model, that is, the operators are defined on functions in a function-space and the functions (of the function-space) are defined on a domain space, but quantum-foam has no valid relation either set-containment of the quantum-system or to measuring position or time in regard to an eigenfunction and its related eigenvalue values and properties in regard to a particle-event's distribution in space.

Note: A domain space is a space upon which the basis for measuring a system's properties resides, because the system is assumed to be contained in the domain space, so that measurable properties of a system, modeled as functions and individual spectral-values in a function-space, need to be relatable to the coordinate containment set of the system.

This cannot be said for quantum-foam, since it is neither quantitative nor a proper containment set.

An energy-spectrum for quantum-foam is implied (by the internal particle-states, to which a set of unitary connection-derivative operators apply, connection-operators whose definitions do depend on [or are derived from] the properties of the quantum-foam), but the spectral-events (associated to indefinably random elementary-particle events) are unstable, so the elementary-particle spectrum of these connection-operators is not well-defined in the context of probability, and thus, again there is no content in this type of a description.

Yet this is "the main feature" of peer-review math and science, (ie meaningless babble, meaningless babble, meaningless babble, babble, etc) since particle-collisions are the main feature of bomb engineering, where making-bombs defines a large commercial interest.

ie it is the structures of abstract math which are being formally applied to the wrong context within which to try to describe the observed patterns which do exist, eg stable nuclei with well-defined (or stable) spectral properties, and these observed patterns are stable and measurable.

But, nonetheless,

These math abstractions, concerning particle-physics, apparently, (or vaguely) fit into empirically determined patterns, but for such a probability based description, the elementary-events are unstable, and thus, it is a descriptive context which is not well-defined, and thus it cannot provide any useful information, in regard to the mysteries which they are trying to describe, eg stable nuclei, so as to compress the information involved in the sets of descriptions of the observed properties of the stable fundamental systems, to which these descriptive contexts are supposed to be related.

Science should be trying to compress information about systems, and trying to fit their observed patterns, into a simple context which is accurately (and to sufficient precision) associated to the properties of a wide range of physical systems, so that the simple context can be used in a practically creative context.

For example, none of the following categories of authoritative description (ie quantum physics, particle-physics, and general relativity) can be used to describe the stable properties of the many-(but-few)-body systems which possess stable spectral-orbital properties and which exist at all size scales, eg nuclei, general atoms, the solar-system etc,

Furthermore the stability of these fundamental systems implies a linear descriptive context.

And,

If information, concerning a linear system, cannot be compressed using one set of descriptive patterns (eg the currently used set of descriptive patterns) then by making new:

assumptions, creating new contexts, and making new interpretations so as to also make a new set of "information-compressing laws" related to descriptive patterns of observed phenomenon wherein models of local measuring are introduced and, subsequently, (newly) interpreted, then the new descriptive context might turn out to actually be useful, instead of only fitting empirical data after a new pattern is detected or observed, as the current model of science is now organized as a descriptive language and context which can only fit data into its language, similar to the way in which Ptolemy's description based on epicycles fit data into its language.

That is, the current description has become far too complicated and elaborate, and it is, essentially, based only on observing empirical properties, ie all of the details of the observed pattern need to be identified and considered, and then the current science can fit it into its descriptive language, where the context of a descriptive state can only be changed by a property which is related to statistical or thermal properties in which it is placed, eg the nuclear bomb is detonated by getting to a critical mass so that the statistics of the critical mass cause the change.

That is, without a valid descriptive context then the description enters into an indefinable random context, where information (eg observed patterns) is not compressible, and within this context there are no stable measurable patterns which can be described, and the description cannot emerge from such a (very random) context, which is based on the random properties of any type, so that all these descriptive efforts become quantitatively inconsistent descriptive efforts (or constructs) which are being applied to unstable (and indefinable) random patterns.

What math properties result from quantitative inconsistency?

That is,

The quantitative relations, in a quantitatively inconsistent context (or in the quantitatively inconsistent manner, which the experts so often and so thoughtlessly and arrogantly use), will lead to

1. a non-measurable context, in which there come to be (emerging from the inconsistent descriptive context), eg

(a) the vacuum state or quantum-foam of particle-physics (where quantum-foam is, supposedly, a state of existence where particles and space-time are not distinguishably separate properties, ie measuring is not possible), where particle-physics emerges from quantum-foam so as to be the model of material-interactions in quantum physics),

or

(b) not being able to extend beyond the 1-body system (which possesses perfect spherical symmetry) in general relativity, etc,

This quantitatively inconsistent context leads to the properties of:

2. many-valued-ness for the function (eg which models a system's measurable properties), ie bifurcations, as well as,

3. the local nearby (or, an expected, continuous set of) relations which might exist between the local coordinates and the system's related measured values (or measured properties) will be chaotic, ie big changes in function values can be caused by small changes in local coordinate values, and the most usual condition in which

4. the (geometric or spectral) patterns, which should define a shape's (or system's) properties, cannot be determined by calculations, which are based on law, but where such a model of "law" is defined in such an (as mentioned above, an) inconsistent quantitative context (eg quantum physics and general relativity, are exactly the descriptive contexts where laws exist but they cannot be used to provide any information in regard to a wide range of general type systems)

[note: the laws of classical physics, when applied to linear and solvable systems, can provide accurate and practically useful information about a relatively wide range of general classical systems to which the context of local measuring of material properties applies (eg position, motion, amount of material, or well defined average values defined over large numbers of components)].

Review of a new descriptive context where the new math context is based-on math patterns which have already been determined (or identified) by the experts, but apparently, these simple patterns are considered to be patterns, which the experts believe to be "too simple" to have any descriptive value, or math interest, since within our propaganda system a math expert is not to be understood, but

The experts are supposed to utter absolute truths, as Yau has claimed, ie that he is sure, what he states, as a professional mathematician, is "absolutely true," since he has provided a mathematical proof, but the only "true"

Which now seems to exist in a "math-world" of "symbols and agreements about symbols," are descriptions of patterns which are a part of an illusionary world, that is there are many, so called, math-truths, which only possesses a delusional content within a world of illusion which the symbols of measuring have now been organized (by the sets of assumptions contexts and interpretations etc) to describe, ie many rigorous math patterns, described in peer-review contexts, are delusional patterns when considered in the world which we experience.

In fact, it is the stable math patterns reviewed below, which have the greatest promise for such a (new) descriptive context to, actually, be able to describe the observed stable patterns of physical systems, but assumptions, containment-sets, descriptive contexts, and ways of interpreting observed patterns (or for interpreting math patterns) need to be re-considered.

Since the current state of physical description is in a such a shambles there should not be any opposition to such an idea, based on the espoused belief by scientists and mathematicians, that all new valid ideas should be considered in science and math

Unfortunately, the current intellectual considerations define intellectual dominance in society which is most notably based on violence so that actions and institutions are arbitrary, so that institutional value provides an institutional personnel with great privileges within society, so the experts would not relinquish such a social position gracefully.

Nonetheless,

So

What is the correct descriptive context?

ie The context in which measurable descriptions for a wide range of general and stable physical systems can remain quantitatively consistent

To do this, to create such a descriptive language based on math patterns, ie patterns about quantity and shape, and set-containment, and (to a lesser extent) random events

Consider the stable structure of compact discrete hyperbolic shapes, which is the biggest category of stable shapes in Thurston-Perelman's geometrization.

In the math literature this is quite often (if not most often) about the spectral properties of the hyperbolic manifolds {either in general, or for shapes of constant curvature} where in the constant curvature case there is the uniformization property, which identifies constant curvature shapes which are isometric-ally (or conformal-ly) related to the three types of, (-1)-, or (0)-, or (+1)-constant curvature manifolds, where the general hyperbolic manifold of constant curvature is modeled as the set of (-1)-constant curvature shapes.

According to geometrization this category of constant curvature discrete hyperbolic shapes is the most stable context in which to describe math patterns which are stable, where math is about quantity (or stable and well defined sets) and shape (or stable and well defined patterns).

[Note: One also wants to consider…, in the context of non-positive constant curvature spaces (and the similar types of stable shapes which this geometric context can contain)…, only "metric-functions" which possess constant coefficients, where such context for measuring (on a coordinate base-space, ie a space within which is contained the system [or pattern] which one is trying to describe in a measurable context), remain consistent with the properties of local linear models of measuring {ie linear measuring models which are also continuously commutative}]

Such hyperbolic shapes can be either bounded or unbounded,

In this paper, it is the bounded hyperbolic shapes which will be of most interest (this descriptive context is very simple)and

It is a context which is related to isometry groups, that is, related to rotations, and translations, and reflections (or 2-dimensional translations),

Where an important invariant property (for metric-invariant maps) would be the type of coordinate transformation, ie rotations and translations and reflections (or 2-dimensional translations), which can be associated to each group element,

Eg

1. there are the (real) 1-dimensional translations (called parabolic group elements),

(note: in regard to accumulation points or limit sets (out at infinity), these would be fixed points of the type of transformation ie related to no fixed points, where the number of fixed-points is important in regard to identifying the properties of a Fuchsian subgroup),

Note: The viewpoint of the Fuchsian group applied to hyperbolic geometry, tends to turn the considerations into an intellectual process which is dominated by the adjustment of symbols, and less about a geometric construct, where a geometric construct has a deeper relation to a description which one might try to make closer to experience.

while

2. there are also the (real) 2-dimensional translations (called elliptic group elements), (related to one fixed point, ie the point at infinity, in regard to the Riemann sphere), (these are also the spatial reflections, or equal-angle reflections)

and

3. there are also the "axially defined rotations" (called hyperbolic group elements) (related to two fixed points, defined between the north-pole and the south-pole on the Riemann sphere model of the complex number-plane) (note: axial-rotations also possess properties of equal-angle reflections, angle transformations can be generated by sets of reflections, in a description where angles have discrete relations, ie when defined on a lattice, then the viewpoint of sets of discrete reflections can be of value)

These transformations are uniform actions on points identified on a shape, which transform between the periodic structure of these shapes, so as to form a (checker-board type) lattice composed of these stable shapes on the coordinate space.

The shape transformed is a fundamental domain, which is then deformed by "cylindrically" identifying the pairs of opposite congruent faces (so that each facial point identifies a circle whose original direction of definition is normal to the facial-surface, ie so as to form a sort-of cylinder shape), into a stable shape identified by its number-of-holes property of its shape (holes surrounded by [or which obstruct "the continuous deformation to a point" of] different dimensional faces, eg curves, spheres, closed cylindrical shapes, etc [within the shape, or within the space]), or its genus.

These shapes are determined in a process which starts by identifying opposite congruent faces (of the fundamental domain) with one another, by forming sets of circles, defined as (these circles) originally being normal to the identified opposite faces, thus, they can be called "circle-spaces."

The fundamental domain shapes for bounded discrete hyperbolic subgroup types of periodic defining by "translations" a set of uniformly fixed points whose "translations" to "what are" equivalent points, are formed in a dimensionally-sequential construct of rectangular polyhedral simplexes (ie simplexes are built of regions bounded by faces which are 1-dimension less than the dimension of the region so that this reduction to lower-dimension faces can continue all the way to the points (or the vertices of the original convex rectangular [or orthogonal] polyhedral simplex):

1. Rectangles of various sizes (but they must be of particular sizes so they will fit together in a lattice) are connected at vertices in a diagonal fashion so as to form a chain of rectangles, then the vertices are pulled apart to form a polygon (this polygon would appear in hyperbolic space,

and relative to Euclidean space, as possessing curved sides) The 1-faces of this polygon would be geodesics when this fundamental domain is moded-out.

2. Right rectangular prisms of various sizes (but they must be particular sizes so they will fit together in a lattice), so that all the orthogonal-faces are rectangles, are also connected together in a diagonal fashion at these rectangular prism's (diagonally related) vertices, but then the rectangular faces of the right rectangular prisms are changed into the type of polygons defined in step 1, so that the number of vertices defined between adjacent orthogonal 2-faces (of the original rectangular prism) are the same (or line-up), then again the connecting vertices of the diagonal orthogonal polygon prisms are pulled apart to form a polyhedron with polygon faces. The 2-faces of this polygon would be minimal surfaces, and the sides of these minimal surfaces would be geodesics, when this fundamental domain is moded-out.

3. The next dimension of right rectangular prisms of various sizes (but they must be particular sizes so they will fit together in a lattice), so that all the orthogonal-faces are rectangles, are also connected together in a diagonal fashion at the (diagonally related) rectangular prism's vertices, but then the rectangular faces of the right rectangular prisms are changed into the type of polyhedrons defined in step 2, so that the number of vertices defined between adjacent orthogonal 3-faces are the same (or line-up), then again the connecting vertices of the diagonal orthogonal polyhedron prisms are pulled apart to form a 4-dimensional polyhedron with 3-dimensional polyhedron faces

The 3-faces of this polyhedron would be minimal volumes, and the 2-faces of this minimal-volume polygon would be minimal surfaces, and the sides of these minimal surfaces would be geodesics, when this fundamental domain is moded-out.

4. Etc,

One can continue this process, apparently up-to hyperbolic dimension-6, after which the bounded polyhedrons with the properties needed to form a lattice composed of bounded fundamental domains no longer exist, but similar sets of unbounded polyhedrons (with finite geometric measures) do exist up-to hyperbolic dimension-10, after which the polyhedrons with the properties needed to form a lattice no longer exist (these are properties of hyperbolic fundamental domains identified by D Coxeter)

Note: The orthogonal structure of the right rectangular prisms though its seems to be lost when the vertices of the diagonally connected vertices of these prisms are pulled-apart, in fact, they are maintained by the deformed shape of the fundamental domain (deformed by the circle-based moding-out process) where the coordinates are either a few geodesic circles, or geodesic-ally determined minimal surfaces (or, in general, minimal faces),

or

the circles, in the circle-based moding-out deformation, become orthogonal sets of pairs of hyperbolically shaped coordinate curves, ie the orthogonal structure is preserved in the moded-out (or deformed) shape of the "circularly deformed" fundamental domain.

Thus, one thinks of the fundamental domain structure as being composed on right rectangular-faced prisms, since it is easiest to think of such a geometric structure as maintaining a continuous locally orthogonal (or continuous locally-commutative) geometric structure, thus, one can use the name of "rectangular polyhedral simplex" for such a construct for a fundamental domain.

The hyperbolic group point-transformations (defined in the fundamental domain shapes, on the coordinate lattice) would define a rotation axis, which would

1. identify a rotation (which is also a point transformation of a fixed arc-length) around (or about) one of the bounded shape's isolated geodesic-circle coordinate structures defined on the [moded-out rectangular polyhedral simplex] shape

(where this "rectangular polyhedral simplex" constitutes the fundamental domain of "the lattice of this shape," where the lattice is defined on the coordinate space, where the lattice identifies the periodic structure of the maps defined by discrete isometry subgroups associated to the coordinate space's fiber Lie group, where the periodic structure (of the discrete isomerty subgroup) is a (stable) shape),

so as to also (or equivalently)

2. rotate between "sheets" of a many-sheeted-covering, where each "sheet" is a fundamental-domain shape,

or equivalently,

3. Rotate (or seemingly translate) between different bounded fundamental domains of the lattice, where the lattice is defined by periodic "rotations, or seemingly translations" of the bounded shape's fundamental domain into the other lattice shapes (ie which define the checkerboard periodic pattern), so that the different fundamental domains in the lattice are identified (by the

(local) coordinate map which is modeled as a discrete isometry subgroup) by means of a rotation (applied to the shape) of a fixed angle (or fixed rotational length),

Where such a rotation of a fixed angle would appears to be a translational shift between equivalent fundamental domains in the lattice, where the, apparent, translational shift is really a rotation by a fixed "circular" length, which is defined on the (or on such a) stable, discrete hyperbolic shape, a shape which can be defined as (or called) a circle-space shape.

(note: it is a "stable shape" since when it is modeled as a differential-form, related to the shape of the bounded fundamental domain, so as to be defined in regard to a general Laplacian of hyperbolic space, then it is a linear, metric-invariant. and continuously commutative shape, ie the general Laplacian operator is separable, where it is continuously commutative (except possibly at a single point) since it is derived from a fundamental domain, which is a bounded rectangular polyhedral simplex, ie it is solvable and stable and controllable). This generalized Laplacian will essentially be the non-homogeneous wave-equation for the electromagnetic-potential 1-form, where the "current" will be related to the faces of the rectangular polyhedral simplex.

An element of an isometry group transforms local coordinates in a manner so that the transformed local linear structure (of the coordinates) remains locally metric-invariant, or for inner-products, after the local coordinates have been transformed by a group element, ie (dx,dy) = (dgdx,dgdy), for group element, g, and where dg is a local Lie algebra element for the group element, g. where the properties of a group element being either a translation, or a rotation, or a 2-dimensional translation (or a reflection) (for each group element) is a property of the group elements, which can be used to identify different subgroup structures within the isometry (or Fuchsian) groups,

There are the fundamental domains of bounded rectangular polyhedral simplexes, which (with discrete isometry subgroups) can be used to define lattices (checker-board patterns) on the coordinate base space

The fundamental domain is transformed in a manner, which is determined by the type of a group element (or by subgroup elements of the same type) so as to define the other congruent (and periodic) regions of the lattice by the group action (or coordinate point-transformations, eg a uniform translation of the fundamental domain) of a discrete (isometry) subgroup

And

where discrete isometry subgroups (can) define the bounded and stable hyperbolic shapes, for the hyperbolic isometry subgroups, ie all the elements of the discrete isometry subgroup are of hyperbolic type, ie each group element in the given discrete isometry subgroup is an axial rotation associated to a fixed translation (or a fixed rotation arc-length)

And

where the bounded discrete hyperbolic shapes all possess a well defined set of geodesics (or minimal faces) where one can think of these geodesics as being defined by hyperbolic discrete subgroups, (ie each subgroup element is hyperbolic, ie it is an axial rotation of a fixed rotational length, ie the "simple" length of the geodesic) of a discrete isometry (or Fuchsian) subgroup,

(or the set of a hyperbolic shape's geodesics can be thought of as defining a hyperbolic discrete subgroup)

A discrete hyperbolic subgroup can be thought of as being defined by the stable bounded discrete hyperbolic shape's geodesic (or minimal) lengths, which, in turn, are determined by the facial measures of the shape's rectangular polyhedral simplex, for a rectangular polyhedral simplex, which satisfies geometric properties which allow a lattice to be defined on the coordinate space, ie an infinite sequence (or series) defined for each of a finite fixed set of discrete "translations" of a fundamental domain (ie a rectangular simplex) so that this sequence completely cover the coordinate space with congruent fundamental domains so as to be without any over-laps.

Where the (conformal-like) Fuchsian group can also be related to the discrete isometry subgroups defined on hyperbolic space

(Then) for the hyperbolic shape, which is bounded, ie the discrete subgroup's fundamental domain (related to rectangular simplexes) there are the natural coordinates associated to the stable shape:

(1) the moded-out faces (ie "opposite" faces of a rectangular simplex identified as being equivalent) which are minimal surfaces within the shape, [or if the faces being moded-out are 1-dimensional, then these 1-faces are geodesics within the shape], and these geodesic coordinate curves (otherwise) fit into (further)

(2) sets of pairs of orthogonal hyperbolic coordinate shapes (orthogonal hyperbolic pairs are defined at each coordinate point), where the geodesics and the sets of pairs of orthogonal hyperbolic coordinates together identify the entire (natural) coordinate structure for this shape, so that the sets of pairs of orthogonal hyperbolic coordinate (though commutative) are locally non-linear, and possess (local) curvature and would (according to general relativity) push any object contained on the shape towards the geodesic (or minimal surface) so that for an object which is not already on a geodesic it would be pushed towards one of the geodesic coordinates, in the

shape's natural coordinate system, and, where the geodesics (or the minimal surfaces) are thus stable, for all the discrete hyperbolic shapes which can be defined by either "allowed rectangular simplexes," or by their set of discrete isometry and hyperbolic subgroup properties.

Since the spectra of discrete hyperbolic shapes are stable, thus, when one considers the spectra of a bounded discrete hyperbolic shape, then it is composed of a fixed and finite number of primary finite spectral-values, which can be counted, so that there are exactly:

(ie The spectral count for a discrete hyperbolic shape)

Number of primary spectral elements =(dimension of the bounded discrete hyperbolic shape) x (genus of the shape [the number of holes in the shape], also the number of toral-components of the discrete hyperbolic shape)

Thus, there are an infinite spectral set for such a bounded hyperbolic shape, where each of these primary spectral values is also associated to all of its integer periods for each such primary spectral-value

Stable spectral properties

This can be talked-about in the context of the generalized Laplacian operator for the hyperbolic shape, where the general Laplacian is a second-order metric-invariant (differential operator) construct, but the analysis is easiest to consider in relation to the above given geometry of (or in regard to) the primary spectra of a given shape

This is the simple construct which emerges from an, essentially, rectangular polyhedron type simplex, which fit's the "two lattice requirements"

in regard to the (discrete translation, or rotation (discretely defined by a fixed arc-length) between the periodic shapes of the lattice [or coordinate partition by fundamental domains]);

(1) the angles defined at vertices of a "rectangular-determined" polyhedral simplex of a stable discrete hyperbolic shape, the Sum of angles at each vertex equals 360, where the set of angles defined at each vertex of the rectangular polyhedral simplex (fundamental domain) in the coordinate space's lattice partition is the sum of fundamental domain's interior angles, where all but one such interior angle have all been discretely translated to each vertex of the lattice-partition of the coordinates and which is defined by the fundamental domain shapes discretely translated (or rotated).… by a fixed arc-length (or solid-angle) defined along the geodesic circles (or minimal

face-geometries) which define the edges (or faces) of a fundamental domain…. to their different lattice positions

(2) All the pairs of "opposite (or equivalent)" sides (or "opposite (or equivalent)" faces) of a fundamental domain are congruent and of equal geometric measure.

Ie That is, in a moded-out rectangular simplex, there are pairs of (congruent) faces…, which define the periodic geometric structure of the new shape (by their discrete fixed magnitude rotations [or translations])…, are identified so that hyperbolic coordinate shapes "fit between" (or connect) the opposite pairs of moded-out faces (or hyperbolic coordinate shapes form between the set of pairs of moded-out faces during the geometric-deformation moding-out process)

The rectangular polyhedron of a fundamental domain are put-together at their vertices, and then each vertex is pulled-apart so that the moded-out faces define toral components, or form circle-spaces, and the coordinates which are defined to be perpendicular to the circles which these coordinates connect by coordinate-curves, so that the coordinate-curves are pairs of orthogonal hyperbolic shapes at each point of any of these hyperbolic coordinates (ie where these hyperbolic coordinate-shapes intersect the pair of local directions of the hyperbola are orthogonal to one another).

The bounded rectangular polyhedron type simplex, {for the moded-out shape's fundamental domain, which defines the shape's "periodic" lattice, defined by the actions of discrete subgroup on the fundamental domain in the coordinate space}, causes the shape to be continuously commutative, in regard to the coordinate domain-space upon which the generalized Laplacian is defined on the natural coordinates, which are defined on such a bounded discrete hyperbolic shape,

Such a local description of the shape's measured properties, where the stable shape's most fundamental properties are the stable spectral-values of the shape, where this shape is naturally defined as a differential-form, which, in turn, is defined on the faces of the rectangular polyhedron type simplex shape (or fundamental domain [within the lattice]).

The spectral (or minimal surface, or geodesic) properties…, associated to the generalized Laplacian…, can be determined from the (shape's) integrals (solved [or summed] on the shape's bounding faces), which are defined on the face structures of the (deformed and) bounded "rectangular" polyhedral simplexes, as the geometric-measures of the differential-forms (measured in a hyperbolic metric-space), where this math construct is linear, metric-invariant, and with continuously commutative local coordinate structures (except at one point, where the local structure can still be organized to maintain it being commutative), so this is solvable, ie the generalized Laplacian is separable, so this is a stable construct and it is controllable.

Within "what context?" of set organization and set-containment, can these stable shapes, and stable spectral sets, which are (also) associated to these bounded and stable discrete hyperbolic shapes, and which are defined for many-dimensions, be used to describe the observed stable patterns of the many-(but-few)-body systems which are so fundamental, and which exist at all size-scales?

Can the new constructs, which are built from the building blocks of the stable shapes, also be used to describe material interactions?

As well as being used to define a unifying global and spectrally hierarchical context, which can be used to describe the properties of living systems, so as to be able to understand the great amount of control which living systems possess?

(note: the new description material interactions, in this new context of containment, also identifies an, apparent, spherically symmetric material-interaction structure for all material interactions observed within the context of a 3-space, ie the new description is not (totally) inconsistent with the angular-momentum properties of 3-dimensional (small-component) material systems in hyperbolic space, ie quantum-theory, especially if the material is free-microscopic-material-components which are interacting).

Note: Since these are rectangular polyhedral simplexes the minimal "surfaces" (or minimal volumes, etc) can be "the equivalent" of a geodesic, however, since these shapes emerge from deformed and moded-out rectangular polyhedral simplexes, there will always exist a relation between the "minimal faces" and the relation of such a face of a general dimension to its 1-faces, or its geodesic-spectral lengths.

Note: An n-dimensional rectangular hyperbolic simplex, when moded-out, will form an n-dimensional shape, which can fit (as a shape) into (or be contained within) an (n+1)-dimensional hyperbolic metric-space.

The allowed set of discrete hyperbolic shapes for all the different dimensional levels, up to hyperbolic dimension-10, upon which such fundamental domain based lattice structures can exist and their geometric properties, within a metric-invariant context, have been identified by D Coxeter.

Thus, there can be defined a set-containment structure, which is based on stable shapes, and which possess stable spectral-orbital properties, and a new math structure of shape and set containment based on a shape's size and its dimension so as to "partition" (or form a dimensional shape-size containment tree in) an 11-dimensional over-all containment set into dimensional subspaces defined by shapes and their sizes, so that the partition can be approximated by considering the different dimensional subspaces of the 11-dimensional space to each have a shape associated to it (the subspace of a given dimension) which is the biggest stable shape of the subspace, and this finite set of largest stable shapes associated to "the number of subspaces of the various dimensions" defines the finite stable spectral set, to which all shapes of all the various dimensional subspaces (or set-containment shapes of various dimensions as well as the shapes of smaller sizes which are contained in any of the stable shapes) must resonate (between any given shape's spectral-values) with this finite stable spectral set of the containing space's new type of dimension-and-shape set-containment partition.

Thus, the set of stable shapes, which compose the new type of spectral-partition, and which define a finite spectral-set, which defines existence for all stable shapes, which are contained in the 11-dimensional hyperbolic metric-space, is the primary context of physical description, where material interactions provide a means to perturb the primary structure of existence's stable bounded spectral-orbital shapes, which define both material components and material-containing metric-spaces.

Material interactions can be related to an interaction being capable of developing resonances with the fixed stable spectral-set of the high-dimension containment set, which when the interaction, eg a material-component collision, has the correct range of energy, the strong resonances (the the containing space's finite stable spectral set) can cause a certain types of new material-component stable shapes to emerge from the (collision) interaction, due to such new (to the interaction) resonances.

It might be hypothesized that such an 11-dimensional containment space, a containment set which contains sets of stable shapes, is a model of a galaxy, where the galaxy really has the property of being high-dimensional.

Instead of this simple description of the measured spectral-values of the bounded hyperbolic shapes which have hyperbolic discrete isometric (or Fuchsian) fiber-groups, so as to be stable shapes formed out of bounded rectangular polyhedral simplexes by a moding-out deformation process which preserves in a continuous manner, local commutative relations of linear approximations to local measuring.

Back to the experts

Instead

All of the various general contexts in which the Laplacian operator can be applied to elliptic, parabolic, and unbounded hyperbolic shapes are considered (by the expert professionals), wherein geodesics do not have such a well defined natural coordinate structure, and the math (instead of being linear) becomes non-commutative and non-linear, and this quantitatively inconsistent descriptive math context is (are) considered, in regard to the spectra of the many different (or other) types of hyperbolic shapes, (other types, which are different from the hyperbolic shapes associated to stable finite sets of spectral properties) and then these other (hyperbolic) shapes are related to the non-commutative math constructs of quantum theory

The current descriptive context (used by the peer-reviewed exerts) is all about empiricism and it is inconsistent so as to support only one dogmatic model about which intellectual domination is defined, yetit is also a descriptive context which is presented as being rigorous and as a description-dependent on deduction from laws, ie partial differential equations,

Consider that the assumptions of: materialism, reductionism, indefinable randomness, and the singularities of general relativity, these are the basic assumptions in the dogma associated to bomb-engineering in commercial-military interests of investment (and molecular models of biological systems)

And

This defines the range of ideas which get peer-reviewed in science (and also in math) professional settings, of either publishing or being part of an academic department (which is get university funding), so that the funding is always related to a fixed set of narrowly defined commercial interests of the investor-class,

Math and science or "the true structure of existence" can be much more interesting than material reduction to point-particles model whose only intellectual intent is for making bombs or to extend the failed molecular models of biological systems.

Appendix

Where an important invariant property (for metric-invariant maps) would be "the set of fixed points" in the coordinate space, but where a particular fixed-point type of (local) coordinate transformation can be associated to each group element,

For example:

1. no fixed points, would be related to translations (parabolic), while

2. one fixed point, would be related to reflections (elliptic), [or, perhaps, "spherically symmetric rotations"]

This would be the single point defined in the "Riemann sphere model" of the complex-number-plane, where the north-pole is the zero on the plane, and while the south-pole is "the point at infinity," so that the one-point, on this spherical model of the complex-numbers, would be "the point-at-infinity," ie the south-pole point, and

3. two fixed points, would be related to "axially defined rotations" (hyperbolic)

This would (could) define the rotation axis between the north-pole and the south-pole,

{where these two points would also be the accumulation points (or limit points) of the lattice which is extended out at the infinite edge of the lattice (or coordinate space), so the points at infinity within hyperbolic space would be the boundary of the circle or the real-line which divides the upper-half-plane, where these are two types of models for hyperbolic 2-space, and in the circle-model it is an unbounded but geometrically-bounded-construct, as a model of the hyperbolic coordinate metric-space}, and would identify a rotation around (or about) one of the shape's isolated geodesic-circle structures on the [moded-out rectangular simplex] shape, so as to also (or equivalently) rotate between a many-sheeted-covering shape, or equivalently, between different fundamental domains of the lattice defined by the shape's fundamental domain, so that the different fundamental domains in the lattice are identified (by the (local) coordinate map which is modeled as a discrete isometry subgroup) by means of a rotation, which appears to be a translational shift between equivalent fundamental domains in the lattice, where the, apparent, translational shift is really a rotation by a fixed "circular" length, which is defined on the (or on such a) stable, discrete hyperbolic shape,

Another form of transformation-invariance would be "a transformed position, in the coordinate space, remains in the same position."

Thus, consider the set of fixed points related to each of the group element's, say, element g,

The set of fixed-points is identified by, ie $gz=z$, where, $z=(x,y)$, and can be solved algebraically by using an inner product on $(gz-z)$ and set equal to zero, where the equation is a quadratic equation.

{ie find the number of fixed-points for each of the g's in the isometry group, where each group element possesses the property of having a particular number of fixed-points in the coordinate space}

So that in regard to the number of fixed points (in the coordinate space) for each g, then (as stated above) g is either a hyperbolic element (two fixed-points at infinity), or an elliptic element (one fixed point), or a parabolic element (no fixed points)

Where the elements in different discrete groups can be either hyperbolic, elliptic or parabolic, or the elements of a discrete subgroup can be chosen to be of only one fixed-point type eg every element of a discrete elliptic subgroup is an elliptic group element, etc,

Chapter 16

Many-valued-ness which emerge
from precise descriptions

(quantitative inconsistencies or stable spectra, (?) chose one or the other)

The social structure of the hierarchical society of the US society is the structure of a collective society…, as is the current configuration of our capitalism-propaganda-spying type society…, and it is very destructive of people, ie all the people become empty and unable to make judgments, in regard to either truth or value.

The many observed stable and fundamental physical systems go without valid technical description, yet, their stability implies they are simple linear systems. That is, the propaganda-education-spying social structure controls language and thought at all cultural levels (even [or especially] the educated-experts are corralled into narrow foolishness). Our, so called, technical expertise is organized so as to only provide military instruments and spying techniques.

For a new type of cultural exploration of a scientific and mathematical truth (a new vision of a technical truth), and it is a (measurable, verifiable, and practically useful) truth which also transcends the material-world,

Consider an article about H Poincare, in 4-14 AMS Notices, where, apparently, the reason Poincare was so "creatively productive" (a relative concept) was that he read so little, so his viewpoints about well-known subjects were his own individual expressions, ie it should be noted that even in the 1870's the propaganda-system was strongly affecting the vision of Poincare about what problems for him to consider, even though he was not an avid reader,

the point of a (strong) society is for its individuals to have a strong sense of their own individual knowledge (and their own relation to practical creativity) which is the reason that law is to be (should be) based on equality, but where equality is mostly about gaining knowledge, as well as gaining new useful creative contexts, (knowledge is not only about what is best for commercial interests, so as to exclude other interpretations of how a "descriptive knowledge" should be organized, in regard to the many ways in which to begin building its [own, ie a descriptive knowledge's own] descriptive language)

Where a descriptive knowledge is best related to creative efforts if it is expressed in a set of systematic (or mathematically) precise descriptive language structures, where measuring is reliable, and the patterns of the description are stable.

That is, the issue of knowledge and practical and useful creativity (which is useful for all of society) is about descriptive truths (as opposed to "the experience of" "seeing the world as it really is" and the direct knowledge to which such an experience is associated, ie knowledge directly related to our "true creative relation" to the properties of existence),

But describing the actual context of human existence is difficult,

For example:

Is the descriptive context (of human existence) materialism?

Or

Is it beyond the material realm?

And

What aspects of (both of) these two different viewpoints can be expressed in a descriptive context of stable math patterns?

Answer: A math model of that which is beyond the material realm can also contain the material realm as a proper subset within the same math model.

The material viewpoint has shown itself to be incapable of describing the observed patterns of "what are assumed to be" material systems, general nuclei, general atoms, molecules, crystals, living systems, as well as the solar-system, all have no valid descriptions (accurate, measurable descriptions of observed properties to sufficient precision) based on what are considered to be physical laws

(eg living systems display a type of entire-system control which is not describable, based on the idea that "all aspects of a living system's material existence being dependent on sets of particular chemical molecular types coming into existence at the 'correct time,' and reacting by random molecular-collisions, or by certain molecules changing their shapes, again 'at the right time,' in the local regions either within or around cells, so as to affect a nearly total control over the entire system including total body regulation biological and emotionally related, but also, in regard to free-will, where there is arbitrary control over the entire system, from total body regulation, to thought, to willed actions,

ie molecules behave in a local regional context, near cells, but their actions, or their processes, are being based (or determined by) on either distant changes or indefinable (distant) properties of the living system's experience (or sense data, or belief).)

Alternative math context for physical descriptions

Whereas, there are sets of stable shapes which can possess many various sizes, and they are shapes which can be defined in various dimensional levels, so as to be defined in similar, or analogous, manner, in regard to math constructs on the different dimensional levels, where the properties of these shapes is directly related to the spatial and temporal subspaces (as well as the discrete isometry or unitary subgroups of the metric-space's natural fiber Lie groups), which a metric-invariant metric-space can possess, where as these particular metric-spaces will have metric-functions which only have constant coefficients, and thus, they would be non-positive constant curvature metric-spaces, in particular the Euclidean spaces and the hyperbolic (or general space-time) metric-spaces, and then the more general set of these type of metric-spaces, which have various spatial and temporal subspace dimensions, etc.

Our hierarchical society with its domineering set of a few (elite) rulers make the claim that only certain types of knowledge are valid (ie the knowledge of greatest interests to the rulers), and it is certain type of knowledge which is superior to any other type of knowledge, which might (alternatively) be related to some other types of creative intent (or creative purpose). This claim of the society's dominant individuals about their possessing superior knowledge is made because the rulers of our hierarchical society only want people (ie the public) to work on the creative projects which would increase the social power of the ruling few.

This is a social construct, built out of the techniques of propaganda,

eg "scientific consensus" is the propaganda property of "individuals (in a group of people) thinking in ways so as to be consistent with the group"

And

Where the set of consistent ideas can be formed through the propaganda system, by using the techniques of substantiating claims by using authority,

Especially, when done in "the media," (which is under the absolute control by the few in the ruling-class, or by their loyal minions) where there is a carefully developed illusion of presenting information as "being observationally objective in its nature," so that authority is used "only if its authoritative claims" have been "well substantiated."

Thus institutional mangers, who along with high-ranking expert-authorities, can channel group-think (represented as being scientific consensus, ie propaganda based-on appeal to authority) into serving narrow (commercially defined) interests, and where authority is developed in regard to academic contests (along with intellectual prizes), which are based on narrow dogmas (where these academic contests form the motivational goals for the [mostly] wage-slave contestants), so that the (wage-slave) experts slowly develop a narrowly defined dogma (and where the dogma defines the rules for the academic competition), and where often the dogma is (can be) based on a partial-truth.

Thus the manipulation of ideas (beliefs) about social-value (held by people in society) can be used to turn science back into a narrowly defined religion.

This is significant in regard to the justice-system, where the judges align themselves with group-think (built from propaganda techniques) so as to "pass judgment" about "a truth" which they have no capacity to determine. That is, it is the justice system which destroys the idea of free-speech, which science needs, in order (for science) to develop in new ways (re-organize the way in which knowledge is being used to interpret observed patterns) so as to be able to provide new contexts for creative efforts, ie new entries into the so called free-markets, but it is really the interference with the so called free-markets, which most caused to investor-class to interfere with science and the propaganda-education system.

That is, the justice system plays the role of the high-priest for the religion of commercial interests, ie Calvinism

Thus the justice system is demanding that the public worship religion of a certain type and in a particular way, ie the religion of materialism as expressed by (big) commercial interests (banking, fossil-fuels, and the military) [the religion of materialism, guided by the commercial interests of the few, and illegally (so as to destroy the development of knowledge) by the justice-system, who act as the minions of the ruling-class].

Then the justice-system colludes with the spy agencies to exclude (or exterminate) the expressions of ideas which are in opposition to this (or their) religion (this has been done, at least, since around 1900, since N Wiener's "Cybernetics," 1948, outlined the context of spying on communication channels (though not explicitly saying the word spying), as a central theme of his discussion about information theory)

Spying is not about protecting the nation, rather it is about "protecting the few" in the investor-class (from risks) ie protecting those in the ruling-class who are driving this fixed viewpoint of knowledge (and its strong relation to), creativity, and trade (ie the artificial control of markets), but this "protection (of the investments of the few)" has rotted-out the, so called, (scientific) belief in correct-practically-useful-knowledge, which has also been a hallmark of the western society, ie the Roman armies built metropolitan regions by (their ability to engineer) using bricks.

The well-educated propagandist, ie the intellectual, is the fiercest defender of the investor-class's dogmas,

While the similarly dogmatic (and similarly arbitrary), often expressed as racism, militant types, ie people of action, who are aided by (and manipulated through, and preyed-upon by) the spying-justice-system, in their acts which terrorize the public, (both groups of people) are used to force society (by physical or intellectual bullying) into the propaganda-developed and very narrowly defined type of group-think

The right claims to support individual rights but supports in a collective manner the rights of the ruling-class to destroy the rights of other individuals

While

The left claims to support peace and humanity (or at least they used to) but they act to support the collective intellectual structures needed for war and doomsday instruments.

Both sides oppose equality and support extreme violence and aggression which is based-on elitism and its related support for an arbitrary hierarchy (the two-sides action and intellect) the result is a regimented narrow society run by an elite who believe in extermination of either intellect or "the many."

Both sides are "turned-around" by both the techniques of propaganda and the hidden terrorism managed by the justice-system, that is, this "turning-around" is done through the illegal rotten-ness of the justice system, which, effectively, only allow one belief and one idea to be expressed (an arbitrary value which defines inequality).

It is the justice system through which the secret-police and the spy-state are created, and this spy-state is used to interfere, in a hidden manner, with individual freedoms, so as to interfere with new ideas, ie it stops free-speech in as "dishonest and unjustifiable manner" as can be imagined, all so that the ruling class can become ever more dominant.

The moral and honest ones (who are a part of this justice system) "whistle-blow" and resign, while those who compose the rest of the system are clearly rotten to the core (they will do anything so as to be given a few crumbs by the ruling-class), and yet it is the justice system which imposes a wage-slave religion, in the form of a required worship of the material-world, on the rest of the population, so that this material-world must be thought about "as the ruling-class wants the public to think about the material-world,"

ie so as to protect the petty investment interests of the ruling-class,

Yet, these people in the justice-system play the role of the society's high-priests, and they sneak into their hidden confessional closets (of the spy-system), so as to be able to destructively interfere with the better development of the society (for a wider range or new ideas about value and truth).

It is this rot at the core of western society, (the subservience of public institutions to the rich, as done by weak and corrupt men, and which is not witnessed by the group-think, wage-slave public), which has resulted in the loss of creative development, the collapse of markets, where markets are (have come to be) too narrowly defined (essentially, fossil-fuels, and military-and-spying equipment) (caused by stupid, selfish, and brutal domination of the public institutions by the rich), so that social power has come to be re-defined as control of only a few types of material resources, and only a few military-ways in which to use and develop technical instruments, as well as the arbitrary ways in which propaganda and hidden violence (spy related interference and violence, ie easily covered-up) can be used to manipulate markets, and politics,

[ie why enter politics so as to be shot-up by a gang of spies, eg the Kennedy, King, Kennedy assasinations], as well as all of society, (but it is a society which has very little to do, ie the dogmas go nowhere, and the people are so empty that they have lost all judgment, and they do not know what is happening to them [hidden actions, and hidden purposes] so the people follow propaganda, rather than "assert, know, and create for themselves").

Alternative viewpoint about both humans and how society should be organized

That is, this (viewpoint about western society) is opposed to the viewpoint that the human condition (is about being creative, and) is to be expressed in a context of equality, in regard to both gaining knowledge and engaging in creativity, and it is only when an (associated) equal sense of free-inquiry is allowed to be expressed, in the realm of practical creative efforts, that the notion of a "free market" might actually exist,

That is, the society must ensure we are each equal creators, ie the society supports everyone, and, in turn, everyone creates things and, thus, develops knowledge over a wide range of possible descriptive languages,[thus people will not feel empty within themselves, and thus not be so easily manipulated]but many restraints on products..., which are traded in a free-market (eg products resulting from a productive team process, or team-effort) which are the, so called, "best sellers,".... need to be considered, since it is a robbery to invest, for selfish gain, in a product which already has a very active market.

That is, "best sellers" need to be turned-into (communal) group creative efforts

(Note: Being a best-selling product does not [necessarily] mean it is a superior product).

In fact, that is exactly what predatory capitalism tries to do, ie make great gains by investing in sure-bets, ie in best-seller products, in the (now) controlled-market, but this type of investing is done in a rigged game, which is controlled mainly by propaganda, by means of possessing control over the media, ie where the media is controlled by the investor-class.

Furthermore, copyright laws are all about lying and stealing for those who..., which our society now allows to..., have total control over the propaganda-education system, ie nearly total control over markets and the education system, so that education is only defined for those categories which have commercial value to those who control both the propaganda-system and the system of investment, where investment, in turn, controls the market (or can be used to control the market).

The society as it is now configured is a collective society which supports the super-rich, in opposition to this type of social organization consider a society where individuals collectively seek knowledge which they want expressed in ways so as to be able to create the types of ideas which they want to creatively develop,

So

The collective aspects of a society (the new type of society, or really the society envisioned by T Jefferson as well as the [Indian] seven-nations or Iroquois) of equal creators, should.., first and foremost.., be "all about knowledge and creativity, but also there should be focus in regard to "successful" products, which have been created," in that, the, market-wise, "successful" products need to be made collective, so that selfish domination does not gain a controlling social position in a society based on investment, ie making money the basis for market trading, and propaganda

However, in a society based on investment and propaganda, there would be an advantage to possessing an unbalanced personality with an interest in domination, so as to be psychopathic, quite similar to many of the psychopathic Roman-Emperors of the Roman-Empire

This inclination towards social domination of the many by the few (ie the western culture is still the [psychopathic] Roman culture of domination), ie the few who compose the investor-class, causes a control over knowledge (by the investor-class), since there is a knowledge-creativity relation, and creativity is all about production and the subsequent control of markets, so that knowledge has come to be (is) restricted to a small set of commercial categories, so that these categories are to be the focus of narrowly defined small set of dogmas (which are sufficient to maintain, and slightly advance, the commercial production in a particular category).

Thus

1. physics is bomb-engineering

2. Math has come to be about fake quantitative structures, applied to an indefinably random descriptive context, so that there are only vague temporary quantitative patterns, so that all quantitative expressions are arbitrary, and only relate-able to an empirically identified (and fleeting and unstable and temporary) quantitative relations (that is, the patterns do not emerge from the deduction of physical law, and they do not emerge from a deductive process in math when the assumptions are that things are random and the math models of measuring are not reliable measuring processes, Note: Feedback systems are based-on empiricism, thus they are tentative and can fail suddenly, though many feedback mechanisms can make them more reliable.

and

3. Biology is about manipulating the protein molecules, which, in turn, are associated (or correlated) with the molecule DNA

There exist "incredible levels of stupid" for the small set of carefully identified, and narrowly defined academic categories, whose properties have just been correctly identified (ie in the above section).

This un-natural narrowness of science has to do with the control of thought afforded to the ruling-class by controlling the propaganda, and allowing (or requiring) academia to be guided (or administered) by narrow viewpoints, ie the categories of commercial interests of the corporate institutions, (this was done using the techniques of propaganda (see below), which was (has been) possible because of the great control the ruling class has had over the media, and this was a result of the [illegal] doings by the justice-system)

ie (furthermore) controlling the personnel in the field (or academic category), by the institutional mangers, so the personnel in the field are autistic, but with the capacity to use language, yet also personalities whom are very competitive, and desirous of being in dominant social positions, so as to be a set of (academic) people who obsessively try to extend the dogma of one of the few categories of expression which is allowed in the professional academic settings, within society, and then turning the education system into a competition, which is concerned only about extending these few and narrowly defined dogmas, which are associated to fixed commercial interests of the investor-class a competition (whose rules are defined by authoritative dogmas) which is only about obsession, as the narrow usefulness of these narrowly defined dogmas has shown,

ie these narrowly defined dogmas, now, have very limited practical usefulness, other than with respect to the narrow commercial use of the knowledge as it was originally used and small extensions of the uses of the knowledge, and thus, creative development has bogged-down, and the resulting narrowness of technical development is causing both no-growth in creativity and the continual use of the same material-resources which are quite often associated to very toxic poisons.

The small number of narrowly defined dogmas are only related to the fixed commercial categories, and often this relation is only tangential, as is demonstrated by particle-physics, which, in regard to its practical uses, is only relatable to bomb engineering, eg it cannot be used to describe the observed spectral properties of a general nucleus, since a bomb's reaction-rate is about the cross-sections of particle-collisions, and this property is totally unrelated to the main problems of physics (which is a clear statement, since the main problems of physics have not been solved using this viewpoint),

Namely, (the main problems being...) determining the stable properties of the observed material systems which surround us, quite often these are the stable many-(but-few)-body systems associated to spectral-orbital properties

eg atoms, nuclei, molecules, and the control of molecular shape, and describing crystal properties, as well as the stable structures of highly controlled living systems, as well as describing the properties of a stable solar-system, etc etc.

The academic subjects of:

1. quantum physics,

2. its, associated particle-physic's model for material-interactions, and

3. general relativity,

and

4. all the other theories derived from this core narrow viewpoint, eg string-theory, are all deeply flawed viewpoints..., where the focus on such topics implies that the intent of the commercial (or product) interests of the investor-class is about basing (or developing) an overwhelmingly explosive bomb based-on elementary-particles and gravitational singularities, ie a doomsday instrument of destruction.

Where

These subjective contexts (subjective, since these categories of description do not answer any of the fundamental questions of physics, and thus they are merely subjective speculations about the nature of reality) which have been instituted, by using the techniques of propaganda:

1. Fear (wage-slavery), and emotion [or desire] (winning the contest, and seeking absolute truths [for domineering personality-types])

2. Authority (establishing a small number of categories for expression and their associated narrowly defined dogmas)

3. Using group-think and (possibly subtle) coercion

This is easy to accomplish by the ruling-class, by both

1. their use of (supposedly, charitable) corporate-foundations,

as well as

2. controlling the instruments of social expression (or social communication), ie the media,

That is, one can continually repeat the same formula (the same expressions of ideas) so as to gain emotional and intellectual control over the population, who are all wage-slaves, and whom are forced to deal with a single-type of a collective form of propaganda which supports (or a propaganda construct which applies to a collective (so called, capitalist) society, whose collective purpose is to support) a narrowly defined market structure, which is highly controlled, and unnatural (ie many "markets" exist only because they have been created by the propaganda system), and associated to a small set of fixed categories of creative efforts.

Social destruction

This total destruction of the social structure of the US society…,

Where, in the US, law, it was claimed (in the Declaration of independence and then supported by the Bill of Rights…, is supposed….) to be based on equality (where the US tried to follow the construct of a nation which was accomplished by the 7-(or 5)-nations of the Iroquois-confederation), but instead (total destruction of the social structure of the US society) was accomplished through the very corrupt justice system, which is so corrupt and rotten, if the nation does not re-form a new continental congress, to do away with the rot, then things will simply get worse.

The main use of spies is, really, to sabotage any new ideas which might be expressed (total spectrum-domination)

The destruction of science

None of the science theories:

quantum physics, particle-physics, and general relativity and their derived theories, should be considered to have any authority, when considered in regard to their descriptive merits (eg accuracy to sufficient precision and over a wide range of systems, and possessing practical usefulness, ie providing new creative contexts which are actually productive, ie can be realized, ie not simply a part of an illusionary world whose only reality is the, so called, "rigorous" words which the descriptive context uses)

and

One can notice that the "progression of ideas" (or the relation, or the correlation) of the ideas in physics (which get a lot of coverage in the propaganda system) to the commercial products,

ie communication systems which are only considered in regard to their military uses, that is, their use for spying on the population, so that no new ideas are allowed to be expressed,and the other military-products to which these half-baked concepts (these (1/10)-truths) of particle-physics apply, these, so called, dogmas of physics, only applied to the nuclear bomb, to the fusion bomb, and now they are being extended to a "gravitational-singularity-elementary-particle-collision bomb,"

That is, other than making these instrumental-objects of explosive destruction, these (what are considered to be) fundamental ideas about physics are, in fact, unrelated to the fundamental questions which should be the main interest of a physics department in a public university (ie describing the stable properties of fundamental physical systems, eg nuclei, atoms, molecules, etc), and along with explosive destructive capabilities, there are also the environmental destruction which is now a necessary conclusion from these efforts to purify these poisonous substances, which need millions of years before they are again safe,

ie civilization is not capable of providing the safe storage for these poisons until they are again safe.

Consider the descriptive basis for quantum physics

First:

Classical physics is continuous and geometric, thus, in the macroscopic regions where measurable physical properties are defined as functions, one wants the functions to be single-valued, thus neither non-commutative-ness nor holes defined in the regional domain space are desired properties for this descriptive context.

Classical physics is based on a local linear model of measuring, so that this local linear model of measuring has an (almost) linear inverse operator, which relates local measures (at points in a spatial region) to geometric measures of boundaries (of that same region), in this context one seeks functions which are single-valued, ie there are no holes in the system containing metric-space,

That is, it is a metric-invariant description which is consistent with geometric measures, and depends on geometric models of material interactions, ie measuring and the physical system are separate (and mostly independent) entities (or the interaction of measuring can be modeled into the system's description in a consistent causal manner).

The local linear models of measuring are associated to the "laws of physics" which identify how local models of linear measuring are related to (partial) differential equations, whose solution functions define the measurable properties of the system.

Unfortunately, this descriptive structure (in which it is assumed there are no holes in the material-containing metric-space) is only "linear and solvable" for either the separable shapes of a physical system, or for the two-body system, ie only for the linear, metric-invariant, and continuously commutative relationships which exist between containing coordinate (shapes) and measurable properties (or function-values) of the system, where the system's object-or-its-components are (supposed to be) related to the local coordinates, and were often the measured system-properties are the properties of position (and position in regard to distant material-geometry) and motion of the system's components within the system-containing coordinates, as well as (other) vector fields, such as force-fields (where vector-fields are often the descriptive context of physical laws).

Here, one is already starting to see problems with the idea of materialism and the dimension of a material-type's containing coordinate-set, eg most classical physical systems determine non-linear (partial) differential equations, including the many-(but-few)-body system, eg "why is the solar-system stable?"

Then (after classical physics):

Quantum physics is based on the set of stable discrete spectra, which are observed for small quantum systems, such as atoms and molecules, but also for some larger systems (with very stable descriptive contexts) such as crystals.

But the systems are assumed to be composed of elementary-point-particles whose properties of motion and spatial-position are random properties,

{However, for an alternative viewpoint, for the small (microscopic) regions where there are separate discrete (spectral) values for the same measurable property which exist, then one wants this small region, where these discrete spectral values are defined, to (possibly) be associated to a shape which has holes in its shape, ie in the geometric region where these many spectral values are defined.

Note: The Brownian motions of these small shapes results in the same properties as a quantum descriptions, ie so as to appear to be fundamentally random.}

The observed characteristics of the quantum system are either about light spectra, emitted by the system, with (assumed to be) particle-components changing their energy-levels in the system so as to cause the light-emissions or about identifying the components of the systems (where the components are assumed to be elementary-particles, randomly distributed point-events in space, which are also contained as material-entities in a particular type of dimensional coordinate-space, ie the domain space for the functions in the function-space assumes the idea of materialism) by identifying point-particle collisions of the system's components (assumed to be in a particular energy-state), at points in space, whose position (or motion) in the coordinate-space is determined by a random spatial distribution structure associated to a spectral-function, where the spectral-function exists within the (spectral) function-space model of the quantum system.

Thus, the assumption has become that "all quantum systems are contained in function-spaces," where each of the individual functions are globally defined over a (or an assumed to be) system containment set of coordinates, ie the assumption of materialism.

The classical system has natural bounds, as well as a natural separation between "the measuring systems" and "the system being measured,"

But

The assumed containment within sets of global functions is an unbounded structure, which isolates the system from all aspects of the outside world, except for internal changes in the system, when the system emits light, ie transitions between the system's set of spectral-functions, (isolates the system from) both external measuring structures and any other set of outside physical systems.

Yet, there are usually many quantum systems surrounding any particular quantum system.

Thus, other than light signals from within the quantum system, there is no valid model for determining the properties of the quantum system's components (ie component-properties), for either finding (measuring) a quantum system's component's position (or motion) within the system or for determining a relation of a quantum-system's components to "distant" geometries of other (quantum) systems.

However, the main problem with this abstract model of a quantum system is that: for most general quantum systems which are (also) fundamental quantum systems eg general: nuclei, atoms and molecules, these sets of spectral-functions cannot be found, by calculations which have been deduced from the, so called, laws of quantum physics.

This means that many other ideas about the nature of the, so called, quantum system need to be considered.

A remarkable set of patterns concerning quantum description is that

1. The solution to the H-atom's radial equation diverges, so this means that Bohr's model of the H-atom and Sommerfled's elliptical-orbital perturbations of energy...,

ie angular-momentum can have a tangent component or equivalently rotational structure can affect the energy-levels of the quantum system, is the more rigorous model, ie more logically and quantitatively consistent math model of the H-atom,and Bohr's model is a bounded (geometric) construct,

2. The quantum system is mostly (thought about as) a closed and bounded structure (though its function-space model is not consistent with this idea, ie its spectral-functions are global functions), which,

3. Though quantum systems are, nearly, always related to charged components, the typical quantum system is mostly neutral, when its properties are associated to other surrounding physical (quantum) systems, though there are many exceptions to this property of possessing charge neutrality.

Yet, there are few valid energy-operator models of these structures, which is basically because the function-space models are built of global independent spectral-functions, and the model gives no method for providing a set of spectral possibilities (or spectral bounds), eg limiting geometric measures upon which to estimate a spectral-value, for the quantum system, whereas

Bohr's model is related to such spectral bounds in a relatively simple manner, yet it is a construct which seems to not be generalized or extended to more general systems ie the many-(but-few)-body stable quantum systems,

4. Furthermore, the only particle-collision model of particle-component interactions which comes close to physical relevance, is in a quantum model of a crystal which is bounded, and it is the phonon-electron collisions in a crystal-lattice, or electron-crystal-vibration interactions, which is an interaction which is supposed to bind two-electrons together in a cooled crystal-lattice (ie two-electrons which are contained in a vibrating crystal),

However, the critical temperature which this model identifies has been exceeded, so it has not been a (truly) successful model,

And

The quantum model of the crystal is simply that of the Euclidean lattice, or rectangular-shape, which, in turn, is moded-out (ie opposite faces to the rectangular-shape are identified with one another) to form a toral geometric structure (so that the internal properties within the rectangular-shape of the crystal's shape are made periodic), though there are many variations of "the fundamental cell of periodicity" for the crystal, there is first the boundaries of the actual crystal's size, then there is the decomposition into further cells-of-periodicity where these small-cells fit into the larger crystal shape as a toral-shape.

That is, the periodicity of the relatively free-electrons in the crystal are being affected by different (other) types of periodic shapes (or periodic distances) within the crystal, but it is an open question as to what these competing or inter-related shapes might be.

It seems that the quantum properties, which are identified as being global and unbounded, in turn, identify an un-natural closed-ness and separate-ness from the physical world, of which they are supposed to be a part.

How is one supposed verify the claim that random electron-events are distributed in space in a particular way. One would need to detect random electron point-particle events in space by means on their collisions with other elementary-particles. So the wave-operator for such a detection system would be the quantum system and a set of other elementary-particle streaming through the region where the quantum system exists in space, but since (1) the original quantum system cannot be solved, and (2) the quantum wave-functions are global, ie defined over the entire coordinate system, and thus the statement "where the quantum system exists in space" seems to have limits as to the meaning of a wave-function for small (eg atomic) quantum systems, thus, this is a pointless exercise

Furthermore, since there are virtually no valid quantum based descriptions of the stable properties of Nuclei, general atoms, molecules, molecular-shapes, nor crystals, it is very difficult to take anything…

which is claimed about "quantum theory being a valid model of the observed patterns of existence," seriously, and since local "atomic" motions are Brownian, and such a (Brownian-motion) context can be used to derive the properties of quantum description (E Nelson, 1957,1967 Princeton), this Brownian motion of atoms and molecules is the likely origin of most of the observed quantum properties, ie the set of observed random properties upon which quantum theory is based.

{So stable shapes of various dimensions and various sizes might be a better model for the observed quantum properties of quantum systems.}

Furthermore, the experimental claims which are made in quantum physics (as well as in particle-physics and in regard to general relativity) are often made "not from experimental evidence," but rather from dry-lab data.

This is how R Millikan dry-lab-ed his results to show that the electron has a fixed quantum of charge associated to itself, and subsequently Millikan won the Nobel prize for this, so called, experimental finding.

That is, the lab results of quantum experiments should not be taken too seriously, the so called lab results are really expressions of authority (used in the propaganda-system), and not real lab-results. The physics is being pushed exclusively into bomb-engineering efforts, due to pressures from the investor-class.

Math and its obsession with quantitative inconsistency

One finds in academic-math peer-reviewed articles, an expression of narrow dogmatic contexts of math, in which the math community seems to only consider some of the most general and abstract descriptive contexts, which the math community considers to be the proper subject matter, in regard to math descriptions,

ie descriptions which are non-linear and/or possess non-commutative (sets of) operator relations

(ie operators defined between local domain values and the local function-value properties) so that this math structure implies a "quantitatively inconsistent descriptive context" for the general and/or abstract patterns which the math community is trying to describe.

On the other hand, they do not consider simple stable shapes as defining a varied enough descriptive context,

Yet, the complicated and very general and abstract shapes (or patterns)…, which they do consider…, are not stable patterns, and the quantitative structures which are supposed to be used to describe these general and complicated patterns are not quantitatively consistent.

And the professional peer-reviewed mathematician uses:

set structures and set-containment contexts which are not logically consistent

eg sets which are too big, are used, eg the set which contains everything, or the set which contains all possible values of (an uncountable number of) points on a line,so that the line, itself, possesses too much information, so that its construct becomes meaningless,

ie points take an infinite amount of information to identify (the number), the axiom of choice does not define a realizable end-result

Also the quantitative inconsistency (eg non-commutativity and non-linearity) of the (local) operators associated to complicated shapes leads to issues of multiple-valued-ness (of the function's values), eg bifurcations,

As well as to, chaotic relations which are associated to initial conditions, or boundary values, ie small local changes can result in very large changes in function-values,

Note: There are four important issues in regard to non-linear (partial) differential equations

1. A (partial) differential equation of a non-linear system has critical points, and associated to these critical points are (often) the geometric-patterns (defined on the coordinate-domain space) of limit-cycles so that associated to the limit-cycles are quasi-periodic quasi-patterns

2. Both of the, so called, solution function (numerical approximation to a solution-function) of a non-linear system there are bounding conditions, which can cause the property of self-organization, as well as the local quantitative conditions which can influence both the properties and the definition of the system's (partial) differential equation

3. Both bifurcations and chaos can suddenly change the "solution-function's" quasi-pattern (or shape) as well as its quasi-periodic response, or both,

4. Either chaos or different conditions in the system-containing coordinate-domain space within which the non-linear system's (partial) differential equation is defined can cause a change in the nature of the non-linear system's defining (partial) differential equation, and thus everything about such a non-linear system can also change (suddenly).

Thus one can ask:

"At what size-scale do stable shapes of various dimensions and in various subspaces (as defined by a dimensional-size-subspace related containment tree, which is defined on an 12-dimensional generalized space-time metric-space, so that the associated metric-space (discrete) stable shapes are characterized by their genus (ie the number of holes in a stable shape), ie characterized by the stable set of spectral-orbital properties, enter into the descriptive context, so as to become the dominant property, in the new "size-scale-dimension" context of "system description?"

These considerations are also related to the issue of finding the spectral-orbital structure (fixed spectral-set) of a system description which is associated to a over-all containing 11-dimensional hyperbolic metric-space (or 12-dimensional generalized space-time metric-space) in regard to descriptions about the galaxy, dark-matter, (apparent) universe expansion, and the life contained in this (math) descriptive construct.

This is about being able to think in a creative context, where the descriptive context allows for control over (stable) systems by imagining in a local-measuring context in which a system's (partial) differential equations are defined.

Note: The stability of the observed properties of quantum systems implies that these systems are formed in a context where the math context is controllable (and not in an indefinably random and uncontrollable descriptive context)

[This note is necessary, since brain-washed and propagandized people will dismiss this idea, based on the failed authority in which they have been led to believe (by the way authority is used in propaganda), about physical description being necessarily based on indefinable randomness. This is a construct of propaganda and group-think, which is also called scientific consensus, but science is (really) developed by individual efforts, and its development is about questioning the assumptions, interpretations, and contexts of an, overly authoritative, precise descriptive language]

Furthermore,

Issues about an inability to identify measurable values, in "what is supposed to be" a measuring process,

And

Stable patterns cannot be identified in such a "claimed to be" math context.

Plague the current overly authoritative and very speculative context of physical description

Holes in space vs. quantitative inconsistency

And

the relation of these two constructs to the property of the many-valued-ness of a function's values

The multi-valued-ness of a function can come about due to its local quantitative structural relation to the domain space (or coordinate space). If this local structure is non-commutative, then this can lead to such a function being many-valued (ie it can bifurcate), or if the local expression of measurable properties, ie a (partial) differential equation in a system containment space (or domain space), is non-linear, then such a function can (also) become many-valued (as well as being chaotic, ie small changes in beginning conditions can cause large changes in the function's values, ie values do not maintain a continuous relation to their near-by values, this could be a form of bifurcation with one value changing a large magnitude while the other value being microscopically small ie an inverse multiple relation),

where both of these many-valued function-properties (as well as chaos) are due to quantitative inconsistency

so that number values discontinuously take jumps (to other values), or pairs of number values simultaneously take discontinuous jumps in value, or triples of…etc, etc.

While

a directly measured property of a coordinate space can also become many-valued (or the value could also be a function) due to the measuring being defined on a space whose shape has holes in its geometry, ie the shape will not continuously contract to a point, since the holes get in the way of such a continuous contraction math process.

But in some space with holes these spaces (or shapes) can be very stable so that the many-valued-ness property (of say a line integral, eg to define the value of the system's energy) would (could) be a stable part of the quantitative structure.

Consider either holes in space or quantitative inconsistency

Continuity and single-valued-ness, of a contractible shape (or space) [ie contractible to a point]

vs.

separate and discrete (spectral) values; of measurable physical properties, may be about the holes which exist on the regional shape (where the shape's dimension might have various, but fixed, values)

where the properties are being defined, ie and where the shape (or space) is not contractible to a point.

That is, issues about path-invariance, in regard to values associated to a curved-line model "of measuring a quantitative value," which is to be defined between two points in space, can be related to a context concerning the existence of holes in the region, wherein the measured values are defined,

(eg this can be the [natural] coordinates of a shape or the value of a system's energy-function, etc) ie on contract-able shapes there can be defined path-invariant values (of measurable properties) which exist between pairs of points, but when path-invariance between two-points is not definable, due to there being holes in the region where the measured values are defined, then there are several spectral-energy paths, which can have different spectral-energy values,

Where each different spectral-value can be associated to a different hole in the shape of the region where they are being defined, so that the energy of a system's component depends on the spectral-path within which the spectral element is a part, it depends on the hole which the system's component surrounds.

This is the fundamental distinction (which should be) made between different types of stable shapes:

(1) the property of path-invariant values ie single-valued-ness of a stable property (the curved path is defined on a constructible shape),

vs.

(2) the property of multiple spectral-values for (different) discrete stable properties for a system's component within the system.

That is, does one want many-valued-ness to represent a natural discrete difference between spectral paths (which the system's components can surround) due to a region's shape (where the shape is stable)? or Does one want to venture into (an assumed to be) a math realm, where the containing region is contractible, but where the quantitative local relations "between (function) properties and their domain space" to be non-commutative, in which case there is neither measuring reliability, nor is there any stable patterns which can be described (due to bifurcations and chaos)? but

Rather, (in the non-commutative case) there only exists randomness, which is always in a state of change (or flux) and it cannot associated to any stable pattern

But

Stable patterns are observed for many fundamental physical systems

Yet

It is the sets of general and abstract and indefinably random descriptive contexts which are the main contexts which the math communities consider, and where it is (still) assumed that such indefinably random (and quantitatively inconsistent) quantitative constructs are considered to be correct descriptive structures by the peer-reviewed math community, apparently, they believe that there are still (stable) math patterns being defined in such a context where (local) measuring is unreliable.

Only an obedient fool would fall for this!

How many levels of stupid does this represent

It is about authority, and about being dominant

It is a social construct (this is about the properties of people, in regard to the social context in which a propaganda system is being used ie their desire to be dominant (associated to the context of a social hierarchy), as well as the affect of the confining social force of group-think (a social force which propaganda generates), and it is about the social construct of authority and obedience to authority (the, apparent, primary strategy of propaganda systems in an authoritative and hierarchical society),

Ie it is not an issue concerning rational attempts to describe the observed properties of the world

Though there are various interpretations of (or conclusions to which one can come to, concerning) these math constructs,

the current dominant conclusion is that (seems to be) that stable patterns must be contained within a space that is contractible to a point, and upon such domain spaces global functions in functions spaces so that each such spectral-function represents one of a system's measurable spectral values, and (this particular spectral function) also represents a spatial distribution of a random point-particle's events (or positions) for the component associated to that particular spectral value when it is contained within the system (which is modeled as a function space of spectral-functions, even though the (operator) math of this function-construct is non-commutative)

Another type of conclusion could be:

The point of math descriptions for physical models are about either local linear models of measuring, which can be related to an inverse process (or inverse operator)…, which relates local linear quantitative relations between the system's properties and the system-containing coordinate space to a sum of (normal, and geometrically determined) vector-field properties…, defined over geometric boundaries of a system-containing region in space vs.

A set of functions in a (harmonic) function-space which represent the system's various sets of spectral properties, where the different functions represent a global model of a system-component's (or an elementary-particle's) random model in regard to their positions in space and time (for each different spectral-function), where the function is related to a random geometric distribution (but an average geometric distribution, defined for an averaging process) for one (of the system's) material-component in space

(ie randomly distributed as point-particle events in space),

ie an average geometric shape associated to the separate detection of a random particle-event in space for an elementary-particle, which possesses a definite spectral value,

Thus, there are properties:

So that the geometric sizes for these random particle-distribution shapes would be the (main, or only) relation of such individual spectral-functions to spectral valuesbut for realistic models of what are often assumed to be spherically symmetric potential-energy term (such as for an atom) of a general many-(but-few)-body stable spectral system, the property of spherical symmetry cannot be maintained, so that such a model cannot provide valid representations of spectral values associated to geometric measures, which are, in turn, defined on random geometric distributions (associated to such a given set of different discrete spectral-values) of such a general quantum system.

That is, one wants to be able to identify spectral-approximations from one's geometric model of the spectral-system, since spectral-values are easiest to conceive in the context of their spectral-lengths.

This leads to an alternative model of spectral-systems in which the systems are stable metric-space shapes

That is, the system's size-scale and its measured properties must be put into a relation in which one compares the system's size-scale to the size-scale of the (material-component containing) metric-space (shape), so as to be able to determine the type containment shape to be associated to the quantitative model of the pattern of interest (ie the stable geometric pattern of the system). However, there is also a change in dimensional-level associated to a material-component (with a stable shape) and its containing metric-space (stable shape).

The set of stable shapes [where these stable shapes are related to rectangular simplexes {or fundamental domains (ie of a checker-board type lattice partitions of the metric-space base-space of a "principle fiber bundle")}] was identified by Thurston-Perelman, ie "geometrization," where the bulk of these simple stable shapes are the discrete hyperbolic shapes.

And (previously) their properties which include many-dimensional properties were elucidated by D Coxeter.

And their geometric-coordinate properties which relate a small set of geodesics on these discrete hyperbolic shapes, to stable spectra, eg geometric-measures (all related to 1-faces) of the different dimensional, discrete hyperbolic shape's rectangular-simplex faces (Eisenhardt, Gromov, etc). Where these different dimensional shapes and models of both material and metric-spaces define the correct context for the descriptive properties of general relativity. These stable shapes are (in regard to their local differential equation representations, or models, of their stable shapes) are linear, metric-invariant, and continuously commutative (except possibly at a single point) differential-forms.

Back to the (difficult to handle) model based on randomness

There are a number of things which the random-based description of quantum physics claims, but there does not seem to be a proof that these claims are true

1. that the randomly determined charges of an atom (or any quantum system) will arrange themselves in a random context so as to be charge neutral

And

2. That the spectral-values of light are related to the geometric changes of average random geometric distribution-shapes of the charges of an atom (or quantum system) (associated to one [or associated to the pair] of the quantum system's individual spectral states) when the atom transitions between the system's spectral states

And

3. That one can determine the geometric distribution of the set of random elementary-particle events associated to a spectral-function (which represents a particular spectral-value for such a point-particle component of the system) in its global coordinate context.

But

4. The main problem with the current theory of quantum physics is that, for the general stable many-(but-few)-body quantum systems, as well as for crystals, (as well as for general quantum systems, as a whole) there are no sets of (actual) solutions in the spectral function-spaces, ie there do not exist sets of operators defined on the function-space model of a quantum system, which, actually, provide sets of spectral-functions, whose sets of spectra are consistent with the observed spectra of the physical quantum systems.

5. That is, the random basis for quantum physics is mainly a way in which to introduce the idea of quantum material-interactions as only being modeled as point-particle collisions between elementary-particles and it is a model which is only relatable, through the measures of elementary-particle cross-sections, to rates of reactions for a transitioning process, which is transitioning in a chaotic and random manner between two relatively stable states

Note: it is the stable states which physics needs to try to describe,

And the random particle-collision context is only relatable to bomb-engineering,

Note: This is provable by induction:

since after nearly 100 years of particle-physics it still has not been able to describe the observed stable spectral states of many of the general nuclei of general atoms

Where is the, so called, rigorous (peer-reviewed) literature which discusses these issues (in a logically consistent and quantitatively consistent context, rather than in an arbitrary context of intellectual dominance, ie as arbitrary intellectual dominance is put-forth in peer-review publications and the illusionary speculative claims about, an assumed to be, material existence, and how a materialistic-reduction to elementary-particles and placed in a context of indefinable randomness are to be put together in a coherent math descriptions, which are both accurate (to sufficient precision) and have practical creative usefulness)

6. The relation of knowledge to commercial interests causes such commercially related "knowledge" to come to be about partial-truths, ie essentially, lies about science and existence, and these lies lead to an inability of mankind to imagine and create for themselves, so that instead their efforts are designed to only serve the narrow (commercially domineering) interests of the few.

Eg doomsday bombs vs. the truth about existence and new contexts for creative efforts

[this is a true seeking of the spirit]

Furthermore, does humanity want there to be a 50-billion (or 500-billion) dollar a year science-program about bomb-engineering, which is trying to produce a doomsday machine, so that this is to be considered the highest intellectual level humanity has achieved,

Or

Does humanity want there to be a science program trying to map the spectral structure of the 11-dimensional hyperbolic metric-space, "partitioned" into dimensional-subspace sets of stable shapes, a mathematical viewpoint which has solved "the stable structure of the many-(but-few)-body spectral-orbital systems for all size scales," so that this gives "to mankind" a further… (beyond simply solving the fundamental problems of the stable many-(but-few)-body spectral-orbital systems facing physics), map of the structure of existence, within which mankind exists, and the range of possible knowledge and creative relations within which mankind can be a part, in regard to their creative actions and knowledge-seeking journeys.

New Creative possibilities (or new creative contexts)

What type of relationships can exist between the many-dimensional structures of stable shapes in a new type of "dimensional-spectral-partition?"

1. The containment of lower-dimensional or the same dimension shapes which are sufficiently small (so as to be contained) within (larger) shapes of different dimensions, where the size and dimensional properties allow such a dimensional-subspace containment tree.

2. Orbital relations between the contained (condensed) material and the shape (and geodesic structure) of the material-containing metric-space, in regard to the metric-space's geodesic structures, as well as an alternative possibility, where a material-component possesses a "perfect-fit" so that the material-component occupies the containing (stable) shape's facial geometry.

3. Spectral relations

(i) a process of continually establishing resonance (eg processes associated to the energy of the existing stable spectral-material-systems)

(ii) bringing new resonances into being (eg collision-energy and the subsequent resonance of a newly formed system)

(iii) component containment, so that the contained components are being organized and guided (or controlled) by resonances between the components and the spectral-shape of their containing metric-space shape, which can cause new shapes to form, and which can be relatable to changes in angular-momentum (and associated Weyl-angles which can exist between the toral components of the stable discrete hyperbolic shapes, eg changing of molecular shapes due to (Weyl) folds between atomic components of a stable molecular shape)

(iv) the changing of internal components (of a living system) due to a controlled spectral context of a stable-shape of the component-containing metric-space shape, where the (rectangular-simplex) faces of the containing metric-space can be used to define the resonances which, in turn, determine what components can exist.

4. Bounds and relations to a containment space

(i) bounds of shapes

(ii) discrete changes

(iii) the relation of unbounded shapes to being contained in a subspace so as to be bounded by the unbounded shape's containing metric-space, ie the bound on the one metric-space shape corresponds to the infinity which the lower-dimensional shape reaches, so that this relation is a relation between shapes ie not so much about limits and convergences (or the limits and convergences are determined by shapes). That is, some of the spectral-energies of the living system's shape can be greater within the living system than the spectral-energies of the external (to the living-system) containment-space on (or in regard to) the components contained within the living-system.

The subject-matter failures of: math, physics, and biology has to do with how the use of "partial-knowledge" which, apparently, is sufficient for commercial interests, but which has no relation to the development of the patterns needed to accurately describe the observed properties of the world,

Where this "partial-knowledge" is used to express social-intellectual dominance (in an abnormally controlled intellectual market-place), where requiring that partial knowledge be used is possible in a culture which is controlled by wealth, and controlled by commercial-market money-flows and not controlled by wonder and the independent and honest search for knowledge, which, in turn, can be used in some individually intended creative use

Through propaganda, and so called philanthropic foundations, with the help of the managers of educational institutions, education has been taken over, so as to express only interest in the categories of commercial interests of the investor-class.

Math seems to only deal with contexts where measuring is unreliable and there do not exist any stable patterns, so the descriptions have no content, or no meaning, so that descriptive information is complicated and not compressible, so that "the use of knowledge" depends on empirical-knowledge, eg empirically determined quantum properties can be used when they are contained within..., or coupled to..., or placed in a context of..., a highly controllable classical system.

Physics only (or primarily) deals with particle-physics the study of "particle cross-sections" and gravitational singularities with the focus on speculative context of producing an uncontrollable doomsday explosion and

Biology deals primarily with the relation of DNA, and its associated molecular context in cells, to a belief that all of life can be manipulated in this context, ie manipulating DNA, but there are no valid models of living systems, so that the manipulation of DNA (if that is ever even desirable) can be placed in its proper context, so that a similar doomsday scenario for the earth's biological systems is to be associated to such molecular tampering.

Chapter 17

Problems with physics in an oligarchy

New ideas, the need for new ideas, the social forces in opposition, and "speculation about bomb engineering" define the interests of the investor-class in physics, and the interests of their selected experts (usually unbeknownst to the experts, themselves), in the (main) subject-matter of physics

This is an issue of free-speech, and about the nature of science and math, and about knowledge and its relation creativity, it is about equal free-inquiry, and the relation of creativity to the, so called, free-market, but it is also about religion, since it is about re-defining classical physics on a many-dimensional context, ie it transcends the material world and re-unifies science with religion, since they are now (in the new descriptive mathematical context) both describing the context beyond the material world, where now religion becomes exploratory and now religion (if one has some imagination) can be framed in a precise descriptive language.

The main problems in physics are concerned about the precise descriptions of stable spectral-orbital physical systems, and there is now a new idea which can be used to solve these problems.

It is a descriptive context which is formal, and in a technically correct mathematical language, so as to be acceptable for its expression at math conferences.

But they are new math ideas which express a totally new context, in which to consider and describe math patterns, in a language where there are stable patterns and the context of measuring is reliable, ie the precise language is describing meaningful measurable patterns (or properties).

Its advantages…, in regard to it (the new ideas about math and physics) forming a new context for the description of math patterns…, are:

1. Difficult problems about the spectral-orbital properties of stable physical systems can be solved,

2. it can provide a more varied and much more precise way in which to consider spectral properties of math patterns, eg eigenfunction and eigenvalue problems, ie the structure of function-spaces can be reconsidered,

3. a new type of containment-tree in regard to a space's dimension and size, ie metric-spaces have stable shapes, and its relation to the properties of spectral-orbital (physical) systems can be explored,

4. a new manner in which the facial-structure of cubical-simplex shapes can affect the properties of the shapes which are contained within any, such, given cubical-simplex shape,

5. the properties of bounded and unbounded sets, which can be modeled within, what would appear to be, a bounded model of an infinite-extent structure, ie considering issues about infinite sets based on the properties of stable shapes (both bounded and unbounded shapes),

6. considering new ways in which to model a living system, and its associated mind.

That is, the new descriptive context for math, which is being described, is a wonderful context in which to consider math patterns in a context in which there exists both stable patterns and reliable measuring properties, where these two properties are central to the organization of the new precise language.

So what are the social forces which are opposed to the expression of these precisely expressed ideas within our (US) culture?

There is a simple reason it is opposed, it challenges the dominance of the existing social order.

There exists in the social context of property rights and minority rule, (where this is an aspect of the US culture which was begun by J Marshall within the justice system (around 1800) and which is a structure of law which, in reality, is outside of American law, but which has been illegally placed there and left in place by a corrupt justice system and a corrupt government, where American law is..., claimed by the Declaration of Independence..., to be based on the idea of equality), an implicit command to be dominant (but which is really only applicable to the minority ruling investor-class). New ideas challenge the social order and the game of domination (see below).

There is a game being played, within the context of the legal structure of property rights and minority rule (which the US government institutions are illegally administering against the public's welfare) and that is the game of domination, where this game is being very seriously played by the few in the investor-class, and it has the most detrimental affects on the US culture, where in this game there are no-rules, and it is played in the unfair context of an already existing economy (the flow of money in a, so called, market) where the one rule of the economy is that the control of money is an expression of domination of society by those who compose the ruling investor-class, and it is the ruling investor-class who control the money-flow and its use within society.

That is the game is a rigged-game, and it a game whose first victor (in the western culture) was J Caesar, the first Roman-Emperor.

The state of "the public" being in "wage-slavery" is an expression of the social domination of the many by the few who play this rigged-game.

In this no-rules game of domination, everyone is an enemy, and the minority who rule use wage-slavery to gain advantages over the other few people in the investor-class, ie the few very rich people who are in the game.

The game requires a deep belief in the idea of materialism.

In such a game the end is to realize complete domination which seems to only be expressible by the extermination of everyone, and where the extermination of the earth is a consequence of the game.

So in this game one must ignore the main feature of being human ie to learn and to know so as to be able to create.

The no-rules rigged-game of domination is a game of regimentation (of the public [regimented] to the interests of the investor-class) and death, but it depends on rules already existing. Namely, that the law is being based on property rights and minority rule, and that people accept the way the economy operates, and that the public worship, in their deepest reverence, the material-world.

Nonetheless, with all the great advantages which the new ideas about math and physics provides to these disciplines, with the only requirement "is to consider these new ideas and to enjoy the new context which they create," one is compelled to ask:

What social-forces seek a math basis which determines a, supposedly, precise language to be meaningless in regard to precise descriptions of patterns (as this is what the, expert, peer-reviewed journals are spending all their time developing)?

The spy-media system (and of course, including peer-review journals) is systematically trying to exclude the expression of these new ideas, which are essentially only being expressed…, at the marginal reaches of the internet media…, for their written expression.

Ideas are only allowed to be expressed, in the society's owned and controlled media, if the person (expressing the ideas) is in a social position of domination, and thus, necessarily, they would express the society's dominant ideas, which are related to the interests of the investor-class.

This is an issue of free-speech, and about the nature of science and math, and about knowledge and its relation creativity, it is about equal free-inquiry, and the relation of creativity to the, so called, free-market, but it is also about religion, since it is about re-defining classical physics on a many-dimensional context, ie it transcends the material world and re-unifies science with religion, since they are now (in the new descriptive mathematical context) both describing the context beyond the material world, where now religion becomes exploratory and now religion (if one has some imagination) can be framed in a precise descriptive language.

The new ideas

In regard to physical description, a new idea of a containment-set…, which is to be used for describing the qualities, or quantitative-patterns, of measurable properties of existing (physical) systems…, is that the containment set be (or is) many-dimensional, up to space-time dimension-12 (or, equivalently, hyperbolic metric-space-dimension-11, where this dimensional cut-off was established by Coxeter, since the highest-dimension [stable] discrete hyperbolic shape is hyperbolic-dimension-10), and this containment-set has a new, "spectral-partition," construct, where both many-dimensions and the new "spectral-partition" are both needed in order to be able to describe the observed set of stable properties of physical systems.

Namely, a new "spectral-shape-dimensional partition" of all the subspaces of all the various different dimensional levels, so that the new type of "spectral-partition" is determined by stable metric-space shapes (which possess stable spectral properties), so as to form a containment-tree for a (dimensionally related) sequence of a metric-space-shape dependent containment-set, where containment depends on both dimension and the size of the (metric-space) shapes into which the metric-spaces (of both the different dimensional levels and different subspaces) are identified as metric-spaces, and placed into a relationship of containment (or not) to one another, where this structure can have a strong dependence on the numbers of subspaces of any given dimensional level.

This new type of "spectral-dimensional partition" also defines a finite spectral set, for this new "partition," where in this new "partition" each subspace of each dimensional level contains both an upper and lower size-limit which exists for (or is placed-on) the shapes which compose the "partition."

Such a "partition" defines a finite spectral-set, since there is a finite number of stable shapes associated to this new "partition," and the spectral-set is defined by the natural geometric measures associated to these stable shapes, which compose the over-all high-dimensional containment set's new type of "partition,"

ie partition based on shape and dimensionally-related sequences of containment subspaces, which in turn are related to the stable metric-space shapes and their sizes.

Appendix

There are two ways in which to identify a finite spectral set on an 11-dimensional hyperbolic metric-space base-space (of a principle fiber bundle) with a new "spectral-dimensional partition" determined by such types of stable metric-space shapes of the different dimensions, by either

(1) defining it on one of the maximal-tori of the SU(11) fiber group (see below) for the over-all containment set base-space, or

(2) by the above mentioned "partition" of the 11-dimensional hyperbolic metric-space base-space, so as to form a dimensionally-related shape-containing containment-tree for the entire high-dimension metric-space construct.

If one follows (1) then the discrete changes in size-scales of the subspaces of a given dimension can change with the discrete changes in Weyl-angles for a fiber group of a metric-space of a given dimensional level

So that for a given dimensional level the size of the spectral set for that dimensional level is given by

(the number of conjugation classes) x (the number of subspaces, in R(11,0), of that dimensional level)

=

{size of spectral-set for that dimensional level}

Then add-up this number for each dimensional level, from 0-dimensions to 11-dimensions.

And/but

for different dimensional levels one would consider the number of conjugation classes (or equivalently the number of Weyl-angles) of SU(11), multiplied by the number of 5-dimensional subspaces (the last set of bounded discrete hyperbolic shapes) in an 11-dimensional space, = {size of spectral-set for all dimensional levels}.

Or

The number of conjugation classes of SU(11), multiplied by the number of subspaces of dimension 5 or less, ie $(2^{11})/2 = 2^{10}$ subspaces, = {size of spectral-set for all dimensional levels}.

This finite spectral construct depends upon the stable circle-space shapes,

eg the torus in Euclidean space, ie a shape which can be associated to a moded-out cubical simplex,

and the existence of holes in these shapes, defined by circles (or other non-deformable shapes), so that there can exist a set of independent stable circle-sizes for each circle-space shape, which cannot collapse (or deform) to a point, where surrounding each such non-collapse-able hole (or set of circles) there are stable geometric measures which might be associated to periodic-flows, which, in turn, can be associated to stable spectral (or energy) properties for the circle-space shape.

The other type of circle-space shapes, different from the Euclidean torus shape, are quite stable shapes so as to be composed of many toral-components, ie one toral-component for each hole in the shape, and is called the discrete hyperbolic shape. These shapes come from cubical simplex shapes attached together and which then are moded-out, by an equivalence relation (identifying opposite faces (or "ends") of the cube), so as to form periodic circle-space shapes with many holes, and so as to be: linear partial differential equations defined on differential-forms in a context of metric-invariance so that the metric-function has only constant coefficients, and has non-positive constant curvature, where the shapes natural coordinates, ie circles and hyperbolas, are continuously locally commutative (so the partial differential equation is solvable, so the differential-form can be found for the circle-space shape) the circles define the shapes geodesics which also determine the shape's stable set of spectral values (and, if needed, their integer harmonic periodic pathways, or vibrations), while the orthogonal set of hyperbolas would push towards the geodesic circles of the discrete hyperbolic shapes.

Resonance and material existence (or stable metric-space shape) after the new spectral-partition is identified

In regard to adjacent-dimensional metric-space shapes, material is now re-defined as a metric-space shape of an adjacent lower-dimensional metric-space (shape), which is contained in its adjacent higher-dimensional metric-space, (which is also a shape, which, in turn, is (can be) contained in its adjacent higher-dimensional containing metric-space (shape)), and where the material-component is "in resonance" with the finite spectral-set, which is defined for the over-all high-dimension containing space, which, in turn, is defined by the (above) "partition."

This adjacent-dimensional and size-dependent containment construct can continue up-to a hyperbolic dimension-10 shape, which is contained in an 11-dimensional hyperbolic metric-space, where the 11-dimensional hyperbolic metric-space is not a shape, and where an 11-dimensional hyperbolic metric-space is (isomorphic ally) equivalent to a 12-dimensional space-time metric-space.

There are discrete, discontinuous transitions, which can exist between the metric-space shapes of the different dimensional levels (as well as discrete and discontinuous transitions between different shapes of the same dimension, where the different shapes represent different subspaces of the same dimension).

When transitioning discretely between adjacent dimensional metric-space shapes, so as to move (by the transition) to the higher-dimensional containment-space, the metric-space's open-closed topology (of the (original) metric-space shape) changes to a closed topological property in the metric-space topology of its (the material-component's) adjacent higher-dimensional containing metric-space shape. The new adjacent higher-dimensional metric-space can have a new discontinuously defined scale change associated to the transition but the lower-dimensional shapes it contains may not be affected by the constant scale-change factor.

This would be the closed topology of a material-component, which is contained in its adjacent higher-dimension (and larger) metric-space (shape) containment metric-space.

This discrete transitioning can also exist between the same-dimensional metric-space shapes (in the same dimensional level), where the discontinuous transition can be accompanied by a discontinuous change in scale between the two different subspaces which have the same dimension, caused by multiplying by a constant factor.

This discontinuity…

(between different dimensional levels, or between different metric-space shapes of the same dimension [but in different subspaces]) and its discontinuous structural properties (which are associated to both measurable and perceivable properties in these metric-spaces, and), which exist between different dimensional levels (or between different metric-space subspaces of the same dimension) causes these other constructs (in our many-dimensional existence) to not be perceivable.

That is, if one changes between shapes of the same dimension then there would be different size-scales, which change in a uniform but discontinuous manner, and there would also be different set-containment relations between the shapes and the different dimensional levels in the new containment metric-space shape,

Thus, new relationships in regard to spectral-orbital sizes would exist, and this would mean new types of perceptions which would suddenly and uniformly, but discontinuously, occur.

That this type of an experience is not common, means that we do not often enter into such processes of transition between subspaces (or between metric-space shapes) of the same dimension.

This is the cause for our deep belief in the delusion concerning the idea of materialism.

That is, we are locked into a subspace topology, in which we can (only) extend past this illusion of materialism with either a correct math model of existence, or by realizing the higher-dimensional construct of that which we "really are" as living systems.

A metric-space can contain within itself lower-dimensional stable (metric-space) shapes which are small enough to fit into such a containment metric-space and which resonate to a (the) finite spectral-set defined by the spectral-shape-dimensional partition of the over-all high-dimension containment metric-space.

But, furthermore,

The material-interaction structure also changes when there is a discontinuous change between dimensional levels.

The discontinuous set of changing properties would include:

1. Size-scale changes, due to a conformal constant factor associated to discontinuous changes between dimensional levels.

2. Lie fiber group changes, due to changes in dimension

3. Weyl-angle changes (as well as size-scale changes of a constant factor) which can occur between toral-components of a many-holed (or a specific genus) discrete hyperbolic shape (where such a shape is a stable metric-space shape)

4. The spatial and temporal subspace structure of a metric-space…

Whose metric-function is related to the metric-space being a non-positive constant curvature metric-space, and the metric-function having only constant coefficients can also change.

Note: The spatial and temporal subspace structure of a metric-space is associated to different types of physical properties which are associated to each of the (these) different types of spatial and temporal subspace structures for a metric-space.

In turn, these physical properties (which are necessarily associated to metric-spaces of both particular dimensions and where these metric-spaces are separated into fixed spatial and temporal subspaces) are related to pairs of opposite metric-space states, and a subsequent spin-rotation between (or cyclically around) these pairs of opposite metric-space states, so that this spin-rotation of pairs of opposite metric-space states, in turn, relates a real pair (or set) of, metric-invariant metric-spaces, to a Hermitian-invariant, finite-dimensional, complex-coordinate space, within which the two opposite metric-space states (associated to a metric-space's physical property) can be separated, ie into the real and pure-imaginary subsets (in the complex-coordinates), and, thus, the base spaces (the complex coordinates, which are Hermitian-form-invariant) would be related to the unitary Lie groups as fiber-groups.

The spin-rotation of pairs of opposite metric-space states contribute to the local inverse structure of discrete (inertial or energy) changes (of stable material-components), which occur in the new construct, for the sets of both inertial and energy changes which are associated to material-interactions, where these two properties, ie inertia and energy, are associated to Euclidean space and hyperbolic space, respectively.

Note: The other changes in spatial and temporal subspace dimensions are related to the introduction of new material-types, and their associated, new physical properties, which can occur (or be defined) in higher-dimensional constructs, where the main focus of this change in subspace dimensions is related to the dimension of the spatial subspaces (of a high-dimension metric-space).

Material-interactions, in the new descriptive context, are now of four different types:

1. The motion of (usually condensed) material (which is contained within the metric-space shape, and) which is pushed, by the (material-containing metric-space) shape, into the

material-containing metric-space's geodesic orbital-shapes, which, in turn, identify stable orbits for the condensed material, on the metric-space shape for the metric-space within which the condensed material is contained.

2. A usually, 2-body orbital perturbation (to the above [#1] geodesic orbits) is caused by an, essentially, Newtonian (2-body) material-interactions, where the geodesic orbital structure is strong enough to negate any further many-body interactions, in regard to the many-bodies which compose the spectral-orbital system, eg in the solar-system this would be the two-bodies of the sun and the one planet in its stable geodesic orbit being perturbed, so that the other planets do not contribute significantly to the 2-body orbit perturbing interaction for any given planet.

An aside about the new interaction construct:

but now the material-interaction has a (slightly) different math construct associated to the material-interaction process, which is a combination of physical properties and their measurable context, which are inter-related by the stable shapes of Euclidean space and hyperbolic space so that this geometric structure is, in turn, related to the geometry of the metric-space's fiber Lie group (for the classical Lie groups).

3. A typical material-interaction (such as the natural interactions of material-component collisions) but the combinations of both (1) the shapes, which form during the interaction process (in the interaction process), and (2) the if the energy-range of the interacting system is "properly defined," can (both together) cause (or allow for) the existence of a new context of resonances for the shape involved in the interaction geometric-process (which form during the interaction) so that a new stable system forms out-of the interaction, due to the new system's resonance with the finite spectral set, which is defined, for the over-all containing space, by the spectral-dimensional partition of the over-all 11-dimensional hyperbolic containing metric-space.

4. A usual material interaction, where neither geodesic structures nor resonances are a part of the material interaction, these interactions are very much like the material-interaction structure of classical material-interactions, though there are now (or there is a re-instatement of) action-at-a-distance geometric-structures (or non-local structures) in this new description.

However, the statistical structures of many-small material-components can also define a valid statistical structure which can also be related by an energy-operator to a vaguely defined wave-function (see two-paragraphs below, E Nelson), where this is a very limited statistical descriptive context based on the many small material components of the very many-component system, ie more like a perturbing structure for classical statistical physics than a valid quantum viewpoint which can logically stand on its own when extended too-far it becomes logically inconsistent.

Note: A shape for a material component, contained in a metric-space shape, can be of either lower-dimension or of equal dimension, as the material-component's containing metric-space (shape), and, if small enough, the material-component can still be contained in the metric-space, but the equal dimension shapes contained in such a metric-space (ie the material-components are equal in dimension as the component's containing metric-space's dimension) would only be "seen" in regard to this shape's lower (adjacent) dimensional faces (or shapes). This would be relevant, especially, for our solar system, as well as for the somewhat amazing and seemingly mysterious math structure which can be associated to the two material types which can exist in 4-spatial dimensions where the fiber-group math structure of the 4-spatial dimensions is an, $SO(4) = SO(3)$ x $SO(3)$, fiber group, so as to divide the two material types into two different 3-dimensional subspaces of $R(4,0)$.

[note: Lie group, $SO(3,1)$, goes with metric-space, $R(3,1)$, ie the usual space-time space, etc]

Furthermore, the interaction (context and geometric) structure, when applied geometrically to many small material components, behaves in a statistical (or Brownian-motion) framework, which is associated to all of the different small material-components which are interacting, in regard to the various properties of neutrality and momentarily charged small material-components which are interacting in a discrete manner, so that this Brownian motion (for these many small components which are interacting) has an equivalent structure to the random structure of quantum physics (E Nelson, Princeton 1957, 1967). That is, quantum randomness is derivable from this new descriptive context.

But

Furthermore, in regard to the stable shapes of the material-components, each shape has a distinguished point associated to its shape, so that material interactions center around this point, so as to cause the appearance given to the random context of small material-components so as to give these small components the appearance of being point-like, when they are detected (ie when these small components are within their random descriptive context).

Note: The many-dimensional containment-set based on stable shapes of many various dimensions and sizes (ie both macroscopic and microscopic sized shapes) so that the shapes each have distinguished points about which interactions are focused, provides a whole new context within which to interpret the properties of elementary-particles, which are observed in high-energy ("particle") collisions, and it is an interpretation which leads into ever-higher dimensions, but the higher-dimensions are mostly irrelevant in regard to understanding the physical properties which actually cause lower-dimensional material components to both form and to be stable in their observed properties.

An important point is that material-interaction structure is a subset of a metric-space shape structure.

That is,

First the material-components are contained in a metric-space, but where the material containing metric-space also has a shape, and this material-containing metric-space shape can affect the motions of the material-components which it contains.

But, furthermore, the set of stable shapes which form as metric-space shapes for any "dimension shape" are (may be) formed because they are in resonance with the high-dimension containing metric-space's finite spectral set, which is defined by the new "spectral-dimensional partition"… of the high-dimensional containing metric-space…. by stable metric-space shapes which, in turn, define a dimensionally related sequence of metric-space shape containment-sets, ie trees of metric-space shape containment-sets.

Within the metric-spaces there are defined "between the material components" the material-interactions, which are related (by new geometric structures) to sets of partial differential equations, similar to the existing math constructs, both classical and quantum, where classical descriptions fail because they do not 1st identify stable metric-space shapes and then define interactions, but

quantum descriptions are (or can be interpreted to be) statistical constructs similar to the statistical constructs of classical thermodynamics, but these statistical-quantum models cannot be extrapolated down to the microscopic-size scale in any systematic (or in any realistic) manner.

That is, the new descriptive context is based on stable geometries, and it is not based on (indefinable) randomness, which is what the descriptive structure…. becomes in its current authoritative, but failed form…. ie the indefinable random context of both quantum physics and particle-physics (as well as general relativity) all become based on indefinable randomness, so as to become a descriptive structure which cannot describe a stable pattern, ie it is a descriptive language which cannot possess any content.

Note: In regard to the relation of general relativity to the new description, first there is metric-invariance and its associated set of stable shapes, wherein the shape of the metric-space which contains material applies to dynamics (for the material contained within the stable metric-space shape) in a most fundamental manner, where the "shape of space" can then be applied to material-components, in the more simplistic context of metric-invariance, ie not in the useless context of both non-linearity and general metric-functions (wherein the idea of measuring loses its meaning).

Note: If a metric-space shape is accelerated, due to a material-interaction between "other material-shapes and the given metric-space shape," eg the frame of our solar-system interacting with distant star-systems, so that this interaction takes place in the solar-system's adjacent containing higher-dimensional frame, eg in regard to the solar-system frame, then the entire (solar-system) shape, and what it contains, would all move in a rigid manner (as all things in a train-car move uniformly when it moves with constant velocity on the train-tracks).

That is, the outside (uniform) motions (acting on the stable rigid-shape) [eg uniformly accelerating motions], are such that they cannot be detected inside the shape, and the shape's rigid (inside) topological structure.

Why new precise descriptive models of existence are needed

(ie social issues, where social issues should not be hampering, a true vision of, science, but they Do!) and

A set of alternative socially determined motives for basing a quantitatively (appearing) descriptive language on the failed context of indefinable randomness and non-commutative math relations.

This new descriptive context (given above), where there are:

1. new sets of assumptions,

2. new contexts,

3. new interpretations, and

4. new ways in which to organize the patterns of math,

where

One wants a mathematical description, because

one wants a description of a pattern of a physical system to be in a context where

1. measuring is reliable, and

2. the patterns used to describe "the observed set of measurable physical properties" are stable

Then one can both identify a context, where the truth about the description, of a system's set of properties, (which) is (more) correct than the currently used context, so that the patterns of the system can be placed into a correct context for use, or for accurate descriptions, which exist to sufficient precision, as "one is measuring," or (there are measured properties) which are distinguished within the system, and the context can be used, so as to place the patterns of the system into a greater inter-related context, so as to build, use, and control, a new system, or control the given system.

The problems with today's precise descriptive languages is that they fail to describe the observed properties of most fundamental stable physical spectral-orbital systems, when their descriptions are based on the supposed set of the laws of physics.

That there are a set of systems whose "observed properties are very stable" this stability indicates that these properties result from a descriptive context (or a true state of existence for these systems) where the system is both linear and solvable, and, thus, the description can be made sufficiently precise and controllable, and systems of a similar kind can be inter-related for building practically useful creative efforts.

The list of these systems (which possess stable spectral-orbital properties) is:

1. Nuclei,

2. General atoms,

3. Molecules,

4. Molecular shape,

These systems are quite often many-(but-few)-body systems, which are very stable in their spectral properties,

Then there are also the many-body systems of

5. Crystals, and

6. Living-systems,

These are stable systems which possess global system properties, ie the properties are not averages of many-components, yet the properties of these stable systems are now (2014) only vaguely related to a descriptive context, yet they also possess stable properties, one needs a more precise structure for these system's stable spectral-orbital properties so as to identify a size and dimensional hierarchy of stable spectral-orbital structures for the system, and a context of control, where the control might be a system's orbital relation to a higher-dimensional stable shape, and there is also the

7. Mind

One needs a new model of the mind,

And then there are again the many-(but-few)-body systems which are macroscopic, and, apparently, very stable, such as the

8. Solar-system

9 and the motions of stars in galaxies

As well as the apparent expansion of the galaxies (this might be a result of group conjugations of SU(11,1), and a resulting apparent drifting-apart motion of high-dimension entities due to the group conjugations)

That is, it is not clear as to what is being observed in the universe, or in regard to very distant galaxies, when supposedly, light comes from such great distances, are physical laws universal, or

Are the physical laws (physical properties, and attempts to relate physical properties to physical law) regional?

or

Are they partitioned into a many-dimensional, and a many metric-space shape, context for a many-layered set of quantitative descriptive containment contexts?

And though there is extensive discussion, in the propaganda-education system, about some of the most irrelevant systems

eg the big bang and elementary-particles,the real issue is to describe the observed stable properties in a linear, solvable math context, so the descriptions are more practically useful for the (above) first 9 categories of stable system properties,

But where there is no description, to sufficient precision, of the stable properties for these systems, where such a description is based on the, so called, laws of physics,

Where the, so called, laws of physics are (now) based on indefinable randomness,

and where the main statement made by this high-brow brand of failed physics, is that "these stable many-(but-few)-body systems are too complicated to describe."

This is, essentially, saying that these systems have an infinite amount of information within them, which cannot be compressed…., in the indefinable random context, which is the basis for the descriptive structure of today's (2014) failed authoritative scientific assumptions and descriptive contexts…. and this leaves the description… of the observed properties of many fundamental physical systems… in a state of total randomness (no information compression).

That is,

The math context in which the laws of physics are now (2014) being presented, require an infinite amount of information in order to describe the observed stable properties of the many, many stable fundamental physical systems.

This is because the current authoritative descriptive context is a failed descriptive context, which is fundamentally based on indefinable randomness, which is only relatable, in a practical context, to nuclear bomb engineering, and it is a descriptive context which sheds no light on the observed stable properties of the most fundamental and simple systems which are so central to our experience.

This destruction of math descriptions by relegating them to

1. indefinable randomness

2. non-commutative relations and

3. subsequent chaos,

4. so the descriptive language cannot describe a stable patterns,

Is done by the investor-class to preserve and stabilize their investment interests which revolve around the material-world

And

the ideas of: materialism, indefinable randomness, and non-commutative math structures, so that the public focus would remain on the material-world, and that the precise descriptions are dysfunctional and development is dysfunctional., in regard to technical development, the way in which our society uses mostly

1. Empirical means of development, [empirical advancement, where data is interpreted through a set of partial-truths] and

2. trial and error methods of development, and

3. enhancing the already existing technical instruments

as the main means of development of the world (along with a practically useful precise descriptive language)..., are (all) controlled by the investor-class, so that the development is slow, and the investment class, who have invested in the material-world, can maintain a control over the relation of knowledge to creativity within society.

Basically, the reason that science is "in" such a "failed state," has to do with, the administration of society and its social institutions, so that these institutions are managed so as to support the interests of the investor-class, so the management is all about both manipulating and preying upon the public,

This managing of society through its communication channels, and the investment-class possessing control over the institutions and their managers, allows for the social control by managers of:

1. Enthusiasm or motivation of the public

2. What it is within society which the public believes possesses the greatest value, ie controlling what is considered to be of high-social-value

3. The small set of categories about which people are to have interest, and through which people can be creative

4. The context in which the small set of narrowly defined categories are expressed, and how observed patterns should be interpreted

5. The language which is used in the categories of social interest, and

6. The personality types which reach the top of the social hierarchy in each category

So thatthe fact that, the personnel system for hiring the authorities whom fill the institutional structure of academic authority, is all about finding the dominant personalities to fill the top authoritative and administrative positions, much as the social system is controlled by a set of domineering psychopaths, which are filling the same type of social position, in our economically identified social system, as the psychopathic-emperors filled such high social positions in the Roman-Empire, while the helpers of the dominant few, are more or less a bunch of obsessive, autistic types, who are not capable of considering the truth of the assumptions, to which they have, implicitly, agreed (by their competing in the academic competitions), on their own…., they are diligent defenders of peer-review…, which promotes they, themselves, being considered by society to be the special people within society.

These obsessive and aggressive and competitive types, personality-types who seek dominant social positions, are a part of an academic setting in which there has been instituted the fixed language structure of formalized axiomatic, instituted by D Hilbert around the 1910's, eg the formal project of axiomatically developed math constructs in regard to a set of books (or a project), called Bourbaki, is a main part of this attempt to fix and formalize authoritative language for the academics.

Thus the technical experts mostly babble nonsense.

A good example of this is economics, where there is no reason for anyone to believe that there are laws of money associated to money flow and markets, since money is really a measure of social domination by the owners of society (the owners of the controlling-stakes of society's institutions) though there are, seemingly, stable patterns related to measuring quantities of money, in reality these are not stable patterns.

This tower-of-babble allows for a greater control by the few (who own and control the propaganda-education system) over both thought and language in society, and, subsequently, it allows greater control by the few over the relation that knowledge has to the society's creative attempts

Where it should be noted that as early as 1910 Einstein was still writing in a very assumption-based technical structure, yet his push was to:

1. the more abstract,

2. the more general, and

3. the non-commutative math patterns, thus

4. leading into chaos (mathematical chaos, ie small differences have very large effects) and

5. into indefinable randomness, which was fully realized as the main context of practicality in regard to bomb-engineering, and

6. the subsequent total control of physical knowledge (control over university physics departments) by bomb engineers, and their administrators, and

7. the subsequent total control of quantitative knowledge by axiomatic formalization,

So that the, so called, peer-reviewed intellectuals babble on-and-on about complicated nonsense, but where their precise technical languages cannot describe a stable pattern, and the context, upon which they base their descriptions, is where measuring is unreliable, and technical improvements are obtained by adjusting the instruments which already exist.

So that, now (2014), the "physical review"… a useless piece of publishing, expressing literary delusions about science, ie it is closer to science fiction than to science…..,

Yet "physical review" is about the most authoritative of the physics publications in society.

(note that all of, the so called, world powers have their people educated about physics and math at US universities so as to place the intellectual class in a context of total social-control within their societies, societies which adopt the basic knowledge of the western culture, quickly become controlled by the west, so as to become a part of the investor-class who collude with one another in their context of social domination of societies based on inequality)

and

Both physics and math's domination by formalized axiomatic language structures demonstrates the total domination over language and thought which has been acquired by those who control the propaganda-education system, where education has been easily controlled due to (1) its historic authoritarian construct in the US and due to (2) the total militarization of the US social institutions immediately after WW II,

The reason that "physical review" has its strongest relation to literary delusions about science, and not so much to science, is because it is the result of formalized axiomatic as the fixed set of dogmas, (which are related to (maybe) a few categories of commercial-investments of the ruling-class), upon which the ludicrous educational set-up of academic-competitions are based, where the winners of these competitions are, supposedly, qualified to work on the technical projects into which the investor-class is investing,

But, in regard to following the narrowly defined dogmas of the formalized set of axioms, by following these fixed formalized rules of language, the practitioner would (might) realize a truth which is only contained within that language, [but when the, supposed, rigorous truth of such statements are applied to an external existence, the descriptions of the language fail] [thus, the great attention which is paid, by the propaganda-education system, to the most irrelevant aspects of empirical science eg particle-physics and the cosmology of gravitational singularities, etc].

Where a precise description's truth can only be determined by:

(1) accurate descriptions which are made to sufficient precision, and

(2) the relation of the identified patterns in language to practical and useful technical developments:

and in regard to this realistic criterion of truth; the, so called, truths (ie deduced descriptive patterns) which result from the formalized axiomatic, (which are applicable only to the context of a formalized academia, about which the professional journals of math and physics, such as "physical review," are based), these "formalized truths" are failing in quite significant ways,

Their construct of assumptions and interpretations are based-on:

1. materialism,

2. reduction to smaller components (detected by high-energy component-collisions), and

3. indefinable randomness (ie the observed events are either not calculable or it is claimed that the elementary-event space is composed on events which are unstable), so that

4. all measurable physical phenomenon are assumed to be the result of solving (partial) differential equations

5. The solution is pursued in a context of using (differential) operators to "diagonalize" a function-space model of a system (within an eigenfunction and eigenvalue setting), but where

6. the model is arbitrary and related to an infinite spectral structure of arbitrary spectral values, {one needs a relation to math constructs which possess stable a determinable spectral properties}

The articles…, of the authorities of fixed formalized axiomatic which exists in the peer-reviewed published professional-club journals…, begin with a technical structure which is far too complicated and it is a math structure which is based on:

1. non-commutativity,

2. indefinable randomness,

3. non-linearity,

4. quantitative inconsistency,

5. many-valued functions (bifurcations),

6. identifying a chaotic context,

which is also

7. logically inconsistent,

Eg

(A) A probability-based descriptive language has within itself a geometric model or point-particle collisions

(B) The local symmetries of particles, SU(n)-groups, (which are related to a system's reduced components) are attached to a globally symmetric, eg energy-invariant, wave-function (which represents a measurable global system)

(C) Non-local properties are attached to a globally defined system, ie there is no bounded pattern for the system, yet the global patterns seems to suddenly collapse, so that the property of non-localness also collapses, and this occurs for no reason which is relatable (attributable) to the descriptive construct of the system, ie the pattern being described as a quantum-system has no validity,

ie it may or it may-not exist, depending on if its wave-function collapses or not,

ie depending on whether the global quantum system couples to some other global quantum system, where this process of coupling is not defined, or it has no cause.

The quantum system needs to be modeled as a bounded system, but this implies a geometric structure, not a global indefinable random structure, where indefinable randomness is defined in a "global manner" over the unbounded sets of the system-containing coordinates.

Note: Material-based models of geometry and randomness are logically inconsistent, where this is due to the uncertainty principle, which, in turn, is associated to any probability based description where the uncertainty relation is defined in regard to a pair of dual (or Fourier transformed) variables which are a part of the descriptive context.

(D) it is a descriptive context where neither measuring nor "system-existence" (or existence of a pattern) are valid constructs.

This is because of both (1) a system's singularity, ie 1/r, energy-term, which is used for charged point-particle collisions, and (2) measuring not being definable in the, so called, quantum-foam of an undefined and mixed context of a space-Fermi-particle-field model of the small-scale structure of a quantum description,

Where (in quantum description) there is a global wave-function property attached to a local (or non-local) set of particle-symmetries associated to the derivative-connection, which is a part of the system-defining partial differential equation.

From this context, in which nothing is definable as an actual quantitative model (or containment set), the partial differential equations, which are picked so as to be worked-on and to get an associated funding are all about irrelevant physics but which are related to the practical relation of both particle-physics and the explosion capabilities of a gravitational singularity to bomb engineering.

(E) particle-physics is applied to quantum systems, and it is claimed to be the correct model of material interactions in the context of quantum physics being described by the particle-states…, associated to particle-collisions (of charged- and field-particles) at a point in space…, which are used to model a perturbation of the energy-levels of the quantum-system's wave-function, where this is done by attaching the particle-states to the wave-function, but this model depends on finding a quantum system's wave-function, but general wave-functions for quantum systems cannot be found,

ie models of quantum interactions are effectively non-existent.

(F) it is in such a type of a, supposedly, precise (or, supposedly, quantitative) descriptive language, in which there is neither stable pattern, nor a reliable model of measuring, and yet it is still considered to be a quantitatively based description.

So that, it is this improperly defined type of a, supposedly, precise language structure, where arbitrary quantitative patterns are introduced, and claimed to be valid, but all quantities are manipulated in a way which is outside of valid quantitative operations, before they are presented, either empirically or vaguely claimed to be predictions (but in fact are arbitrary claims), and the impression given is that, "the descriptions it provides are quantitative and measurable", is the property an arbitrary claim which is, nonetheless, being presented as a predictable and a precisely deducible property, which is used, so as to (fraudulently) claim that the arbitrary claims, which are made in this invalid, but seemingly, quantitative descriptive language, are measurably verifiable claims,

Eg the, so called, predicted "future criminal acts," when the reality is that with spying and active agents and manipulate-able subjects, (subjects who the police-states preys upon) can be used and manipulated so as to cause arbitrary deranged acts, which are manufactured by those secret-agents within the police-state model of a justice-system.

They (the police-state) wants to use this type of (what appears to be) a quantitatively-based language, to provide, what is supposed to be, an irrefutable quantitative pattern (which is really not a pattern at all) in order to justify arbitrary claims or actions, such as predicting a future crime-event in an indefinably random descriptive context.

This can be used by the police-state as a, supposedly, justifiable cause for extermination of the many by the few.

Back to the list of math structures used in peer-review journals,where, in turn, this useless, supposedly, quantitatively-based descriptive structure (which, by the way, defines the descriptive context of peer-review for the sciences and also has the most attention in mathematics) so as to build a precise description upon a context where:

8. "measuring is not reliable,"

9. Patterns are not solvable, and

10. "there are no stable patterns" which this type of a, so called, precise descriptive language is capable of describing.

That is, the descriptive construct (or the professional language) of the professional math and physicist is not capable of identifying stable patterns, so it has no content in its descriptive capacities, ie it is a formalized language which has no meaning.

To prove this, note that there are no descriptions based-on, what is, now (2014), considered to be, physical law, and which describe the stable spectral properties of: a nuclei or a general atom, etc etc, or in regard to general relativity there is no description based on physical law as to why the solar system is, apparently, so stable etc etc etc

Nonetheless, the authority… of these ignoramuses (unable to "discern truth" for themselves, and unable to question authority)…. is unquestioned, where this is due to the way in which the propaganda-education system of the US is organized, it is all about psychopathic domination, which the main authorities of the hierarchical system exercise over the public, where the best example would be the R Oppeheimer, and then "the even worse," E Teller, these are the, so called, experts in administrative positions, who guided physics into (physics) being the study of bomb-engineering, by promoting, in the academic institutions, the autistic, obsessive types who follow rules, but who cannot "evaluate" the truth of a statement for themselves.

Note: the, so called, "quants" on wall-street (the MIT and Princeton math and physics PhD's) are coming-up with (function-space and operator) models of business risks, upon which billion are being bet, but the models (based on the above set of math assumptions) are not true (in regard to the relation that economics has to the "money processes of the material world"), thus leading to very big losses.

But economics is not a valid quantitative model of anything, rather, it is a quantitative expression of the arbitrary domination of society by the very few.

The unexpressed assumptions of economics are that "society must be based on inequality," and "society is a hierarchical-collective, which is needed to support the few rulers of the society."

If one considers the way in which the propaganda-education system presents science to the public, it is…, in many (or in most) significant cases…, about the most irrelevant aspects of science (physics), and it is always given a context of experts, are supposedly, calculating solutions to partial differential equations, which are used to model physical systems, where the case which is most often placed in the propaganda system is about partial differential equations which model elementary-particle properties and gravitational singularities where they are supposedly combined in the model of the big-bang

Consider the big-bang

Its consideration began in regard to the apparent expansive motions of galaxies, which is believed to be true, due to the interpretation of the spectral red-shift of light from distant galaxies, so that there was a theorem about continually extending the assumed to be a spherically symmetric expansion of material in the universe back to the center-point of the, assumed, spherical shape of the supposedly expanding material.

In regard to this, there is a theorem which states that one can continuously extend the material back to a point provided that the extension is continuous.

But

Where is the point in the universe (in "what direction" in the night-sky above) where the expansion commenced? This has been answered as that point being everywhere

Then the center of the expansion being a hypothesized point this makes particle-physics relevant to the description (but where particle-physics is really only about have any practical value in regard to bomb-engineering) where the claim was made that things from the initial explosion…, which came from a point…., discontinuously expanded, suddenly more rapidly, ie inflation.

However, this means that the expansion was not continuous, and thus the extension back to a point is invalid.

And

This means that the point of origin of the big-bang was not a point, but was a region of infinite extent and this region of material broke-up and inflated to give E&M signals whose cause is unknown but it is claimed to be caused by the big bang though there are other models of the origins of these E&M signals.

This is a bit absurd and mostly a bunch of meaningless and logically-inconsistent statements which are backed-up by intellectual bullies (who seem to accept all the baloney) which is focused on irrelevant issues concerning science.

Thus, the real point of the big-bang model is to relate particle-physics to gravitational explosions in relation to bomb making.

How does the media cover these irrelevancies?

1. It requires that the model of our existence as given by the highest and superior intellects is that of materialism and reduction and indefinable randomness and our culture is mainly concerned about bomb-making

2. The society is about inequality and social hierarchy only the expressions of ideas by certain experts are allowed on the media since the media only reports objective truths and only the absolute truths as known to our society's experts

So quantitative descriptions are anything which might appear to be quantitative but it is that which the proper authorities claim that it is true (science is now about the truth based on authority)

3. There is domination of society by both these, deceptively framed as well-meaning intellects, who (apparently) are guided to their beliefs by their pay-masters, so that the experts only consider bomb-engineering ideas in regard to physics (ie science is about the simplest aspects of existence)

But apparently most of these, so called, well-meaning experts do not see their brand of arbitrary and authoritative physics as only being about something which is only practically related to bomb-building, where the bomb being conceived is big enough to annihilate the solar-system, and since it is based on partial truths the knowledge our culture now possesses (or wants to possess) is insufficient to control such an explosion if it ever gets initiated

Thus, the western intellectual efforts are about deceiving its intellects into building a bomb which can create the biblical apocalypse, the experts are in effect building a doomsday-machine, as are the corporations which only profit from fixed ways of doing things ie the bankers oilmen and military corporations, where their destruction and poisoning of the earth has already reached the point-of-no-return, ie in regard to dangerous nuclear material which now exist and have half-lives of millions of years.

Calvinism has come to define god, ie the worship of materialism, and the intellectual efforts are only used in regard to militarism, which has been placed into a fixed authoritative descriptive language based on indefinable randomness, so that no stable patterns are describable by this language, so that there is only the biblical-apocalypse, guided by a narrowly confined class of emperors, ie corporate interests.

But there are other ideas, other intellectual models, about both existence and life and life's relation to creativity.

That is, the corporate police-state is all about a set of irrelevant and non-useful descriptive structures with which one is incapable of describing a stable pattern, it is a descriptive context which is only related to nuclear bomb engineering and the narrow interests of the investor-class, which is being cheer-led by the media.

This is similar to sports events at high-schools where certain sports are endorsed by the school administration (which are related to larger business interests) within which only a few participants are selected, while popularity contests are used to determine cheer-leaders, and people (ie the enthused student-body) then cheer for school victory, but in higher-education the categories are also selected from big business interests, and the cheerleading is more universally framed, so that it is the general media, which cheer-leads a small set of academic categories, thus the researchers are hooked into a competition confined to the dogmas of a small set of categories of interest, which is being led by the big-media so as to cheer-lead for the projects (or categories of interest) which the investor-class wants funded (or worked-on, eg or completed)

For example,

linguists model language in a context of indefinable randomness (eg looking at word-strings), and they place their construct of language development of the (biological) Humans into an also indefinable random context of Darwin's evolution (this will help getting funding),

Thus, such a model cannot compress information, so that patterns are arbitrary, and everything hangs on statistics of a descriptive context based on randomness, but where there is not a valid elementary event space.

But placing the descriptive context into the context of evolution allows for a greater accessibility for funding

This is apparently also a very important aspect of science, namely, that it is improperly pitted-against religion

That is there are the, so called, two sides in society in regard to science and its relation to the, so called, acceptance of evolution (which is really an acceptance of indefinable-randomness, and indefinable-randomness is a structure which destroys science, where one can define science as a description of observed properties of the world which compresses information and makes it practically useable, whereas indefinable-randomness makes what appears to be quantitative description ever more remote from either information compression or practical use)

(1) the intelligent and intellectual adherence to its assumptions of indefinable randomness, so as to get funded for science research, where the descriptive structure of evolution is being fixed and then cheer-led by the media, and funded by the investor-class, but at the same time the religious-right, ie a fake moralistic call to take action, is being funded and covered more closely, by the media, than is the fake-science.

and then there is

(2) the (also funded) religious-right's vision of an independent someone being opposed to the vision of material based science even though the religious-right is, essentially, a pawn in the game, since they are most devoutly adherent to the idol-worship of the material world, since they are being led, in the US, by both, the investor-class and by Calvinism, where Calvinism is the real religion which has been forced on the US public by the investor-class, and their justice-system henchmen, and which is instituted by wage-slavery.(the religious-right tends to be a bunch of bigoted totalitarian violent pawns (deceived) in the game of the media, they are the likely also the set of amoral domineering personality types, which are the spies, or are the agents for the spies, or they are the personnel which the agents manipulate based mostly on emotional methods, and the predatory methods which are used by the justice-system to attack the public and to attack any ideas which might be opposed by the investor-class)

Both sets of people are being preyed upon by the investor-class, whom control the society:

(1) The intellectual types are taught and encouraged to intellectually bully the public

While

(2) The people-of-action are encouraged to spy upon others, and interfere-with others, and use violence and intimidation against the other people, ie they are a team which institutional management protects.

But both groups, which act as operatives for the investor-class, are people who seek to be in dominant social positions, where both are manipulated through funding and cheer-leading.

That is, the people selected to be operatives either the intellectual or the person-of-action are obedient and gullible, but they desire to be domineering and violent (or dismissive) so as to be "on top" of society and to violently defend their top social-positions, as either bullying-intellects or as bullies in a fixed context of righteousness and absolute dogmatic truth,

ie the socially recognized high-value, where the social recognition comes from the management and the media, all controlled by investment

The intellectuals vs. the religious fanatics

But the two types of people, (intellectuals [who compete in, mostly, academic institutions, though some come-from industry] vs. self-righteous people-of-action) which the investor-class exploits, are quite similar groups of people.

They (both groups) are both

1. gullible

2. seek domineering social positions and

3. want to express their self-righteous correctness,

4. are aggressive and competitive, and

5. They are bullies (intellectual or physical)

6. They are, themselves, very easy to manipulate, by the investor-class

Through

(A) the funding

(B) the managers

(C) the media and

(D) through spying and/or the spy-agents (or handlers)

they both seem to follow the hype and high-value which is expressed by

Either

Science (as expressed in the propaganda-education system) (which virtually only supports the small set of categories of commercial interests of the investor class, and they accept and righteously adhere-to a narrow dogma, which gets funded (they are devout believers in Calvinism), though they always pay lip-service to the idea that science can change, but they vehemently insist on peer-review, which only accepts dogmatic beliefs,

ie new ideas cannot be peer-reviewed since there are no set of peers who can judge the dogmatic adherence (of the writer of the new ideas) to the authoritative traditions of the intellectual-class, which the new ideas are expressing)

Or

Religion, and its deep relation to funding, and to the media (ie the same as science)

Or

Through seeking the right-types of people-of-action (through the military and the justice system's penal-system)

That is, people are placed into categories, in regard to enthusiasm, or cheer-leading

People-of-action by the importance of their self-righteous religious and pious beliefs and their also being secret agents which support…, by their being easily manipulated…, the social wishes of the investor-class and

The intellectuals decide to follow the ideas of:

1. materialism, and

2. reduction, and

3. indefinable randomness, so as to not be able to describe a stable pattern, so that their descriptions have no content, other than a vague content about unstable patterns, and the way in which the facts (or descriptions) associated to the scientific descriptive constructs are actually being used for commercial interests (eg bombs, and 19th century physics for: communication systems, and thermal systems, etc, and, whereas the development of chemical systems and mechanical systems usually comes from the technicians, with a context of either rates of reactions or a good sense of how a system's shape affects system dynamics, etc) and

These intellectual attempts, which develop the same categories of descriptive contexts, in which commercial development has already occurred, ie the categories of intellectual dogmas, are being cheered-on by the national media.

While this same management (of the social structure which is directed by institutional administrators, and, subsequently supported (or enhanced) by self-induced (or media-induced) cheer-leading), is also observed at the level of the local high-school.

The local high-school promotes its students to be obedient to the school, and to seek victory in the various school competitions, where this enthusiasm is also based on cheer-leading within the high-school.

So too, the physicists are led to follow the descriptive construct of language, which is being funded. That is, money goes to university technical departments, while they get cheer-led by the national media, and by means of semi-popular science journals, as well as news stories (where the media reports on highly funded technical projects), and (in the media reports) the experts express great interest in the set of observations, which is about stuff which is being cheer-led, a vague…, (vague, since all technical languages are based on indefinable randomness, a language which is easily manipulability, as both the peer-reviewed de-bunking of global warming shows, and predicting a person's future crime acts), theoretical model provides the context in the technical language through which the observed data is interpreted, but which is irrelevant to physics, in regard to the real important problems in physics, but which are being ignored by the physics community, because the real problems are not being funded by both the administrators and investors of society. That is, the issue of funding as being a pre-requisite for creative attempts (or efforts) is an expression of the imposed religion of Calvinism within society.

In the, so called, "real world," (essentially, the social requirement, due to the imposed worship of materialism, of needing to get a job) but more realistically "the real world" is the totalitarian police-state-world run by the investor-class, there are the investors who invest in a fairly limited range of categories, so as to limit investment risks, as new ideas are very risky,

In the context of a small number of categories of investment-interest, there are (undoubtedly) lists of people categorized into skills in which they are (certainly) hierarchically-listed as to their skill level, or

In regard to "new knowledge," and its relation to practical creativity, where

"new" is a misnomer, since academic knowledge follows traditions and authority, especially, as defined by axiomatic formalization, and

Due to axiomatic formalization there is a fixed language which can be easily translated into commercially related categories which have a similar language structure.

There is the investor's commercial investment interest, and then there are the managerial cheer-leaders for the project, where if the new project, supposedly, depends on new knowledge (really only extensions of existing dogmas or simply an empirical solution), then the managers deal with the academic community, where there are academic competitions based on dogmas and fixed language structures, so as to determine the personnel to develop the project, which is considered to be marginally based on new knowledge

It is clear that this scenario is in the realm of nuclear bomb engineering (where the, supposed, new knowledge is about how to relate gravitational singularities to particle-physics processes, ie particle-collisions) where bomb engineering is related to a chaotic context of system transitions, in which the key measurable properties are energy material-component types and probability of component-collisions, where the entire subject of physics is now all about the model of elementary-particle collision probabilities

The, so called, new knowledge, which seems to be of interest, in regard to bomb engineering, and it (ie bomb-engineering) is the relation of particle-physics (material-interactions defined as point-particle collisions) to gravitational singularities, where these singularities are (can be) modeled as being the center-points of spherically symmetric geometries.

The math structure of particle-physics is that of a set of local equation-invariant properties (of internal particle-states) which are associated to global invariant energy wave-functions for a quantum-system (for a quantum system's stable energy states, eg Heisenberg representation, invariance of quantum states, in the Hermitian-invariant (integral) forms of the function-spaces models of the quantum systems).

The local invariance's of a derivative-connection dependent (partial) differential equation model of the quantum-material-interactions is described in a context of elementary-particles, where the interactions are due to particle-collisions, are equation invariant, and they are energy invariant relations

(since they are unitary [both the wave-function and the fixed set of internal-particle-states of particle-collisions are modeled as (or with) unitary operators] and thus, the local changes in particle-state matrices are algebraically-diagonalizable (local-coordinate transformation matrices, but the local-coordinates (in particle-physics) are the internal particle-states), thus the description always reverts to "the set" of elementary-particles, whose spectra is, supposedly, identified by their masses, but new elementary-particles can always be added to the descriptive context [a descriptive context only relatable to particle-collision properties] so as to provide a new type of epicycle set of descriptive structures, which are, supposedly, the absolute "laws of physics" from which "all material-based properties are derivable" [where such a belief is a giant joke, which, unfortunately, no-one "gets", thus leading us-all into our (own) demise and ruin, all based on the partial-truths which are cheer-led as "absolute truths," which are adhered to by a group of domineering wage-slave, so called, intellectuals, who really represent the so called intellectual-top of a very ignorant culture, a culture which is easily being mis-led by the same social forces ass are used in high-schools]),

[note: such a derivative-connection based partial differential equation is a non-linear equation].

So (the technical question)

How do the random particle-events fit into the coordinates of the metric-space in which the material associated to the elementary-particles exists?

Answer: They are attached to the global wave-function in a unitary, or energy-invariant, manner.

But

Finding that wave-function is difficult, since, in general, it is not defined, but

Furthermore, the quantum description provides the probabilities of detecting particle-events (of the particle-components of a quantum system) in space and time.

And the model of a bomb only needs probabilities of material-component collisions in space, which is where point-particle events of quantum systems manifest.

Apparently, there are both (1) the particle cross-sections and (2) the probabilities of a quantum event, to both be realized, but, seemingly, the wave-function probabilities seem irrelevant, since, essentially, wave-functions for general quantum systems cannot be found.

Whereas the gravitational singularity has to do with the math structure of a non-singular point (of the zero-point of the 1/r sectional-curvature term of the general metric-function associated to a sphere), where the metric-function is represented as a symmetric-matrix, so at each (none-zero) point it may be diagonalized in its local coordinates, (ie can be related to a local diffeomorphism group at each point, where the point is not the zero-point) but a general metric-function cannot be parallel-transported (locally) back to itself in an invariant manner, since it is representing a non-linear geometry, and its math structure is non-commutative, thus, it is many-valued and very sensitive to slight changes, ie it is chaotic in its quantitative properties, this means that the so called gravitational singularity has no mathematical representation.

This is true for both descriptive contexts of particle-physics and gravitation [they are both non-linear geometries, and their math structures are non-commutative], yet the indefinable context in 3-dimensions is both spherically symmetric (of a charged field) and relatable to particle-collisions (as high-energy particle-collision data is showing), but

The, so called, experts (really nuclear bomb technicians) are approaching (or trying to approach) a context where particle-collision energies are close to the energies of gravitational-singularity related models of gravitational collapse, though gravitation is Euclidean (and [in the new descriptive structure] the, so called, Higg's particle could well-be interpreted to be a small Euclidean shape, and not a, so called, scalar-field particle) and [in the new descriptive structure] energy is a property of a hyperbolic metric-space (in a context of opposite pairs of metric-space states), and it is (also [in the new descriptive structure]) likely the charged particles would be related to a higher-dimension geometric context, but in regard to bomb technician models "realizing some high-energy cut-off point" might lead to the initiation of a gravitational collapse (or black-hole) type of explosion, which for the partial-knowledge now possessed by the experts (who, mindlessly, serve the investor-class), there is no descriptive context and thus the bomb-technicians have no idea about how to control this (or such a) structure, as the failure to obtain energy-from-fusion has shown,

ie the experts (bomb-technicians) seem to be, inadvertently, racing towards, or enthusiastically working towards, a construct of total annihilation, so as to blow-up the solar-system in an uncontrollable manner (and the media is cheer-leading this process, which society's so called top-intellects are working-on).

This all seems meaningless and not a likely scenario, but the fact that "this is what is being done" is quite irresponsible (though the intellectuals who work on these projects believe these pursuits to be "the highest cultural achievement mankind has ever realized," ie they are "full of themselves," since the media has pumped them up to be so arrogant, [in the new descriptive structure] the charged material, when it reaches a higher-energy, will transition to a new stable shape (as when Dehmelt cooled the electron, ie the cooled electron did not get-still (in the lab), rather it entered into an unexpected relatively stable orbital shape)

Where the stable shapes of physical systems [in the new descriptive structure] might become understood if the physics community, actually, did science, and, actually, tried to understand the structure of stable spectral-orbital systems, rather than speculating and creating expensive instruments so as to experiment about extreme high-energy bombs for the investor-class.

But the social forces have placed the intent on the investor-class, and the intellectuals selected by the investor-class (eg to be physicists) are pawns in the game (of moneyed influence and illusions of high-value).

The need for knowledge about the many-(but-few)-body systems with stable spectral-orbital properties needs to be addressed…, before it can be seriously considered that the physics community can claim that they actually possess the "laws of physics,"….,

This is because the (so called) "laws of physics" (which are endowed, by the media-education system, with absolute truth) are only relatable to the properties of point-particle collisions in high-energy particle-accelerators, and they are unrelated, by their math-use, to understanding the stable properties of the most fundamental of stable physical systems, and, subsequently, and quite mis-guided-ly, they are applied to all-reaches of the universe, so as to assume that these vague (local) quantitative relations about particle-collisions provide a valid model of physical systems, which are claimed to be related to "the set of interpretive patterns, in turn, associated to distant astronomical data."

Thus, apparently, the question which has the investors puzzled is, "How to relate a local particle-physics model of material-interactions to a gravitational singularity?"

That is, "How to relate an indefinably random construct (where with collisions of charged point-particles it also carries indefinable geometric construct) to an indefinable geometric construct?"

But this above question is only about bomb engineering, which is, supposedly, being answered by the ideas of both a Higg's particle (which, supposedly, gives quantum-particles masses) and the, hypothesized, gravitational-field particle, ie the graviton (which seems not to exist, though it is claimed to (already) be validated by particular interpretations of data, eg the recent (2014) wiggles in the, so called, 3-degree Kelvin background radiation), and, unfortunately, but of no concern to the bomb-engineers, is the fact that, the math being used to model quantum-material-interactions cannot carry a stable pattern upon (or within) its structure, where this is of no concern to bomb-making, since the explosion depends on (1) energy-levels, which "for elementary-particles" means their mass and velocity, and (2) reaction-rates which is the only thing particle-physics can calculate, ie the cross-section, which along with particle-density determines the probability of particle-collisions.

That is, the model of quantum-material-interactions is of little (to no) interest to the bomb-engineers (the bomb-technicians simply want masses and cross-sections, the rest (particle-energy and density) is empirical, and can be empirically controlled. That is, their job is to empirically control these properties (ie particle-energy and density) in their instruments).

That is, quantum-material-interactions are a complicated and irrelevant side-show (for the propaganda-education system) in regard to the knowledge of physics, which is of interest to the investor-class.

The, so called, Higg's particle could well be interpreted (in the new descriptive context) to be related to the size scale, in 3-space, at which electron's and protons etc, manifest their Euclidean shapes, so that the mass of these material-components is the shape-size multiplied by a constant.

Whereas the so called wiggle in the, supposed, 3-degree Kelvin, background radiation, can have many interpretations, or causes, but one should note that inflation destroys the continuity needed to realize (in a, so called, rigorous mathematical manner) a big bang, and the, so called, big bang (or a black-hole) is entirely speculation, which is driven by, so called, rigorous math relationships, and interpretation of astronomical data, ie it is speculative science, but it (an, assumed to exist, gravitational singularity) is (or may be) related to rates-of-reactions and explosions, in regard to both many gravitons (which likely do not exist, or they, really, exist as Euclidean shapes of inertial interactions, but they are instantaneous, ie action-at-a-distance shapes) and many Higg's particles.

Furthermore, the measured properties, obtained from astronomy, about big bangs and black-holes, are about (or, it is assumed that they come from) distant regions, whose metric-function and set-containment structure may be quite different from our local regional metric-function and set-containment structure, and the set of partial differential equations which are claimed to be "the absolute truth about the, so called, laws of physics" are dysfunctional (about which the bomb-engineering community does not care, since the model of quantum-material-interactions is irrelevant to bomb-engineering), and thus

(ie since they [the, so called, laws of physics] cannot be used to describe any stable properties of actual physical systems)

They are not likely to be anything close to being "valid laws-of-physics," so extending them out to a fantasy-land context is ridiculous, except that Newton also defines a black-hole, so a "large-massed object" might be gravitationally collapsible, in a way which is relatable (in 3-space) to a spherically symmetric interaction field, and thus the particle-collision model of rates of reactions for bombs might be relatable to such a collapse in 3-spatial-dimensions.

Socially, the population is preyed upon by the investment class, their spies, their field-agents, their managers, and the enthusiasm-controlling media (where the media includes: publishing, education, and politics), where there are certain people which the investment class can manipulate to its, supposed, advantage.

Chapter 18

Intellectual revolution

Understanding that "science and math have failed" is of deep importance in regard to realizing the need for an intellectual revolt which needs to break-away from the US totalitarian collectivist society, where now narrow intellectual dogmas (followed by both the academic institutions and defined by the narrow ideas which peer-review requires in regard to being admitted to the professional intellectual class) have come to define all types of intellectual activity which are used within society, but it is too narrowly defined.

That is, if one "wants more jobs" then one needs a much wider set of creative contexts than the set which the academic and industrial experts provide to society, since these so called experts, only deal with the creative context which serves the investor class,

Note: The Powell memo (from the Associate Supreme court person of the 1970's, Powell) expressed an interest in requiring that all ideas expressed by the public be put under surveillance, apparently, to protect business (or investor) interests.

But, furthermore, it has been "recently" revealed, by documentation, that a totalitarian police-state exists, so as to be a deeply invasive, and personally interfering, where such an arbitrary interference is allowed based on arbitrary ideas which support the interests of the investor-class, and where this interfering is based on a person (a spy) having a petty relation to arbitrary social-influences, basically the agents are to attack any aspect of the public expression which might slightly interfere with the narrow interests of the investor-class, where this is done in a context of improperly identified, but which nonetheless was "a very authoritative type of expression" that the authorities of the academic institutions already possess an essentially absolute truth, ie acting with impunity for selfish advantage of both the ruling few and for the spy-agents, themselves,

(which has, for a long time, been the norm of the American political process, where justice and propaganda are wedded-together, in the name of an arbitrary-form of minority-rule (US law is based on property rights and minority rule), but apparently the politicians and publishing editors [the propaganda-education system] can no longer control the cover-up of the corruption of the ruling-class)

The failure of science is not identified in our propaganda-education institutions, and new ideas about math and science are excluded from the media, because of the mythology that the media (the entire-set of the propaganda-education system, even all of the, so called, serious internet-sites) is only about providing the public with "the authoritative and, supposedly, objective truth," but (apparently, a fact which is never actually internalized by those who claim to be serious about truth) this is all about the truth which the investor-class has provided to the collective society, and this truth is "the truth" which is being defined in our, so called, public academic institutions, and this truth is the truth which best serves the interests of the investor-class,

It needs to be noted that the, so called, laws of physics are sets of (partial) differential equations associated to a small set of fundamental assumptions,

ie materialism (a fixed dimension for a description's containment space), reduction of material to elementary-particles (all material interactions are supposed to reduce to collisions between elementary-particles, the only reason for such an idea is that it is easily related to rates of reactions in nuclear explosions), and the property of (indefinable) randomness (the world is related to a quantitative model only when the descriptive context is to be only about fitting measured data to fundamental random events),

But where the (partial) differential equations are only of any value…, in regard to the practically useful properties of both accurate descriptions (over a wide-range of systems) and control over a description…., if these equations are solvable, where they are solvable, effectively, only if the equations are: linear, metric-invariant, and continuously commutative (ie local matrices of the function's have relations to the system's coordinates are (continuously) diagonal, where these coordinates indicate the system's shape within the system's containing coordinate space, a set of properties which are also called "separable"). Unfortunately, none of these physical laws are solvable, except the box and the harmonic oscillator, so the laws of physics keep the attention of the scientists on an irrelevant descriptive context, which essentially has no practical uses, but which causes the math structures, upon which the scientists focus to be quantitatively inconsistent and chaotic, thus enforcing the notion that "the main property of the world is that it is indefinably random." furthermore, the non-linear (partial) differential equation is a natural part of feedback systems which the military wants developed.

Instead of trying to identify the observed stable physical systems by using partial differential equations to model material interactions between point-particle models of material, which the partial differential equations try to describe, but this viewpoint has failed, to be able to identify the observed set of stable properties of the world

Rather,

Instead one should try to identify a context for existence based on the stable shapes, and their natural association to stable spectral-orbital properties,

And,

Furthermore, from this new context, then identify a (new) structure for material interactions, and from the new context (for interpreting existence, and identifying a dimensionally-related containment-set structure for existence)

Derive the, apparent, randomness of the microscopic systems from a (new) model of material-interactions (note: the new model is quite similar to the old model of interactions), [and where the point-like properties of quantum models of existence are (in the new model) a result of each stable shape being also related to its own distinguished-point (on its shape), where this distinguished-point becomes the center of material-interactions, such as material-component collisions] but where the main attributes of existence are the stable shapes,

which can model both

(1) metric-space containment sets (for material), allowing a stable solar-system to be understood, but the stable shapes can also model

(2) the stable material-systems, allowing the stable, and usually neutral, quantum systems to be understood, Where a material-system (a stable shape) is contained in an adjacent "1-dimension higher" stable metric-space shape, and where the set of stable shapes, and their (new) interaction structure, are now placed in a many-dimensional context, for all of existence.

It needs to be noted that a valid expression of revolution which is completely consistent with the original American Revolution and the (expressed) ideas of T Jefferson, and by the courage of the Quaker, W Penn, ie the original leader of the American Quakers, so as to provide both relatively good relations with native peoples (in Pennsylvania) and building the most intellectually energetic city of the colonial Americas, Philadelphia, the true cradle of a Revolution based on equality, where the original idea upon which the American Revolution was based is about: organizing a society around protecting the individual within a society of equal creators, and with the need to express equal free-inquiry in regard to developing knowledge related to one's own creative intent.

These are the main expressions of the Declaration of Independence (law based on equality), the preamble to the Constitution (the basic spirit of the law and governing), and the Bill of Rights (the relation of individual creators to knowledge and creative intent, and it is about curtailing an intrusive state whose goal is to help the selfish intent of a minority-ruling-class).

European societies have, since the Roman-Empire, always been collective societies, where the public supports the ruling-class, but it has only been the American Revolution (as well as many native American societies) as expressed by the Declaration of Independence and the Bill of Rights, which has been a true social revolution about structuring a knowledgeable and creative society around all of its equal individuals.

Developing knowledge individually, new ideas which, in fact, put the current descriptive ideas about math and physics to shame, since the current ideas are failing so badly, but (the current ideas) remain so narrowly authoritative, so as to only apply to a few commercial interests, while the propaganda-education system beats the drums for the absolute truth of these failed ideas of modern (or currently accepted) science,

(unless the findings of a science institution opposes the interests of the ruling-minority, where it then attacks the statistical basis for the new science findings, but all alternative ideas are excluded, so as to only allow the science ideas which are used in relation to investment interests to be expressed by society's official expert-science class).

Consider the new, revolutionary, viewpoints about math and physics, which are based on both many-dimensions and on the idea of stable shapes, so as to form a new descriptive context where both measuring is reliable and the patterns being described are stable math patterns, where these stable math patterns (ie stable shapes of mathematical descriptions) form the basis for the spectral properties of the spectrally stable physical systems composed of many-but-few-bodies (or many-but-few-subcomponent systems), rather than being defined (as is now being done in the professional [or peer-reviewed] math and physics journals) by an arbitrary spectral-set, where the arbitrary spectra are associated to harmonic function-spaces and its related set of harmonic-analysis processes, which determines an indefinably random structure, which has no meaning.,

ie it is unrelated-able to some specific (observed) spectral-system, in particular, when placed in the descriptive context of quantum physics, eg the wave-equation for a general atom, except when bounded in some artificial (or externally determined) manner,

eg artificially truncated, ie effectively the math context is discontinuously changed in a way which is not based on math principles, but rather based on "the wish to fit the description to known data," (ie of the H-atom), or bounded by the properties of some controllable electric circuit.

These new ideas put knowledge.... of the physical world... into a valid descriptive language, where the fundamental properties of the many stable systems in the physical world are, in fact, describable, and so that the nature of existence can be appreciated in a new context, in regard to practical creative concerns, as well as one being able to now (in the new descriptive context) place the context of the human-spirit (ie living-systems contained in a many-dimensional context) in a more realistic, and much more relevant, set of relationships between (material) existence and the spirit, where the descriptions of these math models of physical constructs are provided in, the new books, so that the structure of living systems are related to stable shapes, which have the property of being of the shape-type of an odd-dimension and odd-genus shape, and (these living systems are) contained in higher-dimensions, than the idea of materialism allows, but where the hidden structures of these living system models are about (or are due to) topological properties of metric-spaces and the bounded shapes of (stable) metric-spaces (of constant non-positive curvature), placed in a dimensionally dependent sequential-tree of set-containment of a dimensionally partitioned many-dimensional space (dimensionally partitioned by the stable shapes).

The simple ideas in math upon which the new ideas are based

Math is about quantity and shape, where quantities are (math objects, or measuring standards, which are) based on stable uniform units of measuring, and a subsequent construct of counting (in this uniform unit),

but such uniform units of measuring can be consistently placed on either the shape of a line (or line segment) or identified from some fixed point on a circle, and measured around the circle (or so as to measure around the circle).

This consistency of quantities on the two shapes of both lines and circles is the basis for the number systems of both the complex (or planar-numbers) and the real (linear-numbers) where the two number-systems have the same set of algebraic (or number-operation) properties, where the algebraic number operations are defined in regard to the two (main) number-operations of counting and grouping (where grouping may also be thought of as re-scaling, ie changing the length of the uniform unit of measuring), or adding and multiplying, respectively.

But "measured values" are usually related to (or thought of as) uniform units of measurement defined on a line, so as to partition the line by uniform line-segments, where the uniformly spaced partition, identified by the integers, begins from some fixed point (usually the zero-point) on a line.

Measured properties are most often about relating a (measurement on a) physical set with a set of numbers, through a process of measuring, ie so as to define a (single-valued) function, often (especially in physics) the property of a position determined in an object's physical containment set, where this position is given in relation to some arbitrary fixed point (in the containment coordinate space) so as to be put in relation to the position of a center-of-mass point associated to a (the) physical object, so that changes in position can be determined from measuring between the two distinguished points in the coordinate containment set.

This is elementary, but somewhat difficult to describe, but it is also fundamental to how everyone should think about measuring and numbers, so as to put (or place) measured properties into quantitatively consistent, and logically consistent, descriptive contexts, and placed into a well defined quantitative construct concerning the (measurable) properties of an external world (ie measured properties are often best thought about as values of functions defined upon (in) the coordinate (domain) space where measuring processes can be related to uniform units of measuring).

[where indefinable randomness is the most common context, in which numbers are most often placed, when the idea of measuring is applied to the world, so as to be placed into a context of probability (or randomness and uncertainty), and are, subsequently, used to form improper math models of measured properties, or counting within improperly defined sets of random events (ie or improperly defined counting) {ie an elementary event space of stable and well defined events, must be the basis for properly defined probability} (so as to form quantitative contexts which actually possess no meaningful content, since there are no stable patterns which are being referenced)]

The derivative is the natural construct for a local linear measure, which can be defined on functions, and it is defined as a local linear map from the domain space (ie the system's containment coordinate space) to the space of numbers of the function's values, ie the function represents a system's measured values, or measured properties (where the {physical} system is contained in the (or its) domain space).

The derivative is best considered to be a local linear model of a "measuring system" defined in regard to a set of variable function-value (or measured properties), where the function's values are number-relations, so the function-values are to be uniformly (or consistently) linearly related to the uniformly defined set of quantities which compose the domain variables, ie the quantitatively consistent (locally linear relations to the domain variables) (where the domain space is the system's containment-set, ie "set of domain-space measuring attributes").

However, one might ask:

1. Is the derivative about continuity?

Or, alternatively (an equivalent alternative),

2. Is the derivative defined on a domain-space, set of containment-coordinates, that, in turn, defines a continuum?

Or, a different alternative,

3. Is the derivative a discrete operator, where the containment set is also partitioned into a discrete set of stable shapes, where these stable shapes (or resonating with these stable shapes)…, which define the stable properties which the containment space can describe…, and where the coordinate metric-spaces are also partitioned into two sets of opposite metric-space states, which are defined in regard to a fundamental (physical) property associated to the coordinate metric-space.

Where, in math (or in the math-models of the physical world), it is assumed (in the new descriptive context) that quantities can be measured in a linear real-coordinate model, and the fundamental properties of these measured values are (can be), simultaneously, spin-rotated between a pair of opposite metric-space states, which are defined for each metric-space, so that (or where) locally (ie in regard to the derivative model of measuring), there can be discretely distinguishable opposite (or inverse) fundamental local relations of the function-values, in regard to the changing values of these measured properties (identified by the function-values), defined in relation to the set of fundamental properties of the (or associated to) metric-spaces, so that these changes are defined on functions, whose domain-space is the coordinate space.

The only patterns which can be distinguished in this math structure, used for a quantitative description (or math model) of a physical system, which are measured locally, ie related to (partial) differential equations, are the local linear, metric-invariant, and continuously commutative, properties of stable systems (or stable patterns related to physical systems).

These are also the main attributes of the main set of shapes defined in geometrization theorem of Thurston-Perelman (see below).

There has been a map of the set of stable patterns, or stable shapes, which can exist in the various dimensional levels…, in regard to the non-positive metric-spaces of constant curvature (where the metric-function can only have constant coefficients), ie the general these are the Euclidean and hyperbolic metric-spaces…., given for the stable hyperbolic shapes which exist in hyperbolic metric-spaces, and which are defined from 1-dimensional up to 10-dimensional sets of discrete hyperbolic shapes, where this map (or list) into the higher-dimensions was given by D Coxeter, and where these are the stable shapes…

[doughnut shapes and shapes made from strings of doughnut components connected together]. along with the cylinder, the cubes (or rectangles) and a small set of a few other stable shapes, are the entire set of stable shapes, which also compose the stable shapes defined by the geometrization theorem of Thurston-Perlman. The discrete hyperbolic shapes are the main set of stable shapes. The stable shapes are the non-positive constant curvature metric-space shapes, for metric-spaces whose metric-functions possess only constant coefficients, ie the discrete Euclidean shapes and the discrete hyperbolic shapes where the discrete hyperbolic shapes have a dimensional bound of 11-hyperbolic dimensions. The stale hyperbolic shapes identify the stable fixed-energy shapes for material components, while the discrete Euclidean shapes are a part of the geometry of the spatial-displacement interactions, which exist between the discrete hyperbolic shapes.

This is simply a re-expression of the more fundamental idea, that only lines and circles can be put into quantitatively consistent (algebraic) geometric relations to one another, right rectangular shapes and circle-space shapes, and (in some cases) cylinders.

That is, if one wants to describe a pattern which has a set of stable measurable properties, then it needs to be related to this particular type of a set of math patterns of stable shapes, placed in a many-dimensional containment set, where the shapes can be of:

1. either even-genus or odd-genus (where these shapes only exist if their own dimension is odd, ie the dimension of the shape itself is odd, and where these odd-dimensional shapes fit, as geometric shapes, into their adjacent higher even-dimensional containment [hyperbolic] metric-space),

or the shapes can be

2. either bounded or unbounded, except after hyperbolic dimension-5, ie all the discrete hyperbolic shapes dimension-6 up to dimension-10 are all unbounded (stable) shapes, and there are no known discrete hyperbolic shapes which exist for 11-hyperbolic dimensions or higher (D Coxeter).

The odd-dimensional and odd-genus shapes oscillate (and generate their own energy) and define a set of new material types in the higher-dimensions, in hyperbolic-dimensions 4, 6, 8, and 10, while the unbounded shapes are either light (or field) shapes (with distinguished-points) or neutrino types of shapes, with a relative relation to being bounded or unbounded depending on the metric-space containment set's properties of being stable shapes which are themselves either bounded or unbounded (if of hyperbolic dimension five or less).

This is the most interesting set which should be of greatest interest to math, especially, if one wants the descriptions of the measured properties to be about stable patterns. That is, the description of math models only has meaningful content, in its descriptive set of symbols, if the descriptions of measured properties are about a context in which the set of measurable properties being discussed is related to this relatively small set of stable shapes, but this description can be defined over many-dimensions, and it would have shapes which would have new "material" properties, ie the odd-dimensional and odd-genus shapes, ie a set of high-dimensional (or, seemingly, hidden) stable shapes which can be used to model a living-system. A mathematical description of what appears to be a hidden math structure, whose apparent hidden properties have math explanations about this seemingly hidden structure so as to fit new mathematical models of living systems which fit into the math context of stable shapes defined over many-dimensions.

That is, it is a much more interesting descriptive context than the context which is based on materialism, ie arbitrary restriction of the dimension of a system's containing domain coordinate space.

Unfortunately, instead of focusing on the set of contexts where these types of stable patterns can be inter-related in the context of a containment set for the descriptions of stable patterns, instead the professional math community focuses on irrelevant descriptive contexts (eg the laws of physics are to be based on partial differential equation models of material-interactions which take place between point-particle components, but where there are no stable patterns and in which there are only contexts where measuring is unreliable and patterns are unstable and fleeting, ie (thus) the description has no content, or (in non-linear feedback systems) the content is really about the stable structures which compose a containment set (ie feedback systems) the interaction is to be adjusted within the context of the system's containment-space's geometric properties [but where the true nature of the containment set (and its set of stable properties) is not identified, but the (an effectively) un-nameable stable set, is really the relation of a metric-function to external shapes (eg the distant geometry which is the cause of the feedback system's non-linear interaction structure), so that local measures of a model of an interaction are (can be) adjusted, by feedback, to (try to) conform to the relatively stable external properties, in a way in which an external observer desires]. (excluding the context of feedback) for "free systems," this is a quantitative structure which allows for the description of any arbitrary pattern, so that the descriptive structure is designed to fit the data, obtained from any arbitrary set of patterns, so as to be put into its vague and inconsistent structure, so as to validate the arbitrary patterns as being a pattern which exists,

and

The, subsequent, unstable pattern is identified by data, where the data is processed in the set of inconsistent math structures, which can be adjusted in arbitrary ways.

This is essentially, the math structure of quantum physics, where quantum physics cannot describe the stable properties observed for general atoms or general nuclei or molecules or crystals etc etc

but nonetheless the "science authorities" proclaim "quantum physics has never been shown to be wrong," well, in fact, it has obviously failed, but since it is an mathematically inconsistent structure, which is used to fit data to arbitrary patterns. Note: The epicycle structures of the Ptolemaic system could have equally, as confidently as today's quantum scientist cheer-leaders, have proclaimed their epicycle system is never wrong, compared to the Copernican model.

With such a math purpose (to fit data within an indefinably random context), it is difficult to say anything, at all, about its truth, ie it is a math structure which is truly designed to "try to fit data" from sets of unstable (or indefinable) patterns (which are non-commutative) so as to fit-to arbitrary sets of unstable and fleeting patterns, ie it is a descriptive language which has no content.

But it is descriptive pattern which (when the propaganda-education system identifies as being valid science, as proclaimed from the experts themselves, the experts which the propaganda-system so reverently worships, so as to provide society with the absolute truths, which, in turn, the propaganda system proclaims) can be used to justify arbitrary claims, eg it, supposedly, can be used to identify future criminal behavior, (behavior which is allowed for the rich, but which is not allowed for the poor) and this, inconsistent description of nonsense, can be used to justify the pre-emptive extermination of the subject's, so called, future criminal behavior, and thus it can become an authoritative, so called, scientific basis for the rich to exterminate the poor.

The set of ideas (or descriptive contexts) upon which the Science and math professionals focus is being determined by the investor-class, since the investor-class controls, through the propaganda-education system, that which society believes possesses high-intellectual-value, as well as how language is used, where the categories of technical language are being controlled by peer-review, a totalitarian model of science, which most certainly excludes any new alternative ideas or interpretations of science such as Copernicus provided to the intellectual-class in his age, but Copernicus was excluded from the intellectual class's language based vision of truth in that age, and these categories of thought, and their associated descriptive contexts, are most often related to the commercial and social-control interests of the investor-class, eg oil, the military, control of communication channels, and control over technical language so as to get a, so called, scientific validation for any arbitrary model which is placed into a probabilistic model of description (note: {to re-iterate} but arbitrary probability-based descriptive constructs cannot be used to even identify the observed stable properties of physical systems..., as quantum physics and particle-physics, as well as general relativity [since general relativity is, in general, non-linear and cannot be fit to (or be solved for) any stable pattern, except a 1-body perfectly-spherically-symmetric system, and which is unperturbed] ... demonstrate..., due to the inability of these descriptive contexts {which are based on indefinable randomness and/or non-linearity} to be able to describe the stable properties of: nuclei, general atoms, molecules, crystals, (living-systems), and the stable solar-system, where such an inability to describe these fundamental properties so clearly demonstrates the failure of these descriptive contexts)

Note: This is the main problem with a collectivist and hierarchical society…namely, it is hierarchical, where actually, it is difficult to imagine a collectivist society which is not also very unequal as "the collective" is going to be directed by a social-hierarchy, which is going to be providing direction to such a group-centered society.

The main point of the American-European experience, in regard to the noble American revolution, is about allowing the individual to act on their own, the main point of the first amendment, while the second amendment is about there being no standing-army, and subsequently, no set of amoral spy agents and no set of criminal spies interfering with individuals. However, the point of supporting the individual, over a collective, is about an individual's relation to knowledge and creativity, and not as much about the individual gaining social power for themselves, but European society was/is all about (aspiring to) the individual-power of the Roman-emperor in a collective society.

Whereas, many of the societies of the native-peoples of the Americas were about allowing an individual to be themselves, so as to not let a fixed way of using language to become a dominating influence in people's lives.

It is remote speculation, by the highly controlled obsessive and dominance-seeking small set of the, so called, top-intellectuals, who have been led to follow authority and intellectual traditions, that the sets of non-commutative, and quantitatively inconsistence, and both unstable and fleeting shapes, which are (only) slightly related to physical mathematics, are in fact to be considered the most relevant aspects of all the math structures in their relation to physical descriptions, this is all-about trying to provide a so called form of scientific validity for the substantiation by data of arbitrary claims by the technical-arm of the ruling-class ## other than the indefinably random and chaotic physical contexts which possess a relation to the rates of reactions through particle-collision probabilities, which exists (or where this context exists) in the physical transition processes which take place between two sets of relatively stable states, where this chaotic descriptive context is used as a model of a nuclear explosion, ## however, there is no way in which these illogical, and quantitatively inconsistent descriptions of unstable properties actually possess any meaningful content.

Section about alternative viewpoint as to the tenuous validity of modern science and math

Hoaxes perpetrated on the technical communities (but what the hoaxes really reveal)

Deceptions of submitting to professional journals, and the subsequent (in regard to these journals) publishing by these journals articles which are fake-authority, or only "authoritative sounding gibberish," is really a demonstration that the, so called, "real authoritative articles" published by these professional peer-reviewed journals are, themselves, without any (useful) content, other than (maybe) being consistent with a formalized axiomatic language. Formalized axiomatic, implemented (around 1910) into academia, has made math and science ineffectual, that is, there have been no new creative contexts to consider by the intellectuals.

That is, one might seriously question what the underlying content actually is in the highly technical peer-reviewed articles in professional journals, where high-academic or high-industrial social positions are determined by being published, in these peer-reviewed technical journals especially, in regard to the hoaxes recently perpetrated on these professional authoritative journals,

Recently, as reported by Sigma Xi, 2-15-14, on the "Institute of Electrical and Electronic Engineers" journal, (as well as the old, Sokal's hoax, in regard to a professional academic cultural-journal), in which articles which are "apparently, similar in content" (when read) to "the authoritative sounding articles" of the, so called, "real authoritative articles," which, in turn, are published in professional journals, so that "the hoax is," that the submitted "similar sounding types of authoritative" articles (which are accepted for publishing in the authoritative professional journals) are, in fact, articles with either arbitrary meaning, or no meaning, or gibberish (ie computer generated gibberish), so as to be "an authoritative sounding" sets of gibberish

However, in fact, this would certainly be true of many, if not most, of the articles in the very highly regarded Physical Review professional-academic journals, and the subject matter of most of the academic math journals, which are full of abstract general nonsense, where the peer-reviewed professional experts, are trying to describe the state of unstable fleeting patterns in a context which is no longer a valid math…

(or reliable quantitatively measurable [or containment])…. context, the correct interpretation…, in regard to the hoax-related articles, which are published in these professional journals…, is that this really means that, in fact, the authoritative articles themselves possess no actual content, and this is because they are, essentially, irrelevant in regard these…, so called, "real authoritative articles"… actually possessing, within themselves, either accurate descriptions of physical properties, or related to practically useful technical development.

That is, the actual authoritative articles, themselves, only communicate useless information, which is unrelated to the actual properties of the, often called, real-world, where this context of communication is related to a fixed language determined by a set of formalized axioms which, in turn, define the range of academic competitions in this meaningless formalized professional language structure.

That is, in the context of formalized axiomatic language, the (so called) real authoritative articles are those published articles which correctly state the formal statements which are consistent with the formalized language of the professional languages of either in math or in physics (where physics is now very axiomatic) but which have no valid relation to either accurate descriptive capabilities or to practically useful models,

The point is that it is relatively easy (when either formal axioms are followed or when formal-sounding language is adopted) to put something which sounds authoritative (or is considered to be professionally authoritative) into an authoritative journal, whose professional language is being determined by a set of formalized axiomatic language structures, where this formalized languages has no content in regard to its relation to the real, or physical, world.

For example, the quantitative validity of economic models should always be put into question, but when it is not questioned, then the authority of simply "sounding quantitatively economic" carries far more weight than it ever should carry.

That is, a formal language comes to be totally random, or which is based on being totally random, then in terms of meaningful content, there is none.

This means that there is no compression of information into a set of statements, which are being provided by peer-review journals, and which, thus, these non-compressible statements have no meaning, ie the set of statements are irrelevant to either accurate descriptions or to practical usefulness.

Thus, it is (would be) easy to trick the people who participate in the "acceptance process," if a submitted article sounds like the formal language, especially, since in both cases, either the "completely consistent article" or the "fake article," (neither of which) have any content, so this means that both articles require the same "large amount of information" in their expressions (which are empty of any content, that is, both the real and fake articles, in relation to either accuracy or practical use, have no content) and thus, they express the same context of being completely random expressions of symbols, so they are indistinguishable except in the meaningless context of getting the exact formal language exactly correct (this is the essence of the peer-review process). That is, the most fundamental and stable many-(but-few)-body systems are claimed by the current science dogmas to be too complicated to describe. The current descriptive context cannot carry within itself any content, ie it cannot compress information, it cannot provide any compressed information it can only describe the (indefinable) random context upon which it is based.

This is the natural end-result of information, which is being channeled into a formalized axiomatic construct, as is the language of the professional math and physics (or in science "in general") journals today, ie it can provide information which is only random and incompressible.

Though C Ventur can, supposedly, make exact replicas of DNA-molecules, in an essentially classical domain of the nano-world, but these molecules only "work," as actual DNA, when they are processed (within living cells) in a context which is outside of scientific description, or outside the descriptive capabilities of authoritative science and math.

In elementary information theory, the property of being random implies the use of ever-more information, so if a language has no content then this implies it is random, so it requires ever-more information (to identify a random state), thus the authorities can be tricked by either gibberish, or by expressions which have no content, in relation to either accuracy or practical use.

(the whole exercise is a meaningless exercise in using formal language, which has no meaning, in regard to having a practically useful (or even accurate) relation to the observed properties of the world).

The "deception," and the, so called, real thing (both) have the same information content, and both are totally random, in regard to the practical usefulness or accuracy of what they... (the two articles, submitted to authoritative, peer-review journals).... are describing.

Thus, the fact that it is so easy to deceive the publisher into publishing (both the fake and the real authoritative stuff) is a proof that the so called real-authoritative-stuff is as useless and random (and without content) as the so called fake-stuff.

.... that is, without the above considerations about the stable properties which a math description can actually describe in a meaningful way to balance what has been published in the professional math and physics journals since about the 1910's, that all of the professional-journal published material since the 1910's should also be discarded, since it also (or mostly) has no meaningful content, ie it is also gibberish.

Social comment,

The types of people who are called society's leaders (obsessive and seeking domination over others) and the nature of the propaganda-education system which exists

Where the behavior of Obama shows the moral development, and lack of character, of the people running the society, a bunch of empty suits, seeking only good-fortune for themselves.

The issue which the elitist press is working so hard to cover-up (eg Democracy Now, or Rush Limbaugh etc etc) is that the blanket-spying is a practice which has, effectively, been occurring since the Pinkertons, 1850's, and it was also before that, as G. Washington had his own spy-ring, and it is all about protecting the rich from the poor

(the elitist-press is talking about a fake-story about the freedom of the elite press being intruded-upon, the elite-press is effectively saying, "do not the elite politicians and judges know that the elite press is also 'immune to being called criminals' and to be protected"), where the rich and their politician-media-propagandists are lying, stealing, and murdering, (an observation made by M Twain)

But, where this is (mostly) being done is by the palace-guards, or the secret-police, but where the palace-guards (or secret-police) are in reality the types of people that J E Hoover was.

The palace-guards and/or the spy-agents are racist, opportunists, bigots, whose main life-interest is psychopathic domination over others. This use of secret police and wire-tapping into the information system was instrumental in forming and manipulating the Lee H Oswald, Sirhan Sirhan, James E Ray, etc etc. all either assassins or stooges or patsies, who were being manipulated by a form of blanket spying in a very confining social structure, which can be much more confining to some people than to others, so that these people preyed upon and set-up to be chumps, so as to be directed and controlled by handlers,

And thus,who is directing the handlers of these chumps, where these chumps are being preyed upon within the institutions of which they are (or were) a part?

There is the, so called, revelation that the spy agencies are allowed to attack individuals based on the spy agency not liking the ideas of the individual, this would be about the public not possessing ideas which are consistent with the ideas of the investor-class (wage-slavery was motivated so that the investor-class would have more control over knowledge and its use), this is the typical example of the justice system not upholding the law, but rather to use the justice system so as to attack the public, and to serve the interests of the investor-class, ie to help the investor-class, to lie, to steal, and to murder.

The knowledge and related creative structure of society is set-up to make social-power flow to the investor class, but new ideas and new creative contexts could, or would, interfere with this flow of social power to the investor-class, so this seems to be the justification for attacking individuals by using a spy agency, because the ideas of the public might interfere with the power of the ruling-class

The constitution was itself a conspiracy which was only allowed to be instituted if there was a Bill of Rights, but if the Bill of Rights is not enforced then the constitution is no longer valid. The bill of Rights was never enforced, as the "Washington put-down" of Shay's rebellion demonstrates this constitutional government's unlawfulness, note: where Shay's rebellion was about a group of revolutionary war veterans whom Washington did not pay.

It is clear that the society is disintegrating because there are no alternative ideas in science and math, but this is mostly because any such new idea is being excluded by the secret police, that is the attacks on the individuals with new ideas is primarily being done through the secret police because of the use of the secret police to stop any new ideas, that is, the justice system and its association to the secret police, which began early in US history and was an important instrument in maintaining the power of the ruling-class, and is the main cause of the way in which the half-baked partial knowledge which mankind presently possesses (about the nature of the world) ie its failed and very incomplete scientific knowledge, which is being used to maintain the power of the investor class, in the so called highly touted system of "democratic" free-market capitalism where "democracy" means the justice system allowing the propaganda-education system to be absolutely controlled by the investor class, so that politics is about controlling the language and social discourse of the public through the media ie these morons (the investor-class) are being allowed (by a corrupt justice system) to interfere with beliefs and thinking of all people in society by means of:

1. controlling language,

2. controlling the public as wage-slaves and

3. by controlling science by means of axiomatic formalization which has resulted in a totally irrelevant form of science, but this irrelevant science which is now institutionalized is completely related to the commercial interests of the investor-class, so that

4. this way of using a failed knowledge is leading to the destruction of the earth and

5. it has caused markets to be destroyed

So that it is easily seen that M Twain is absolutely correct the ruling-class are a bunch of lying, thieving, murderous thugs, and Jefferson was correct that the nation needs to re-visit the Declaration of Independence in a revolutionary manner from time to time.

A way out of this failed-state of our nation, which would also be consistent with the US law, is to discard the constitution and go back to the continental congress, and to revise the law of the continental congress, so that law is based on equality, as stated in the Declaration of Independence, so that all the judges in the courts are dismissed (put in jail), that the government based on the constitution is dismissed and also put into jail so that the new governing body is the continental congress so that the government is to support the common welfare where each citizen is addressed by the law as an equal creator with the right to equal free inquiry and no one is a wage-slave and this would mean that the banks are to be seized, the banker's casino pinched-off, and the bankers also sent to jail, property is to remain as it currently is, but all the vacant housing seized and subsequently, given to the public so as to ensure the common welfare of the US's equal people, but where each person can contribute to the market by their creative efforts (which may be realized through the existing institutions, or factories), but a market which has "too big of a share" needs to be managed by the state, so as to be sure the common welfare of a bunch of equal people is maintained. The military would have to be temporarily left, but without its spying capabilities, so as to be wound-down over a few years (5 years at the most), and so that the public once again becomes the militia, ie no standing armies.

The main point concerning the freedom to believe and the freedom to express one's beliefs is not so much to convince others to follow an idea about society but rather to gain knowledge and the have new creative contexts lead to a wide ranging market place based on new creativity, so that everyone is allowed to express themselves as being a creator.

Otherwise the investor class and the spies which they control and the set of psychopaths (obsessive domineering personality types) which they control within the institutions of justice and education and propaganda will lead the nation to ever greater ruin, and to ever greater arbitrary violence, their current use of partial knowledge, so that they also insist on leaving knowledge so as to only be and remain in a state of being partial knowledge, is criminal.

Note: this type of activity is most likely quite prevalent in regard to the spying agents,

That is, these people (spying-agents) are the personnel which are being used in the spying operations, so that, these spies may behave in the same manner as the villain behaved in the movie, "Something about Mary," Ben Stiller and Matt Dylan, namely, getting personal information about women and their suitors by spying, so as to interfere in both of the people's lives, whom they are spying-on,

so as to take advantage of the woman, where this can be done with impunity by the (clever) spying agents, furthermore, these spying agents must possess the personality types quite similar to that of J E Hoover obsessively seeking socially dominating social positions.

The American propaganda-system is, supposedly, highly concerned about protecting women, but where they show this by their willingness to "throw the American women to the wolves" (ie to the spy agents, however, the spies are rich…, so in our Calvinistic proclaimed idol-worship-of-money, imposed by the justice system by requiring the public to have the social position of being in wage-slavery, where this imposition is done by using great violence, and that due to Calvinism…, such behavior, ie the interference in the spied-upon people's lives by rich spies, is OK).

That is, it has been reported that the police-state takes all information available on the society's communication channels, and they can do a great deal more than that,

Furthermore, any of the spies (with access to the storage of such an all encompassing information grabbing operation) can access anyone on the channels, then it is reported that the police-state organizes people to interfere with any of the sources, which communicate ideas, which are opposed by (or disturbing to) the investor-class (ie it is an organized party whose purpose is attacking a person who has been black-listed, note: R Ragaen became president for telling-on "the people in the movie-industry" who were on the black-lists)

And where this interference is about planting false information and discrediting the person and much, much, more.

This is being done within the justice system.

That is, the point of the class warfare is about how the ruling-class "and those who help them" can prey upon the public, while the public must pay their bills and become prey for the ruling-class.

Thus, there is "sure to exist," so called, US scientists, like the scientists at the German concentration camps, that experiment on the public with impunity.

This is possible due to the way in which the propaganda-education system is organized, and due to the wage-slave status, which the justice system forces onto the public. The propaganda-education system is about the right, which cheer-leads for the rich, and expresses the idea of violence to protect the social order while the left are the intellectuals who are arrogant and support (without any questions) the authoritative dogmas which are used by the ruling-class in their technical development.

The ideas expressed by the arrogant intellectuals cannot be new ideas. Exactly similar to how the ideas of Copernicus could not be expressed so as to challenge the dogmas of the church, where the church was used as a propaganda-education system in that age.

Thus, the left is all about the intellectual structures which support the ruling-class, and the right is all about the needed violence, which is required to maintain (with great violence) the arbitrary social and intellectual structures of society, upon which the ruling-class has based their social power.

The intellectual-class, in the context of the technical methods, always deal with physical models of:

materialism, reduction, and indefinable randomness, where local measured properties of systems are mathematically modeled as partial -differential equations, which, in turn, model material interactions, whose solutions are supposed to provide the quantitative models of the (a) physical system's (observed) properties, but his context of partial differential equations is always: non-linear, non-commutative, with arbitrary models of measuring (for both general relativity and particle-physics), in a chaotic and indefinably random (context), which is either related to feedback systems, or is related to categories of system types whose properties are not describable "except by collecting data on these systems."

That is, the descriptive structure is either stable in regard to the purpose of an object's relation to its surrounding (relatively stable) geometry in regard to feedback, or using the failed math techniques (used in the system's math model) to distinguish, empirically (distinguishing features), the categories of system behaviors, or system properties, which are distinguished in the vaguely limited set which has been defined by the descriptive context of indefinable randomness, so as to try to place a vague notion of an "elementary event space" into (onto) a context where there are no distinguished (or calculate-able) stable patterns, ie the context of indefinable randomness most often modeled as a partial differential equation and/or a harmonic function-space (which is not relatable to sets of operators which allow the function-space to be diagonalized).

Since, the partial differential equation models of physical systems have failed, the use of a system's properties is to be about "placing the unstable (system) structures into an information system," with the hope that vaguely distinguishable categories (from which a non-linear system might suddenly switch) can be helpful, in some further information context, which might use the system's data-information.

This is the information theory model of materialism and indefinable randomness.

This is all about staying within the indefinably random state, so that information is not compressed, but

Possibly it might be used in some other technical context (mostly unrelated to the eigen-function spaces of a random model) The method of spectral function spaces deals with both arbitrary definition of a context's spectral structure and also the arbitrary manipulation of these spectral values, since there are no stable systems, or events, associated to the containment context, ie differential operators acting on function spaces which is assumed to contain a context which has no valid definition eg the observed spectra are not calculable and events are simply unstable and/ or vaguely distinguishable patterns.

Note: the ruling-class is already in control of the markets which the current technologies define, so the ruling-class only wants the public,

(ie the top-intellectuals, supposedly, identified by educational-competitions where the competitions are, in turn, defined by the fixed dogmas, associated to technical languages, which support the interests of the ruling-class), to do things "in the setting of high-valued research," which extend the ideas, which are already being used to support the technical interests of the investor-class. That is, new ideas are to be excluded, since they create new contexts for creativity, and interfere with the investment structures which already exist.

Chapter 19

New expressions about, truly, new ideas, concerning math and physics, independently expressed, in a totalitarian age

The new descriptive context for the physical world, as well as for math descriptions, which the 5 new books by m concoyle propose…, in regard to both math and physics…, can be summarized in a few paragraphs.

The new ideas about math and physics are in agreement with some aspects of the old ideas about math and physics, namely, the idea from general relativity that, "it is the shape of space" which is the ultimate cause of both stable system structures and dynamical motion (of material-components which are contained within metric-spaces), but this old context is given a new more meaningful (mathematically) descriptive context wherein geometric patterns are stable, and, either stable systems can transition discretely between various other stable shapes, or the material "which a stable (metric-space) shape contains, and whose primary orbital-motions are being determined by the shape," can have its motion perturbed within the stable spectral-orbital structure of the stable shape, where this perturbation is due to a further material-interaction (structure) whose process is determined within the descriptive context of partial differential equations and the associated idea of a connection-derivative, but the new ideas also require a many-dimensional containment-set, within which significant macroscopic geometries exist [for example the higher-dimensional shapes can be as big as the solar-system, ie the new ideas are not about (or are not derived from) string-theory], and in higher-dimensions, these (new) stable macroscopic geometries also possess significant physical properties (ie metric-space types are intrinsically related to particular physical properties [due to invariances of these physical properties under certain types of coordinate changes]),

ie they are mathematical models of a measurable, physical world, where the physical descriptions are not based on the idea of materialism.

(note: string-theory and particle-physics are both based on the idea of materialism.)

In the new descriptive context, dynamical motion needs a stable containment-set (or exists within a new context for geometric descriptions), eg note that physical description needs a relatively elaborate laboratory-frame (an inertial frame relative either to the fixed stars or to an unbounded distance which is associated to a stable shape, where measuring is reliable and the lab's properties remain stable), but it is a new description of the physical world, which is not based on the idea of a set of "locally coordinate-transformation-invariant" "partial differential equations," which (in the old theory) determine both [stable* (see *note below)] (1) dynamics and (2) stable systems [where, supposedly, it is assumed that these stable systems come into being due to material-dynamical-interactions]), but where (in the new descriptive context) such a set of "invariant partial differential equations" is a subcategory (or subset, or a perturbing structure, or a perturbing process) which is subordinated to the main category of stable (linear, continuously commutative) shapes, which determine both motion and stable-system-components, where in this (new) stable-shape-category, for both containment and existence, [which are both defined relative to (hyperbolic) stable shapes]...., each such shape possesses a set of stable spectral-orbital geometric properties (ie properties which are associated to the "holes in the stable shapes," eg a doughnut has a single hole in its shape)..., and the spectral properties of its containment set determine the context (of stable spectral-resonance) by which these stable shapes come into existence (through the stable shape's resonance with the containing set's spectral properties).

That is, the stable shapes come into existence by being "in resonance" with a finite spectral-set defined for (or upon) the "new" many-dimensional containment set, ((where such a stable geometric structure (as a (new) model for stable material-components) exists for material systems which, in turn, exist at all (bounded) size scales, where this construct is applicable for all of the myriad of stable spectral-orbital "material" systems, which exist in regard to all the various dimensional levels of existence), and where the stable geometric structures determine both "material-components" and the many hyperbolic metric-spaces, which contain the stable material-component-shapes, and for which (both categories: material-component or containing metric-space) have the property of being shapes, where the containing-space is at least one dimension higher than the dimension of the component-shapes which it contains [ie the set of discrete hyperbolic shapes define a set of hyperbolic metric-spaces, which, in turn, contain the (observed) stable material systems]), and it is within this containment-set wherein the stable hyperbolic shapes..., which define both material-components, and the metric-space containment sets (of the material components)..., where a new type of partition-set, or a "spectrally-partitioned" 11-dimensional "over-all containing" hyperbolic-metric-space..., is defined, but where this (highest-dimension) 11-dimensional hyperbolic metric-space has no shape of its own, however, all the lower dimensional hyperbolic metric-spaces, which it contains, do have stable shapes, and these shapes are the very-stable discrete hyperbolic shapes, though these shapes may be either bounded or unbounded shapes,[at least up to and including hyperbolic-dimension-5, and all the

6-dimensional through 10-dimensional stable "discrete hyperbolic shapes" are unbounded shapes (D Coxeter)].

There are no known discrete hyperbolic shapes above hyperbolic dimension-10 (D Coxeter).

*Note: stable dynamics would be defined in regard to material-motions of (what is usually condensed) material which is, in turn, contained within a metric-space shape, so that the stable material-motions are mostly being determined by the condensed-material interacting with the metric-space's shape, and such motion is only "a bit perturbed" by the subset of material-interactions which are defined (or can be identified) by sets of partial differential equations, which can be, in turn, associated dynamical ordinary differential equations.

That is, such stable orbital motion is being determined by the shape of the material-containing (hyperbolic) metric-space. The condensed-material, being contained in a metric-space which has a stable shape causes the condensed material to move in stable orbits.

This is, apparently, the only way in which one can describe the stable orbital properties of the planets of the solar-system, as well as the stable spectral-orbital properties which are observed to exist for the (usually) very stable many-(but-few)-body quantum systems, eg nuclei, atoms, molecules, etc,

That is, shapes are fundamental... to both material-systems and to metric-spaces..., while dynamics... defined by partial differential equations.... are incidental to the shapes of metric-spaces, where the shapes of these material-containing metric-spaces dominate the stable (orbital-spectral) dynamics of condensed-material. Yet, there is also a local material-interaction structure associated to alternating tensors, which, in turn, are related to the symmetric metric-function 2-tensor, ie differential-forms defined in a context of metric-invariance, ie related to the classical Lie fiber groups, eg SO(n) and SU(n) Lie groups.

That is, the currently accepted descriptive structures for math and physics (2014) are based on the wrong "shape of space," where in the current descriptive structure "the shape of space" is mostly considered to be spherical-symmetry (with an irresolvable singularity-point, eg 1/r [for r=0]), as well as a deep failure of the currently accepted descriptive structures, namely, that the current descriptions are not capable of identifying, based on its own laws, the observed stable spectral-orbital properties of the most fundamental of physical systems, namely, the many-(but-few)-body systems, which exist at all size scales: nuclei, atoms, the solar-system etc. Furthermore, the current descriptive construct cannot identify the correct many-dimensional context of an actual existence, wherein stable physical systems do exist and they can be described in a (mostly) linear context. The currently accepted descriptive construct is based on materialism, reductionism, and (indefinable) randomness; within which its (ie the current, authoritative math and physics descriptive languages), failed, descriptive laws are based.

This is the essential viewpoint of the new, revolutionary, viewpoints about math and physics, which are based on both many-dimensions and on the idea of stable shapes.

Some issues about the currently accepted descriptive dogmas for math and physics

(Note: Internal particle-states of elementary-particle-theory and their associated string-theory geometries, which fit into an arbitrary notion of materialism, are descriptive constructs which are: illogical, quantitatively inconsistent, non-linear, non-commutative, chaotic, meaningless, and practically useless descriptive constructs, namely, they are descriptive constructs which are not capable of describing the stable spectral-orbital properties observed for the many-(but-few)-body systems which exist at all size scales, rather they are descriptive structures which are designed to describe a chaotic transition process, whose transition-rate is determined by probabilities of random particle-collisions, as the system transitions between two relatively stable systems, but descriptions for the stable states do not exist in such a chaotic and random descriptive context).

Topology

In the new descriptive language the stable shapes of the hyperbolic metric-spaces possess open-closed topological properties for the many different "topological metric-space structures," which are used in regard to describing the new form of a dimensionally specific type of materialism, which is associated to "each dimensional level being contained within the coordinate structure of the stable discrete hyperbolic shapes" of hyperbolic metric-spaces (ie hyperbolic metric-spaces are generalized space-time metric-spaces, hyperbolic n-space, is an (n+1)-space-time space) where these hyperbolic metric-spaces (can, and do) possess the shapes of stable discrete hyperbolic shapes (in this new model for the context of physical description).

Partition

These stable hyperbolic shapes can be defined for each of the varied metric-space sets which can be characterized by both (1) the various subspaces of the same dimension, and (2) this can be done for each of the different dimensional levels, so that the metric-space shaped subspaces fit into (or are contained within) the 11-dimensional hyperbolic metric-space), [note: the last definable discrete hyperbolic shapes has hyperbolic-dimension-10, (D Coxeter).]

note: the different subspaces of the same dimension are used to count, in a finite context, the set of different hyperbolic metric-spaces, which can exist as discrete hyperbolic shapes, these metric-space shapes can move between subspaces, but each such metric-space shape would be constrained by their individual context of metric-space containment (within an open-closed topology), ie it is a bit of an "artificial" way in which to get a finite number of stable shapes at each particular dimensional level.

It should also be noted that when a metric-space shape of a given dimension is within an adjacent higher-dimensional hyperbolic metric-space, then "the (adjacent) lower-dimensional shapes" (or the lower-dimension discrete hyperbolic metric-space shapes, which are contained within the higher-dimensional metric-space), are seen to possess boundaries. That is, one must "look-down (in regard to dimension)" in order to "see" (or to perceive) the boundary of a component-structure, which exists for a discrete hyperbolic metric-space shape, yet all shapes of all the different dimensional levels are contained in an unbounded 11-dimensional hyperbolic metric-space. Thus, beyond our topological confinement in our bounded metric-space shape, (where evidence for the existence of such a metric-space containment set being a bounded stable discrete hyperbolic shape (evidence for this to exist) is the stable orbital properties of our solar system)

…, light can extend into an unbounded subspace (or light can extend toward us from an unbounded region, ie seeing the distant stars) and

It should also be noted that, there is also a simultaneous containment of our, apparently, material existence within an Euclidean space, wherein both (1) spatial positions of a material-component's center-of-mass exist, and (2) dynamics are defined, relative to both bounded and unbounded stable shapes.

That is, in the new descriptive construct, there are many different dimensional metric-spaces, for each of the different subspaces in each of the different dimensional levels, within the 11-dimensional hyperbolic metric-space, where this containment context of different dimensional "metric-space shapes" for each of the different metric-space subspaces can, in turn, be characterized by a dimensional-spectral sequence of increasing (spectral) sizes of the stable metric-space shapes, so that the metric-space-shape partition of the 11-dimensional hyperbolic metric-space can be defined as the set of largest-sized shapes for each subspace of the same dimension, and for each (or all of the different) dimensional level. That is, if one counts the stable shapes for a given dimensional level then one way in which to make such a count finite is to restrict the number of biggest-shapes to the number of subspaces of a given dimension in the 11-dimensional over-all containing hyperbolic metric-space, and to do this for each of the different dimensional levels.

This (new type of) "spectral-partition" defines a finite spectral set, which in turn, can be used to identify what type (or what shape) of metric-space-shapes can exist, by requiring that all such metric-space shapes (contained in the 11-dimensional hyperbolic metric-space) be resonant with this finite spectral-set, and so that these metric-space shapes can be contained in an adjacent higher-dimensional metric-space shape, which is "big enough to contain" the different sizes of the lower-dimensional shapes, so that these "contained-shapes" define material-components, with definite stable spectral-orbital properties, which exist within a (or that) particular dimension metric-space, which is defining a particular subspace, but the subspace is relative to the geometric-topological containment structure of the material-containing metric-space (in turn, being contained in a higher-dimensional metric-space).

This very restrictive context for containment makes the finite number of "allowed shapes" in "the count" of these stable geometric-spectral entities, where this "count" is similar to counting all the n-faces (for some fixed, n, n<11) on an 11-dimensional (Euclidean) cube, and then summing for all the different, n's, and then dividing this sum by 2, so as to get all the subspaces of the different dimensions in the 11-dimensional space. This has the same pattern of dimension and containment of material-components and their associated metric-space containment sets of this new descriptive structure, so seems to represent a good model in which to identify the new construct related to a new type of a finite "spectral-partition" of a high-dimensional hyperbolic metric-space (over-all) containment set.

Though the stable shapes can drift between subspaces, the topology and the metric-space containment structure of this model of different dimensional shapes suggests that "the counting of faces," on an 11-dimensional cube, is a good approximation to the number of shapes of different sizes which also possess the given dimensional properties, ie the number of pairs of faces of a given dimension on the 11-cube can define a "dimensional-subspace context" for the different sizes, which can exist in a finite context, for the given dimension, ie each of the set of different facial-pairs defines a different size for a metric-space shape of that dimension.

There are several ways in which to get-at this containment context, which is now (in the new descriptive context) "being related to (the over-all containment metric-space possessing) a 'finite spectra,'" which, in turn, is related (in the above new context) to a "finite number of 'biggest shapes,'" forming into a "spectral-increasing-size-sequence-partition" for each dimensional level, and then "as the sequence also increases dimensionally" so as to include all the different dimensional levels, (but the only dimensional levels associated to well defined spectral properties are the dimensional levels, namely, one-dimensional through five-dimensional levels, since all the stable hyperbolic shapes in dimension-6 through dimension-10 are unbounded shapes),

Of (or in regard to) the over-all containment, 11-dimensional hyperbolic metric-space, space.

That is, for each dimensional level, each subspace is given a largest* (bounded) shape upon which there can be defined a finite sequence of such sizes of stable shapes. That is, the set of subspaces do not need to be fixed subspaces in an 11-dimensional metric-space, but rather represented as sets of metric-space shapes of the same dimension which are (bounded) shapes which together define a finite sequence of shape-sizes whose sequence-length is equal to the number of subspaces of that given dimension in a hyperbolic 11-space, where these stable hyperbolic metric-space shapes possess an open-closed topology.

*note: Each such distinctly identifiable subspaces (for each of the different dimensional levels) may also be associated to a smallest bounded stable shape.

For example, one can begin from the higher-dimensions...,

[ie there are 11, 10-dimensional subspaces in the 11-dimensional over-all containment hyperbolic metric-space, and due to the requirement for the containment of shapes in some higher-dimension hyperbolic metric-space (all of which possess the property of being shapes, except the 11-dimensional metric-space)], and then one can then proceed sequentially downward to lower-dimensions, [where there is also the property that "all 6-dimensional and higher-dimension discrete hyperbolic shapes are unbounded (D Coxeter),"] etc. etc.

The real issue..., in regard to a precise descriptive language through which an actual and meaningful description of content about stable, measurable patterns can be expressed..., is that an 11-dimensional containment space (which is a hyperbolic metric-space) must be divided, or spectrally-partitioned, so as to model a containment space which defines a finite spectral-set, or there cannot exist a precise descriptive language whose content (or meaning) can possess stable patterns, note: this can also be accomplished on the maximal tori of the fiber-group (where the math of the fiber group is a bit easier to understand in the descriptions which are associated to the special-unitary fiber-groups defined over complex-coordinates).

Without such a descriptive context then

"What math and physics can now offer as a descriptive language is ludicrous, which is, at best, only capable of fitting observed data in an undefined context of probability so that their current (2014) descriptive language is not able to (causally) describe a stable pattern (either mathematical or physical)."

Reduction to smaller components, ie reduction to elementary-particles (a fundamental assumption of quantum physics), is to "now be understood" in regard to the dimensional-spectral sequence, and where (now, in the new descriptive context) the elementary-particles are related to "the vertices" of the dimensional-spectral shapes...,

{Note: vertices are related to these shapes when these shapes are modeled as cubical-simplex structures, which, in turn, define a lattice for the metric-space coordinates, within (or upon) which they are defined},

..., where (to re-iterate) the dimensional-spectral shapes define both (1) the stable material-components (which fit into an open-closed topology of these shape's containing metric-space) and (2) the stable set of metric-space containment subspaces of a (or this new) high-dimensional construct,

[the new descriptive context is, primarily, based on stable shapes].

Material-interactions

The metric-spaces (defined in the adjacent above paragraphs) are all hyperbolic metric-spaces, but material-interactions (and its inevitable relation to inertial-momentum) depend on these components (or shapes) also being (thought of as being) within Euclidean metric-spaces, and the, subsequent, interaction geometry is related to both

(1) the non-local (or action-at-a-distance) discrete Euclidean shapes, which are defined between the stable discrete hyperbolic shapes of stable material components, [which, in turn, these discrete hyperbolic shapes can be approximated, during the interaction, as discrete Euclidean shapes], and where, in turn,

(2) this Euclidean shape's geometric relation to the local geometric properties of its Euclidean fiber group, SO(n), for an n-dimensional Euclidean space, eg the fiber-group geometry of a maximal torus defined within the fiber-group (which, in turn, identify the local group transformation of the local coordinates of the material interaction).

This identifies the local structure of the partial differential equations which are involved in material-interactions. These material-interaction constructs are mostly non-linear, except for the constructs which determine the stable discrete hyperbolic shapes, and discrete Euclidean shapes.

Thus, there are both (1) the stable shapes of material components, and (2) metric-spaces, which, together, form the stable scaffolding upon which the partial differential equations descriptions can be applied. But "partial differential equations" are used to mostly perturb the motions, in regard to (condensed) material interactions, whose dynamics is mostly defined by the condensed material's relation to the orbital structures defined by the shape of the metric-space, within which the material components are contained.

That is, materialism and reduction to elementary-particles, all placed in a context of randomness, so that the description is dominated by the idea of partial differential equations, is quite wrong, both as a model and in regard to what such a descriptive language structure is capable of describing.

The new descriptive context is discrete, where discreteness is identified between (or related to discrete):

0. the discrete discontinuities associated to the spatial properties of unbounded shapes,

1. dimensional changes within a descriptive context, and

2. discrete Weyl-angle changes, and

3. discreet time intervals (associated to the dynamics of material-interactions),

where the discrete time-intervals are defined by the periods of spin-rotations between pairs of opposite metric-space states,

Where to identify "metric-space states" then one needs to consider that the metric-spaces (hyperbolic space and Euclidean space) are identified with the properties, which E Noether's symmetries have defined, that is, time and spatial-displacement symmetries (or invariances) are associated to both energy and inertial (or momentum) properties, respectively..., as well as the pair of opposite metric-space states which these properties define, eg (+time) and (-time) for hyperbolic space [fixed stars and rotating stars for inertial frames (or Euclidean-space), etc].

note: energy is associated to hyperbolic space while inertia (or momentum) is associated to Euclidean space.

Continuity of motion

The descriptive context is discrete, but in lower dimensions..., ie in the hyperbolic metric-spaces of the different dimensional-shapes from 1-dimensions to 5-dimensions..., the dynamic structure is continuous, where this is due to conservation of energy in each (topologically open-closed) hyperbolic metric-space (which is a discrete hyperbolic shape), but continuity is mainly defined in regard to the motions of the adjacent lower-dimension discrete hyperbolic shapes, which a hyperbolic metric-space contains, ie continuous dynamics exist in regard to the "material" components which the metric-space contains.

That is, the discrete structure of dynamics depends on identifying a position in space, and defining pairs of opposite metric-space states, usually of equal energy, associated to the two opposite spatial displacements associated with the two opposite time directions of a hyperbolic metric-space of a material-interaction's dynamics. It is these pairs of opposite metric-space states, which are local inverses to one another (accelerating or decelerating), which identify the continuity of motion.

(note: most of these types of motions are non-linear, it is only (primarily) in the context of being contained within stable metric-space shapes that material dynamics can be put into a stable (dynamic) context.)

It might be noted that in higher-dimensions than 4-hyperbolic dimensions the dynamic geometry has a greater relation to tangent forces, ie torques, so rotational dynamics might be the predominant type of motions

(as opposed to translational motions and its associated "radial geometries" of (what are assumed to be) spherically symmetric force-fields, but these radial geometries are really mediated by toral shapes, within Euclidean space, which define spherical symmetry, where this spherical symmetry is realized, in the new descriptive context, by the vector relation that local tangent Euclidean toral geometries have to the local geometric properties of either SO(3) or SO(4) = SO(3) x SO(3) Euclidean fiber groups)

Unbounded shapes

It should be noted that, the unbounded discrete hyperbolic shapes, which exist for all dimensional levels (wherein discrete hyperbolic shapes are defined), are models of both light and "neutrinos," where neutrinos are the lepton counter-part to the hadronic neutron, where the neutron is a bounded shape. Both the electron-cloud and the nucleus are mixtures of "charges and neutrons or neutrinos," where, in turn, the neutrons and neutrinos are mixtures of equal numbers of electrons and protons, but electrons can couple with the unbounded neutrino.

Can the, apparently, unbounded neutrino be totally contained in a higher-dimensional metric-space which is itself bounded, thus providing a bound for (or a bound imposed upon) the neutrino? How would charge distribute itself in such an unbounded-yet-bounded context?

We have trouble perceiving higher-dimensions than the 3-dimensional spatial-subspaces, which we do "so easily perceive," where this is due to open-closed topological properties of the 3-space containing hyperbolic metric-space, and due to the size of our containing metric-space, which is as large as the solar-system, but without such a structure one cannot describe the stable orbits of the solar-system, ie neither Newton nor Einstein was able to solve the problem as to "why we exist in a stable solar-system," whose basic orbital shape was determined (in the history of our western culture) by Copernicus.

This problem, about the cause of solar-system stability, ie the many-(but-few)-body problem, is an old, but most fundamental, problem of physics and math. And the new ideas expressed in this review, and as pictured in the diagrams provided in the 4 new book's, by m concoyle, describe the first known solution to this problem.

Where the solution can be given in diagrams, but the descriptive context is new and quite different, but nonetheless similar, in the "similar way" in which Copernicus's ideas were similar in some ways to those of the Ptolemaic system.

Further social comment

The propaganda-political-economic-militaristic-oil-education system is driven by investors, and it (all aspects of this social system) serves the investor's interests.

The main idea presented by the propaganda-education system…,

ie the system which represents high-value and truth for society, is the idea of inequality, and thus for a hierarchical social-system based on individualism, the idea of individual freedom, [ie what should be "equal" individual freedom], needs to be made consistent with a social hierarchy, where this social-hierarchy is used to express social-dominance, so the idea (or language construct) of individual superiority must be cemented into the language of the society by its propaganda-education system.

Thus

The main right-left "political" dialog of our (the US-European) propaganda-education system is:

Individualism (right) vs. statistical collectivism (left)

Which is incorrectly reformulated as:

Spiritual (right) vs. materialistic (left)

or

Religion vs. science

(and this is, mostly, incorrect, since what is worshipped in the society's religious construct is the material-idol of money. This required worship of money is violently forced on people by the state, by its justice system, in support of the interests of the ruling investor-class)

Furthermore, according to the right, individualism is not about equal individuals and their individual relation to knowledge and creativity, rather the individualism envisioned by the right, is the individualism of a few who possess the social properties of being self-righteously superior (or a few self-righteously dominant over the many),.

That is, the very rich investor-class are the very few people who have been elevated to near saint-hood by Calvinism, which expresses the dominant protestant religious tenant, that "the rich are those who have been favored by God."

But

where such elitism is ultimately based on lying and stealing, which, in turn, is based on (or brought into being and, subsequently, maintained by) violence.

In this context of violence the intellectual-left is maintained (by educational indoctrination) as a group of aspiring wage-slaves, where the justice system has coerced the population into being wage-slaves, where wage-slaves can aspire to social heights, through their indoctrination within the education system's contest, so as to become an authority of a narrow dogma, ie so as to become an aspiring group of academics, which intellectually supports (without question) a statistical model of the sciences, which includes a statistical model of physics and math, where (in physics) this is about the models of materialism and randomness and reduction of all material systems to elementary-particles, and

subsequently society's structure is also supposed to be dependent on this viewpoint of indefinable randomness, which, in regard to a materialistic-scientific model of life, where life is assumed to be based on the chemistry of DNA, where the superior members of the species, and their DNA, determine the direction of life's development by being more successful in their environment.

That is, life is believed to be reducible, by Darwin's evolution, to the material structure of DNA, and a random statistical structure, which is governed by the "rule" of "survival of the fittest."

One sees again a focus on "a chaotic transition process, which is defined between stable states of life-forms."

That is, society's structure is ultimately to be explained by the Darwinian-and-DNA model of "mutation and survival-of-the-fittest," which is, in turn, is interpreted to be the scientific proof that "life is statistical" and, subsequently, that the (through arbitrary) categories of society (which define the selfish interests of the ruling, or investor, class) and measuring performance-qualities (eg good-and-bad) within these (arbitrarily defined) categories associated to the statistics of normal curves, (where, in statistical models which are based on finding many samples of statistical averages within a category, then the statistics of these many samples (of averages) define a statistical normal curve), where, in turn, these normal curves can be interpreted to mean that an "elite set of people," the, so called, "superior set of people" can be identified by means of the qualities of performance, which are in turn, defined on these normal curves, (of arbitrary categories of actions associated to either material or process-related products), which, in turn, are defined on (within) arbitrary types of categories (and using the statistics of many samples where averages are measured for each sample),, so that it is claimed that "it is proved" that society is to be, inevitably, ruled by an elite set of people, ie the ruling-class.

This is (all) interpreted to mean that society is a statistical-collective, which supports the ruling-class.

That is, "the right" (in the right-left dialog) claims that their belief in individual 'freedom" must be translated, (through the, so called, scientific viewpoint of Darwin's, that is, the survival of the fittest, or equivalently, the superior members of the species naturally "rise to the top" and rule (or determine the development of a species), where this is all due to a random context, within which it is assumed that life and existence are contained, and onto which the environment forces an evolutionary direction), to mean that only "the superior individuals" of the species can cope with individual freedom in a socially meaningful manner, so as to become the ruling-class, which the rest of society must collectively support due to their inferiority.

This is a very materialistic viewpoint, which is manipulated to imply a natural hierarchy for life (or life's purpose, ie to survive, and, subsequently, for people to be a part of a [arbitrary] social hierarchy upon which social-value is defined).

while (for some religions) the religious viewpoint (in particular of the Abrahamic religions) also assumes a natural hierarchy for life, so that (especially Calvinistic) Protestantism has interpreted this to mean that "those who gain wealth" must be a part of this natural, God-given, social hierarchy, where, in fact, this social-hierarchy worships the material-idol of money, ie a man-made construct in an established but arbitrary social hierarchy.

That is, high-social-value is given to "arbitrary sets of ideas," which can be logically reduced (in the language of the realm of the investor-class) to material-based ideas, eg property rights and the control of material-instruments and the investments of banks, but all related to "the worship of money" by the rest of society (ie idol-worship).

By agreeing to a fixed dogma (ie if one agrees to a fixed dogma) then, all those in agreement with this dogma, become manipulate-able by the social hierarchy to which the dogma is a part.

In particular, the dichotomy of: materialism vs. "the spirit," has no meaning (it is a dichotomy without any real content), if (or because) there do not also exist math-models of the possible patterns of existence which transcend the idea of materialism. That is, math-models of "the spirit" are needed to give the: materialism vs. "the spirit," dichotomy, any meaning.

That is, it is in the interest of the oligarchy to make-sure that all the other "oligarchic societies" with whom the US oligarchy has agreements, also agree to the fixed dogmas of the US oligarchy, since these are the fixed dogmas which best serve the interests of the investor-class.

This allows the propaganda-education system to possess such great influence over society.

The dogma of materialism has failed to solve the fundamental problems which it has identified, ie describe the properties of the observed "stable material-systems."

That is, the dogma of materialism is a failed dogma, but it is still the basis for the US oligarchy.

Thus, to have faith in "the dogma of materialism" is a mistake, and this dogma has been cleverly transformed into an idolatry-worship of money, but where this worship of material things (ie worship of money, or "in an equivalent propaganda form" "the workshop of money") is forced, by violence (through the justice system), upon the public, but due to the scientific viewpoint being based on materialism, this idol-worship is expressed as a scientific necessity.

For example, the, so called, quantitative science of economics, but economics is an expression of a rigged-game, ie the oligarchs have all the (extra) cards… they need… up their sleeves.

All intellectual viewpoints are circular, in that, they end-up expressing the idea that "inequality is a scientific law," and thus there is a societal "need for a ruling-class."

It is through idolatry worship (eg worship of money as wages) that categories of commercial interest also become categories for intellectual dogmas, and all commercial categories are put into a material context.

Bombs are related to physicist's models of material interactions

Economics related to counting "paper representations of money"

Medicine related to cell-like organisms and their relation to DNA, and subsequently related to chemistry,

Communications; to electrical engineering and computer programming, (isolating common instruments from public development and from public control, by placing their study in competitive academic institutions)

In this regard the local rules of measuring are desired, ie partial differential equations, so that (usually) the global properties (of the subject of a category for measured descriptions) are not prominent considerations,

eg in economics the oligarchs have all the money so it is a rigged-game (ie this is the global structure of economics), where risk is managed by essentially keeping society fixed (eg technical development is mostly entirely dependent on 19th century science, ie classical physics. Thus, there is all the propaganda about "how everything is changing so much."

(But local measuring, partial differential equations, must be quantitatively consistent, otherwise the described properties, which are deduced form local structures, are not reliable patterns. But partial differential equations are not quantitatively consistent in the math contexts of: non-linearity, non-commutative-ness, arbitrarily (or improperly) defined randomness, etc etc, and these are the math contexts within which academic math and science persons are trying "so hard" to extend, ie the contest to extend the failed dogmas of the traditional authorities of these subjects, but really "the traditional authorities" are the, so called, authorities which have been designated as authorities by the investor-class, where these "traditional authorities" are usually "housed in" academic institutions.)

That is, the society revels in partial truths represented as intellectual dogmas, especially, if the current state of knowledge is directly related to commercial interests of the investment-class

Thus one sees the destruction of the earth based on the, so called, practical use of "partial truths," associated to fixed dogmas which have failed, eg GMO's (altering the DNA of existing organisms)

and accumulating large amounts of fissionable material, spraying of insecticides, dumping toxins into the environment, etc etc.

The intellectual left is duped (by educational indoctrination, and the experience of "winning" in a competitive education-system) into supporting high-intellectual value, where this high-intellectual-value is being defined by the fixed dogmas which are used to support the interests of the investor-class.

Whereas, the left, apparently, sees its own high-intellectual-value as being the "true" God-given intellectual discoveries, which, apparently, the left believes, are a part of the social hierarchy, which the left also mistakenly sees as being related to "a steady progression to truth."

This, so called, "truth" is also modeled (for no apparent reason), by these same intellectuals, in terms of "a consensus," apparently based on the number of indoctrinated, so called, intellectuals, which support these ideas in "competitive" academic institutions.

Yet, the nature of deduction (or rigor)…, about which physical and mathematical "law" are, supposedly, based…, is that it (finding properties from axioms and definitions) stays within the language itself, and it is not an averaging-process.

Perhaps, when knowledge exists in a chaotic context, wherein, perhaps (in some cases), the idea of "forming a consensus (or a best-bet)" is implied.

But usually (or always) best-bets are related to the interests of the investor class. So if, the so called, scientific "best-bets" are proclaimed (for a [usually, improperly defined] random descriptive context), then the complexity of any such scientific statistical model (which might be relatable to a consensus) can be used to show that there is statistical ambiguity.

However, in regard to physics, any such efforts to un-track the statistical structure of material-interactions as a consensus…, which should be questioned due to both the description's failures as well as the description's complexity…, are never considered.

(note: the statement that "the stable many-(but-few)-body systems are too complicated to describe" cannot, in-any-way, be considered to be a valid statement, there should be very many "very deep criticisms" of this statement, than there are criticisms of the global-warming statement)

That is, the questioning of an intellectual dogma associated in a "directly functional" manner to a commercial product, eg bombs, never occurs.

That is, "a physicist with such questions" would simply not become a physicist, ie they have not mastered the correct dogma.

Thus, through idolatry worship, the apparent dichotomy of material vs. spirit is resolved within society due to the ruling-class creating idolatry worship of money and subsequently, forcing the so called intellectuals of society to accept "formalized axiomatic intellectual structures" which have been formed into (or these formal-axioms are used as the basis of) intellectual dogmas, which are (can be) well defined within competitive academic institutions, where intellectual wage-slaves compete within the complexities of these dysfunctional dogmas.

But, in regard to global warming (or the deadly poisoning of the earth by pollutants), any possible threat to human safety should be interpreted in the context of the precautionary principle, where the safety of the earth and life should triumph over concerns of the ruling-class about their commercial (or investment) interests and an associated relation by the public (created by the ruling-class through the justice system) to the material worship of a false idol (ie the public, forced into wage-slavery, has come to worship money). That is, the human being is an intelligent species which is capable of adapting to new contexts, so society should change so as to accommodate even "dangers" which might only be "possible dangers."

Thus, "the right" supports individual freedom, where within society this means "supporting a collective oligarchy," since the oligarchy defines the superior individuals, who, in turn, define "superior culture," and thus, the ruling (investor) class define the high-social-value within society,and

The intellectual-left supports the high-intellectual-value which the oligarchy has defined for the intellectual-left.

For example, the intellectual-left supports the idea of randomness, which is not definitive, rather

Such "knowledge," based on improperly defined randomness, is only arbitrary, and/or ambiguous.

The vision of "equality" expressed by the elitist-intellectual-left, is that, society should take care-of the "inferior people."

That is, the left is blind to (or ignorant of) the failure of the intellectual constructs, which it (obediently) supports, ie the left mistakenly sees themselves as members of the upper 5th or 10th percentile on the normal curve, whereas, in fact, they are mostly stooges, ie chumps, who actively participate in the oppression of the many.

But (despite the indoctrinated ignorance of the "superior-intellectual-class") in fact the oligarchy is arbitrary and is upheld by extreme violence, and it is those violent forces which the right controls by means of:

1. spying and manipulating personalities, by

2. Or through the justice system, eg implementing the religion which requires wage-slavery, and by

3. War, the expansive control of societies by the investor-class, and

4. economic growth (in a context of wage-slavery), and by

5. Propaganda (incessantly beating the drum for social inequality, so that the wage-slaves will compete to be a part of the failed dogmas).

It is upon the "heights of social power," upheld by violence, that the ruling-class, in turn, defines the intellectual structures which support their selfish interests, and it is these intellectual structures, which the intellectual-left, in turn, supports.

In regard to expansion, of the economy, this is a natural part of empire, but in globalization, the expansion of the empire is now defining an expansion of "(little) islands of, so called, economic development," ie the narrow set of categories which define the set of investments of the ruling class, in turn, defined by (a narrow set of products in regard to) product production, defined in a context of social hierarchy, in turn, defined by the investor class, and defined by the, so called, economically easy (efficient) methods of such expansion, eg building instruments, which are used to support the social hierarchy, and using the (so called) cheap energy sources.

The issue is, that the expansion is artificial, and based on a narrow set of creative efforts (or creative products) which are the instruments (the institutions) through which the ruling-class maintains its power

Unfortunately, the material-random context of knowledge and creativity is failing. But its disintegration is designed to result in the same violent context through which it is upheld. That is, idolatry worship and the subsequent misguided intellectual efforts result in intellectual inertia, ie an inability to change contexts, so that the violence which really holds the systematic oligarchy together is allowed to act first so as to be directed within the context of the ruling-class, the intellectuals only have a delusionary context of the role which their beliefs play in the scheme of society (usually, they claim that they are progressing to a greater scientific truth, by trying to extend the fixed dogmas, within which they compete on an intellectual level. But Godel's incompleteness theorem, ie that precise languages have great limitations in regard to the range of patterns which the assumptions upon which such precise languages are based, shows that such a competitive and arrogantly superior belief is misguided, so as to lead to delusion, and not lead to a practical, wide-ranging truth.

The Romans conquered and built with bricks, the structures which defined their domination, but this was only temporary, since "the locals" soon learned to build with bricks, and, subsequently, became more independent, eg Japan soon learned western technology, to which only slightly better technological advantages (were possessed by the west) and apparently (also) more abundant material resources led the west to winning "the war," ie WW II, but now there are no such advantages of knowledge, but only fleeting strategic advantages.

Where, it might be noted that "in WW II," Hitler failed to take Moscow with his Blitzkrieg, at the beginning of December 1941 (12-2-41 to 12-5-41), ie the Nazi's could not "take over" the communists, and it was at that time that the US entered WW II (on 12-7-41). That is, when the Blitzkrieg failed against the Russian communists then the US immediately entered WWII.

(the Russians had then gone-on to beat the Nazi's by the end of 1942).

But the US re-instated the corporate-state model of governing, as fast as WW II ended, as evidenced by "the use of Nazi spies," by the US, immediately after WW II, as well as the obvious way in which the model of the corporate-state was already well established during the gilded-age, wherein M Twain's social criticisms are being re-iterated in the social criticisms expressed in this paper, (ie a political-ruling structure based on lying, stealing, and murdering), but with a new analysis expressed (in this paper) about the "religion of materialism," the idol-worship of money, upon which such a hierarchical social order…, as exists in the US society…, depends.

The proclamation…,

Public (there is a new game in town)

"Get your guns (or bring your high-brow intellects)!"

"There are some inferior people that 'Need to be put in their place,'" (those who "play-ball" in this context become well-rewarded wage-slaves), is a very good example, by which there is implemented an arrogant (and confident) social-hierarchy, and is an example of the basis "in violence," by which the US oligarchy is upheld.

In fact, the fixedness of knowledge, and its formalized axiomatic structure (where its "being fixed" leads to failure), and its, subsequent, obvious failures, eg its inability to describe the properties of a general stable spectral-orbital, many-but-few-body, physical system, can only lead to the conclusion, that it is better (for "the west") that the societies, which are outside "the west," (ie outside the investment structure of western banks, or the outposts of empire) should learn the narrow set of dysfunctional knowledge, which is possessed, and used, by "the west," so that any new form of knowledge which might be found by these other societies, is, in fact, not found by these other societies, where such "new knowledge" might lead to an advantage of possessing a "new superior knowledge," (for those societies outside the west) over the "fixed western knowledge," ie for those societies which are external to the banking-west (ie the investment context might be disrupted), in the current social-model of a "'barbarously elite (and quite confident oligarchy)' model of human-life on earth." (essentially, the same social-model as expressed by the Roman-Empire, but now (2014) with the investor-class representing the emperors, and material-based science defining the new religion of idolatry [money-worship])

That is, the lying stealing and murdering by a person who then "gets away with it," is defined to be the person who is "the fittest," and it is these few-people who then define the high-social-value for society, and, subsequently, the intellectual left is duped into supporting the intellectual foundations of the institutions which allow (and support) this type of a social structure to exist,

ie (of a social state of an individual being considered to be a superior person, a property of an individual of a species often referred-to as "being the fittest," ie so as to be a fit member of the few people who "get away with lying, stealing, and murdering").

Namely, the (wage-slave persons whom are formed into the) intellectual-left support:

0. Property rights,

1. the idol worship of money (in both economics and in the law), and

2. the belief in randomness, which, in turn, allows

3. the survival of the fittest to be defined (biology and the law), and

4. the belief in "the random theory of materialism"

so as to promote:

5. arbitrary money worship and its relation to the central issue for banking, namely,

6. investment risks (Calvinistic Protestantism, and the law), thus leading to the belief in

7. Keeping the social structure fixed, so as to lessen investment risks,

8. Which implies axiomatic formalization of the academic categories, where these intellectual categories are all related to investment interests, and

8. A belief in meaningless randomness, and

9. its, so called, (apparently, minor) relation to the development of life (biology), and

10. the knowledge-base (of indefinable randomness) upon which weaponry is so dependent,

eg the probability of particle-collisions being related to the rate of reactions (and subsequently, to bomb explosions), and then applying the associated "transitory randomness idea of particle-collisions" to defining, by means of the idea of material reduction to elementary-particles, all material interactions (physics and math), and subsequently,

11. the cultural knowledge of society pivots upon the fake, "material vs. spirit" dichotomy, which is represented as "the core issue of existence," but it is a fake issue, since the people of society worship the material-idol of gold and money,

12. When randomness is placed in a descriptive context which is based on materialism, ie science, and on the randomness of:

(a) material, and

(b) the life-of-the-fittest, and

(c) investment risk

then the "meaning of language" can be adjusted, as anyone (ie the top people in society) might like language, [in the propaganda-education system] to be adjusted, where

13. all of this arbitrary context for language is upheld by extreme violence (depicted in the movies as the person who can apply extreme violence to achieve a purpose defined by material relationships, eg investments, ie the God-like socially meaningful creative context of human creativity, which is all being controlled in our, western, society by investments).

The ruling-class has what should be a transitory relation to knowledge, but which is fixed and very official where the official nature of knowledge is expressed and maintained by academic institutions which work on creative projects into which the investor-class have invested where their projects depend on a narrow region of knowledge and it is these narrow categories of commercially related knowledge which are canonized as an, essentially, absolute scientific truth by means of university departments and their associated set of peer-review academic journals.

Where the ruling class organizes its social relationships (to society) through the types of knowledge and the narrow categories of knowledge which are related to their commercial interests so that these interests are related to descriptive contexts of randomness (or more correctly indefinable randomness, ie randomness is not correctly defined) and quantitative inconsistency which can also be a math context of irresolvable quantitative structures, eg chaos and bifurcations, ie many-valued (supposedly) measurable properties,

And these relations are both commercial and structural, that is, it is claimed (within the propaganda-education system that) society has been organized so as to be consistent social-Darwinism, ie a scientific justification of a hierarchical society which is called a meritocracy (to be defined on normal curves of behaviors associated to narrow categories of knowledge), and of course chaotic randomness is directly related to the narrow commercial-military interests of nuclear weapons of the ruling-class, but this has led to a complete failure of physics and math. The strategy of the ruling-class is to simply control the expert language, that is, keep the wage-slave, articulate, elitist-intellectual-left properly indoctrinated. These experts are aspiring, obsessive, aggressive, competitive, set of intellectuals, and this is another example of how personality types, in a relatively fixed set of social institutions, are so easily manipulated by the ruling-class.

That is, life's meaning is attached, in any arbitrary way, to the idol worship (of money in controlled markets), within a context of randomness, so that "pursuing money through violence," and the social power this gives a few people, in regard to controlling language for all of society, is supposed to be (has, apparently, come to be) the (highest) meaning of life-on-earth in our hierarchical society.

And the few who have this power, apparently, are using it for self-aggrandizement.

As long as they can regiment society so that the world population continues to grow, then they may be able to hold-on to their social power. But this is, apparently, leading to the collapse of the earth as a structure which can support life, as the earth is ever-more poisoned and the life which the earth supports continues to decline.

While, in fact, the mathematics of stable shapes, and its natural many-dimensional context, in a measurably reliable and quantitatively consistent context, indicates that life does have purpose, in relation to knowledge and creativity, in the context of stable shapes, but it is a many-dimensional context.

The point of many religions, eg Buddhism Taoism western spiritualism and native American religions, has been to store energy so as to get-at the many-dimensions, of which we are, naturally, a part.

But the math structure of the many-dimensional context makes it seem (appear to be) quite hidden, especially, when our word structure so ruthlessly asserts the narrow viewpoint of a material-world (based on [indefinable] randomness), and whose sole purpose is to survive (in the material world), within what has become a culture of great violence, instead of a culture of great knowledge and many individual creators who honor earth's balance.

Chapter 20

Information

The descriptive organizational structure of:

containment-set, context, measurable properties and processes, and operations or functions defined between sets, in which things (or physical systems) become stable, fixed, and controllable.

Information is about organized (electronic) signals and/or symbols which are compressed (reduced to an information equivalent, so as to compress a signal in which some of the (whose) information-properties of the signal are redundant, [after it has been placed in a "correct" descriptive context, both correct for

(1) an accurate "compression" of information, and also correctly coded so as

(2) to be introduced (or introducible) into an electrical circuit of switching circuits, (which has been organized in both its switching sequence, and in the inter-relations associated to the inter-related switching circuits, which are a part of an (or in the) over-all electrical instrument, which is processing the correctly "coded" "in-put information")], to a smaller signal-set (or a smaller set of signals, or smaller set of coded in-puts) so as to be sent..., at a fast-rate (or in an effective manner)..., through a noisy (or dispersive) channel (or electrical-line), and then received (at another place), so as to be re-constructed, so that the organized information which was obtained through the channel has an equivalent information content as the original signal.

Information is (or can be) measured so as to be based-on the (average) number of different (and/but identifiable) signals, eg where a "different signal" can be any of a set of distinctly distinguishable voltage-levels... (eg on-off voltages, or bits, and which exist on a signal transmission channel, or sets of waves with different frequencies)... which are needed to distinguish a symbol, of the many symbols (of) which the message (or the information set) is composed.

Information theory, in electronic instruments, is about sending and receiving information (or messages, ie voltage signals put into a code) so the information remains consistently accurate (where accuracy is compared to the original set of sent information). This is about compressing, sending, and re-constructing the set of information (of interest) accurately.

Descriptions of math and science are mostly about "information compression" of (or in regard to) descriptions of measurable properties, and their associated descriptive contexts, which are:

(1) measured in a reliable context, and

(2) stable patterns, and

(3) controllable processes,

…, which, in turn, are patterns relatable to sets of signals, which might be represent-able by only a few symbols, and possibly with only a single symbol.

While

Patterns which are random require the most number of symbols (which are needed) to identify a random state's distinguishing features, where in a "valid probabilistic context" this would be the entire list of the stable and well-defined elementary events which compose the probability system's elementary-event space.

One of the great fallacies of our, so called, expert technical culture is that "invalid random contexts" are being claimed to be useable (where the invalid probability contexts would be related to "the containment descriptive contexts" in which the events of the random system's elementary-event space are either not stable or not well-defined, or the claimed stable events exist, eg the observed energy-levels of general atoms, but these stable events cannot be calculated in the descriptive context within which it is claimed that they exist, [eg the energy-levels of a general atom so that the description is given to sufficient precision "as the observed patterns can identify," or (more technically) even the fundamental energy-levels of the H-atom, which have been found by calculations which are based on a valid set of math operations, associated to the, so called, laws of quantum physics]) so as to describe stable patterns, this is, supposedly, the content of quantum physics, and in this same context it is now being claimed that a random context of description can be used to confine and identify stable states (or stable patterns).

Note: Unless all of the stable properties of general: nuclei, atoms, molecules, crystals, etc can be calculated to sufficient precision by using the, so called, laws of quantum physics, eg containment of the quantum system in a function space, and a subsequent set of operators (a complete set of commuting [local] Hermitian operators)…, which can be used to actually [globally] diagonalize the function space…, are found, then it (ie the confinement of a (claimed to be) stable pattern [which is fundamentally an indefinable-random pattern] because the indefinably-random pattern is claimed to possess some vague and indefinable relationship to probability) is not shown (or has not been shown) to be valid basis for description of fundamentally indefinable-random descriptive patterns.

That is, if the discussion which is provided "about a descriptive structure's random context" is incoherent and devoid of any meaningful content;

and the main statement about the (observed to be) stable patterns are that "either they are indescribable, or they are too complicated to describe;" then there is no compression of information for such a random pattern, ie there is no confinement of information,, ie all the randomness which the system possesses must be expressed in order to be able to describe (with symbols) the stable state (or stable pattern).

If, nonetheless, this invalid descriptive method is deemed to be theoretically true (based on a set of invalid assumptions, within a fake descriptive context of formalized axiomatics) then….(this can have dire implications)

This then becomes a licensing-technique, based on fallacious expertness, to be used for petty-tyrants to act in amoral and atrocious manners, so as to express the great evils associated with arrogance or self-importance.

That is, the meta-data associated to a vague context of language, and its relation to social-behavior, which is, essentially, being held fixed by the propaganda-education system.

It is being claimed, (meta-data) can be used to form an (invalid) probability model of a person's behavior, so that it can be further modeled that a probabilistic model of this person "doing a crime in the future," can be calculated, based on the (invalid) laws of probability, so that this model justifies a "pre-emptive strike" on the person, so as to murder that person, before they can do the {(invalid) probability-hypothesized} crime.

Of course such a model is arbitrary, and would be used to exterminate any person, which the secret police might want exterminated, ie we are living in a concentration camp.

On the other hand, there actually is a more realistic model of the propaganda-education system, which, more like a controllable electric circuit, can have its language, almost, entirely controlled, where the controlling mechanism is, essentially, continual repetition of ideology and word-strings placed in a behavioral context in which a person's motivation structure is based on "wage-slavery and the associated forced idol-worship of money," ie a social construct violently forced onto the public by the justice system, [the value and reality of wage-slavery is made clear by force, where this is effectively the context of a concentration-camp, ie a reservation into which a few crumbs are tossed], where in the context of language control of the propaganda-education system, a system which only expresses the values of the ruling investor-class, a political propaganda system in which the right expresses support for the ruling class, and all of its violence and arbitrariness, while the left expresses support for the intellectual foundations upon which all the creative contexts into which the investor-class has made investments are based

The politicians are about those scaly-wags who want to make money for themselves by being a part of the propaganda-system (in a manner which is entirely similar to how the, so called, artists seek entry into the propaganda system),

Thus, voters are voting in a context in which the society, and their own place within that society, has been entirely misrepresented to them by the society's information system, ie the media-education-information system

That is, in the propaganda-education system there is a definite effort to control "what is stated" over the communication channels. Thus, there is a causal relation, not a probabilistic context, ie an intent to communicate in a particular highly information controlled context, where in the case of the propaganda system information is about a causal context of "describing the real state of existence" be it "within existence itself" or be it "a personal existence" "within a society controlled by its propaganda system" so that people will behave as the ruling-class wants them to behave, based on limited knowledge and a limited set of allowed actions within society.

Within this type of a casual model of information, ie communication within the propaganda-education system, the spy agency can be very active, so that word-strings which are in opposition to the intent of the propaganda system's goals of limiting the public's behavior, can be easily identified and there source identified, so that a great deal of interference can be related to excluding these types of expressions in a "land of the brave" where there is "freedom of belief" and "freedom of expressing one's beliefs" particularly, if those "beliefs and expressions" are, fundamentally, all about "valid knowledge, and the relation of a valid knowledge to the creative efforts of human beings."

It is at this juncture, where the organization of the propaganda system around the categories of expression which are directly related to the creative interests, into which the investor-class has invested, and the, subsequent, relation which these "categories in language" have to the university system of expert intellectual efforts can be considered.

Namely,

The propaganda system is about editors, ie the gate-keepers of the propaganda system, and the categories of intellectual efforts, which are so narrowly related to the commercial interests of the investor-class, (where this) it is basically the relation of the intellectual efforts (which are dogmatically expressed in the education-system so as to be formed into intellectual contests), to the "complicated sets of instruments" into which the investor-class has invested, so that the, so called, intellectuals have been indoctrinated, in an intellectually competitive context, so as to possess the correct model of research, which surround the instruments and institutions, which form the structure of social domination, put in place by the investor-class.

But, furthermore, the intellectual categories are intimately related to the commercial structures of monopoly and social domination, due to both

(1) the products which the investors control and

(2) the way in which society is organized around some of the products,

eg nuclear weapons and the spy-structure of communication systems, are used to define a police state called a national security state which, in turn, defines a narrow fixed way of organizing and managing society so that new intellectual efforts are not allowed to be expressed so as to not "de-stabilize" the traditional efforts, or the current investment contexts, of instruments and institutions within which society is organized.

It is Rome which, so effectively, designed society so as to get the

(1) propaganda system the

(2) technical system (then, brick-laying), and

(3) the structure of social violence to work together, so as to form a social organization which was (has been) designed to support the ruling-class, ie the Roman-Emperors, and it is exactly, this type of social structure which is in place today, but now (2014) the emperors are the investor-class, and now it is the hand-ful of investors (maybe 12 or less) who own the controlling interests in "a majority" of the world's monopolistic commercial institutions.

Information:

In regard to the random occupation of a system's energy-states by the system's components,

1. Entropy, which is the number of possible energy-states into which a fixed-number of components of a physical system can be fitted (or filled), [in a closed system].

Or

In regard to randomness,

2. If an n-decimal-place number, identified as a base-2 number, ie n-bits, is determined in a random manner, then 2^n is the amount of information to which such a number can be associated,

If the number is identified systematically, then the information the number contains is less, yet the systematic process of limiting the number also has information in its construct,

That is, in a context of both great diversity, ie many possibilities and thus existing in a state of great randomness then any of the set of random events has a great amount of information associated to itself, while when there is knowledge and rules which exist within context which limit the set of possibilities then a distinguished property has less information within itself.

That is great randomness implies great arbitrariness and little capacity to evaluate, while more order allows for greater evaluation.

But perhaps the order comes from a new context, so that the randomness of the old context still exists, but that old context is not important (or not as important) in an evaluation (or a limiting of information) process.

Laws reduce information

If everything is random then anything goes.

Knowledge is about compressing information, where perception and intent can also do this, ie compress information, where one is perceiving in new contexts, beyond the material-world.

Using information to manipulate and deceive. The perils and problems with language, or descriptive knowledge.

But perhaps the material-world is not (or should not) be the main focus of a human's attention, and not the context where knowledge (or [information limiting] rules) can be best utilized, eg knowledge of the material world has little value in regard to some other context for a creative purpose.

3. The average amount of information, eg the number of bits, needed to identify a meaningful symbol, eg a letter, then one would need a certain number of bits of information for each letter, in order to identify each one of the 26 letters, etc.

4. The ability to compress information, by using knowledge of (stable) patterns, ie reliable descriptions of properties from contexts which can be used to compress information, so as to form an algorithm.

How can symbols compress information?

Or

Can symbols be used to compress information?

About communication systems, or information systems

Information systems only perform well when placed into a "language construct" which, in turn, is based on valid knowledge, ie the organized symbols and assumptions provide a valid model of the observed events, or a valid model of a context which is being considered (or described), within which there is to be action taken, or processes initiated and/or concluded.

That is, the probability needs to be properly confined to a well defined elementary event space, or a properly confined relation (a representation of knowledge) is (can be) associated to a symbolic category, or

To a context (certain types of processes [or properties], which exist within a containment-set).

One can see that information is deeply related to probability, but compressing information is about organized knowledge, or organized sets of symbols, and this depends on different set-containment contexts, as well as depending on different viewpoints.

Chapter 21

Review of society, and the place
of knowledge in society

Though some propagandists, such as C Hedges, are correct about the delusional nature of people's viewpoint about life, in society, due to propaganda

Nonetheless

These propagandists do not get the great depth of the absurdities embraced by the public, as well as the, so called, expert-class, or better stated as the illusions which are being forced upon both the public and its "experts," where all of the public has been forced into wage-slavery,

It should be noted that there is a short, (2½ -page) section* of this paper, which is succinct and elementary, yet it is a rigorous technical discussion about the quantitatively inconsistent context into which the current descriptive languages of math and physics end-up, so that, it (the current quantitative descriptive structure) becomes a descriptive language which loses its capacity to describe any meaningful content about the physical world (by means of the math structures which the current descriptive language uses).

That is, the current context of physics cannot accurately (to sufficient precision) describe the stable spectral-orbital properties of the general types of the most fundamental such systems from nuclei to the solar-system and it does not contribute anything in regard to the development of new practical creative contexts so as to expand the practical context of invention, ie it is not true and it is not practically useful, (technical development today (2014) is all about extending the ideas and constructs of 19th century science, in a manner similar to how the Roman armies expressed their technical prowess by building their engineering projects with bricks, engineering projects which defined the cities and towns of civilization for over 1000 years [but what creative expressions of knowledge, and related creative new contexts, were suppressed by violence and propaganda for those 1000 years?])

That is, the type of physical systems which the current language is incapable of describing, are the bounded and stable many-(but-few)-body physical systems which are observed to possess stable spectral-orbital properties, and in regard to these most prevalent and fundamental physical systems the current language-construct can only describe, in a very limited way, a (an elliptic) 2-body bounded physical system

And then the discussion (in the next part of the same section* of the paper) explains the very simple (but many-dimensional) math structures within which physical descriptions can correctly identify meaningful physical content, so as to describe the system-binding context of the stable bounded many-(but-few)-body spectral-orbital physical systems,

or the discussion explains the elementary math context within which (in regard to) the descriptions of stable bounded physical systems must be framed

These new ideas can also be extended so as to form new models for both life and mind, where the extensions depend on both the genus of (or number of toral-components in) ..., a discrete hyperbolic shape, as well as the various properties of the discrete hyperbolic shapes for the different dimensional levels, which were uncovered by D Coxeter.

In western culture, {where the western culture is a collectivist culture, but the propaganda-system proclaims "it is a culture based on individual-freedom" (such a claim is an obvious absurdity, only those few in the ruling-class have individual-freedom, since they dominate all of society)}.

That is, individual freedom is defined for an individual within the public as their ability to dominate others through bullying (in a manner similar to how the ruling-class uses the public institution of the justice-system to bully and terrorize the public), where an individual member of the public is allowed to bully those whom "the bully" believes are inferior to their-self, by means of violence, or (authoritative) persuasion.

But if an individual within the public uses violence (or persuasion) against the ruling-class then this is opposed with greater violence which emanates from the military-justice-spying-police-state or by excluding such persuasive discussions.

The delusions of the western culture (the main delusion of our western culture is our arrogance and our belief that our culture is in the possession of absolute truths and our lives are related to an absolute context, which is needed to determine value)

The short list of delusional choices which are forced onto the public (the wage-slaves) who have been induced into a sense of arrogance by the propaganda-education-indoctrination system which is controlled by the ruling-class

(1) One can either be an idiotic fool (authoritarian expert, all of whose intellectual-models are the models which best support the interests of the ruling-class, eg physics is primarily about detonating a [mythological] gravitational singularity)

or

(2) one can be an accessory to crime (eg a person of action in any of the many institutional structures) in regard to the spying, lying, robbing, and murdering of the public by the ruling-class, in a one-sided class-war waged by the ruling-class against the public, and the world,

eg copyrights are about stealing from the public in order to support the ruling-class,

Through "capital," ie moneyed-exchanges for value and controlled by the banks, and "law," backed-up by "the force of society," ie the military and the governing and justice-system structures which relate the public to the military, ie the intentions of the investor-class (or bankers-oilmen and military-investors [and their organization of "perverted set of spies [the wolves to whom the women have been thrown]"]) are backed-up by the governing-and-value-determining-institutions of society (the instruments of the communication system), the public is made into wage-slaves, resources are stolen, and a narrow set of categories of behavior and creative efforts are imposed on society, where the limited number of categories and the social-value associated to these narrowly defined categories (of allowed social actions by the public) are controlled by both "capital-investment" and "the communication instruments for society," which "capital" owns and controls. And then the public is spied-upon, so as to cause them to be prey, for the very-rich, where these very-rich are clearly empty within themselves, they want to prey upon a set of people, whose culture has been reduced and it has come to be defined in the narrowest and most meaningless social contexts, namely, the few categories of behavioral action related to the monopolistic businesses of the bankers-oilmen and military-investors.

Science and math are highly controlled categories (controlled by formalized axiomatic [see below]) where science and math are to be used in very limited and narrow ways, ie for the military and to control technical instruments [they are placed in a similar social context as the brick-laying of the Roman armies, where their engineering defined a social context in a very fixed and stable way].

However, due to the possible social instabilities which new knowledge might cause in society, thus new ideas are to be excluded, ie by peer-review. There is the criminally-tyrannical-authoritative claim that truth has no validity unless it is peer-reviewed.

This is a totalitarian viewpoint about truth, yet this is the way the, so called, progressives and liberals argue to support their viewpoints, whereas it is the, so called, "right" which takes both sides (either the authoritative side or the non-authoritative-side), in regard to their claims about truth.

"This is how insidious the US political system is," where the political system claims to possess two-sides, but which are both supported by the same investor-class.

The so called "right" wants totalitarianism, ie they very deceptively stand-with the investor-class, eg deceptively attributing the freedoms of the investor-class to the freedoms of the public, and the left implicitly "hands" a deep agreement with a "totalitarian viewpoint about truth" to the "right," "on a silver-platter," but where the "right" simply takes the claims of the investor-class (either expert or non-expert) as the totalitarian-truth.

This absurd social relation to "trying to determine the truth about existence," which is a relation which excludes society from the truth, is all controlled by "capital," and the management (or selection) of personnel by the institutional administrators, so that the personnel of an institution reflects the interests of the investor-class.

The relation between knowledge and creative efforts is quite guarded (by the ruling-class).

This class-war and (colonial) expansion is framed as the set of a few superior people who resolutely in opposition to the inferior people (who possess less [or no] authority), where social-value is being determined by "capital," through their propaganda-education-indoctrination system (social-value is measured on an artificial and arbitrary and meaningless value-scale).

This is about arbitrary inequality and social domination, which is fundamentally based-on the idea of materialism, and material-value, but material-based science is a failure, ie it is all arbitrary, and it is solely upheld by violence and "obedient inertia" of the manipulated public, eg manipulated institutional experts.

This is about a big deception, put in place by the propaganda-education-indoctrination system, which is run by the investor-class, so that the propaganda-system defines the range of political discourse (as well as the range of knowledge and creative contexts for all of society) and this structure of deception is used to steal the resources from the public, and the world, and it was also used to turn the public into wage-slaves (historically it is those people who do not own (real) property), where the wage-slaves, in turn, then seek to demonstrate their value to the investor-class, in arbitrary fixed categories and in an arbitrary measure of social-value and social-worth, wherein the investor-class controls the state-run instruments of violence

This criminal destruction of the world and its people, which is the apparent function of the ruling-class, is a far cry from the touted expressions (in the western-media) of "equal people who are free to believe and observe and to express their own thoughts,"

Where this has been reduced to the public has an equal opportunity to play a rigged-game, whose only "moves (in the game)" are to help the ruling-class, so as to "prove their social-worth" to the public, where the public must only act so as to serve the interests of the ruling-class (this is the absurd state of equality for the citizens in the western-culture).

But

Where a social context of equal people who are free to believe and observe and to express their own thoughts would satisfy the principles which best relate humans to their intrinsic inquisitive and creative natures.

The criminal destruction of the world (by poisoning), and its people (badgering them until no thoughts by people in the society have any validity or any creative value, so that there is nothing left for the ruling-class to steal), a state of action, so that the action is an expression of social-domination, to which the western-culture has descended, is a culture which is based on arbitrariness and violence

Where this is the current context of western society, a western culture which violently upholds a material-based society which, in turn, is used to define and uphold the arbitrarily determined value of a, so called,

Calvinist religion,

ie materialism and "the personality-cult worship" of the rich, this is the arbitrary base from which western culture violently advances (in 2014).

This is done through police-state terror, so as to force the public to effectively worship (against their will) materialism and "the personality-cult worship" of the rich, this is the arbitrary base from which western culture violently advances (in 2014) upon the world, its university physics departments are dedicated to trying to detonate a gravitational singularity, apparently, to blow apart the solar-system

The Puritans (Calvinists) made religion serve monopolistic empires much better than the Roman-church

The central issue of the culture is identified in the context of inequality, and its expression as the domineering behavior which is, truly, only successfully expressed by the ruling-class, but which is encouraged for the ditto-head "small authorities" within the public, who mimic the arbitrary expressions of dominance which the rich express

Data

Supposedly there are 85 corporate-individuals who own and control 60% (?) of the world's wealth, so they clearly have the controlling-share of any public-ally traded company, but it is likely that only 5 to 10 "individuals", ie from oil, banking, or military monopolistic businesses, control the world by means of economic strings (controlling percentages of stock) of institutional control, and they own and control the propaganda-education-indoctrination system, ie politics is expressions of propaganda, since "how society is organized and run" is being determined by these 5 to 10 "individuals," ie the politicians are pawns in the game.

The inequality and dominant behaviors, of the fairly small ruling-class, is mirrored in the public, and it is certainly mirrored in the administrative managers of, so called, public institutions, ie aggressively and violently domineering, and in the "institutions" organized and controlled by the spy-police-state system, eg militias and churches and schools etc

The definition of narrow social categories of allowed behavioral actions (allowed for the wage-slaves) was initiated by the engineering which was achieved by the skilled Roman brick-layers (during the Roman-Empire) so that these engineering efforts defined the structure of the social institutions so as to have cities be formed in a fixed relation to the technical structures, ie categorized and compartmentalized so as to most easily use the technological conveniences

Food and tools got categorized and subsequently defined bounds on creative actions (or creative behaviors)

Bankers used this fixed structural relation between technology and a community, and this could, subsequently, be related to large-scale investment, and monopolistic control of the small number of socially acceptable categories of creative actions, and within this narrow range of allowed societal behavioral actions the public was corralled as wage-slaves by means of the bounding institutions and technical structures, and with subsequent, terrorist-or-secret-agent actions directed against them,

Those without "social-value," in regard to the investment structure, are to be pushed to the edges by the social institutions, where all the social forces conspire to exterminate them, the unworthy

Corralling intellectual efforts

Where this same context of a limited number of categories are defined for intellectual attention or endeavor

And

This also turned into intellectual actions related to absurd delusional, so called, intellectually based descriptions (at least attributed to being intellectual by the propaganda-education-indoctrination system)

The succinct section*

The contexts of formalized axiomatic based languages, which in math and science, divide sets into (1) containment of a system (which is, at least, descriptively bound together), eg either coordinates of a metric-space or (well defined) random elementary-events, eg events of point-particles in space, and (2) the set of measurable properties of that physical system (geometry or spectrum, ie coordinates or function-space, respectively), and (3) the methods, processes, and/ or sets of operators, which are used to identify the actual models of local measuring in regard to the measurable properties of unified physical systems (or, at least, described physical systems) [which are contained in event-coordinate-"metric-spaces"], and (4) the set of, so called, physical laws, which, in turn, are used to identify the sets of (partial) differential equations, which are to be solved in order to find the system's measurable properties, [that is, the solution functions to the system defining (partial) differential equations possess the function-values which are the measurable properties of the physical system],

Formalized axiomatic assumes that (1) the quantitative structure of each local coordinate of the local coordinate properties of the metric-space satisfies the axioms of the real numbers so that the coordinate space is quantitatively consistent, and (2) the quantitative structure of each local function-value of the local set of function-values, and for each function-value satisfies the axioms of the real numbers so that the set of function-values is quantitatively consistent, with the function-values seemingly possessing more quantitative variety, thus if the one operates upon the function-domain pair with local operators which model a local measuring construct then both (1) and (2) maintain their quantitatively consistent properties in an independent way within each of the two sets, but this is not true, that is, neither (1) nor (2) can maintain quantitative consistency if either the local operators in which the system defining (partial) differential equations are non-linear or non-commutative and/or the set of spectral functions in regard to a {quantum system's} function-space model of a quantum system are non-commutative or the so called randomness is actually "indefinably random" (eg the elementary-events are unstable, or they cannot be determined by means of calculations)

but

That is, formalized axiomatics assumes that each set, ie both the containment sets; and the set of the system's measurable properties; {as well as the set of maps from "the system-containing coordinate metric-spaces" to "the set of the system's measurable properties,"} (each set) in an individual-way, is consistent with the axioms of the real number systems, and it is assumed in the formalized axiomatic that the operator structures (which are defined by converging-sums and limit processes) is defined on numbers, {as well as the maps (from the system-containing coordinate metric-spaces to the set of the system's measurable properties)} and thus these improperly defined operators (ie non-linear, and/or non-commutative, and/or indefinably random operators [or maps]) do not disturb these two individual set's relation to quantitative consistency

But This is not true,

The operator structures, which act on the maps (which are the transformations defined from the system-containing coordinate metric-spaces to the set of the system's measurable properties) are improperly defined, ie they are either non-linear and/or non-commutative and/or indefinably random, and though each set, in an individual-way, is consistent with the axioms of number systems,

nonetheless, when these sets are inter-related by maps and operators which are either non-linear and/or non-commutative and/or indefinably random, and the subsequent solution functions are placed into a context where their domain spaces are quantitatively inconsistent and their function-value spaces are quantitatively inconsistent so that there are no stable patterns to which such quantitatively inconsistent sets can be related, (or) by moving (or mapping) into (or forming) a descriptive context of locally definable models of measuring, in which the local models of measuring are: non-linear, non-commutative, and/or indefinably random, then this means that the quantitative structure in the system-containing coordinate metric-space set, as well as in the set of the system's measurable properties (ie the solution-function's set of function-values) are both quantitatively inconsistent (which means that: the quantitative scale, the number property of continuity, and the function-structure of single-valued-ness are all lost in [such] a quantitatively inconsistent number-system), and thus, there is neither reliable measuring properties nor can there exist (stable) identifiable patterns, in either the coordinate metric-space or in the set of measurable properties (ie the solution-function "function-value space") which can be described for the system, which is supposed to be contained in the metric-space, where measuring is supposed to be reliable and stable patterns are assumed to exist, but, in fact, the descriptive context possesses no meaningful content (ie there is neither reliable measuring nor do there exist stable math patterns).

But along with quantitative inconsistency, there is also, for each value of a number-component, (ie for either coordinates of the metric-space or for function-values), there are the real-numbers (or complex-numbers) being modeled as a continuum,

but

This model of the real-numbers as a continuum is "too-big of a set" so as to result in the quantitative sets (thus) being very large sets, which allow for the definitions of limits and convergences which result in the formation of separate and distinct set structures within which the convergences exist (or are defined) where these separate and distinct set-structures are logically-inconsistent with one another,

eg the reduction of material to a point-particle geometric model of a (point-particle) collision, which defined random point-particle events in space, but this is defined in a (ie an assumed to be) random-based descriptive language, so such point-particle collision-geometries are outside-of what the uncertainty-principle allows.

So (in the new descriptive context) what seems to also be about "fitting data" (finding spectral-values so as to fit the stable discrete hyperbolic shapes) but, in fact it is not about fitting-data...,

(as a probability based description is all about fitting-data in regard to a random descriptive language, where fitting data to a probability based description is exactly analogous to Ptolemy's epicycle constructs and its data-fitting relation to planetary-motions), rather it (a finite-set of spectral-orbital values) is (about finding) the needed information in regard to the construct of a number-system, which is needed so as to properly define a finitely generated number-system (generated by multiplying each of the values of the finite spectral-set by the integers), which is "not too-big," and which is to be used in both the coordinate metric-space-set as well as in the set of function-values (in regard to the system's measurable properties) so as to define a quantitative-set of coordinates within which we are (or our measurable experiences are) contained, and where measuring is reliable and the math patterns being described can be stable patterns and the quantitative-sets can remain logically consistent.

It should be noted that

By geometrization, the only quantitatively consistent math patterns which can exist in the context of the sets of: (1) containment and (2) maps to measurable system properties, (3) which is being described by local operator models of local measuring (or geometries which can also be related to stable spectral-orbital values) (ie the only math patterns which are quantitatively consistent in the currently accepted (2014) context of physical description) are [briefly] the: cubes, cylinders, tori (doughnut-shapes), and discrete hyperbolic shapes (ie strings of put-together tori-components, and possibly folded at each of the points joining the toral-components).

That is, the only quantitatively consistent math constructs are geometric, where it also needs to be noted that the apparent random and reduction-ist context, which is the basis for of quantum description, can be derived from a geometric interaction construct which acts on small components whose local geometric properties of charge distributions are always changing, due to particle-component collisions so as to model local Brownian motions amongst the small material-components (see E Nelson [Princeton] publications in, 1966, 1967, 1979, where it has been shown that, the context of Brownian motions of small components are equivalent to quantum randomness) and each of the material-components, modeled as small stable shapes, possesses a distinguished-point, which would be emphasized during material-component interactions, thus we see the set of random (of what appear to be) point-particle events, but are the distinguished-points defined on geometric models of material-components, but one also wants a quantitatively consistent math construct, where the quantitative-sets, eg local real-number-line approximations for each of the coordinates, are not too-big, so that local measuring is reliable and the math patterns are stable, and they are patterns (of a precise descriptive language) which are logically consistent. This can be accomplished by generating a number-system from some large finite set, such as the finite-set of spectral-orbital values, which can be defined on (or within) an 11-dimensional hyperbolic metric-space containment-set (containment in regard to the set of discrete hyperbolic shapes), where the number-system generated from the finite spectral-orbital set, can be related to all the real-quantitative-constructs of stable systems, which are a part of our measurable experiences, and so that relations of our experiences to the higher-dimensions can be identified, and subsequently, realized so as to be used in a practical manner (but "practical use" might be related to an experience which is outside the material-world), eg the stable solar-system is evidence for the existence of a stable 3-shape in 4-space

Thus, one can make the claim that by E Noether's symmetries relating "metric-space-type math properties" to physical properties, where the hyperbolic metric-spaces are the spaces upon which the physical property of energy is defined, and thus, the discrete hyperbolic shapes are "the stable forms of energy which can exist," and then, according to D Coxeter's theorems on discrete hyperbolic shapes, these discrete hyperbolic shapes exist in (hyperbolic metric-space) dimensions one through ten inclusive (all of which can be contained in an 11-dimensional hyperbolic metric-space, but [it should be noted that] the 11-dimensional hyperbolic metric-space is not a stable shape), thus, this many-dimensional context for the existence of forms of stable energy-structures (in regard to the entire set of discrete hyperbolic shapes) which, in turn, define finitely generated number-systems, is the correct descriptive context for both science and math, if one wants to describe stable patterns in a context where measuring is reliable.

The new descriptive construct also models local linear measuring and the basic difference between the two descriptions is the context in which the stable bounded elliptic systems interact and subsequently form in a containment context which is controlled by a system's resonance properties with a finite spectral-set.

This new context is great for physics and life can be re-considered in new mathematical forms, and religion is given a context which is associated to direct "physical" exploration, but this new many-dimensional context, to which life is connected, seems to have a mysterious distant relation to the proper subset structure of the material-world to the actual set-structure of existence.

End (not quite)

Where it needs to be noted that, it is the failure to provide the descriptions of the observed stable patterns of the many-(but-few)-body spectral-orbital physical systems which exist at all size-scales, and which now (2014) go without valid descriptions based on the, so called, laws of physics, which demonstrates that modern physics, based on quantum physics, particle-physics, and general relativity, has been such a great-big failure (but where this obvious failure is so successfully ignored by the propaganda-education-indoctrination system, but it is ignored since science and math [ie the failed context of math and physical science] are currently (2014) serving the interests of the investor-class so well, ie physical science is about developing the communication-system technology associated to 19th century science [and to Tesla], while modern science [ie quantum physics, particle-physics, and general relativity] is being used by the military-investor-class for bomb engineering)

Everything (all social-value) is arbitrary, and society is controlled through the propaganda system through a context of either ignoring things or placing ideas into the attention of the public by continual repetition of the same ideas (the same messages). It is the usual "con" of the "stage illusionist."

That is, or

Furthermore, the core of modern physics and math are about quantum physics, particle-physics, and general relativity, where each of these descriptive constructs is deeply related to one or more of the quantitatively inconsistent set of local math operator properties of: being non-commutative, or non-linear, or indefinably random, and, therefore, this means that these descriptive languages (also) have no meaningful content.

Furthermore, particle-physics is a descriptive language, which is deeply related to the mathematics of quantum physics, and which is only related to determining the cross-sections of material reduced to particles and thus they are descriptive languages which are principally (or only) related to rates of reactions in explosions, (the model is about trying to detonate a gravitational singularity, it has no other descriptive relations to the physical world) this is how the monopolistic military businesses control language and thought, so that virtually all scholarship into math and science is about either bomb development (and communication instruments) or it is meaningless baloney, which is mainly used by the, so called, educated-scholars to intellectually bully, the other people in society, who are always assumed to be "the lesser minds of the world."

It is an expression of domination (it is intellectual colonialism) but it is not an expression of being rational, nor is it about truth, it is more related to the "bricks" of the brick-laying Roman engineers.

It is ironic that the, so called, highly educated scholars do not (apparently, do not) have the mental capacities to "determine truth for themselves." Perhaps it would help if they: questioned their assumptions and their contexts and the interpretations and also if they actually assessed the actual successes of the viewpoint which they "parrot," where the viewpoint that they do "parrot" is consistent with the propaganda-education-indoctrination system.

This is more about the language and psychological effect which people with high-social-positions have in regard to engineering a propaganda system, than these, so called, scholars having any relation to curiosity or seeking truth or seeking to be practically creative.

Furthermore,

The set of local math operator properties of: being non-commutative, or non-linear, or indefinably random, also [then this] leads to a condition in which the system cannot be solved (or it is not integrate-able) in the sense that there cannot be found a global-function which identifies the system's measurable properties, and/or which can be controlled to some extent by controlling the system's initial conditions or its boundary conditions when one is setting-up (or building) the system (to perform some task) rather local quantitatively inconsistent "solutions" are claimed to be pieced together, but doing this "piecing together" with only math adjustments of the numbers leads to a chaotic context where no math patterns exist,

So the "piecing together of number-values" is done by using adjustments so as to "fit the numbers together into a pattern" of "what is to be expected," ie epicycle reconstructions, and this non-math process has no relation to any type of deducible or controllable math pattern,

Eg using a geometric particle-collision model in a descriptive context which is based on randomness (this is logically inconsistent), eg the re-scaling of the numbers is the structure of renormalization (which Dirac would have nothing to do with this, so called, re-normalization or re-scaling process) [this type of adjusting, such as by re-scaling, has nothing to do with measurable controllable math patterns {but it should be noted that atoms with the same number of components always have the same spectral properties, so this means that they are forming in a controllable and measurable context}, thus, to say that "they (ie the particular number of components) form into a stable system in a non-linear context," is simply an absurdity, since such a claim cannot be deduced from a quantitatively consistent context, ie it is a statement of faith, so it is the same as saying ((in a sincere voice)) "I believe in the tooth fairy" or in any other noun (or belief), which has no relation to"reliable measuring and using stable math patterns to build some practically useful thing,"

ie descriptions which are not consistent math structures, and which are neither accurate nor of any practical creative value (in regard to their descriptive context) are simply nonsense,

Eg this type of descriptive claim is exactly analogous to the use of the epicycle structures in Ptolemy's model of the solar system, although they were somewhat accurate, and they could (sometimes) be used for navigation, but they are of no value in regard to navigating a satellite through the solar-system (not in the context of a satellite orbit about one body, but rather in the context of getting the satellite to go where one wants it to go in the solar-system),

but this is a parabolic-hyperbolic (see below) type (partial) differential equations, which model material-interaction, and which is also non-linear, and thus, it is dependent on feedback (in order to achieve its externally provided purpose), which, in turn, is based on knowledge of the planetary positions (and possibly, the positions of the moons around a planet), but it should also be noted that this feedback is done in an action-at-a-distance model, ie it is done using Newton's model of gravity]

It should be noted that, all of quantum physics, particle-physics, and general relativity are neither accurate nor practically useful

That is, they neither accurate in regard to their application to general quantum systems nor in regard to their application to the set of (many-body) gravitational systems, and they have no relation to developing new creative contexts; and

note that particle-physics only applies to bomb-engineering (a very narrowly defined application in a very chaotic descriptive context eg the context of probabilities of random particle-collisions during a transition between stable states of matter)

It becomes so difficult, and absurd, to express an actual rational viewpoint in a public discourse, even in the very primitive language of math and science, because the propaganda-education-indoctrination system has colonized so many minds. The meaningless and arbitrary content which should be attributed to: quantum physics, particle-physics, and general relativity are, in fact, taken quite seriously by everyone who reads and believes the education-indoctrination system, as M Twain said "I try not to let my schooling get in the way of my education."

It might be noted that

Both...

"how Copernicus changed the assumptions and context as well as the interpretations of observed patterns" and that Godel's incompleteness theorem implies

that due to the limitations as to the set of patterns a precise language is capable of describing that the assumptions, contexts, and interpretations, as well as the organization of concepts into a new descriptive language need to be continually changed and questioned, especially, when a precise descriptive language cannot describe the patterns which are being observed

Thus these, are two examples about changing the technical language of science and math, where such considerations about changing assumptions and contexts are (supposed to be) fundamental to the development of both science and math.

So that when the observed patterns of physical systems are, nonetheless, continually placed into the same (failed) descriptive language then this means that the science institutions are in opposition to good science practices

This is an example of the differences between schooling and education.

And where the control of society requires only such a small percentage of fully indoctrinated types, so as to ruthlessly dominate the others through the institutions which serve the ruling-class

Where only a few can dominate the many since, according to the propaganda-education-indoctrination system, those (indoctrinated and authoritatively domineering personality types) who get the high-social-positions in institutions are, supposedly, superior types of people, ie they have been indoctrinated into the world-view of those who "truly do" hold the social power

Freedom of intellectual belief, and equal free-inquiry, and freedom of expression, can (all) be excluded, under a society ruled by a culture whose first consideration is about inequality and explicit domination, where the use of the society's instruments of communication is directly related to this inequality, and in a society where the people are governed by both wage-slavery and the propaganda-education-indoctrination system which serve only the interests of the ruling-class, so that the main behavior which is highlighted and rewarded is the expression of arbitrary and aggressive domination.

It is exactly the same issue in regard to the individual right to birth control, but where the ruling-class wants to colonize each body of the public, so the ruling-class supports the idea that a particular religious idea should be dominant, so as to allow (force) a religiously based belief to be required, ie "it is to be required in society" that everyone believe such an arbitrary (either moralistic or intellectual) idea. For example, the media expresses the idea that "people cannot use birth-control, due to one (media domineering) group of people's, so called, religious belief," so as to justify a forced invasion of technical probes into each of the individual bodies (of the public) based on the religious belief of one group of people, subsequently, this can then be the "legal" basis (based on a, so called, superior religious belief) so that technology can be used to invade and occupy each individual body, of any body which is contained within an inferior social-class, when the individual social-class of any particular person is compared to the superior social-class of the ruling-class.

Similarly, failed intellectual beliefs are forced onto the public, where this is done through the propaganda-education-indoctrination system, but where the forcing is implemented by the wage-slave nature of the public so that only those who express the correct indoctrination are allowed into the public and private institutions so as to be rewarded for their loyalty to (and faith in) their indoctrination

This is true for both intellectuals and for people-of-action, but people-of-action are indoctrinated with failed arbitrary beliefs, which are often arbitrary moralistic points-of-view, which are to be imposed on others in a dominant way

Or consider that,

When the local models of measuring, [which are operators which act on (assumed to be, solution) functions (which map between metric-spaces and the physical-system's measurable properties)], are: non-linear, non-commutative, and/or indefinably random then this leads to a quantitative structure, for both the coordinate system containment space and the set of the system's measurable properties (or solution-function-values), which is quantitatively inconsistent,

ie where quantitative inconsistency includes:

(1) the number-scale is not well defined,

(2) the number-system is discontinuous, and

(3) the numbers (or function-values, eg coordinate functions) become many-valued, and subsequently, the description becomes nonsense, within both the containment metric-space (which contains the system-containing sets of coordinates) and in the set of solution-function's-values.

That is, there cannot be any stable pattern defined in the coordinate space or within the physical system's set of measurable properties, (there cannot be any stable and reliably-measurable patterns identified) within such a descriptive context (or construct).

The context of solvability for these descriptions based on models of local measuring (of either geometry or of spectral-values) requires the properties of local measuring of their being (1) linear, (2) metric-invariant, (3) continuously commutative, and (4) parallelizable [or locally commutative parallel transport of any coordinate-direction around any local closed curve (a curve determined by the coordinates)] [that is, if 1-3 do not already include the property of parallelizability?]then in this case a global function which models the system's measurable properties can be found (or solved) and subsequently somewhat controlled by boundary conditions or initial conditions

Also

Second-order linear (partial) differential equations (or pde) whose different terms (of the pde) are only related to constant-factors are (always) solvable. This is the context of most of modern (2014) technology eg electric circuit control and control of some thermal systems etc.

Descriptions of physical systems, and material-interactions which are involving measurable properties or events, are (usually) second-order (partial) differential equations which can identify three different types of dynamic systems

[Or in quantum-physics:

are defined by a parabolic wave-equation (or energy-operator)]

The three types of (partial) differential equations (or pde) for dynamic systems are

1. Parabolic (or free, in that the boundary conditions can allow the projectile [or material-component] to escape its bounds)

These are the projectile paths

2. Hyperbolic system-types where this identifies a collision between material-components

And

3. The Elliptic systems, where these are the bounded systems, defined by stable orbits and stable spectral properties

It is the elliptic (or bounded-orbital-spectral) systems which are so prevalent, eg nuclei, atoms, molecules, crystals, even life-forms, as well as solar-systems and (apparently) motions of stars in galaxies where these stable elliptic systems are typically: stable many-(but-few)-body spectral-orbital systems, but the only context in which these many-body systems are described is in regard to the above descriptive context of local measurable properties which describe the elliptic pde of material interactions,

But these elliptic pde can only be solved for the 2-body systems, but where the solution to the radial-equation for the H-atom in a quantum-description diverges

That is, this entire descriptive context of elliptic pde is a big failure; that is, locally measurable properties of material interactions cannot be used to describe the most prevalent and fundamental of the stable systems,

ie the stable and bounded many-(but-few)-body systems

Furthermore,

Quantum descriptions cannot really describe the H-atom, and in this case, then

quantum physics can only be used to describe (or solve) either (1) particles in a box, or (2) a harmonic oscillator modeled for a spring-constant, (say) k, in any believable detail,

But---

the stable geometry which is associated to the physical conditions in regard to the physical context within which the harmonic oscillator is applied (as a physical model) as a (or in a) descriptive context--- so that in this context there is the question as to "what stable geometry actually allows such rigidity of motion for a small component, in regard to its oscillatory motions?" is seldom considered, and is not ever answered.

That is, the quantum harmonic oscillator is more of a fiction, than a valid model of a true physical context.

However,

It should be noted that,

The box is as about as good a model as exists for the properties of crystal energy-levels. In general, the more detailed type descriptions of crystal properties based on the local structure of a crystal lattice-site for the periodic crystal lattices do not identify a well defined method, which consistently yield descriptions of the observed crystal energy-level properties.

That is, local models of measuring, in regard to either material-interactions, or material probability distributions, though they work fine for the parabolic and hyperbolic dynamic material-interactions they seem to have no value in regard to their (methods and contexts) being able to describe why many material-components can bind together to form stable systems, so that the same number and type of components for some specific quantum system implies exactly the same spectral properties of the bound-system as for any such system with the same number and type of components in its composition.

A new descriptive context can describe a greater range of these system structured new context is about a many-dimensional context of containment, organized around stable shapes, so that the entire high-dimensional space (the over-all containment set) identifies a finite set of spectral-orbital values, where this finite spectral-orbital-set is based on the stable shapes of the "discrete isometry or unitary shapes" for the different dimensional levels, within a metric-invariant model of a containment set.

In this new context the stable shapes form when they are in resonance with the finite spectral-set defined on the over-all high-dimension containment set. These stable shapes are models for both stable material-components and they are also the models for the metric-space containment-sets of the lower-dimensional material-components for a "local" dimensional level, so that the material which is contained within these metric-space shapes is material which forms as condensed material, which is pushed onto the shape's stable orbital geodesics, which are defined on (by) the metric-space's shape,

ie the stability of the solar-system is understood with the same type of stable shapes used to model stable material systems but the dimension of the metric-space is an adjacent higher-dimensional space than the dimension of the material-components which it contains

That is, the organization takes place due to the different size shapes which determine the many-dimensional levels, where the stable metric-space shape of each dimensional level has an open-closed topology, which causes the different dimensional levels to be independent of one another, and mostly isolated from the other dimensional levels, and/or from the other stable metric-space shapes of the same dimension.

The dynamics is similar, in each isolated metric-space, in regard to the same type of parabolic and hyperbolic (partial) differential equations, which define the material-interactions, but the context which exists in regard to different systems (different components) binding together... is related to the new viewpoint about existence (or... defines a new context) of the above mentioned many-dimensional levels organized around stable shapes of different dimensions and sizes so that a finite spectral-set is defined for the limited number of identifiable stable shapes which are used to organize the many-dimensional structure of existence.

(one uses the model of local measuring, unless algebraic models are used to inter-relate measured-values, as in thermodynamics, but where in this case, the measured-values are (can only be) defined over "average-values of a large-numbers of components, which are confined in some way," ie thermal physics. However, thermal physics is best organized as a physical descriptive context around the ideas of differential-forms, where the energy and entropy functions are assumed and then defining a first-order exterior derivative on either of these two assumed to exist functions, where, in turn, these exterior derivatives (from which differential-form emerge) are defined over the thermal variables eg component-numbers, volume, and either energy, or entropy) so that these operator structures are non-linear, non-commutative, or indefinably random structures which are associated to, supposedly, measurable models based on operations (or math processes)

Note

Geometrization is a statement that the physical-system containment "metric-space set" must be related to stable patterns of coordinates, where the coordinates are related to the stable metric-space shapes of the discrete isometry or unitary subgroups, ie lattice structures associated to orthogonal simplex structures, of the non-positive constant curvature shapes, which can be parallelized [due to the orthogonal structures of the simplexes] (or which are parallelizable).

Governing institutions are based on illusion and distractions created by the propaganda-education-indoctrination system which is owned and controlled by the ruling-class so that the ruling-class can lie steal and murder with impunity so that the social-collective social construct of western-capitalism supports such a de-lusionary set of actions for society, so as to be able to base the system on the atrocious crimes of those controlling capitalism

Politicians are selected, by business managers, to fit into the propaganda system, and where politicians, essentially, see "politics" as an opportunity to cash-in by being a part-of the "business based" propaganda-system.

The baloney that the corruption of politics is all about (or has been a result of) "how money is used in politics" is mostly a red-herring (ie irrelevant), especially, in regard to "how the political institution has been organized" so as to be a part of the propaganda-system, which is controlled by the ruling-class

That is,

In western culture one (a person) can either be an idiotic-fool (ie an expert who is dominating in their intellectual authority over other professionals, and who is worshipped by the public, but most of whose intellectual models have either failed, or are out-right nonsense) or an accessory to crime (ie helpers in all the other institutions, which are different from the education-indoctrination side of the propaganda system) where all institutions are designed to serve the arbitrary and selfish interests of the... lying, thieving, murdering... ruling-class

In such a society the people must adhere to (or be given to) delusions

Chapter 22

Physical descriptions are distorted by social forces

The science of the western culture is about quantitatively based descriptions which can be verified by measuring, but Ptolemy's epicycle model of planetary motions fit this criterion, so a better characterization of science is that it is general, ie widely applicable with quantitative descriptions which are fairly easily formulate-able and solvable, so that the description, ie the solution functions of the system formulated equations based on physical laws, (which are related to both the system containment-set (coordinate metric-spaces) and to a system's measurable properties ie the solution functions are the system's measurable properties), so that the descriptive properties are accurate (to sufficient precision, ie related to the precision of the system properties themselves) and they can be easily related (through geometric properties) to practical useful inventiveness

Note: Once a subject matter has been described in a quantitative context, so as to provide this type of measurable information which has been placed into a set of stable descriptive patterns, so that it (the precise descriptive information, ie quantitative-values placed into a descriptive context) is of practical useful value, or this type of stable measurable information is available, due to a descriptive construct, such as both Newton's mechanics and Newton's gravity, or Faraday's electromagnetism, where Maxwell summarized and added one term to Faraday's language (and Maxwell was given far too much credit, this is a part of class warfare, the wage-slave public feel that they are engaged in class-warfare with one another [or with the other people in the public], where this is all about arrogance intrinsic to a hierarchical social structure, and the constant repetition that the rich are superior people, and that the poor-people can also be rich, if they have value, but that value is to only be expressed in a particular type of social organization (built around narrow categories of production), and in that particular social context, "about material production or social organization," there is the repeated claim that people are not equal) but it might be better to recognize J W Gibbs as the person who best summarized Faraday's descriptive language for electromagnetism, with the differential-forms which Gibbs helped develop by their application to physical descriptions [about a metric-function's relation to other local geometric-measures within a coordinate system]) as well as the usefulness of thermal physics and the statistical models of these thermal systems (where the statistics are based on the atomic hypothesis), then the inventiveness associated to the descriptive context should be considered to be part of the commons, just as the descriptive language is a part of the commons, (but this is not how the justice system works, ie the justice system works in conjunction with the ruling-class and it helps the ruling-class wage class-warfare on the public, so that the, so called, great economic benefit are only realized by the ruling-class, eg the public are turned into wage-slaves (who must help the ruling-class paymasters) so that whatever is in the commons (or in the culture) which can be identified as possessing value (in regard to production, and which fits into the interests of the ruling-class) can be stolen and controlled as property is controlled by the ruling investor-class so that what comes to possess value in the, so called, commons is used for profit by, and for, the ruling-class, where huge profits of monopolistic businesses are one of the main basis for social power of the ruling-class within a society, where the society is engaged in class warfare (the ruling-class is always waging class-war against the public), or the ideas (which might be expressed in the, so called, commons) can be kept from wide dispersal, and/or kept from being accepted as being true by the public (by the control of the ruling-class over the propaganda-education-indoctrination system) so that only the ruling-class is allowed to use ideas and those ideas are used by the ruling-class to help the ruling-class, where this can be very much to the detriment of the public, where some "new ideas" could greatly improve the public, and the social position of the public, but not help (to increase the power of) the ruling-class, so much)

That is, everyone in society is empty, this is because no-one is allowed to develop knowledge (the experts are engaged in delusion, befuddled by the nature of axiomatic formalism, in a descriptive context which is so vast it is very difficult to navigate) and no-one is allowed to create on their own, they only create in very narrow contexts that which the investor-class wants created, a person is only allowed to assert their dominant social position, but if this assertion is considered to be out-of-place (ie one is acting uppity) then the class warriors come to attack the justice system and the propaganda-education-indoctrination system which develops narrow minded information zealots they see their mission as attacking any idea which differs from the ideas of the academics who are indoctrinated to only think about, or to believe-in, ideas which support the interests of the ruling-class, ie ideas cannot be placed onto the media which do not serve the interests of the investor-class, where any such ideas are excluded in "the name of" being accurate (truthful), ie the rules which the media propagandists are required to follow. This is the same role that the church played, in regard to controlling intellectual efforts, in the age of Copernicus.

The way to resolve the issue of "truth," (the so called, objective truth which is the duty of the professional journalist to protect [only peer-reviewed experts are allowed to express ideas] is to allow all ideas to be expressed but to require that the discussion most often remain elementary, ie at the level of: assumption, context, interpretation, ways of organizing language, etc and to put on a freely-publishing-internet web-site and classify the expressions based on their sets of assumptions, and other elementary discussions. Einstein was one of the very few who discussed many ideas and quite often he expressed these ideas at an elementary level, but what now (2014) passes for "truth," is a rigorous discussion about illusionary patterns which are only related to the set of assumptions, where it is required that an academic completely believe-in the set of assumptions of the formalized axiomatic, that is, the experts must express (in peer-review publications) only the delusional ideas associated to formalized axiomatic, ie the narrow set of technical ideas related to the incremental development of already existing instruments, instruments into which the ruling-class already has a history of investment, eg the development of nuclear weapons etc (Note: Einstein was excluded from the development of nuclear weapons).

When considering physical descriptions there is a lot of attention paid to the context of material interactions, defined by Newton, even though it has been a descriptive context which has only been relatable to the solution of a stable and bounded two-body system, where it is assumed that all stable physical systems both "form" and are held together within a context of material-interactions, where these material-interactions are defined in relation to a model of measuring which is local, so as to either relate changes of motions to force-fields, where the force-fields are generated from geometries of distant materials, or energy structures of spectra related to geometries of force-fields which are dependent on potential energy geometries, or

Dependent upon a set of walled potential barriers which confine (or bound) a set of material components which when bounded compose the system, and whose measurable properties are discrete spectra (different from the set of measurable average-values, eg temperature, pressure etc, of the descriptions of thermal physics, which also describes the average measurable properties of confined system), and wherein (in the spectral model of the box-system) material is reduced to random point-particle spectral-events in space.

Apparently, because of its limited descriptive range (of quantum physics), in regard to describing the measurable properties of stable bound physical systems, the discussions are abstract and esoteric, a context for such discussions which demonstrate a lack of questioning, ie a lack of interest in the descriptive context, and are presented as a sequence of memorized and complicated math processes, which seem to be held in place by the authority of the highest-worshipped experts, since critical questions are not asked.

There is a lot of concern about invariance's and symmetries, in the abstract discussions about physical law, but where these physical laws can only be related to a narrow range of types of physical systems, that is, the interaction construct which is built upon physical law has only one success, in regard to describing stable material systems ie the 2-body system for 1/r^2 spherically symmetric force-fields (existing between the two interacting bodies), the subtle and abstract notions of the related constructs of physical law, are always the central focus of the physics community, but the discussion is never diverse and it is always dogmatically authoritative and often it is about the inability to mathematically deal with the r=0 singularity of the (1/r^2)-form for the spherically symmetric force-field geometry, especially, for the point-particle models of the material to which material systems are assumed to reduced, so as to always lead to divergences in regard to the math models based on the dogmatic authority of current descriptive contexts of physical law

The natural limiting descriptive properties of physical interactions, in the quantum descriptive context in which randomness is fundamental, the spectral-functions are assumed to represent a probability distribution in regard to finding a particle-component within a quantum system with a particular spectral-value and in some region of space ie random spectral-particle-event in space, and then this viewpoint is adjusted so that a spectral-function is to be viewed as both an occupation-number and so that the function comes to be modeled as a set of creation and annihilation operators, but which are conceived to be primarily about the invariance of an equation, and the carefully considered "invariant ways" in which a particle changes its internal particle-states (ie models of creation and annihilation transformations), especially, during a field-particle (random) collisions with a system's Fermion material-components, and where the state-changes of the particle during collisions, supposedly, represent some physical law, which is, supposedly, the basis for all many-bodied material-systems existing in states of spectral-orbital stability, but the descriptions of the measurable properties of these stable many-body systems based on physical law are non-existent.

That is, there does not exist any wide range of accurate applicability in regard to the so called laws of quantum physics particle-physics, as well as general relativity

(where general relativity is also about the invariance of an equation when it changes between some set of allowed frame changes, general relativity is, supposedly, about changing between general accelerating frames, but it is always represented as changing between sets of constantly accelerating frames, which are claimed to be expressible (or modeled) locally, where this local representation (associated to the idea of frames differing by [or possess the same] constant acceleration(s)), where this local model of frames is always represented as being local space-time construct (being in constant acceleration), actually implies (or is best interpreted) that the system is a part of a stable hyperbolic metric-space shape).

The interaction between materials, where the interaction is based on forces, is about changing motion, or changing the inertia of a constant straight-line motion, or the changes of a constant rotating-state of material (ie the existence of torques, or tangential rotation-rate-changing forces), rotating about an axis of (rotating) motion, and so the set of system-containing coordinates…, which are also metric-spaces, where metric-spaces are needed for the geometric measures on the system containing metric-space…, are defined in regard to a frame of a uniformly-moving component, which defines either a set of moving coordinates, or a locally constant acceleration of a material-component's moving coordinates (moving in a manner of a locally constant acceleration) so that the equations of motions, or the equations for force-fields, in the coordinate metric-space frames, which the component defines, are both invariant (in regard to the form of the equation) to changes in these types of coordinate frames (either uniformly moving or constantly accelerating) and they are to relate to some (local) property of the metric-function in an invariant way,

Eg either

The "general metric function" is found by solving a form-invariant equation ie invariant to changing frames which are moving in constant accelerations (ie transforming between frames with different constant accelerations), where one can then define parallel transport (or the path of a geodesic on a general shape) once one knows the form of the local metric-function, where the physical law of material-interaction, (or the law of motions) is that material-component follow (local) geodesic paths (the problem with this formulation is that the shape itself must possess a well defined orbit in order for the this description of material-interaction being relatable to the descriptions of stable orbits that is, the shape where a geodesic is defined must itself already be a stable geometric pattern in order for such a geodesic model of material-interaction to be able to define a stable orbital structure within such a math construct of modeling material-interactions as being based on material-components following geodesic pathways,

Or

The equation which identifies a physical law is invariant to changes in coordinates, ie changes in frames of uniformly moving coordinates (frames with different uniform motions [or different uniform velocities]), which (also) leave the metric-function invariant, this is the construct of Newton for inertia (or mass) in Euclidean space, and the construct of Faraday-Maxwell-Gibbs for charge in space-time, thermal physics is in either Euclidean space, or represented in statistical-physics it is symplectic space (ie phase-space of position and momentum).

In regard to the stable spectral systems of many-bodies, there are issues about what equations, or what sets of operators, are applicable to modeling a system's local measurable properties, so that these local measuring constructs relate to the system's observed stable spectral properties.

But both methods of either

(1) limiting how the equations of a measurable description are to be determined, and/or

(2) how these (or some set of) equations are related in a meaningful way to a stable system's observed (or described (ie supposedly deduced from physical law)) properties, and in dong this they (the current reigning experts, or the set of dominant intellectuals (who, in fact, hide behind the skirts of the bankers-oilmen-military monopolistic business owners, eg this is peer-review, which is simply an expression of authoritative dogma no different in its relation to science than was the relation of the church to science during the age of Copernicus, science is individual interests in knowledge (this is what learning is), it is not the collective-enterprise of building consensus (a "consensus" whom work on particular instruments), that consensus is being determined by the investment-class), which is how science authority is modeled in a society which is organized around inequality and social hierarchy, essentially, based-on materialism and personality-cult, ie the Calvinist model of empire) (they, the current experts) miss the subtle point about requiring the following set of properties in a math description, which is mostly related to metric-spaces, and the associated set of related metric-space shapes, which are to play a central role in regard to the true and solvable models of stable and bounded physical systems:

(that is there is needed a new context for physical description which is brought on by both the Thurston-Perlman geometrization theorem, as well as Coxeter's classification of all the different dimension discrete hyperbolic shapes), systems are stable metric-space shapes, defined within a many-dimensional context, where these dimensions extend up to a hyperbolic metric-space dimension of eleven (or equivalently, up to a general space-time metric-space dimension of twelve))

Namely, the following considerations about math are required, when math-modeling the properties of physical systems:

1. linearity,

2. metric-invariance, and that metric-spaces are to be assigned with physical properties, and that the natural set of opposite physical properties defines the spin-rotations of metric-space states (of a metric-space's physical property and its opposite property)

2b. The symmetries identified by E Noether concerning both spatial displacements and time displacements associated to inertia and energy, respectively, are fundamental to the "mass equals energy" formula of special relativity, but they can be interpreted to mean that metric-space types, eg Euclidean space-time or equivalently hyperbolic metric-spaces, are related to particular math and physical properties and the natural discrete metric-space shapes of these spaces identify the stable, or conserved, types of system-properties, eg the discrete hyperbolic shapes are the natural shapes of stable energy systems while Euclidean spaces identify inertial properties of material systems etc.

3. continuously commutative local matrix (or operator) relations of the local models of measuring,

So that these first three properties allow a math model of a physical system to be solved and they are also properties which are required in order for the descriptive context to have the property of being quantitatively consistent, so that the descriptive context can possess meaningful content in regard to what it does describe

3b. The natural coordinates, which follow the continuously commutative local models of measuring, and which are also consistent with a commutative metric-function, are locally commutative in regard to the parallel transport of any of the natural coordinate's local directions about a local closed curve defined by these coordinates, ie the natural coordinates of the stable shapes are parallelizable for the (entire) shape, [this is a property which should be derivable from the first three principles concerning math which must be followed if the quantitative structure of the description is to be quantitatively consistent,

And (other issues about math which need to be re-considered and/or re-interpreted, so as to re-consider math models of physical systems so as to be within a math context where there is quantitative consistency in regard to math operations and math processes which are a part of what is supposed to be a measurable description which is to be modeled mathematically)

4. the deep relation (due to properties 1, 2, and 3 above) that the non-positive constant curvature stable metric-space shapes, ie discrete Euclidean, hyperbolic and unitary groups, have in regard to the stable properties, including the geodesic-based orbital structures of metric-space shapes which contain material, especially, material-structures of material contained within metric-spaces which require material condensation or material-accumulations and subsequently the deep relation that:

5. holes in these stable metric-space shapes have to the set of stable spectral-orbital either dynamic or "occupation number" properties in regard to the sub-components of the stable material-component shapes or stable metric-space orbital shapes (for the condensed material which the metric-space contains) as well as

6. action-at-a-distance toral shape of Euclidean inertial constructs of force-field interactions, and the close geometric relation that tori (or discrete Euclidean shapes) have to the toral-component structures of the more stable, and thus more prominent, discrete hyperbolic shapes, which compose both stable material systems and stable metric-space shapes whose geodesic properties are upon which the stable orbits of condensed material planets depend in a stable many-(but-few)-body system

7. And that these stable shapes exist in a many-dimensional context; from hyperbolic metric-space dimension-1 to hyperbolic metric-space dimension-10, inclusive, so that the whole context can be contained in an 11-dimensional hyperbolic metric-space.

That is, if the so called experts do not consider in a critical manner the issues of

1. stability,

2. solvability,

3. stability of shape,

4. alternatives to spin properties, which are not simply abstract math properties, which possess complicated properties of invariances (or complicated symmetries)

5. metric-spaces which can be modeled as (stable) shapes,

6. the failure of the material interaction construct,

7. The deep need for quantitative stability,

8. an alternative interpretation for the mathematical role of holes in stable shapes, and

9. the relation of holes in shapes to stable spectral-orbital properties,

10. the need to avoid the issue of a necessarily non-linear model associated to the spherically symmetric model of material interactions, and its associated 1/r potential energy pattern, which leads to divergences,

11. an erroneous attempt to re-define physical constructs by some form of arbitrary convergence (or divergence) [which are defined in quantitatively inconsistent containment-set context] {eg the relation of point-particle to wave-packets} which are allowed to be defined because these infinite convergence processes are being defined on the continuum-model of the real-numbers, a set which is too-big,

12. to deeply re-consider the; action-at-a-distance, a property, which has been confirmed by the A Aspect experiment in regard to EPR,

And

13. a needed-desire to extend past the unnecessary confinement of the idea of materialism, as well as

14. a willingness to embrace the ideas of E Nelson (Princeton) who related to Brownian motion to quantum randomness, ie to try ideas outside the notion that "randomness is a required physical law,"

15. where it is necessary to re-consider randomness, because the stability of the many quantum systems implies that they form in a controlled geometric context, and

16. the only stable math patterns (so that the math patterns can be contained in a quantitatively consistent containment-set construct) are the very simple stable... "discrete isometry or unitary subgroup"... shapes, which are defined in regard to non-positive constant curvature metric-spaces.

The problem is that stable systems do not form when they are described in the context of the measurable forms of material-interactions and the associated laws of physics

(in the currently accepted descriptive contexts of) Either geometry and motions or

probability motivated function-spaces where the goal is to find a "complete set of commutative Hermitian operators" which act on these functions spaces, so as to determine the sets of stable spectra, which are observed to be associated with the physical quantum system, (ie where the quantum system is modeled as a function-space), either canonical quantization (ie associating classical measurable properties with analogous sets of operators) and most fundamentally related to the geometries of potential-energy functions (eg material-component confining boxes which are walled-off by infinite-valued potential-energy walls, the 2-body H-atom with its $1/r$ potential energy term (but whose "solution-function" to the radial equation diverges), and the mythological harmonic oscillator with its kx^2 potential-energy term [these are the highly coveted, by the authorities, mythological quantum systems, whose energy-forms are obvious fictions, but these fictions are integral to their formulation]) or creation and annihilation operators (which are claimed to be related to harmonic oscillators, but the physical cause of the kx^2 potential energy term seems to not exist) which act on an occupation-number context, which is, supposedly, ultimately related to a vacuum-state and elementary-point-particle internal-states, which change during the random interaction-point-particle-collisions, but the type of occupation structures and their energy properties in regard to quantum systems seem to only be occasionally determinable in a systematic context and this has always been done by first using canonical quantization to determine the energy-levels of the (above mentioned) mythological quantum systems and then determining the occupation properties in regard to these energy structures

The problem with the formulation of physical systems based on modeling the systems by means of function-spaces is that the math property of being commutative (or the math property of commutativity) is a required property (for the description) but it is a property which cannot be realized, so there are only a few systems which are both formulate-able and solvable.

The Dirac operator (or the Dirac equation) is a first-order linear operator which incorporates "sub-operators" as a part of its operator structure, where these sub-operators result in the Dirac operator not being commutative with itself (thus, whatever system to which it is applied the system is not energy-invariant) and it introduces states of a supposed, solution function which are different from the solution function being a scalar, eg wave-functions are (at least each component is a) scalar-functions and energy-functions are scalar-functions, and these wave-function cannot represent a spectral-value which possesses a constant energy. That is, because of these new sub-operators ie different matrices introduced for each different variable into a particle differential equation, and new wave-states the Dirac operator does not commute with itself so its solution functions cannot describe stable patterns.

That is, the descriptive context of the Dirac operator is not a wave-equation or energy-equation but rather the quantum descriptions of occupancies for spectral-values which cannot be calculated by the occupancy model of a quantum system. That is, the energy levels of a quantum system, to which one applies the Dirac operator, must be found by other methods and then the Dirac operator applied in a context of quantum occupation numbers and the annihilation and creation operators are thus associated to a wave-function, where the wave-function can now (in the context of a Dirac operator) have various internal particle-states, but this interpretation seems to not go anywhere, rather it is only empirical not predictive, ie there are not descriptions of stable patterns associated to the applications of the Dirac operator, though the literature is full of mis-representations of the properties of the Dirac operator, and subsequently, fake solution functions, supposedly derived from the Dirac operator, are expressed about systems, like the H-atom, but the Dirac operator does not commute with any angular-momentum operator which incorporates the spin-matrices in its construct.

Note that the Dirac operator (or the Dirac equation), though it is supposed to be Lorentz-invariant…, but because it is a new type of a mixture of operators an algebraic square-root of a metric-invariant operator, ie similar to the wave-equation operator for space-time, but which is represented as a scalar-product of space-time momentum 4-vector with itself, and which is represented as a mass equals energy relation, which introduces new types of operators into the set of partial differential equations…, but which (it, the Dirac operator) cannot actually be classified as being Lorentz invariant.

That is, (there is no (single) Lorentz transformation which applied when to the Dirac operator...., in regard to all factors for each of all the terms in the operator..., leaves the Dirac operator invariant.

That is, these square-root matrices are a part of the Clifford algebras, but not a part of the spin-cover group of a space-time transformation.

They are most closely related to the SU(2) x SU(2) double spin cover-group of the SO(4) Euclidean 4-space, R(4,0), where the Clifford matrices of the Dirac operator either multiply the separated SU(2) matrix components (within a larger dimension Clifford matrix) either by (1) or by (-1), and so they are not as related to the Lorentz fiber group SL(2,C) = SU(2) + iSU(2), or SU(2) x iSU(2) double cover of R(3,1), ie space-time.

Thus, they are mostly about mixing space-time, ie the variables of the Dirac operator are space-time variables which are being mixed with Euclidean space, and a new interpretation of the math patterns, in which there are new sets of pairs of metric-space states, which are associated to any (real) metric-space.

The simple-connected property of the fiber (isometry) group is all about the model of material-interactions in which locally the dynamic changes of a material-interaction are the inverse properties of the two opposite metric-space states, so that in a local neighborhood of a discrete change associated to a local model of a material-interaction, so that the inverse of one action in one metric-space state is the (direct) action of the opposite metric-space state material-interaction spatial and motion changes in a dynamic process, ie pairs of dynamic functions defined in a context of a simply connected fiber group are each locally the inverse of the other, and almost involution (where an involution is a function which is its own inverse).

That is, there are two functions which are a part of a dynamic process..., which are in the two opposite states which are a part of each real metric-space..., so that one of these functions is the local inverse of the other function, there is a two-state dynamic process, where the dynamic process can be related to a dynamic function, and where each state is the local inverse of the other opposite-state dynamic function.

The double cover for a simply connected fiber isometry group is the spin group, where this can be defined for various metric-function signatures,

Ie (where) for R(s,t), (s,t) is the metric-space's signature, note that space-time is R(3,1), and it is a math pattern which beckons into the higher-dimensional constructs of metric-spaces, ie to follow Coxeter's expressions of the math patterns of the discrete hyperbolic shapes up to 10-dimensional stable discrete hyperbolic shapes, where the hyperbolic metric-space is isomorphic to space-time, where a general space-time is R(n,1) which is an (n+1)-dimensional space-time. And this (n+1)-dimensional space-time is (isomorphically) associated to an n-dimensional hyperbolic metric-space. That is, for hyperbolic metric-spaces one projects-out the time component.

Also

Note that, the Dirac equation (or Dirac operator) does not commute with itself, but a quantum system's Hamiltonian (or the square-root of a Hamiltonian) must commute with itself, if the energy of the system is to be conserved, thus, the Dirac operator cannot be used to identify the properties of a bounded stable system, its descriptions of a, supposedly, stable system's spectra are both "not stable" and "the descriptions (predictions) are not reliable."

Thus, the Dirac operator has been placed into a context where the number of material-components, which are occupying a quantum system is always changing, namely, the context of occupation-numbers and creation and annihilation operators of unstable random point-particle events, which occupy incalculable spectral-states, since the canonical quantization of a stable many-(but-few)-body spectral-orbital quantum system cannot be formulated, furthermore, the process of counting sets of unstable random point-particle events is a context where counting is not reliable and so it cannot be related to a reliable probabilistic description, and

So even though there is a lot of effort put-into showing that the Dirac equation is Lorentz invariant, and that the spinor-operators, which are incorporated into the Dirac equation (and are the cause that the Dirac operator does not commute with itself), are in fact related to the spin angular-momentum of Fermions, which are, assumed to be, a part of a relativistic quantum system,

Nonetheless, sense the Dirac operator is not commutative with itself, this means that the Dirac equation will be a part of a quantitative containment set which is quantitatively inconsistent, so its descriptions are meaningless, ie the quantum systems which are formulated using the Dirac operator cannot be solved, so as to be related to any type of stable spectral patterns, and the measuring structures within the descriptive context are not reliable.

It (the Dirac operator) is (and also, the Dirac operator, as well as virtually all of: quantum physics, particle-physics, and general relativity, are all) only related to a context of empiricism, which has less value, in regard to science, than did the descriptions of the epicycle structures used by Ptolemy to "predict" the motions of planets in the heavens.

All of these, so called, highly rigorous scientific descriptive languages (quantum physics, particle-physics, and general relativity) are only related to the probabilities of particle-collisions in regard to the detonation of bombs, where general relativity is related to bombs in that it is about trying to detonate a gravitational singularity)

But only the "box" of component confining infinite potential-energy walls, and harmonic oscillator have been calculated, but/and where even the "highly touted" H-atom's solution function to the radial equation (of the H-atom) diverges, and it is not clear "what the actual physical context is?" in regard to "how a harmonic oscillating object can be physically realized?"

ie what physical context exists wherein a small material-object has a linear relation to force, as does the classical spring harmonic oscillator?

The reason that the laws of material interactions are not important, is because during either the interaction processes, or during collisions, (either classical or quantum), it is not the interaction-process through which a many-(but-few)-body system enters into a relatively stable system in a context of a bound state for the system components, but rather (in the new context of physical description) it is the relation (of resonance) which both the "shape of the interaction structure" and the "energy properties of the interaction-shape-spectral-complex" have to a finite set of spectral-orbital values [where the finite set of spectral-orbital values are a part of the new mathematical context, in which there is a finite spectral-set, associated to a high-dimensional containment set, and it is this finite spectral-set upon which are generated the set of quantitative structures upon which the description is built (and which are related to measuring)] so that an interaction which leads to a new material-system is a result of an original interaction process, in which an interaction-shape-energy-complex, and so that this complex also satisfies a set of: energy, spectral, and angular-momentum properties, so that when the complex enters into resonance with the finite spectral-set then it can (geometrically) deform and shift energy-spectral-occupation structures, so as to form a new stable geometric shape, which is a new stable material system, which is in resonance with some spectral-values of the finite spectral-set which is defined for (or defined upon) an over-all high-dimension containment-set,

And (or that is)

Where this new stable system's shape is consistent with the spectral-orbital properties associated to the finite spectral set defined within the 11-dimensional hyperbolic metric-space, which, in turn, is organized around stable shapes of various sizes and organized around shape and dimension and size containment relationships in regard to the various different-dimensional subspaces, and various subspaces of the same dimensional level, which exist within a high-dimensional set

When this resonance does not exist, during an interaction process, or does not occur because the set of energy spectral-set and angular-momentum properties are not satisfied, then the interaction, or collision, reverts to a regular classical interaction,which, nonetheless, may also be interpreted as a quantum relationship, for small entities, where this interpretation of an interaction as a quantum relationship is due to the relation of Brownian motions to quantum probabilities which E Nelson's has described

The interactions which are used in modern physics today (2014), and their carefully related invariance's are not all that central to physical descriptions, especially, in regard to the descriptions of the formations of stable many-(but-few)-body spectral-orbital systems (of all size scales), whereas the relative unimportance of the carefully identified interactions (which characterizes modern physics) is mostly due to the fact that most of these equations ((partial) differential equations).

Which are associated to the descriptions of material-interactions are related to the mathematically untenable conditions of these equations, being either non-linear, or non-commutative, or indefinably random, where any of these properties (for the equations associated to the implementation of, so called, physical law) leads to quantitative inconsistency (ie the number-scale is not defined, the quantities have properties of being both discontinuous and many-valued, ie the numbers are not well-defined), and which, in turn, causes such math descriptions to not possess any meaningful content,

eg the interactions are not the descriptive contexts through which stable material systems come into being, that is, this type of careful description of interactions has not resulted in any valid descriptions of the observed properties of stable many-(but-few)-body spectral-orbital systems.

Note that, general relativity is formulated not in regard to isometry fiber groups, but rather in regard to a diffeomorphism fiber groups, where the diffeomorphism group allows for arbitrary metric-functions, which can possess any signature,

Ie (signature of a metric-function) R(n) = R(s,t) where s is the dimension of the spatial subspace and t is the dimension of the temporal subspace and where s + t = n, so that (s,t) is the signature of the metric-space (or sometime, s-t, is called the signature of a metric-space's metric-function, but then n must also be given)

Thus, if it is claimed that a metric-function of a general metric-function, which is supposed to be related to the laws of general relativity and thus a general metric-function can be related to any signature, but then if the general metric-function is also claimed to be related locally to a (3,1) metric-function signature, either

(1) so that if one claims that this local property of the metric-function's signature is also needed in order for the geodesic to exist on some general shape then this can only mean that the shape upon which the geodesic is defined is a discrete hyperbolic shape, ie where a discrete hyperbolic shape is one of the main shapes in the geometrization theorem of Thurston-Perlman.

or

(2) For a claim, such that, a general metric-function, in the context of general relativity, is related locally to a (3,1) metric-function signature, so that locally (in a small neighborhood of a point in space) the acceleration can be linearly approximated as a constant, and thus, the description has a local general frame where the acceleration is a constant-value (where, it is being claimed that, the equations of material interactions [or finding a shape's (local) general metric-function] need to be invariant in regard to local coordinate frames which possess a constant acceleration) but a local metric-function with a (3,1) signature can be related to a discrete hyperbolic shape, but then there cannot be any general metric-function.

Indeed this is the type of discussion which results from a descriptive context in which the central (local) equations are formulated so as to be: non-linear, non-commutative, and indefinably random, in which case the descriptive language possesses no meaningful content, thus, the meanings of words (or the meaning of quantitative patterns) become arbitrary, and without any meaningful content, ie nothing can be resolved when the containment set is quantitatively inconsistent.

That is, the so called authority of the intellectual-class, and their so called leaders,

ie the very highest stars of the personality-cult associated to authority,

Express their ideas in an intellectual context of arbitrary abstractions, so that their, so called, rigor is only related to a world of illusions, and it is not true rather it is arbitrary, and

There exists a fixed core of the focus associated to the entire descriptive context (to only be focused) on particular instruments, which are related to the technical interests "needed" from science by the banker-oil-military ruling-class...

That is, the way in which the propaganda-education-indoctrination system is related to the technical interests of society is that it is fixed and dysfunctional and in this respect it is no different from the arbitrariness of possessing a racist viewpoint, which is OK or often desired, if such a viewpoint serves the ruling-class

That is, the authority of the intellectual class is as arbitrary as is the viewpoint of racism, and the intellectual-class of indoctrinated, or educated, people are the same sort of bullies (intellectual bullies) as are the emotionally motivated physical bullies, which the ruling-class also manipulates, and it is the arrogance of the big set (perhaps 5% of the population) of indoctrinated people, which try to exclude new ideas, so that in their arbitrary blind support for the educational-indoctrination...., which was provided for them by the bankers..., is the intellectual context which best supports the ruling-class and

It (this intellectual viewpoint, which best serves the banker's interests) is used (by the, so called, educated 5%) to oppose new creative contexts, where this arbitrary opposition is based on a false sense that the so called (indoctrinated) authorities are people who seek truth, but this is not true, rather

They (the highly-touted experts) are all either indoctrinated wage-slaves, or members of the ruling-class, and they are rewarded for expressing the ideas which support the ruling-class

This is the underlying assumption or viewpoint of the, so called, objective news-reporting, those who dispense the "objective news" (only the facts, and only what the experts say is true) are also required to shill for the expression of the idea that the (exceptional) highly-touted highest-ranking in the intellectual-class are the guardians of an absolute objective truth, and they (and the investment-class, whom these, so called, high-intellects work-for) are guiding the society into the best possible social context, for which there are no other viable (in regard to the dogmatic truth of the bankers) alternative possibilities, ie the claim of no viable alternatives, associated (in the media) to M Thatcher.

To believe in, and to follow, the authoritative intellectual-class of any particular time period is all about expressing the idea (which is placed by the propaganda-system onto M Thatcher) that there is the only way there are no alternatives, and it is the way in which the expert-intellectual-class is leading society, but the patterns of intellectualism (eg Godel's incompleteness theorem) have already uncovered the serve limitations of precise language, namely, that a set of assumptions and contexts and interpretations can only be related to a limited set of patterns, and the conclusion of this claim, is that precise language should inhabit not complexity (which has no relation to practical development) but rather simplicity at the elementary level of assumption, context, interpretation, new ways in which to organize patterns, etc etc. that is, both expertness and the context of learning need to be elementary and they need to be about equal free-inquiry where there is freedom of belief and freedom of expression but the context of the efforts is about developing practical creative contexts and "products" which are given to the earth and the community so as to help and they are given as gifts of appreciation. That is one wants the expression of arbitrary ideas to be placed in the place of relatively unimportant expressions of ideas while new ways of using language aimed at new creative contexts needs to be the focus of discussions in society

Intellectual efforts, especially efforts in regard to math and physics, are individual efforts, which are often about ideas (or conceived operations) which seem quite different from what the propaganda-education-system has us (the public) believe (or have been led to believe), is (are) normal viewpoints, but only a small set of fixed language relations which are related to particular instruments, are rewarded, so that truth is about extending the profit structure of the investor-class, where an instrument's relation to "profit obtained from investments of the ruling-class" define the fixed categories of acceptable intellectual social behavior, especially, behavior related to intellectual efforts and arrogant claims (by the, so called, experts within society) about truth (ie the set of experts who express all of their ideas (all their intellectual efforts are) about the development of the particular instruments, which, in turn, used by the ruling-class to derive their social power over the public

The society needs to be based on equal creative persons, which the society needs to organize so as to support the efforts of each person in an equal manner, so that the society is not a collective, which only supports certain sets of behaviors in a small set of fixed categories of doing using and behaving (all of which support the social-power of the ruling-class).

The laws of society need to be based on equality, and not on property rights and minority rule

Education needs to be based on equal free-inquiry and "the individually built relationships" between knowledge and creativity (is good to be expressed)

Trade is OK, but when certain products are popular, then collectivist structures can be built, but there needs to be bounds placed on collectivist behavior, so that each person's individual creativity is not opposed (however, all the selfish and perverted and the establishment of domination (or dominant behaviors) forms of creative efforts need to be opposed, that is neither value nor truth have absolute structures, especially, within society) and/or language development is free and equal, and all creative efforts which extend knowledge and extend new creative contexts, or whose motivation is to help others, are to be aided by the society.

One finds amongst the, so called, expert-class it is fashionable to portray oneself as an expert as being skeptical of the so called paradigm of the day, but nonetheless when such an expert discusses the subject matter over which they are authorities that they establish in their speech all of the fundamental sets of: assumptions, contexts, and interpretations of data, etc, which form the basis of the ruling group of the intellectual-class, the, so called, group expressing the scientific consensus, and furthermore, "the expert" most often both expresses skepticism, yet, nonetheless, validates the basic uses of language by the marketing-class, or the propagandists, in regard to the use of expert language by marketers as a basis for sales and social manipulations. That is money reaches deep into the behaviors of the experts so that they do (to) support the investor-class

This is very much similar to how the politician appeals to some of the interests of the public but who once in office never acts so as to help the public

The big business interests are always supported at the expense of the public, and the system is set-up so that when the public does add new value to the culture, such as a new alternative viewpoint about physical description, then these ideas are incorporated into business interests so as to shut-out social advantages for the originator of the new ideas so that the business interests are allowed to legally steal the development of actual meaningful cultural advances from the public so as to serve the ruling-class, eg Tesla, who seems to have had a different viewpoint than the dogmatic science community of the time, yet the ruling-class's propaganda-system put the obscure and mostly misleading Einstein as the dean of the illuminati or the intellectual-class, though Einstein had "difficult ideas about society" and he was excluded from the new illuminati, though it was only Einstein who expressed a wide array of different ideas which were often expressed at the level of assumption rather than technical babble, while the new illuminati were presented as the people who "always have the correct answer," ie those who could discern truth (namely, the truth, which the ruling-class wanted developed) and run by first, the wishy-washy Oppenheimer, and taken-over by the totalitarian scientist Teller, and whom (as a class of intellectual super-humans) worked to develop the bomb, ie the particular instrument to which the fake-illuminati focus their: narrow, obsessive, and memory-driven intellectual efforts, which have no meaningful content, where this is because the math structure upon which these pursuits of the further development of bomb instruments depends in non-linear, non-commutative, and indefinably random, and thus, the quantitative structure (within which it is assumed that the description is contained) is quantitatively inconsistent and possesses no stable patterns and measuring is not reliable.

These are issues about whether there is some other way which also deals in a rigorous manner with truth and provide an alternative, but where the message always given by the expert, who is a person who has been indoctrinated so as to have beliefs consistent with the ruling-class, and the claim made is that the set of alternative viewpoints are not rigorous, the other claim is that descriptive language built from different assumptions contexts and interpretations of patterns and data need to be peer-reviewed but where peer-review is about being published for being consistent with the established authority so as to extend the current authority and to further relate to the development of the particular instruments which a particular authoritative dogma is related, eg quantum physics particle-physics and general relativity are all related to the development of bomb technologies in particular the detonation of the mythological gravitational singularity

But

These authoritative dogmas, ie quantum physics particle-physics and general relativity are not true (where a more valid measure of truth, rather than a descriptive context being related to a particular instrument, is described below)

That is, rigor is the development of language based on sets of assumptions and contexts which are can be related to assumed truths which are not consistent with the world we experience but rather identify truths in a world of illusion

The point is that quantum physics particle-physics and general relativity are, supposedly, designed to describe the most prevalent and generally numerous sets of physical systems which possess stable measurable physical properties, such as the stable many-(but-few)-body spectral-orbital physical systems which exist at all size scales, and which exist in a wide variety of slightly different but nonetheless general contexts, but which have no valid descriptions based on the, so called, laws of physics which are associated to the so called scientific disciplines of quantum physics particle-physics and general relativity. If the laws of physics cannot be applied in a general manner so as to provide a valid descriptive structure of stable physical systems, which is both accurate to sufficient precision (where this phrase "sufficient precision" needs to be re-considered, in regard to the wide variety of data which is possessed about many of these stable and general systems) and so that the description lends itself to practical creative development, not only to the development of one particular instrument, such as quantum physics particle-physics and general relativity only seem to be relatable to the one instrument associated with a high-temperature explosions, and otherwise their descriptive contexts have no relation to the practical development nor to any new practically creative contexts, but rather these overly authoritative descriptive languages, which serve military monopolistic businesses so well, are only related to creative literary contexts which deal with a rigorous description of an illusionary world.

Chapter 23

Quantum physics and particle-physics

What began in quantum physics with an attempt to model (small) physical systems with harmonic (or probabilistic) function-spaces, and subsequently, Dirac operator's relation to unitary-invariant internal particle-states (spin), as well as a historic interest in the internal states of nucleons, was moderated (or adjusted) by general relativity's claims about the types of partial differential equations which can be used to model physical systems in a, so called, valid context, ie partial differential equations derived from the extrema of action-functions, where the action is based on gauge-or-fiber-group-invariant scalars, so that the fiber-gauge-group associated to (nucleon) internal symmetries, could define a derivative-connection which was defined to be associated to changes in "internal particle-states," and be a partial differential equation which can model quantum systems in a, so called, valid (equation) context,

so this was promoted so as to be the natural progression of the development of physical law

But

(1) Using the partial differential equation to model material-interactions so as to be the basis for describing stable, bounded many-(but-few)-body spectral-orbital systems, ie the elliptic types partial differential equation (which exist at all size scales), has failed, since the valid descriptions within this descriptive context only exist in the solvable context in regard to partial differential equation models of physical systems, and only the 2-body system of interacting material-components can be solved (separated) for a bounded stable system placed into the context of partial differential equation models of material-interactions

And

(2) internal particle-states are very local, and they have divorced themselves from possessing any type of (external) geometric relations, due to their only determining particle cross-sections, and their, primarily, being related to the properties of nucleons, and if there is geometry in the descriptive context, then it is the geometry of either electromagnetism or gravity, which are both

partial differential equations based on the geometry of the field-sources, and geometrically related field-particle densities, and these statistical field-particle models either only mimic the results of electromagnetism (but this seems to, actually, be a result of data fitting, the main attribute of probability based theories, and new adjustments are continually found until the probability based description can fit the data) and gravity, as represented by Einstein's general relativity, which also cannot describe a system with two-bodies, so it is irrelevant, but, the nucleus has no valid model of sources of fields, so as to allow a geometric model for the sources of the field-particles of particle-physics (upon which to base an ordered statistical structure of flows of field-particles in a system).

The point to be taken is, that particle-physics and all of its formal appearing rigor is, in fact, non-linear, non-commutative, and indefinably random (thus its point-particle collision models are not really allowed by the uncertainty principle, but this is ignored) and thus, it is quantitatively inconsistent, so it has no valid meaningful content, measuring is not reliable ie re-normalization, and there are no stable patterns upon which it is based or which the, so called, solution-functions…. of the partial differential equations of particle-physics…. can describe

And

The only reason for the study of particle physics is because it is related to the chaotic and non-mathematical context of a bomb's explosion, a transitory state in which there is a changing between two stable states of material, where this change is characterized by collision probabilities between the material-components which the unstable (transitioning) thermal state has caused the breaking-apart of the initial-states stable material structures and such a context only has a descriptive relation to particle-cross-sections, so particle physics is used in bomb engineering (the goal is to detonate the mythological gravitational singularity)

Quantum physics

The discussion of quantum physics begins with the belief in both (1) reduction of all material systems to elementary point-particles, whose random "point-particle-event identification in space" is associated to a spectral-value, to which the point-particle (or material component) belongs (as an energy spectral-level in the system into which the particle belongs), and that all point-particle events in space are determined by (2) "random distributions" of (or in regard to) the positions of any point-spectral-particle event in space, ie a belief that physical description must be about both reduction and randomness of point-particle-spectral events in space.

Nonetheless, the basis for description is about finding geometrically-motivated classically measurable properties (or variables) and representing sets of operators where these operators are (to be) based on these classically measured properties, where these (classically-measured motivated)

operators act on a function-space…., which, in turn, has become the way in which a (random) quantum-system is modeled…., by these classically motivated measurable properties so as to find the ("diagonal-set" of) spectral-functions (which solve the partial differential equations which are defined by the commuting set of operators) of the harmonic function-space (which models the quantum system), and the associated spectral-values (which help define the quantum system's partial differential equations), where the spectral-values are the measurable properties of the quantum system.

That is, quantum physics is still fundamentally geometric, and calculations, which are dependent on this geometry, are supposed to identify the quantum system's set of measurable spectral-values.

That is, To find a physical system's (both) spectral-values and "random event distribution in space" the quantum system is modeled as a function-space, whose spectral-functions, or infinite dimension diagonal representation of the function-space, is to be found, by first identifying a "complete set of commuting Hermitian operators" whose operators act on the function-space, which, in turn, determine sets of partial differential equations which model the quantum system, and whose (ie the set of operators) algebraic properties allow for (or result in) the function-space being diagonal-ized, ie one function (eigenfunction) for each spectral-value (eigenvalue), where the set of spectral-values are assumed to be the system's set of measurable spectral-values, which supposedly can be calculated, and, thus, identified.

The complete set of commuting Hermitian operators are found (determined) by first determining a set of classical measured properties, which (might) apply in regard to determining a quantum system's measurable properties…,

ie its measurable spectral-values associated to random point-particle-spectral events, and then representing those measurable physical values (or measurable physical properties) as quantum operators,

Where classically measurable properties can be related to quantum operators by means of Fourier transformations of the classical variables, usually the direction vector components of the classical linear momentum variables are Fourier transformed into becoming a partial differential operator in the direction of the linear momentum vector component, where the context of sets of pairs of Fourier transformed variables which are being related to operators are, in turn, related to the property of pairs of variable associated to defining the physical property of action, which is defined by a multiplicative-product of an action-pair (of variables), where this property of action is usually defined on infinitesimal elements of measuring, in turn, these pairs of factors define dual pairs of variables-operators,… in (or through) which the uncertainty principle of quantum physics is defined (on these dual pairs of variables which are related to the physical property of action)…, so that the dual pairs of variables are related to one another as being operators (in a quantum operator descriptive context) through Fourier transforms, eg so that the Fourier transformed momentum variable becomes a partial-derivative operator and in this way the classical variables are transformed into quantum operators. Furthermore these dual-pairs of action-related operators do not commute. This non-commutative properties of the fundamental set of classically-and-quantum-mechanically measurable properties (along with the (1/r) geometry of material-interactions (the potential-energy term) in the context of point-particle models of material) could be the source of the failure of this method (of quantum physics) to provide a formulate-able and solvable descriptive context for the most general and wide-ranging of the quantum systems, namely, the stable many-(but-few)-body spectral-orbital systems, and thus, the great failure of quantum physics as a valid descriptive context. That is, the descriptive context of quantum physics is fundamentally non-commutative (and thus quantitatively inconsistent, and thus, it cannot be used to identify stable patterns).

That is, the so called canonical quantization of a quantum system is done by turning classical measurable properties into (quantum) operators by means of the Fourier transformation of the classical variables, where these (newly defined) operators are related to the quantum system's measurable spectral-properties, through the algebraic principles which might be associated to the finding of a set of a quantum system's "complete set of commuting Hermitian operators," through which the function space is diagonal-ized, and subsequently, the set of spectral-values associated to the quantum system are found.

But for the wide range of general, stable many-(but-few)-body quantum systems, which possess measurable sets of spectral-orbital properties, this descriptive context cannot be realized, for these types of quantum systems the system equations cannot be formulated, and the system's (hypothesized) partial differential equations cannot be solved.

The only quantum systems which can be both formulated and solved are:

1. the box with infinite potential walls,

2. the harmonic oscillator (but it is not clear what, actual, linear model of a force which exists, and which can cause such oscillations for a particle-component which exists in any particular localized position),

Note: this (the harmonic oscillator) is the context where annihilation and creation operators are first defined, and, subsequently, occupation-numbers are related to bound energy-levels (originally) defined by (or within) a parabolic potential energy geometric-shape.

That is, the physical context is not well defined for a harmonic oscillator, and this becomes a basis for an arbitrary empirically determined energy-level, where there is both energy-level occupation and the numbers of point-particle-components in the energy-level can change numbers, ie the energy-level is arbitrary and the occupation stability of the any energy-level is not defined.

and supposedly,

3. the H-atom but the solution-function to the radial-coordinate-component of a separated (Hamiltonian) wave-equation for the H-atom diverges, but by artificially truncating the diverging series solution function of the radial-equation, the energy levels of the two-body H-atom seem to be identified fairly well.

But this is not math, it is not a valid math model.

That is, something is wrong with the description, based on solving a (parabolic type partial differential equation) Hamiltonian energy-operator, which is the H-atom modeled as point-particles for a two-body system of point-particles related to one another through the 1/r geometry of the potential-energy term in the Hamiltonian wave-operator, thus, the r=0 point in space, is a point in space which can be occupied by point-particles, and this is (seems to be) the main cause for all the math problems with this model, as well as all the models derived from it, eg particle-physics, and subsequently, string-theory.

But quantum descriptions were also motivated, at an early stage in its development, by the particle-collision model of probing a nucleus, by E Rutherford, where the nucleus is also modeled as a point-particle, even though it is composed of many different point-particle components, (and the, subsequent, properties of the particle-collisions came to be the main focus of particle-physics so there came to be a model of the particle-types (of the nucleus) which compose the many-(but-few)-component small systems, so that certain particle-types could be modeled to change between themselves, so as to form inner-particle-state vectors, which change between the different particle-types (which the particle-state vector identifies), so as to do this in a manner (or due to a transformation matrix) in which energy of the system is conserved, and for the unitary-energy-invariant transformations of quantum descriptions, based on energy-operators, this was the sets of operators which acted on the infinite-dimensional function-spaces, but (it was considered that) this could-well be (unitary) operators acting on finite sets of particle-states, which would be transformed by the finite-dimensional unitary matrices, and this has been the model chosen for particle-physics, and placed into the context of a force-field related derivative-connection structure within the energy operator,

Thus, there was developed the nuclear-force-field model related to changes in the nuclear-component particle-states, ie the main idea was reduction to point-particles, even though these point-particles (seen coming from nuclear-particle collisions) were mostly unstable.

An unstable event is not a valid basis for a statistical (or probabilistic) model of anything (ie unstable events cannot be counted in a replicable manner, ie the counting of, so called, distinguishable "events," which may or may-not exist, is not a valid basis for building a number-system). Yet, this was pursued, because it was a context related to particle-cross-sections, and subsequently, it is related to the context of an exploding bomb, and, thus, the changes in internal-particle-states was a means to model the rebounding particle-products produced from particle-collisions, so as to be able to determine particle cross-sections, though (again) these particle-collision products (resulting set of reflected particles and their angular distributions) are mostly unstable events, ie they decay in very small intervals of time, that is, they behave (ie their unstable event structure) as if they are ghost-particles, which are "not really there," and these "highly speculative models" of the rebounds of particle-collisions, are used to determine particle cross-sections, for the, so called, particles, which are really unstable (fleeting) vaguely-defined random events is not really science.

That is, though such speculation is a valid activity of science, but when it is a ("math") model, which is not mathematically sound [ie non-commutative, non-linear, indefinably random, for speculative unstable events, in which the main set of descriptive attributes pertain to hypothetical-objects which cannot be observed] it is certainly not science,

Where science is supposed to be about valid math models, for observable (and measurable) properties, which if, so called (peer-reviewed), science is only about hypothesized unobservable properties, then the description must be relatable to some wide-range of useful practical contexts, for both description and practical creativity,

But where none of these criterion are actually met by particle-physics (or quantum physics, or general relativity, or any derived speculative theory), thus, there needs to be a great many of other alternative ideas expressed about this same subject matter, eg about why the nucleus is stable, and why there arc so many-many stable many-(but-few)-body spectral-orbital systems which exist at all size scales.

The, so called, (peer-reviewed) scientific context is now (2014), formally, only about the models which interest the military big-business investor-class, and the, so called, (uncritical) scientists (who help this investor-class), and the, so called, consensus which is claimed to be reached by the (thoughtless, non-critical) wage-slave science community, is not science, but rather it is an expression of religion, ie the faith that the, so called, science-community..., ie paid scientists working in science institutions, ...has (possesses) in the military investor-class,

ie the religion of personality-cult

ie worship by the institutional scientists of the ruling-class and their faith-in their paymasters.

That is, the data of elementary-particle tracks was obtained from the particle-accelerators, and these tracks seemed to be characterized by their emanation from particle-collisions, so that particle-states were invented to coincide with particle-collisions, so as to model the particle-angular rebounds (particle-scattering-angles), "which are observed between colliding-particles" in the detection-chambers of particle-accelerators, so that these changes of particle-states were modeled so as to be associated to particle-collisions and the angular distributions and particulate constitutes of the elementary-particle products of (high-energy) particle-collision, (but these collision geometry schematics can also be derived from elementary orthogonal faced simplexes which are the fundamental-domains of lattices defined on metric-spaces, ie related to the discrete subgroups of SU(n)'s), so as to form models material-interactions based on particle cross-sections related to elementary-particles modeled through their particle-collision properties and, their particle-collision rebound structures, which exist between field-particles and the material-particles (or Fermions, though some field-particles can also be Fermions), where the algebraic structure of "changes of internal particle-states," associated to "collisions and angular scattering particle structure which is "observed" between Fermion-particles with field-particles," (where "observed" has quotes since the "observed" properties are for unstable (quickly decaying) (not necessarily) events),, was put into the (descriptive, or measurable) context of derivative-connections, (in what is really a speculative non-measurable descriptive context), where derivative-operator terms were found in (or put into) energy-operators (or into Dirac operators), so that the rebounds of the scattering elementary-particles, were modeled to be (or are due to both) particle-collisions and changes of internal particle-states, which lead to the final "reflection-angle structure" of particles rebounding (or scattering) from collisions, in regard to the various patterns of changes in particle-type; from before and then after a random particle-collision event, so as to determine the cross-sections (or probabilities of collisions) of (unstable) elementary-particles, and thus....,

Note: This is not science.,

To plunge deep into a (into an invalid) statistical model in a descriptive structure, which is non-commutative and indefinably random (ie the elementary-events used as the basis for the probabilities of particle-collisions are unstable events), so that the descriptive structure cannot describe stable (math) patterns.

That is, the particle-collision model of material-interactions is only related physically to an unstable and indefinably random context of chaotic particle motions (or in regard to a classical field, possibly a regular flow of motion of the field-particles from a geometrically modeled source (modeled from electromagnetic-theory), though it may only be an impulse of a force-field, ie a solitary (or spherical-shell) wave-front, ie another geometric-model, that is, the statistics (of random particle-collisions) only has relevance if there is an associated geometric-model to which the statistics can be related) and the collisions between (unstable) particles can (only) be statistically related to particle-cross-sections (measures of collision probabilities for a particular particle), and thus it is a description based-on determining the statistical magnitudes of field-interactions, based on the number of such field-particle-Fermion-particle collisions, in a non-commutative and indefinably random basis for descriptions. The meanings of the description are arbitrary, and solely related to fitting data, and related to the cross-sections of particles, where the cross-sections are to be primarily used in models of bomb explosions

Though the energy of a collision with a, 550 nm wave-length, light-wave is, according to the de Broglie matter-wave-relations, hf=E, about 2 eV, and the binding energy of an electron in the H-atom is about 14 eV, ie they are comparable energies, but the spectral-energies are less than the binding energies of the system, but in regard to classical force-fields, particle-physics does not add anything different from regular quantum physics, that is, to say that the statistical field-particle-theory only has meaningful statistical structures if is first depends on a geometric model upon which to base (within itself) an ordered statistical structure, where such statistical-order is needed for geometric descriptions, ie the statistics of particle-physics is manipulated so as to fit with the observed data, (but) furthermore, the operators of particle-physics are not assumed to conserve energy, in regard to the system which they describe, ie the Dirac operator does not commute with itself, thus, the energy of the system, which the Dirac operator is being used to describe, is not conserved, and the unstable context of elementary-particle collisions also means that the "individual elementary-particles are unstable" so the whole of (all) particle-physics descriptions is (are) only about a small region associated to (or surrounding) a particle-collision,

ie it is only a description which possesses an unstable relation to the cross-sections of unstable elementary-particles, which are related to models of unstable forces, which are very short-ranged constructs.

It is (or the descriptions of particle-physics are) unrelated to classical geometry, except the geometric model of particle-collisions (a geometry which is inconsistent with the uncertainty principle), and how particle-state changes affect the geometric-angles of the particle-collision rebounds or scatters, (eg of the Fermion and the field-particle [or any other elementary-particle] with which it (usually the Fermion) collides).

But the statistical properties of particle-collision-interaction is related to Fermions which are, supposedly, already contained in spectral energy-levels of a quantum system, so it is an interaction-property, which can only adjust the energy-structure of the quantum-system within which the Fermions are contained (based-on unstable collisions with (ghost) particles in the immediate region of the Fermion), ie there is no systematic relation to the system (ie to the system's energy-levels), rather only local statistically vague relationships to particle-collision cross-section, when the assumption of continual particle-collisions in a quantitatively inconsistent descriptive context needs much more than (simply) assuming such a local chaotic state exists when one considers how little the model of particle-physics actually explains (other than rapturous expositions, by the expert propagandists, about the beauty of math patterns, in a quantitative context which does not allow stable patterns to exist).

That is, the particle cross-section cannot be related in any meaningful way to determining the stable properties of a material-system (because its descriptive context is the vague statistics of collision-probabilities in a chaotic state) rather it can only adjust things at a statistical level (ie the statistical-level of particle-physics) which is less well defined (ie defined in a very small region) than the statistics of Brownian motion, where Brownian motions form a (the) statistical context of regular quantum physics, ie the sets of commuting operators which act on harmonic function-spaces, where the (algebraic) aim is to identify a quantum system's spectral-energy levels,

ie the statistics associated with (occupying?) the harmonic-functions form a subset structure to the more dominant sets of measurable properties (or sets of commuting operators)

ie the statistics of the particle-occupancy which is associated with (occupying) the harmonic-functions form a subset structure to the more dominant sets of measurable properties (or sets of commuting operators)

And

Furthermore, the changes are mediated by random collisions of field-particles with the material or Fermion-components, where the Fermion components are contained within energy-levels of quantum systems, where the field-particles may of may-not possess a vector (or internal particle-state) structure of their own

But this descriptive context has not been successful.

The primary and wide ranging sets of stable many-(but-few)-body spectral-orbital systems, though, are observed to be stable and uniform, so as to only depend on the number of components which compose such quantum systems, but otherwise, when the number of components is fixed, then these systems all have exactly the same spectral-properties.

This means that these systems are forming in a casual and solvable, geometric context, not a random, chaotic, and quantitatively inconsistent context.

The new context

In turn, this means that the descriptive context is the very limited context of the very stable geometric contexts of the very stable discrete hyperbolic shapes, which are defined over a range of 10-dimensions.

This geometric context needs to be organized into a language so that physical description can be based on these stable shapes defined over many-dimensional levels.

This has been done by: M Concoyle, G Coatimundi, D Hunter, B Bash

(see these books at Trafford.com).

Otherwise

Physical description sinks into irrelevant complications, which exist in a quantitatively inconsistent context, that is, the math structures of quantum physics, particle-physics, and general relativity, are non-linear, non-commutative, and indefinably random, and in this math context quantities: (1) do not have a valid scale defined and (2) the quantities are discontinuous and (3) the quantities, eg coordinate functions, become many-valued, and this all means that such descriptive context cannot possess any meaningful content,

ie the descriptions are arbitrary, and they are based on randomness, so they become arbitrary, but apparently quite complicated, models for fitting data, just as Ptolemy's epicycles were used to fit data, and the discussion is about delusional contexts,

Eg particle-state transformations are being used to adjust the energy-levels of a wave-function… (where there are no energy-functions of any important quantum-system which actually exist which can be adjusted,

eg the general and stable many-(but-few)-body systems,

ie it is a discussion for which no substance (or no subject) exists) (adjust the energy-levels of a wave-function) by after lengthy careful descriptions in which the quantum system's component's cross-sectional properties are adjusted, so as to fit the observed energy-level data,

ie a probability based description is always about adjusting the description to better fit the data, rather than talking about substantial (or stable, and measurable) patterns which can be identified, and controlled and measured and used, instead there is a discussion of "complex data fitting techniques" which are unrelated to either stable patterns or to reliable measuring, but one really wants to fit stable system-properties together in a control manner, in a new creative context, instead of there only being a vague discussion about uncontrollable and un-useable probabilities of unstable properties, where these complicated discussions are very much similar to considering "how many angels can fit on the head of a pin," ie it is a delusional (formally complicated) and meaningless discussion.

To elaborate on the complicated descriptive context of particle-physics (a math context which has no capacity to possess any meaningful content)

This model of internal changes of particle-state was brought into its formal context through the Dirac equation or the Dirac operator

Dirac tried to take the square-root of the "energy subtract the kinetic-energy equals the mass" operator, where this is an operator which is represented in its (classical) quadratic (squared) metric-function for the 4-momentum local vector structure, so the energy spectra of this energy-operator would be the mass, so as to get a first-order linear differential operator, which is Lorentz-invariant in the local 4-momentum coordinates, but the square-root was itself an (Hermitian algebraic, or Clifford algebra) operator structure (whose transformations are related to a unitary group, ie [SU(2) x SU(2)]) so that the square-root operator does not commute with itself (due to the Hermitian algebra matrices which are a part of the square-root operator), which means (can be interpreted to mean that) its energy is not constant,

ie it fits into the empirical occupation-number model of a quantum system, and the fiber groups (for the Dirac operator) are {both} ([SU(2) x SU(2)] x {and} [SU(2 x iSU(2)]) which are the spin groups (or the simply connected covering groups) of [(SO(4)) x (the Lorentz group)].

So the Dirac operator cannot be called Lorentz invariant, since it depends on two fiber groups, and it is not based only on the local 4-vectors of space-time (or the 4-vector of momentum), ie it is outside of the context of only the Lorentz group, but it led to both the empirical "occupation" and annihilation and creation operator model of quantum states, and it led to more complicated fiber group structures for the principle fiber bundle models of quantum systems, where the local base-space, associated to a fiber group, (what is usually considered to be a metric-space) are now (in particle-physics) the sets of internal particle-states of the particle-components which compose the quantum system, where these particles are operated-on by the creation and annihilation operators.

However, the context of the fiber bundle especially, in regard to how to organize and interpret, and, subsequently, how to model the observed property of spin, what is its correct descriptive context how should its math structures be organized in relation to the physical properties of metric-spaces of various metric-function signatures, that is what is the (its) relation between a metric-space and its simply connected covering group, in regard to Euclidean metric-space structures which relate to both spatial positions and spatial displacements, and its relation to dynamics, and its relation to the square-root of the generalized Laplace-operator here (where the generalized Laplace-operator is {here}, [defined to be the metric-invariant partial-differential operator for any signature metric-space]), so this would mean that the Dirac operator is the square-root of a generalized Laplacian related to the local coordinate transformations of R(3,1), ie of space-time.

It seems that the square-root operator is more closely related to SO(4) than to SO(3,1), and this can be interpreted to mean that the simply covering space of the fiber-group is most naturally related to Euclidean space, where the most fundamental type of opposite metric-space states can be locally identified in regard to dynamic properties which are most elementarily defined in the context of local spatial displacements.

That is, one might conjecture that this is in regard to pairs of opposite spatial displacements which relate to inertial processes of change, and also being related to pairs of opposite metric-space states, being the real relation to the locally measured properties of a metric-space so that there are two states of each dynamic change, which are opposite, and determine "inverse dynamic constructs" inverse spatial displacements in regard to the time associated to the period of a spin-rotation of the opposite metric-space states of each metric-space, where these opposite spatial displacements are locally connected, so they relate to the simply connected model of the metric-space's local transformations in a dynamic interaction related to geometrically measured properties(this is the type of elementary discussion about spin-properties which is needed in physics, and it is a context which needs to be discussed by 7 and 8 year-olds, since their opinions about these issues are as important as any experts opinion, especially, since the experts have driven physics into meaningless babble)

Back to the particle-physic's interpretation

The complicated pattern of several separate, seemingly, unrelated fiber groups, associated to the same base-space, as the operator and fiber-group structure of the Dirac operator are interpreted, was developed in regard to, the momentum-related partial-derivative in the square-root Dirac operator, which was (can be) related to a connection-derivative, associated to the electromagnetic-field, and this was a 4-vector potential-energy construct (so that this electromagnetic-model of a connection commutes), but this idea for a derivative-connection was extended in particle-physics, so as to model a collision-interaction of a system's particle-components with field-particles which, in turn, is associated to internal-particle-state changes of the system's particle-components, but this model of a connection is non-commutative and it is non-linear.

The fiber groups of the internal particle-state construct are related to these non-commutative derivative-connection constructs, which model field-particle-collision material-interactions in an occupation-number and annihilation and creation operator model of a quantum system, which is supposed to be the occupation-number of some wave-function for a stable quantum system

ie but this can only be either

(1) the fake model of the 2-body stable H-atom

or

(2) a quantum-box model, but

In these two quantum systems only the spin-property manifests, except for photon-collisions, and, possibly, phonon-collisions, (furthermore, the weak-force is most often ignored for the H-atom). However, the angular-momentum operator for the H-atom cannot have a spin-operator context, since such an operator would not commute with the Dirac operator. Otherwise, none of the internal particle-states manifest in either space and time, or in space-time, in these two systems so the inner-particle-component model of a wave-function has no measurable physical meaning in either the H-atom or in a (quantum) box.

Particle-component structures only have some vague meaning in particle-collision experiments in particle-accelerators, or in nuclear quantum-systems, where in particle-accelerators the uncertainty principle of quantum description does not allow the observed particle-tracks, which are seen in detection cloud chambers, and where the spectral-values, ie the masses of the particles, of the particle-interactions cannot be calculated.

The claim, which seems to be made, is that the cross-section of a particle is affected by particle-collision interactions, in regard to internal-particle-states, so this change in a component's cross-section in a random particle-collision model of material-particle-interactions can affect forces between electrons and protons due to changes in the probabilities of field-particle collisions with a quantum system's material components, due to the changed value of a component's cross-section.

Note: The particle cross-sections (or probabilities of collisions) is the only quantity which seems to be to be of central concern, and in particle-accelerators it can be empirically calculated, and it is of central importance in particle-physics, and it is a statistical property (models can be adjusted to fit empirical evidence), and it is the only property of any significance in regard to bomb engineering.

In a descriptive context where there is only occupation-numbers and annihilation and creation operators which are most often (only) associated to collision processes, in turn, (where the components, upon which the collisions are defined, are [empirically] placed within an energy-level of the quantum system, so that collision-interaction-processes are) associated to empirically determined energy-levels for a quantum system, and the interaction structure is based on random-collisions between Fermions (material components) which occupy the energy-levels of the system, and field-particles, then the magnitude of the force between particle-components would be related to "the probabilities of particle-collisions" and this is determined by both the particle cross-sections and the density of the flow of field-particles at the position (or at the energy-level) of the Fermion, where, supposedly, the energy-level determines the types of field-particles, which are related to the description of a Fermion which is occupying such an energy-level (of that particular energy-value).

But this descriptive structure of material-interactions cannot describe how a system originally comes to be bound together, Instead it is only related to a state of chaos, where there are unstable transformations associated to broken-apart stable systems, and the rates and energies (or temperatures) of material-component collisions, either as the original system is made unstable so as to disintegrate, and thus to give-off energy, or as the system cools so as to resume a new stable state.

That is, cross-sections as a model for material-interactions, which result in a system being bound together is a much more primitive descriptive context (or a description based on even less information than a classical..., or a classically motivated identification of energy-operators..., descriptive context)

That is, particle-physics provides a more empirically determined descriptive context than do the other descriptive contexts of physics (ie classical physics or regular quantum physics), and these other (ie classical physics or regular quantum physics) descriptive contexts are more sophisticated descriptive contexts, but even these descriptive contexts can only relate the 2-body systems wherein material-interactions are being modeled, so that the bounded property of the system is being caused by the descriptions of material-interactions.

The descriptive language of particle-physics is less capable of compressing information than the other types locally measurable models of physical properties of material components.

Particle-physics is a description of randomness (in regard to): motions, positions, collisions, and internal particle-states; the particle which is being described (which is assumed to be a material-particle, and thus, it is a Fermion), in a descriptive context of particle-physics, is a particle which is, supposedly, occupying a particular energy-level, but there is no reason that it is in any particular energy-level, so that the main property of the point-particle-component in an energy-level is its cross-section (which apparently is affected by its internal particle-state structure, which manifests during particle-collisions), since it is this cross-section property which connects the particle-component to the random structure (of being related to random collisions with field-particles) of which it is a part, in the vague descriptive context of particle-physics.

That is, particle-physics introduces more components to the descriptive structure, and the information is more difficult to organize, or to cause a necessity for the system to logically reduce (the description) to a descriptive context of a system entering into a stable state, especially, when the causes are all random events

This statement is proven by the fact that particle-physics has not been used to describe any stable system.

That is, such descriptions have little meaning when there are stable systems which imply a geometric context, where geometric patterns give a useful descriptive context, ie geometry allows information about (the descriptions of) physical systems to be compressed, whereas randomness can only be manipulated to fit observed data.

That is, time going forward and backward, and focusing on particular types of particle-collisions are aspects of physical description which can be manipulated so as to fit data.

Yet, there is nothing which is understood about any of these particle-states, including a particle's spin properties, except that they are based on a fictional (or a hypothesized un-measurable) model to find elementary-particle cross-sections, which can be manipulated in arbitrary ways, so as to fit data, eg re-normalization and identifying non-existent particles adjusting non-existent mathematical-models of stable spectral structures of quantum systems, namely, the very stable many-(but-few)-body quantum systems of observed spectral properties which exist widely in our experience, (the elementary-particles are known to be unstable and they are, really, unstable facial-geometric aspects of broken-apart stable shapes) associated with particle-states, with non-existent (or no valid models of) forward and backward time-states

In particle-physics, there is a lot of discussion about the form-invariance of the equations, [as well as symmetries of time-direction charge-sign (charge conjugation) and spatial orientations (parity symmetries)], where these equations are based on operators which have been, essentially, obtained from the Dirac operators (eg derivative-connections, which are really associated with decay processes (as in the context of explosions) (and they are not about random changes of internal particle-states except in regard to determining a particle's cross-section)), but (or and) since the context of the equations is: non-linear, non-commutative, and indefinably random, this means that the quantitative containment context is quantitatively inconsistent, so that, in turn, this means that there are no stable patterns, which can be related to such a descriptive context, and there is no reliable measurements, which can exist in this context, and furthermore, the events are unstable, and the spectral-energy-mass values, in regard to each of the (hypothesized) various internal-particle-states of the (given) component (but where very unstable decay patterns are taken as some sort of proof that the hypothesized particles are actual entities) and whereas predicting the masses of the various particle-states should be of central importance, but they cannot be calculated.

Thus, there is very little reason to deeply consider a descriptive context, where there are no stable patterns, which can be described, and only a vague relation, through a component's cross-section, to the random events surrounding the component, where a material-component is assumed to be a Fermion.

That is, a Fermion exists in an energy-level of a general quantum system (which cannot, in general, be calculated) and is an assumed to be material-particle which is a member of a set of its own internal particle-states (or possible particle properties of itself) (each of which possesses a mass, where the different masses (would) identify the spectral structure of internal particle-states, but where the masses of each of the different particle-states cannot be calculated [by using (or solving for) the energy-operator for the system which the Fermion is a part]) the position (in space) and motion of the Fermion is only relatable to a vague probability distribution, (but where the, assumed to exist, functional property of probability distributions seems to not be available for careful descriptions based on the, so called, laws of physics) and the Fermion is, supposedly, colliding with field-particles, where the type of field particles, apparently, depends on the energy-level of the Fermion, and the strength of the material-interaction due to the collisions with the field-particles, where the number of collisions depends on both

(1) the density of the field-particles at the Fermion's position (and its motion? And/or its system energy structure, apparently, a property which seems to primarily to be about the Fermion's kinetic energy) and

(2) the Fermion's cross-section, ie probability of colliding with a field-particle,

(3) where particle cross-sections are the main (and only) focus of particle-physics, in particular, in regard to the details of models of elementary-particle-collisions, both the field-particle models, based on a collision-determined material-interaction, and their relation to changes of a Fermion elementary-particle's hypothetical changes of their, assumed to exist, internal particle-states, where these constructs are hypothesized so as to fit the particle-collision data seen in particle-detectors in particle-accelerators

That is the only property which particle-physics focuses upon is the cross-section (or probability of collision) of elementary-particles which are related to the so called particle-collision model of material-interaction processes including the changes of particle-states during particle-collisions which is the basis of material-interactions in particle-physics where changes in particle-states during particle-collisions is the mechanism by which both particles re-bound, or glance-off, one another during a particle-collision,

So as to be able to count both(1) collision numbers and (2) "collision reaction (or re-bound) angles" during a particle-collision

The number of various sized-angles of collision-rebound-angles depends on the size of the, supposed, sphere-model of a particle which is a part of a (Fermion) particle-collision of a collision's material-component target, eg more wide angles means bigger cross-sections.

That is, the so called scattering-matrix associated to both "existence of" and "changes to" (hypothesized) particle-states during particle-collisions... is a model related, primarily, to identifying particle cross-sections.

The issue of being able to identify all the particle-tracks in bubble-chambers (or particle-detectors) of data obtained from particle-accelerators that exist in regard to collision energies of particle-accelerators.

It is not clear that this model can be claimed as being verified, since the events are nearly always unstable (random) events (ie they are events which should not be counted in order to determine a set of probabilities for sets of well defined events) and since the, so called, (hypothesized, elementary-particle) quarks are unrelated to direct measurable properties, furthermore, "as many" particles of high-energy are gluons (as there are types of the hypothetic quarks) and gluons are the hypothesized mediators of the so called strong-force (which it is claimed holds the nucleus together but there are no math models based on the model of the strong-force which can actually describe any property of a nucleus which has more than (say) 5-nucleons, ie beyond a stable helium-atom's nucleus.

The gluon supposedly interact with one another in a multi-component context of gluons the, so called, color-interactions of the, so called, strong-force but one does not hear about the detections of this relatively big set of various types of gluons or are these gluons always virtual as the photons hypothesized in collision-particle-changes-of-state photons (which are also never seen)

The descriptive context of particle-physics is about a chaotic transition process, characterized by random collisions of "Fermion and field-particle" particles, which are related to average temperatures of this chaotic system, where the temperature is related to the rate at which stable components are broken apart (the stable components, which are from the original stable system-state, which has, subsequently, become unstable so that random collisions characterize the chaotic transition state, this is the model of there being a system through which one can detonate a bomb, so that the chaotic state of collisions and dispersions is what cools the initial rise in temperatures, this is all statistical and closely related to component cross-sections,

But

Now the stable system is also being modeled as being in a state of random chaotic particle-collision probabilities based on a state of collision-rate determined transitions which is supposed to result in a stable context for a system, but this only makes sense in regard to a cooling and dispersing context

But

How can such a model of interactions as being in a continual state of random transitioning processes also remain in the same state of heat-energy? What is confining a random and dispersive process which either liberates energy or cools down the random component motions

It is claimed that this confining or binding of components is caused by the strong-force or by electromagnetism

But electromagnetic forces, modeled as particle-collisions, are only related to small adjustments to energy-levels, in regard to the methods or processes, hypothesized by particle-physics, where the stable energy of the quantum material system's Fermions are assumed to be in a stable energy-state established by regular quantum physics

But, essentially, there are no such descriptions of stable energy-structures for quantum systems which exist within regular quantum physics there are no descriptive structures which successfully apply to the general stable many-(but-few)-body spectral quantum systems.

Thus, the descriptive context, I. which was originally geometric and causal, in classical physics, was reduced to

II. A semi-geometric descriptive context, in the descriptive context of quantum physics, where geometry is related to the potential energy term, which is modeled after measurable physical properties of classical physics,

and where the point of the descriptive context was to determine the energy-spectra of the quantum system.

Note that the quantum description is less causal (depending, passively, on finding the set of algebraic operators), which can be used to identify the system's spectra, but this depends on being able to find sets of commuting operators, rather than (as in classical physics) simply identifying a physical system's geometric context and then identifying a local coordinate context for measuring the geometrically motivated measurable properties,

However,

As in classical descriptions there are very few examples of identifying a bounded system's natural interaction context for separate components binding together to form a stable system in regard to a description of material interactions based on geometry with a 1/r (or 1/r^2) binding potential-energy term especially for a system-model where the components are point-particles, these types of interaction force-fields can only be solved for a 2-body type system or for some clear geometric pattern related to a system's components being bound together in a system eg a harmonic oscillator with a parabolic potential energy function translated below the x-axis of an (x,y)-graph. to now

III. in particle-physics, where there is a Fermion, which occupies and energy state, which is not calculable, and this energy-state, apparently, is related to the type of field-particles which are present (in the system of which the Fermion is a part), while the Fermion has associated to itself an internal particle-state structure which is activated, in a local descriptive context, by collisions with field-particles, so that the strength of the field...., in a random field-particle-collision model of a material-interaction between a Fermion in an energy-state and field-particles, which, in turn, relates the Fermion to its internal particle-state (intrinsic) structure [where the strength of the Fermion-(particle-state)-structure]...., is determined externally by the density of the field-particle-flow around the Fermion, and internally by the Fermion's cross-section.

Note that the geometric relation of the field-particle source to the Fermion is not an intrinsic part of this "physical" model, rather, it is a part of the wave-function model of the quantum system which is supposedly provided in II, but this cannot be true, since II identifies a context in which it is assumed that both the energy and the component structure of the quantum system are conserved, while particle-physics is based on everything changing in a chaotic context, with no geometric structure, certain types of field-particle flows (where the field-particle types assumed to exist is to be based on the energy-level of the description) from unknown sources, and only vague relations of collisions between the Fermion with these field-particles based on collision probabilities, in turn, based on the particle-cross-sections of the Fermions and the particular-type of field-particles, and internal particle-states (of the Fermion and [often, also] the field-particles) which a formula defines, a formula which is all about continual changes to the system's energy and component structure.

This is a description about the changes between stable states ie initial stable state and a final stable state, in a chaotic transition process, which is based on particle-collision probabilities, in a context where the structure of the original stable state is being broken-apart by the high-energy particle collisions, in regard to the particles liberated by the original set of destabilizing particle-collisions, which are related to average values of random (number of) events, ie it is the description of an explosion, and it has nothing to say about the structure of stable physical systems, except in some random context which can vaguely relate this description of an explosion process to some physical property eg particle cross-sections.

The solvable aspects of classical physics, which are in a context of geometry control and causality, are the preferred descriptive context, and the existence of stable spectral-orbital systems composed of many-(but-few)-components implies that it is only in a solvable (and/or geometric) context where there is geometry and control, and there is a context in which a stable set of patterns for physical systems can exist.

The proof of this is simply that the current descriptive structures have failed despite there great relation to the propaganda-education-indoctrination system, and thus there is a lot of high-valued, so called, intellectuals (who dominate the propaganda system, and) who will not let go of their worship for the set of illuminati, which have been put forth to them (as representing high-intellectual-value by the propaganda system

But the Socratic method or equal free-inquiry, is a method equally accessible to anyone and when expressing ideas at an elementary level of assumption context interpretation it works better than does a system based on the aristocracy of intellect it would be better to have the 7 and 8 year olds (who have taken an interest in these matters) lead us, rather than the fake-science and math of the illuminati and the banking-oil-military big business interests whom these, so called, illuminati serve.

In general relativity, the, so called, local symmetries, ie the local invariance of a partial differential equation ie equations which are invertible at a single point for arbitrary frames of motions, means that the physical properties can be very different in different regions of the descriptive domain space, eg the tangent vectors at one point are quite different from the tangent vectors ate a distant point in the domain space (or the base-space).

But

In particle-physics local symmetries, ie the local invariance of a partial differential equation, ie local gauge invariance (or local unitary invariance), but the system's properties do not change in distant regions of the domain space, ie there exist the same changes of the internal particle-states at distant points, but where the model assumes that the quantum state which the collided-with particle occupies is at a higher energy-level than are the energy-change relations to internal particle-state interaction constructs, so this is a local invariance property, and not about its (the occupying particle's) relation to distant energy levels, and not about its relation to distant regions in the domain space (or base-space), where the same relations to a distant (or different energy-level) occupying particle would (also) exist, even though the invariance is local, as the invariance in general relativity is also local.

Since particle-physics is primarily about the nucleus,

ie it adapts its statistics to the electromagnetic geometry,

And in the nucleus,

the material-interaction is modeled as a point-particle-collision, this means that the description has no "place" within the nucleus from which the force-fields of particle-physics can emerge as sources, other than the source of the force-fields being the particle-collisions themselves.

This is because the point-particle-collision implies a place for the collision within the nucleus (about 10^(-15) m) which by the uncertainty principle (or deBroglie wave-relations) estimates identify an (kinetic) energy for a nucleon of about 10^(-10) J = 10^9 eV, which means that, within this particle-collision model of interactions, the energy levels within the nucleus seem to be only the (kinetic) energy associated to the nucleons themselves.

However, the nucleus does possess stable energy-levels, related to high-energy electromagnetic spectra, eg gamma-ray spectra, the energy-level ranges can be from 2-40 x 10^6 eV, (data from nuclear physics text) where these energy-levels are assumed to result from some box-like potential-energy structure associated with regular quantum physics, but this requires a cause (a descriptive reason) for such a box-like potential-energy structure, which the particle-collision model cannot provide. That is, the nucleus is a stable many-(but-few)-body spectral-orbital system, and thus, requires a more detailed description than a mythological box-like potential-energy geometry, ie a box-energy-geometry which would be shredded by point-particle models of the nucleus (due to the uncertainty principle).

On the other hand there are very good (or much better) alternatives to the current physical description dogmas, especially, concerning any stable many-(but-few)-body spectral-orbital systems

{while confinement of physical models to stable geometric shapes, within which there can exist both stable orbits, and a surprising charge-distribution (or orbital arrangement), "on a 2-dimensional topologically-closed shape (when viewed from 3-space)"}

While (within particle-physics)

Electromagnetism is defined to have a global-symmetry relation in regard to its U(1) fiber-group, so that electromagnetism maintains its classical context when described in quantum physics, ie the U(1) connection in a field-particle context, is no different from (provides no-more information than does), say, the Dirac equation with an electromagnetic field connection, which (also) identifies the geometric context of classical electromagnetism, (but the Dirac equation does not require that the system it describes "to be at a constant energy" [since the Dirac operator does not commute with itself] as do either the Schrodinger equations or the Bohr equations (ie these two equation-types {Schrodinger and Bohr} do commute with the energy-operator])

When solvable formulations of electromagnetic systems either classically or in (regular) quantum physics exist, then these two descriptions can provide some useful, ie both accurate and practically useful (or at least, fairly accurate), information about the system

But

These instances are rare, ie models of material interactions based on partial differential equations provide too limited of a range of useful descriptive information, and those based on probability are both less accurate, and they are almost entirely non-useful in regard to practical creativity

And

Neither general relativity nor particle-physics provide any useful information about any systems, but the most irrelevant systems in regard to (wide ranging) general accuracy and practical usefulness.

Interpreting the fiber groups of particle-physics (the order based on unstable point-particle-events) in relation to stable shapes

The symmetries of the internal particle-states are relatively simple, ie internal states (of a random model of material) are "invariantly" transformed by, U(1) x SU(2) x SU(3), and thus, these (claimed to be) unstable states of material, are apparently really related to the simple stable geometries of the 1-dimensional, and 2-dimensional, and 3-dimensional rectangular faced simplexes of Euclidean space, ie SU(n) is naturally related to SO(n), where SO(n) are the Euclidean metric-invariant local coordinate transformations, where SU(n) relates the simple rectangular faced simplexes to a set of moded-out orthogonal circle-space shapes (because complex-coordinates are based on lines and circles, thus, allowing (or demanding) the geometries of moded-out rectangular-faced simplexes in the complex-coordinates to have a geometric and quantitative relation to circle-space shapes), but whose sizes can exist in a continuous context, thus the particle-collision empirical measurements can be interpreted to mean that the fermions and/or photons exist in a context of toral components which do possess particular sizes, ie particular masses (or for photons in lasers particular frequencies) and that certain high-energy collisions can break-apart these otherwise stable shapes (but where the oral-components are actually toral-components of the very stable discrete hyperbolic shapes, and the particle-states are related to the different sizes, or the geometric-measures of the rectangular simplex's lower dimensional faces, of the toral-components of the discrete hyperbolic shapes of the material-component being collided in a particle-accelerator.

Rather

There is some interest in the problems of the square-root of the Hamiltonian (or Dirac operator) in regard to the nature of the square-root operators, as Clifford algebra elements or as SO(4) operators, and the relation of a metric-space, itself, to the properties of spin-rotations of metric-space states, so that components which are stable metric-space shapes can themselves possess the property of spin and furthermore that the property of metric-space states may be deeply related to the relation of dynamics to Euclidean space and the relation of space-time to an equivalent hyperbolic metric-space so that the stable shapes of the discrete hyperbolic shapes, ie the shapes of most importance in geometrization of Thurston-Perlman, are the natural stable forms of energy into which either metric-spaces or equivalently material-components can be organized.

Note

The Dirac operator cannot give stable energy estimates of the H-atom's energy-levels, due to it (the Dirac operator) not commuting with the angular-momentum operator, which includes spin-angular-momentum term, rather the electron can be assumed to be in a Schrodinger (or Bohr-Somerfeld) determined H-atom energy-level

Chapter 24*

An alternative viewpoint about chemistry and lattice structures

An alternative viewpoint about chemistry and lattice structures

[which can be related to both material-components, and to the lattice structures associated with the states of matter (eg solid, liquid, or gas, or condensed-matter, and the fluid-material)]…, but our totalitarian society does not allow these types of highly informed expressions of new ideas,they are new ideas which are excluded by the institutional intellectual-class, who serve (or worship) the "high-value" defined by the investor-class (our totalitarian society does not allow these types of highly informed expressions of new ideas, they are new ideas which are excluded by the institutional intellectual-class, whose high-salaries and their media-image of possessing high-social-value are contingent upon their serving "the social structures and high-social-value which have come to be defined by the investment-class.")

The set of institutionally identified intellectual-class does not define science as a method of inquiry,

ie equal free-inquiry as Godel's incompleteness theorem demands for an axiomatic based language, rather they define science as form of authoritative power, supposedly, defined by a, so called, "consensus of ideas" into which the investor-class has invested.

(begin) An alternative viewpoint about chemistry and lattice structures

Chemistry is about atoms and molecules, their thermal context, their thermal context is traditionally described in regard to lattices, and the reactive context of atoms and molecules (or their context of formation),

ie their formations, and the contexts of their functioning as independent components which can react,

Ie in regard to their reactive context of, lattices, and their thermal properties.

The Chemistry of atoms and molecules (or chemical reactions) are modeled as being based-on there being a collision relation between chemical components and their formation (or their chemical reactions and their relationships to the how the thermal-properties of the system affect chemical changes), where the collision relation (or collision-rate) is associated to both the thermal properties of the system, so as to be associated to chemical-reaction properties by means of the rates-of-collisions of the chemical components, as well as the energy of these collisions, where, in turn, these thermal (or statistical) properties are being related to reaction-rates, but chemistry (can) also have a simultaneous relation to a [higher-dimensional] lattice structure, whose dimension is the dimension of the (original) metric-space, within which the chemical component-shapes are contained, (the lattice-dimension [as well as the dimension of the metric-space] is a higher-dimension than the dimension of the chemical-component-shapes, which are dimensionally related to the lower-dimensional facial-components of the given energy-lattices which are associated to a thermal system which is composed of a certain type of material)

There is a (clear) deep relation between (crystal-like) lattices and the atomic hypothesis, ie nuclear, atomic and molecular structures (as well as crystal-structures) are based on energy-lattices, where the relation between the energy-lattice and the chemical components of the thermal system will [needs to] be made into a math model (in this paper), where (in the new descriptive structure) the different properties of the chemical components (which are related to the faces of the lattice) of the thermal system, and their lattice origins, are mainly being determined by two different adjacent dimensional structures, as well as the different topologies (eg simple connectivity, as well as the dimensional level which allows the existence of a shape, or a component) in the different dimensional spaces of such a new type of a component-lattice model.

This relationship between chemical-components and lattices is traditionally seen in regard to:

1. Crystal-lattice structure is seen most definitely in crystals, whose lattice structure, traditionally, emerges primarily due to a model of the crystal being within a mythological box geometry (of a potential energy term representing the geometry of the potential-energy walls of a box, in an energy-operator) which, in turn, is modeled as a (moded-out) periodic crystal-box, math structure (where moding-out a lattice to form a shape, is about the topology defined by an equivalence relation between the faces of the box, which, in turn, is hypothesized to contain the crystal),

where

Then, to refine the description, the crystal is modeled as a set of local elementary-cells, which define another periodic-structure, where these local crystal-cells are claimed to be defining the natural periods (repeated patterns at (of the) crystal lattice sites) of the crystal, where this extra periodic-structure refines and adjusts the crystal's energy-level structures, so as to be adjusted from the more uniform box-potential-energy solutions,

yet

the big-box seems to give essentially, the same math model for crystal-lattice properties (or used to provide answers to questions about the properties of the crystal-lattice), particularly in the case of a free-electron gas within the crystal. But for either the box potential or the different set of bounding relations defined by the crystal's elementary-cells, the math result is dependent upon there being several (or two) sets of standing waves each associated to a (slightly) different periodic structure

The basic assumption, used in regard to describing the properties of crystals, is that there are at least two (slightly) different (periodic) lattice structures (or energy-partitions associated) to a crystal, which determine its energy-structure and subsequent physical properties of conduction, heat capacity, etc.

But, why will a crystal organize itself into these particular periodic structures in the first place?

Is it primarily about the boundaries of the crystal, and then a slightly different local-regional size of the crystal's components, or is it caused by some other fundamental aspect of the organization of material-components, which is causing the elementary-cells to exist,

ie are the boundaries of the crystals forcing the local crystal-cell structure to possess the particular local crystal-cell properties, or is the local crystal-cell structure fundamental, and the crystal's shape then conforms to the local (periodic) crystal-cell structures, or can it be the result of various combinations of these two main ideas, ie sometimes the macro-size of the crystal is the determinant cause and sometimes the local regional size is the determinant cause and sometimes there are several energy-lattice structures which exist in a (general) crystal?

2. But, when modeling an ideal gas, the probabilities of the components, which compose the gas, are traditionally required to fit into an imposed hypothetical (energy) lattice structure (an arbitrarily imposed energy-partition), which defines the set of energy-levels into which the, hypothesized, chemical components occupy, where the occupation is (traditionally) based on Fermi or Bose statistics, where this model is consistent with the observed thermal properties of an ideal gas system (ie sets of average measured values, which are averaged over the many-components, which, in turn, are hypothesized to occupy the thermal system made-up of a closed box which contains an ideal gas) of an ideal gas for the component-types of Bosons or Fermions.

The math model is the model of a system containing box which defines a rigid potential-energy function which is an infinite wall so the energy-levels are determined by the integer-related periodicity within the walls of the box, so that, an energy-partition is then defined based on partitioning the (positive) integers. Then the energy-partition is filled, essentially, from lowest energy (smallest integer-values) to highest energy, but the harmonic-wave solution of the box does not (necessarily) imply there is a well defined idea about a lowest energy state in the box, rather, this is an assumed aspect of the partition of the integers.

3. The chemistry of fluids (or liquids) is often modeled as an ideal gas, and the usual possesses the periodicity of a system contained within a box potential-energy (shape), and

Where long axial-molecules, which can be used to define a liquid crystal, also (often) align themselves in lattice formations.

But

The chemistry of life, which often occurs in a liquid, is baffling, since the idea that (or the research process in which) one first identifies molecular-components, and then one tries to correlate these molecules to certain types of life-system-processes is first questionable (ie it is at best a random correlation) and

Second, it is a process which simply leads to infinite complexity, but the living-system, in turn, seems to exist within a highly controlled and precisely functioning system

Where this precise control really implies a simple controlling context for the entire system.

What is the state of the so called identified chemical components (which are being correlated with life-processes)? Are they stable chemical entities of a local region, or do these chemical entities…, which are a part of that same local region…, change rapidly? Are they transported, or are they suddenly not as prevalent in a local region without any relation to their being transported?

Life

That is, the simple context (of identifiable chemical-components in a local region [of a living-system]) can be related to a lattice structure from which chemical components seem to both suddenly emerge, and then suddenly disappear, perhaps on "command" from an even higher-dimensional lattice structure, but where living-systems (seem to) need to be modeled by (an at least) a 3-dimensional shape, which exists in a 4-dimensional metric-space, within which mass, which is related to either inertia or spatial-position, is to be modeled as a Euclidean math-construct,

ie the inertial properties and spatial-position properties of a living system are contained within R(4,0), with fiber group SO(4) = SO(3) x SO(3), which, in turn, acts on R(3,0) x R(3,0),

Where there is a common "2-plane" in both R(3,0)'s, and which is a part-of (or distinguishes) the two sets of inertial properties, of either one of the two R(3,0)'s, but (it can be hypothesized that) each SO(3) is associated to a different form of material-type, ie each different material-types is defined in a separate 3-space so that the separate 3-spaces are each associated to different types of energy-lattices, which, in turn, are associated to two different material-types,

(1) one material-type is inert, and

(2) the other material-type is naturally self-oscillating (generating its own energy), ie the shapes possess the property of having an odd-number of holes in their shapes which when all the (related) 2-faces are occupied with charge causes a charge imbalance so as to cause a natural oscillation to occur.

In this structure there exist the set of 2-dimensional facial-components of the (energy) 3-lattices, but now there are two different sets of 3-lattices, each associated to a different material-type, where the inert-type of material-components are what form the molecular-components, which are being detected (and then vaguely correlated with processes in the living-system), while the other type of material can generate its own energy and thus the different spectral properties of its different 2-faces can be related to causing resonances from a selected set of spectral values so that this control over spectra, by means of the directed flow of a shape's energy (still the mystery of living-intent, ie a mental construct which is associated with a control over directed energies), can influence both sets of 2-face system-shapes through resonances.

The inert material-components can be influenced by the other (or activated) spectral structures of the energy-generating type of material, and this indicates that the chemical components can emerge from, and be absorbed back into, the lattice properties of the 3-lattice's 2-face structures, and there may be many more inert-material lattices present than "the assumed set of two," which is the basis of the traditional descriptions of solid-state physics. That is, there can be many sets of 3-lattices associated to the two-sets of different material-types and which are all associated to the living-system. Thus, the 2-faces which can emerge from such a descriptive context can, as 2-components, seem to join together in many ways and when doing this could actually affect the structure of the sets of 3-lattices which compose the living-system, but they could also temporarily join-together to form a needed shape and then when the process (which needs the particular

2-component-shape in order to function, or to allow a process) is finished then the temporary mixtures of 2-component-shapes can return undisturbed to their original 3-lattice structures, due to the untangled threads in regard to the new model of the neutrino or the new model of an electron-cloud.

But there is another aspect of the descriptions of living-systems which can be considered in this new model of the solid-state (or new model of the lattice structure of material) and that is about the relation of the stable shape of a cylindrically shaped DNA molecule. Namely, that the cylinder-segment-lengths along the cylindrical DNA molecule define the physical properties in relation to the 1-lengths which, in turn, define the spectral properties of the 2-faced related 2-component-chemical structures in the living-system, which emerge from the life-system's 3-lattice structures so that a living-cell has its own relation to the spectral properties of the sets of 3-lattice structures which are hypothesized to compose a living system, where within the cell there is the structure of an oscillating (or energy-generating) 3-lattice's fundamental-domain, so as to as to transmit lattice-spectral properties between the two types of material which composes the living-system

This hypothesis about cylindrical-segment-lengths of a DNA molecule relating to the spectral properties of the set of two different types of material 3-lattice structures is supported by the mysterious fact that though the molecular structures can be identified and composed in the lab the initiation of the functioning of these molecules only occurs after there is both embedding within a cell and the subsequent starting-up of the molecule to function only after electricity is applied, a clear sign of a dual set of living-structures.

Living systems and their energy-lattice context (to re-iterate about life-forms)

The model of a living system, and the relation that its living-system-processes have to chemical components, might be better modeled as the living system's relation to an extensive set of energy-lattice structures, where the chemical components can both emerge from and retreat into, various aspects of the sets of energy-lattices, as these chemical components are needed at the various time-periods of these living processes, and this is done in a context of non-locality, ie the chemical component structure is not so much about getting chemical component to the right place at the right time, but rather activating lattice properties at the right position in the system, at the right time, so that the chemical components can emerge from the lattice at the right place and not need to be transported there.

Where these activations would be done through an energy-body. An energy-body for a living-system results from a living-system actually being a 3-lattice whose shapes are realized in a 4-space, so that the material-inertial properties of the chemical components, associated to these 3-lattices, ie the thermal properties of the living system, exist in R(4,0) or Euclidean 4-space, where the fiber group of Euclidean 4-space is SO(4) = SO(3) x SO(3), so that there is a common 2-plane to these two 3-space local transformation groups, where these separate Euclidean 3-spaces compose 4-Euclidean-space, but one has the third direction emerging from the common 2-plane being associated to the idea of inert materialism, while the other direction in 4-space (emerging from the 2-plane) is associated to a new material, a material which naturally oscillates and creates its own energy, so that this energy can be used to activate different energy-lattice types associated to the inert-material structure of the life-form.

Thus, there are two sets of 3-energy-lattices, which can both be associated to a living-system, and which exist in 4-space, but because the 4-space splits into two separate 3-spaces we are left with an illusion that we only experience 3-space, ie the 3-space containing inert material.

However, since these become different 3-spaces (separate and independent, except for the common 2-plane) then each 3-space would also have a relation to an independent energy 3-space.

Note that in 4-space the two sets of 3-lattices would be simply connected in either 3-space or 4-space and thus,

when 4-space is viewed as a pair of different sets of energy 3-lattices, so that the neutrino thread-like component connections (explained below) can give the component 2-faces of these 3-lattices a non-local context, yet these 2-dimensional components can still be in a simple shape relationship with the 3-lattice, because the threads of the unbounded shapes stay untangled due to the fact that the 3-lattice is simply connected (as a 3-space) (or in 4-space) thus the 2-components associated to the 2-faces of the energy-3-lattice can stay in a simply geometrically connected relation with the 2-component's original energy-3-lattice, which can be expressed immediately, because the thread-like connection to the 3-lattice is simply connected.

though the two 3-spaces are not really independent, since they share a 2-plane, which could have implications about the relative sizes of the 2-components of the inert material and the (size) of the 2-components of the oscillating material shapes (relatively stable shapes).

Non-local properties of material systems

Furthermore, it should be noted that, in traditional (solid-state and regular) physics, there are several examples of components which seem to couple to one another from a distance

The, so called, electron-pairs of a superconducting material, which by being coupled become Bosons, so that they can occupy (or fall into) a single lowest energy-level of the energy-lattice, so as to flow freely in the crystal, or so the model claims, [but this model of superconductivity identified a critical temperature above which superconductivity is not supposed to exist, but this (predicted) critical-temperature has been surpassed, thus negating the model, and attempts to save this model, in turn, predict other properties which are not seen]

And

The so called, entanglement quantum constructs associated to both Bell's inequality and to A Aspect's experiment, which showed that things can exist in non-local relations, and physical (quantum) properties are transported instantaneously in regard to the use of information about the quantum system's state so as to connect a non-local relationship ie a state seen here requires, if the states are still connected, a state at a non-local place (it demonstrates action-at-a-distance within Euclidean space).

Are there other alternative models of this phenomenon of non-local coupling?

Yes

One can model a: nucleus, atom, molecule, crystal, as well as a fluid, and a metric-space containment set, as being related to lattices, where each of these sets can be modeled as a 3-dimensional energy-lattice, but since the metric-space containment set is the energy-lattice which gets represented as a (stable) shape (whose shape is identified, or defined, in a hyperbolic metric-space of dimension-4, within which this shape can interact with other material of the same dimension and of about the same size) then this means that only stable shapes associated to these types of 3-dimensional lattices which are, say, within an order of magnitude of 4 of the size of the solar-system's lattice's fundamental domain, ie the size of the original metric-space containment set, are the only sizes of these shapes which can interact with the solar-system's metric-space shape, and this is interpreted to mean that the smaller lattices tend to keep their lattice structure, rather than become 3-dimensional shapes contained (as 3-shapes) in a 4-dimensional metric-space.

However, the 2-faces of these smaller 3-lattices associated to: nucleus, atom, molecule, crystal, as well as a fluid, can appear in 3-space as (or turn their 2-faces into) 2-dimensional component-shapes, but

one of these 2-shapes would (could) be an unbounded shape, which is a (new) geometric model of a neutrino (in the lepton category of stable components), where such semi-unbounded shapes (as such a model of a neutrino) must be a part of a system which is composed of both electrons and protons, apparently, the stable 1-dimensional shaped (positive and negative) charges which can be a part (as a face) of a 2-dimensional discrete hyperbolic shape, can be related to an unbounded 2-dimensional stable shape, and this is a natural model of a neutrino which, in turn, is associated to an electron-cloud, where an electron-cloud allows a material-component, which is composed of the two charges, to interact with the other components contained in the 2-shape's adjacent 1-dimensional-higher material-component containing metric-space

The shape of the neutrino is semi-bounded, where the bounded aspect of this unbounded 2-shape represents the set of bounded faces of a material-lattice, which, in turn, represent the set of 2-dimensional component shapes, whose shapes are contained in the original metric-space, while the unbounded aspect of the neutrino is, in turn, bounded by the boundaries of the fundamental domain of the system's 3-dimensional lattice structure, so that within the 3-lattice the topology is simply connected (ie there are no holes in the space).

Thus, the 2-face components can separate along the different sets of toral components, associated to the bounded parts of the 2-face, and yet still be bound together in 3-space, so as to be bound together within the original simply connected 3-dimensional metric-space, or simply connected 3-dimensional energy-lattice (or a local metric-space defined in relation to the system's 3-lattice)

Thus, there can seem to be sets of independent components of the system, which are in fact bound together in the original 3-lattice within a simply connected metric-space, by the strands (or unbounded threads, which extend to the boundary of the simply connected metric-space) of the neutrino's geometric connection to both (1) the bounded part of the lattice and (2) its metric-space containment set.

In such a non-local context in which a system is both a set of separated 2-components, which are nonetheless connected together in their 3-lattice, by means of the geometrically connecting strands which are, in turn, associated to the boundaries of the system's containing metric-space 3-lattice. Thus, the 2-components can move into and out-of the lattice, so as to still remain a unified system, but whereas they can, apparently, also divide into seemingly independent components, which are nonetheless still connected to their material-defining 3-lattice, so as to be able to return to its (equivalent) lattice structure, where the simple connectedness property means that when the 2-components return to the 3-lattice along the geometric "threads of connection" they will not be tangled threads, ie the 2-components can return without any geometric changes to the geometric relation to their defining 3-lattice.

The structure of material components, to which material systems can be reduced, are the stable shapes of the discrete hyperbolic shapes which appear to be a set of unbounded strings of (an infinite extent) torus, whose toral-components may be folded, where these folds are based on a discrete structure of change, which can be associated to the Weyl-angles of the fiber-groups, and they are easily related to the set of semi-stable shapes of the discrete Euclidean shapes, or tori (or doughnut shapes).

The main categories of these reduce-able material components are the 1-dimensional shaped charged material shapes

eg Dehmelt's attempt to hold an electron stationary resulted in a solitary orbital electron shape, where the protons, are the components of the closed and bounded nucleus, while the electron are associated to a, so called, electron-cloud, which is defined about a nucleus, but both the nucleus and the electron-cloud are related to the neutral component shapes of the neutron and the neutrino, respectively.

It should not be allowed that this data to not be related to (or not be interpreted to be related to) quarks, since quarks are outside of observable science, and their only relation to physical descriptions is their relation, through models of elementary-particle cross-sections, is to nuclear-bomb engineering, and this is in regard to the determination of elementary-particle cross-sections for elementary-particles, which, though momentarily identified, are unstable events, so they cannot be related to the descriptions of stable patterns, and it is to the descriptions of stable patterns which both physics and math should be primarily concerned.

One cannot take seriously particle-physics when the elementary-particles upon which it is based are either unstable, or un-observable eg the quarks are unobservable, and where the main issue of particle-physics is to determine elementary-particle cross-sections, which, in turn, relate only to the single-point-particles which exist in a solitary manner within an (incalculable) energy-level, so that the system is to be defined at only one point, where it is modeled to interact through particle-collisions whose primary particle-scattering properties are exclusively related to determining particle cross-sections, so as to only be statistically related to the force-field interactions by their probabilities of particle-collisions, but in which there is no systematic statistical basis for a field, ie the force-fields are unrelated to geometry (except the derivative-connection of an electromagnetic field), so (otherwise) they possess no statistically identifiable properties, ie particle-collisions are randomly applied to a single-point, this is only about a reaction-rate, in regard to a set of arbitrary particle-collisions at a single-arbitrary-point in a box, where the reaction-rate determines an explosion, a chaotic transition process which possesses no attributes of geometry, and otherwise it is a descriptive context which is useless, yet, it is being used as a model of material-interactions most particularly for the forces within an atom and nucleus, but if there are no descriptions (predictions) of stable properties of the nucleus, or (similarly) for many-particle stable atoms, which emerge from such a descriptive context then the descriptive context is useless.

Instead try to relate an energy-lattice to non-local component relations, which, nonetheless, emerge locally from an energy-lattice, and then can be absorbed back into the energy-lattice,

where this is a more interesting model of existence, which is consistent with stable math patterns, which are quantitatively consistent patterns.

In chemistry, within a liquid (or fluid), where the energy-lattice is not so fixed, or rigid (as one would expect in a crystal), then a 3-energy-lattice can have sets of paired-2-faces, which can be attached to the 3-lattice by untangled neutrino-threads, so that the components associated to the 2-faces can be somewhat independent within the liquid-system..., so as to behave chemically different depending on the 2-face structures which emerge from the 3-lattice...., but the 2-components are always non-locally (and in an untangled manner) related to the 3-lattice, so as to be able to change their 2-face identities at the site of the 3-lattice, to become different component-types in the liquid, and function differently chemically, but to return to the 3-lattice, so as to be able to undergo yet another change and, again, function differently as

The components, related to elementary-particles, which need to be accounted for, and thus they need to be modeled as stable shapes, are the:

protons,

neutrons,

electrons, and

the neutrinos,

where the proton and neutron are related to closed and bounded (or compact) system stable shapes of the nucleus, while the electron and neutrino are related to a semi-bounded set of stable shapes.

The neutrino, with its small mass, is related to an unbounded type of discrete hyperbolic shape though it may be bounded by the metric-space within which the system of which the neutrino is a part of an electron-cloud type system, whereas the photon (or light) would be an unbounded shape which is not bounded by the metric-space, within which it may be temporarily contained.

It is likely that the neutron and neutrino are examples of systems composed on at least a pair of a proton and an electron, where in the hadron-case the shape is bounded, but "in the lepton case" the shape is semi-bounded

The relation of the neutrino to the boundary of the system containing metric-space (or metric-space's boundary) could be used to create a model of the nucleus where the charges of the electron may be both at the center of the nucleus, as well as at the boundary of the system containing metric-space, ie the 3-dimensional lattice so that the protons are in (folded) orbits, whose shape has high-energy, being pulled within the nuclear system to both the electron center and the electron boundary of the stable metric-space shape

While for the atom, the nucleus is at the center, and the leptons are bounded by the metric-space defined by the solar-system shape of the stable spectral-orbital structures of the atom, where at the boundary of the this shape (it could be that it) also defines the boundary of the neutrino, where one might hypothesize that the charge at the neutrino's boundary, ie on the system's 3-lattice, there would be the positive charge that is within the neutrino,

But

The molecule can have a similar, but a different set of folded and orbital structures, so that the neutrinos are still relating the charge distribution of the molecule to a positive charge at the boundary of the 3-lattice which defines the stable metric-space shape of a molecule.

That is, the component-shapes, which may be related to the unbounded neutrino, ie unbounded when it (the neutrino) is identified as a 2-face of the system's stable shape, but which is bounded by the system's 3-dimensional containing metric-space shape, but this metric-space is to be thought of as a 3-lattice, which is spectrally consistent with the 3-lattice of a stable molecule, which can be quite a diverse, and there can be various sets of many different molecular-component shaped 2-face structures, which the neutrino binds into a stable system shape of its (the molecular-shape's) 3-lattice by means of its semi-bounded properties.

In this section shapes are identified in their very simplest form ie cubes instead of polyhedron for which all of whose faces are rectangular shapes, etc

If existence possesses order

(as opposed to arbitrariness, as in the math contexts where the math patterns are based on… either a set of interaction partial differential equations, or a set of operators representing a set of the system's measurable properties, and they are operators which act on a function-space-model of a quantum system, so that these descriptive contexts of a physical system's measurable properties are (or happen to be):

non-linear,

non-commutative, and

indefinably random

set of math operators (or math patterns) which are being used to describe the physical system's measurable properties) and the order, which is observed to be a part of existence, is describable in quantitative context, where the stable quantitative math patterns extend beyond the quantitatively consistent shapes (consistent with the axioms of the real numbers) of lines circles eg the complex-numbers are the stable shapes, then such a quantitative and measurable description is to be based-on simple cubical and (associated) toral shapes (along with shapes composed of toral components, as well as cylinders) and such an existence is also modeled to be many-dimensional, [because otherwise]

(1) so that the descriptive context is capable of describing the observed stable patterns of existence, and

(2) Coxeter has shown that the very stable discrete hyperbolic shapes, the basis for a math description of stable patterns, are defined in the dimensional range for hyperbolic metric-space from hyperbolic dimension-1 to hyperbolic dimension-10, ie in hyperbolic metric-spaces.

Then there exist relations between cubical lattices, and their manifested (or associated) set of toral shapes, where these shapes are contained in the "one-dimensional higher (ie adjacent metric-space in regard to dimension) metric-space," wherein the lattices are contained,

ie the 2-square can be mode-out to form a 2-torus (or doughnut shape) which fits into 3-space,

so that, as lattices, the math properties of the metric-space (within which the lattices are contained) is simply connected (ie the space does not contain any holes), and this property of being simply connected also applies to the lattice,

ie $R(n,1)$ is simply connected (or equivalently, an $(n+1)$-dimensional space-time is simply connected), but the space-time metric-space, $R(n,1)$, is also equivalent (as an isomorphism) to an n-dimensional hyperbolic metric-space, which is also simply connected.

But the moded-out shapes…, associated to the lattices…, where the shapes associated to the lattices are defined in the adjacent higher-dimensional metric-space (shape)… lose their topological property of being simply connected when viewed from the metric-space within which the moded-out shapes are defined (or are contained), ie the torus has a hole in its shape which causes it (the torus) to not be simply connected in the metric-space within which the toral shape is contained.

The lattices in the metric-space within which they are contained, are simply connected spaces, while the moded-out shapes, associated to the lattices, are not simply connected within the adjacent 1-dimension-higher metric-space containing space for the shapes.

That is, when the property of the simple connectedness of the moded-out shape is determined in the 1-dimensional higher containing metric-space (but which is also a moded-out-shape), then the (moded-out) shape is not simply connected.

The metric-space within which a set of small lattices are a part, is a much smaller lattice…,

ie the fundamental domain of the larger lattice is much bigger than the fundamental domain of the smaller-partitioned lattice, so that one can conjecture that when the fundamental domains of the same dimensional lattices are either 4-powers bigger or negative 4-powers smaller, then the associated moded-out shapes whose sizes (ie sizes which are within this range of sizes), are comparable to the size of the fundamental domain of the moded-out metric-space shape…, and all of these different sized moded-out fundamental domains are all contained in the next higher-dimension sized metric-space…., then the shapes in this range of sizes will interact, in regard to a significant force between the shapes of such sizes (as the sizes of these moded-out shapes, which are within this range of sizes, and all contained within the next-higher-dimension metric-space shape), but the smaller fundamental domains which are outside of this size range, will not mode-out and subsequently, interact in any significant way with these larger sizes.

So it is natural for these smaller sized lattices to stay in their lattice form.

ie only shapes of a comparable size will actually interact (as the equivalent of material-components) to any significant amount.

These small lattices are of the same dimension as the (original) metric-space within which they are contained, so that they keep their lattice properties within this original metric-space, and one of these math properties is that, as a set of lattices of cubical-related shapes, they are (all) simply connected (as lattices in their original metric-space).

But

The faces of the lattices, eg the 2-faces of a fundamental domain of a 3-dimensional lattice, can manifest as shapes within the original metric-space.

The shapes of these moded-out faces can be related to unbounded shapes which might appear to be shapes which possess a set of 2-dimensional sets of unbounded strings, associated with what would otherwise be (if associated to a bounded lattice) a toral-components, but where there can also be a bounded part to the lattice so that the bounded part of the 2-face could be moded-out to form a set of connected toral-components, which have been folded, in a discrete folding-process where this folding is related to the fiber Lie group's Weyl-angles, which, in turn, are (angles) defined between the set of distinct maximal-tori which are in the fiber group (and this set of maximal tori are conjugate to one another).

The toral components of the lattice, and their different sized tori, which are the set of folded toral components, can define stable spectral-orbital properties, as well as possessing various relations to different systems, where these different systems would have different angular-momentum properties, depending on the way in which the toral components are folded.

That is, these 2-faces can be of two parts a bounded part related to a set of stable spectral-orbital properties of a shape composed of toral-components, and an unbounded part of the lattice, which when moded-out possesses the appearance of a set of threads which extend to either infinity, or to the boundary of the metric-space, which contains the 3-lattice from which (or the 3-lattice which, in turn, contains) the elaborate 2-face whose shape is composed of both a set of toral-components and a set of unbounded threads (seemingly extending out to infinity)

These 2-shapes associated to the 2-faces of a 3-dimensional lattice can be related to:

Either

1. positive and negative charges, or

2. The shapes of stable nuclei

3. Atomic shapes,

4. Several atomic components (or a molecule) related to one another through a new shape for the system's charge distribution, or

5. the components can be related to the 3-lattice as they are so related in an ideal gas, or

6. they can be more rigidly related to a lattice as in a crystal (or in condensed material)

The first type of system to consider would be the neutron and the neutrino, which are both hypothesized to be composed of an equal number of positive and negative charges, but the neutron is a bounded shape, while the neutrino is a semi-unbounded type of discrete hyperbolic shape, but the 2-dimensional toral components are contained in the 3-dimensional metric-space, which can also be related to the 3-lattice whose 2-faces are related to the toral-component shapes so that the unbounded part of a neutrino could be bounded by the metric-space, within which the shape is contained, as well as being bounded by the 3-dimensional fundamental domain of the lattice within which the 2-faces are contained.

It also needs to be noted that the 3-lattice is simply connected, so the neutrino can bind together 2-dimensional toral-type components in 3-space by strands which are bound by the 3-lattice, so these strands (threads) bind together the 2-dimensional components by strands or threads which, in turn, are bounded by the metric-space which contains the 2-dimensional component shapes, but these component connecting threads will not tangle, since they are contained in a metric-space which is simply connected, and thus the 2-dimensional components can be directly related to their 3-lattice positions so as to keep their stable properties of the their shapes which are associated to their facial relations to the 3-lattice from which they emerge, and can be re-absorbed in a direct untangled manner.

Thus, seemingly disconnected and separate 2-dimensional toral-component shapes can be bound together by thread which do not tangle.

So that this can be the model for pairs of charges in crystals, ie the Cooper-pairs of electrons in crystals allowing the electron pairs to be Bosons

Or

It can be the model of so called quantum entanglement where the threads connect distant components in a non-local manner

The piece of data which needs to be considered is that charges combine to form stable spectral-orbital systems, which are often, charge-wise, neutral, but for protons (hadrons), these positive charges combine with electrons to form neutrons, and one can also hypothesize that they also combine with neutrinos, so as to form, what seems to be, a closed bounded stable metric-space shape of either a neutron or a nucleus,

ie a discrete hyperbolic shape (or a stable shape composed of many folded toral-components), which is a closed and bounded metric-space shape, which possesses a positive charge, ie a nucleus,

While, in turn, this system of positive charges combines with electrons and neutrinos to form a (charge-wise) neutral, so called, electron-cloud and a nucleus system, which seems to be a semi-bounded stable spectral-orbital shape, and which is neutral, and

Thus, it is a system which is likely to be a closed metric-space shape, ie the charges and the electromagnetic-fields get trapped within a closed metric-space, but which seems to extend to the boundaries of the metric-space, within which the atom is contained (and its charge-neutrality defined),

ie to the boundary (when viewed in the next-higher-dimension metric-space) of the metric-space within which it seems that our experiences of the material-world are contained, [or to the boundary of a 3-lattice from which it (the atom) emerges], where this metric-space (within which our experiences of the material-world are contained) has our stable solar-system as one (ie one pair) of its 2-faces, but the actual shape of our solar-system is a stable discrete 3-dimensional hyperbolic shape, which is contained within a 4-dimensional (hyperbolic) metric-space [where this 4-dimensional hyperbolic metric-space is also a stable, closed and bounded, discrete hyperbolic shape].

Whereas

A molecule is composed of sets of atoms, which are combined together, in a way in which the positive charged parts of the molecule seem to keep their localized (compact, or closed and bounded) positively charged properties, but the entire charged system must integrate (or consistently combine) with the (unbounded shape of a) neutrino containment context, so that the entire set of charges, both positive-nuclei and electron-clouds, define a stable shape (a closed discrete hyperbolic shape), which extends to the boundary of the system-containing metric-space (which, in turn, is defined by the shape of our solar-system).

Thus, these charged components can fit into various sets of discrete hyperbolic shapes in regard to various types of charge distributions and related folded spectral-orbital stable structures (or shapes).

Furthermore, the various components of a molecular system can seem to be independent and separate of one another, and yet still be bound together by the stands (or threads) which extend to the edge of the metric-space, and yet, they remain untangled, so that the system can stay consistent with the facial-orbital context of a molecule which can be defined by a 3-lattice.

But such small 3-lattices do not mode-out to form a 3-shape in a 4-space, rather they remain 3-lattices in which their 2-faces can form into 2-shapes which are contained in 3-space.

In a crystal (or condensed matter) the relation of charged-material to a 3-lattice seems to be the main attribute in a fixed lattice structure, but again the charged 2-dimensional components can seem to be independent, locally, but still have a fixed relation to a lattice, thus there can be non-local properties associated to some of the 2-components, which from in the crystal lattice.

Where for gases, this same relation seems to still exist, but the independence of the 2-dimensional components seems more definite, but it is the relation of these components to a 3-dimensional lattice which seems to be the basis for the thermal law of the conservation of energy.

That is,

Is conservation of energy due to material-components being confined in a box?

or

Is it due to the fact that the (unbounded) neutrino geometries keep the system's components unified to a lattice in a more fundamental way?

Considering the traditional quantum viewpoint vs. the more natural viewpoint of both existence and material naturally defining lattice structures which are associated to stable shapes

Why, in the traditional descriptive context, are lattices so important

That is,

Why is the energy-lattice structure of a gas so demonstrate-ably statistically related to the thermal properties of ideal gases?

Yet this hypothesized lattice structure does not appear as a solution to a set of operators which act on a (quantum)-system's function space (as in the case of either the harmonic-oscillator or as in the case of the H-atom), rather, it is an ad hoc (ie try this and see if it works) math context, ie the energy-partition of a lattice which is related to a gas contained in a box, ie the partition is actually added-on to a set of solution-functions of one-component quantum-states (or quantum energy-levels) ie the partition implies both (1) the counting states and (2) the existence of a context of high and low energy-values, but the solution to the box does not imply this structure. Yet, the existence of this partition is fundamental to the description, and the way of filling the quantum states.

The alternative expresses the idea that the 3-energy-lattice is always existent in our 3-metric-space, but a container with rigid-walls pushes on this natural 3-lattice, with the rigid 3-lattice associated to the container walls, where the 3-lattice is always there because this 3-lattice's fundamental-domains have such a small size compared with the size of the fundamental-domain of the metric-space within which the 3-lattice is contained, that there is no point to becoming a 3-shape in 4-space, ie the smallness of the fundamental-domain means that the material associated to such a (small) lattice structure will be defined by its relation to condensed material, which mostly exists on the planets of the solar-system, where the solar-system is the shape which defines the (size of the) natural metric-space, within which our, apparently, material experiences take place.

There are other fundamental questions which one can consider, questions, such as,

Why are... "the prevalent very stable many-(but-few)-body spectral-orbital systems which exist at all size-scales"... so stable? The alternative answer to this question (as opposed to claiming that this stability is based on particle-physics, which is an absurd foolish statement) is that these stable systems are also based on stable shapes, which, in turn, are related to the same set of 3-lattices as related to the descriptions of the states of matter (ie solid, liquid, gas) but as shapes they are focused upon the fundamental-domain's relation to its "moded-out stable shape" which is associated to a hyperbolic metric-space (shape).

These are the most fundamental questions in physics which are not asked, whereas, asking why, in a useless (quantitatively inconsistent) model of elementary-particles,

1. How elementary-particles get their mass, and questions about

2. the nature of gravitational singularities (and about their mythological gravitational field-particles), which

3. are all descriptions which only relate to one point, ie the point of a point-particle collision, and where

4. the only property of interest in this description, is the particle-cross-sections, ie probabilities of point-particle collisions which take place a single-point (thus, this is all about asking how to model and measure particle cross-sections of elementary-particles which are unstable so that these types of (foolish) questions are of prime importance to the investor-class), are these the only type of questions which are considered by university physics and math departments to be the only valid questions in physics?

Answer: Because the only thing of interest to the military-banking-oilmen, and their investment and their solitary and singular control of the propaganda-education-indoctrination system, has resulted in only knowledge being considered in university physics and math departments which is knowledge which is to only be about detonating a mythological gravitational singularity, ie a doomsday machine with which the military-banking-oilmen can terrorize the world.

But, one can also ask,

Why does modeling an ideal gas as an energy partition (or energy-lattice) work so well, in regard to using such a math construct to derive descriptions of the properties of thermal and statistical systems, but also useful in regard to its relation to stable spectral-orbital systems?

Is it because the material components are best modeled as being in a box?

and

Thus, these components are related to a quantum energy-operator so that the energy of a solution wave-function is related to independent directionally motivated integers and then arbitrarily partitioning the 3-dimensional (directionally motivated) integer-space into unit volumes?

But from a solution-function, which depends on integers for its energy-values, to the act of partitioning the (only) positive integers into unit-volumes there seems to be no motivation,

ie "why would such an energy-partition exist?"

and "why should individual components in a box fill-up such an energy-lattice of integers?"

Answer: The claim would be for the components to seek the lowest energy-levels of the box.

But there is no natural lowest-energy for a particle-component in a box, the idea of a lowest component-energy for a component within a box is only implied after the energy-partition is made, and there is no reason that a energy-partition for a box would exist.

Is this partitioning of energy simply done to fit data, to a probabilistic model of a system?

Where the probabilistic model of the system is only, vaguely, related to physical measuring through sets of operators, which are used to diagonalize the system's function-space, or to find the system's set of spectral functions, where the quantum system is modeled as a function-space of probabilistic harmonic functions which seem to be coincidentally related to a parabolic energy-(or-wave)-operator.

There is nothing compelling in such a descriptive context, rather it is put-together in a way so that, if it is determined that some relation may exist then try this math structure and see if it can be used to fit data,

ie it is all done in a ad hoc manner (ie do it if it works, manner).

Where this ad hoc math context is a descriptive context which can only (really) model and solve (1) the box, and (2) the mythological harmonic-oscillator, where the harmonic oscillator is mythological since there is no valid context of the harmonic oscillator in which the forces on an independent component (in an actual physical context) would be in any way equivalent to a spring, with a linear spring-constant.

Furthermore, the H-atom actually provides a diverging solution function for the radial equation.

That is, the quantum energy-operator has a very limited range of descriptive capabilities, and

Furthermore, the fact that "it is defined as a probabilistic context" means that the description is all about fitting data, and it never really predicts anything new, ie where "new" means either "new properties which are accurately described" or "the identification of controllably useful physical properties."

That is, to solve an energy-equation for a single component, which is identified by the single-component existing within a box. Then, in this description, the component will have whatever energy which it does have, and there is no condition of a negative potential-energy identifying (or causing) a natural lowest energy.

Rather, it is only after one does the extra and independent assumption that the energies in a box can be related to a partition of energy, associated to a partition of the integers, that the context is then set to be able to talk about the lowest energies in the partition, then there is the further assumption that the energy-levels, defined by the partition of the integers, fill-up the energy-levels of the box from lowest to highest, but this is an extra assumption,

That is, only a negative potential-energy function (and the associated potential-energy term in the quantum-wave energy-operator) would make this a necessity, whereas the box only identifies component-confinement, ie the binding of a component to a system by a box's potential energy term does not relate a component's energy to its state of bounded-ness so that there is no natural energy-level which identifies the component as being bound by the greatest energy, ie all energy-levels in a box whose potential-walls are infinite are equally bound within the box.

That is, one does not solve the harmonic oscillator, and then independently define an energy-partition for the harmonic oscillators "solution function's quantum numbers," instead the energy partition is intrinsic to the solution functions and the system's operator constructs.

Furthermore, the energy-levels of a crystal are also essentially determined by assuming the binding potential-energy terms of the crystal are the infinite potential-energy walls of a box, but which also must accommodate the crystal-cell periodicity of atoms and the periodic occurrence of the nuclei of these atoms.

And there are the extra assumptions about energy distributions,

ie (again) filling the lowest energy-levels to the highest energy-levels (so as to be consistent with conservation of energy), but there is no math context which pulls a component to the lowest energy level of a box.

The alternative, based on stable math patterns

On the other hand, the great usefulness of an energy-partition, and subsequent energy distributions within that partitioncan be interpreted to mean that the energy-lattice is a not a result of a solution-function to a particle differential equation, and then these single-particle energy properties arbitrarily related in an independent manner to an energy-partition, but rather an energy-partition (or energy-lattice) is an intrinsic part of existence, in which defining highest and lowest energy-levels becomes a natural attribute of the system's energy and its energy-distribution, and furthermore, the energy-lattice is at the core of the stable aspects of physical descriptions, ie the structure of stable many-(but-few)-body spectral-orbital systems it seems can only be described in regard to energy-lattices

And being bound within a lattice is an expression of relatively low energies for the system's components when compared to other lattice-sizes (of energy-lattices) within the same dimension, where, in turn, the larger energy-lattice sizes are related to stable shapes, which can naturally interact as shapes with other shapes, rather than a system having material-components emerge from, and/or to be absorbed within, the system's energy-lattice, within a system whose physically measurable properties are defined, not as much by the system's shape, but rather defined by the (relatively small fundamental-domains of the) system's energy-lattice, ie where a larger sized energy-lattice (with a higher-magnitude energy) results in the ability of the higher-energy-magnitude-lattice's larger fundamental-domain (of the lattice) to be able to contain more energy within itself, ie it can contain more energy-components (which may be either bound or free, smaller-energy components).

That is, (Does the idea that the usefulness of an energy-partition should be interpreted to mean that)

"Existence (already) has a deep relation to energy partitions (or equivalently, energy-lattices)," or equivalently,

Existence itself has a natural lattice structure, which is defined in the context of energy, where a 3-dimensional energy-lattice structure, possesses a, subsequent, relation to its 2-faces, and subsequently, their relation to 2-shaped components (where 2-shapes are contained in 3-space) and where the folded orbital-structures of 2-shapes can be composed of stable many-(but-few)-bodies, which, in turn, are contained within a (stable) spectral-orbital structure related to a 3-dimensional energy-lattice (this is something quantum physics is not capable of describing, by means of partial differential equations and solution-functions, which are associated to general harmonic function through sets of operators), and it (this energy-lattice structure) is also related to a non-local structure for existence, where this non-local structure exists in an energy-space which possesses an extensive lattice structure.

Note:

(our totalitarian society does not allow these types of highly informed expressions of new ideas, they are new ideas which are excluded by the intellectual-class, whose high-salaries and their media-image of possessing high-social-value are contingent upon their serving "the social structures and high-social-value which have come to be defined by the investment-class.")

This is an intellectual-class which defines science as authority, so that science has become much like a dogmatic religion, which is based on personality-cult. Everyone has come to claim authority for their opinions based on what is now (2014) considered to be science. That is, the institutionally identified intellectual-class does not define science as a method of inquiry, ie equal free-inquiry [as Godel's incompleteness theorem demands for an axiomatic based language], rather they define science as form of authoritative power {these institutional-intellectuals adhere to, and support, a "new" form of totalitarian science}.

If one opposes: dogmatists, arbitrary authority, and racism, then one also needs to oppose the set of dogmatically authoritarian institutional intellectuals. That is, it is: dogmatism, racism, and intellectual-elitism (which can only be defined in a [fraudulent] context of an absolute authority), which define the idea of an exceptional-ist form of imperialist domination by the few..., ie in the investor class (whom define high-value for society)...., over the many.

Chapter 25

Re-iteration

Physical science is about materialism, and either geometry of material or randomness of material which has been reduced to the random event structure of material point-particles in space and time, and the (partial) differential equations which are used to determine the measurable inter-relationships between material components (or material geometries) and system structures into which the system components form, where in both cases (geometric or random) the material interaction, ie the (so called) cause as to why material forms into stable systems, is modeled by means of a partial differential equation either in regard to (1) a force field (related to material geometry), or (2) a geometric structure determined by the system-shape's local curvature and related metric-function, or (in the case of quantum systems) (3) a combination of both energy-operators, forming a quantum system's wave-function, and sets of internal particle-states, which are attached to the wave-function, including the, supposed, sets of "energy-level related" elementary-particle-families where the non-linear properties of these particle-states are supposed to have the effect of perturbing the quantum system's energy levels (of the elementary-particles which occupy the quantum system's wave-function).

That is, in all three cases, where the properties of a physical system (based on either geometry or on randomness) are being quantitatively modeled, the center of the model is the system's defining (partial) differential equation (or pde) (where partial differential equations model the local [linear] measurable properties of the system which have been put into relation to the containing set's measurable coordinate-values).

[Note: Where parenthesis are put-around the word (partial) since in Newton's model of forces affecting the inertial properties of a system's composing material the differential equations defining the material's motions are ordinary differential equations.]

It has only been the solvable systems…, which exist almost entirely within the set of classical models of physical systems (material in a context of geometry and force-fields)…., which provide both accurate useful information, and the descriptive structure identifies a useful context for practically creative efforts in the physical world, as well as the (pde, or partial differential equation) "solution model" of the system's properties being physically controllable by means of controlling the boundary and/or initial conditions of the system, which are implemented in a classical context of measuring.

Though, often it is quantum-type physical properties which can be coupled to a controllable classical system which allows for many practical creative designs which work, but this has little to do with the information about the quantum system obtained from the so called laws of quantum physics, ie finding a complete set of commuting Hermitian operators defined on the quantum system's function-space model of set containment, where such a set of operators can seldom (if ever) be found for general quantum systems. That is, the quantum properties are measured (classically), and through the same measuring context, the quantum properties are coupled to the classical system, and then used in a controlled system.

That (both) (1) the quantum model is based on randomness, and (2) the (seldom found) operator structures…, used to solve the system's spectral properties…, are operators which model classical measuring properties, which together these two properties cause this type of a "random-descriptive and classical-coupling" context for precise descriptions to exist.

But this descriptive context of materialism and partial differential equations (except for the few solvable cases in the classical context) leads directly to an operator context of either; non-linearity or non-commutative, and/or indefinably random, which implies the property of quantitative inconsistency, which is, subsequently, induced into the descriptions quantitative containing sets, that is, this is because the operator (or local equation) context is: (1) non-commutative, in regard to either (1a) locally (over some regions of a geometric containment set), or in regard to (1b) a function-space model of a spectral system, and/or (2) non-linear (pde), and/or (3) the system is indefinably random, ie either the random system's elementary events are not stable, or they are not determinable (eg not calculable).

Where quantitative inconsistency means (1) the scale of measuring is not well defined, or (2) the scale of measuring (or the measuring process) involves discontinuity of structure, or (3) the measuring scale becomes many-valued, so in any of these cases the quantitative sets are no longer properly (or logically) defined, and the properties described in such a (non-commutative, or non-linear, or indefinably random) context and which are contained in a containing-set which is a quantitative-set but where the quantitative-sets are quantitatively inconsistent, are not valid, ie the properties are not stable and/or the descriptions have no meaning.

This means that since : quantum physics, particle-physics, and general relativity are all non-commutative, and thus the properties…, which are, supposedly, being proven in their descriptive contexts (but where the descriptions possess these non-commutative properties)…, are meaningless (or invalid) properties, since they are all properties which are related to a quantitative containment-set, but where "in this quantitative containing set" measuring has no meaning, ie the quantitative sets associated with non-commutative math relations are quantitatively inconsistent (discontinuous, many-valued, and the scale of measuring is not defined).

The problem is that these descriptive structures are all, trying (or should be trying) to describe the very fundamental and very stable many-(but-few)-body systems,

eg general nuclei, general atoms, molecules, solar-systems etc, since these systems are the most fundamental systems which compose our "physical" existence and they have stable spectral-orbital properties but the stable (and internally controllable) properties of these systems go without any descriptions based on the, so called, physical laws.

And the context of materialism and partial differential equations is not the correct context.

In such a descriptive context only the 1-body (in general relativity) and the 2-body (classical) systems can be described and solved.

That is, the idea of materialism identifies the wrong containment set-structure, and (apparently) the partial differential equation can only be used to perturb, at most, two bodies within a stable many-body system.

The correct containment set is both many-dimensional and each of its various dimensional levels are composed of stable shapes, ie the very stable discrete hyperbolic shapes which are defined from hyperbolic-dimension-1 to hyperbolic-dimension-10.

The discrete hyperbolic shapes are used in the many-dimensional structure to define both hyperbolic metric-spaces as well as stable material components which are of a lower dimension than the metric-space within which they are contained (usually the dimension of a material component is, an adjacent, 1-dimension lower than its containment metric-space).

That is, consider that in conceptually "moving between" classical physics to the quantum physics, then in regard to the descriptive context the basis (of description) changed from "geometry and local measuring of motions and fields" to randomness and the context of measured properties being related to their containment in a function-space set-containment model of the quantum-system so that the system's measured properties are to be identifiable by the correct set of commutative Hermitian operators which act on the function-space model of the quantum-system, eg differential and integral operators which represent physically measurable properties.

Now the model needs to change from randomness function-spaces and sets of commutative operators (where the change is necessitated since the set of commutative operators for the fundamental systems, eg general atoms etc, have never been determinable in this descriptive context)

To

A containment set which is both many-dimensional, and each of its various dimensional levels are composed of stable shapes, ie the very stable discrete hyperbolic shapes which are defined from hyperbolic-dimension-1 to hyperbolic-dimension-10 and, where for each distinguishable subspace of 5-hyperbolic-dimensions in an 11-hyperbolic-dimension over-all containment set there is a largest spectrally valued stable discrete hyperbolic shape (while there already is a minimal spectrally valued stable discrete hyperbolic shape in that subspace, where this has the effect of causing the independent, five- and lower-dimensional hyperbolic metric-spaces to define a finite spectral set, which, in turn, causes the quantitative structures…, which are needed to measure the system properties…, to be finitely generated, and thus the quantitative sets cannot be "too big," and thus, the logical consistency of the identifiable patterns in this descriptive context can be assumed with greater confidence.

That is, in the context of stable shapes of many various sizes defined over many-dimensions, where once inside a 3-dimensional or 4-dimensional discrete hyperbolic shape where materialism is well defined it is very difficult, due to topology and the dimensional-structure of both material and their interactions, to identify anything outside of the apparently confining 3-dimensional or 4-dimensional discrete hyperbolic shape material-containing (hyperbolic) metric-space,

The stable shapes are both {{seemingly}} closed metric-spaces and stable material systems, but the descriptive context for these shapes is not material interactions, but rather these very stable shapes are related [relation of the shapes is] to the Lie group fiber group of the principle fiber bundle within which these shapes are defined, but within this new context (of the set-containment of existence), in which to understand the stability which is observed in the world, ie the very stable discrete hyperbolic shapes, these stable shapes are (or can be) related to partial differential equations, which, in turn, are associated to differential-forms, where the math structures of these shapes (differential-forms) are: linear, metric-invariant, and continuously commutative (except possibly at one point) and these shapes are related to rectangular-generated convex polyhedrons, which, in turn, are related to the (flat, within themselves) discrete Euclidean shapes, or the Euclidean tori (ie doughnut-shapes), where the tori are related exactly to rectangular (prism) and continuously commutative shapes.

The main point

That is, there is both quantitative consistency, and a finitely generated quantitative containment set-structure…, ie a quantitative-set (which is used for measuring) whose properties are closer to the rational numbers than to being related to the real-numbers…, in the new descriptive context, where the real-numbers (in regard to how their construction is related to the axiom of choice) define a quantitative set-structure which is too-big, (and thus it [ie the real-number based quantitative containment set] leads to logical inconsistencies).

Back to arbitrary descriptive structures

To say that "particle-physics is being verified, due to observed particle-properties being associated to vaguely correlated math patterns" is an absurdity, since particle-physics is a set of descriptions which exist within a quantitatively inconsistent containment-set structure, so the patterns it describes are, essentially, meaningless,

Except (there is an exception, namely, where the descriptive context does not possess any distinguishable patterns other than the sets of random component-collisions, in regard to the components which have been broken-apart due to the high-energy component collisions which initiated the explosion's detonation) in the chaotic context, which is the descriptive context of a "rate of particle-collision model" of a nuclear-reaction (or, equivalently, model of a nuclear explosion).

That is, when the, so called, laws of particle-physics can be used to describe, explicitly, the stable spectral properties which are observed for (relatively stable) general nuclei (which are related to the atoms in the periodic table) only then can one say that particle-physics has any validity.

Otherwise

Particle-physics is only a story about the very limited context of bomb detonation (it describes only patterns which exist in a chaotic nuclear explosion) (wherein the properties of both measuring and the exploding system are transitioning in a chaotic manner (between descriptive contexts where measuring [and stability] does make sense), and between two relatively stable states [ie (1) the state before detonation, and (2) the state after the explosion has cooled]).

The so called experts have tried (in a, seemingly, disingenuous manner) for over 100 years to describe the stable patterns of physical systems, which are observed, using the math properties of: materialism (which is a math statement about set-containment) and partial differential equations but where this descriptive context possesses the properties of: non-commutative-ness, non-linearity, and indefinable randomness, and subsequently, they have consistently failed to be able to do this.

The efforts of these so called experts only result in: (1) the adjustments of certain types of instruments based on 19th century science, and (2) the building of horrendously destructive war instruments

But

Their only statement about describing the stable spectral-orbital properties of any of a myriad of the many-(but-few)-body systems, is that; "these systems are too complicated to describe," apparently, or especially, when one uses the descriptive context of; materialism (which is a math statement about set-containment) and partial differential equations, but where this descriptive context possesses the properties of: non-commutative-ness, non-linearity, and indefinable randomness, so as to result in trying to describe these stable systems within an implied quantitative containment set which is quantitatively inconsistent, and thus it is an invalid descriptive structure.

Wherein the only statement one can make is that "within such an invalid containment context, all systems are too complicated to describe."

That is, the intellectual class…

{those thoughtless and dogmatic, but, nonetheless, tagged-with (identified with being of) "a high-intellectual-value" by the propaganda-education-system, and (these obsessive people, driven more by their memories of traditional and authoritative "high-valued" dogmas, are being) placed in the, so called, higher-institutions of education, so as to tend-to the narrowly defined set of instruments, which the ruling-class wants them to tend-to, ie the technical instruments about which their monopolistic businesses are organized and built (where the point-less, authoritative traditions of great generality and abstract complexity, are being used as a propaganda-instrument of great importance to the ruling-class's propaganda-system and their propaganda-strategies about placing the false image of the existence of intellectual-inequalities within the population)}, only deal with the (above mentioned) very limited overly authoritative, and overly traditional, context of math, a math context…

(or math structures and methods which are related to math models based on),

ie a (failed) math context which mostly leaves the main issues concerning math quite arbitrary,

eg its inability to solve the very stable many-(but-few)-body physical systems,

ie issues about describing stable patterns… where, in the descriptive context, measuring is reliable, but instead

Where (now, in peer-reviewed math journals of 2014) the main focus of math revolves around the following math methods, or math operations, and math processes which are:

1. non-commutative (either locally in regard to geometry or, in regard to sets of operators acting on function-spaces)

2. non-linear, and

3. indefinably random (either elementary-events which are unstable, and elementary-events which are not determinable through calculations),

Where these types of math properties lead to quantitatively inconsistent containment-sets, ie where quantitatively inconsistent means: many-valued, discontinuous, chaotic, with no ability to describe stable patterns, and a descriptive language without any meaning.

And, furthermore,

They are… (ie these above listed, 1-3) math methods (or math structures)…. also placed into models of quantitative-sets which are "too-big (as sets)," so that this results in math models which have no logical consistency, ie the property of convergence is used to fit together quantitative properties (or structures) which are logically inconsistent with one another, eg point-particle collisions are a central property of the particle-physics model of material interactions but it is also a geometric model within particle-physics but particle-physics is fundamentally based on the idea of randomness of the wave-function, of particle-collisions, and in regard to the implied randomness and chaos of a non-linear connection term in the particle-state-transition model of particle-collision material interactions (but geometry and randomness are logically inconsistent, or logically incompatible with one another), thus forming a failed mathematical containment set, because the model of the quantitative sets in the model are sets which are too-big, and the use of these types of math models also has a great deal to do with the failure of the intellectual-class

Note that, quantitative inconsistency is an elementary property of mathematics.

There are three examples of quantitative inconsistency, which can clarify the issue:

1. (simple quantitative inconsistency) The uniform unit of measuring does not stay uniform, so if one measures with a (12") ruler, the measuring scale will keep changing, so instead of 12" one could see 33" or at another instant one could see 4" at the end of the ruler etc,

2. (discontinuity) when measuring with a ruler; the ruler would suddenly split-apart and separate itself, so as to form measuring-gaps along the direction of the ruler, so that the end of the ruler, which represents the ruler's measurement, misses (measuring) the gap which was discontinuously formed between the (two) separated parts of the spontaneously split-apart measuring stick, ie one cannot get an accurate measurement with such a discontinuous property associated to "elementary measuring" processes.

3. (many-valued) when one tries to measure the end of the ruler one finds that the ruler has become several different ruler-ends, so that each ruler-end provides a different value for the measurement. Thus, there is not one-value for a measured quantity, but rather several (or multiple) values, so that one cannot distinguish between the true measurement of the different values.

That is, in regard to quantitative inconsistency this small set of (above mentioned, three) measuring irregularities automatically occur, in the measuring process, when the quantitative sets, which one uses in a mathematical model, have the property of being quantitatively inconsistent.

That is, in such a quantitatively inconsistent structure (or model) for quantitative sets…

(or a quantitatively inconsistent set-structures which is used for the purpose of "modeling mathematical pattern-containment within measuring-sets"), there are no (stable) patterns which can be measured in any reliable manner.

That is, the quantitative description becomes meaningless, or equivalently, the description loses any meaningful content.

That is, one can speculate about such descriptive contexts, where meaning is apparently (must be) considered to be irrelevant, so that in such a descriptive context, where there is neither quantitative consistency nor a capacity to describe a stable pattern, one needs to ask, "what type of content can actually exist within such a descriptive language?" {There cannot be any content.}

Apparently, essentially, there is no content which can exist within such a descriptive structure.

That is, the determination of "What aspect of a "quantitatively inconsistent"-based mathematical model possesses any value?" especially, when the math model is placed within a meaningless descriptive language, is a central question, which needs some answers, in order to justify that "such a meaningless descriptive language (based on a quantitative containment-set which is quantitatively inconsistent)" should be "the dominant focus of research" for the academic institutions of the, so called, public universities.

That is, this type of judgment about the singular use of a precise descriptive language (eg physics is to only be considered in regard to its sole relation to bomb engineering)--- (but where the physics principles used in bomb-engineering cannot be used to describe a stable pattern)… in regard to the rational inquiry of the, so called, experts, whom are given positions in a public institution --- where such a descriptive basis (of physics principles used only for bomb-engineering) seems to have no justification what-so-ever,

ie and where such a descriptive basis (for bomb-engineering), which possesses the property of being quantitatively inconsistent, and where this quantitative inconsistency exists because the math structures being used in the math-description are (locally) non-commutative, non-linear, and indefinably random.

Deluded by one's obsessions

Nonetheless, these types of meaningless descriptive structures are embraced by the experts, where the experts seem to believe this is because of:

(1) these contexts seem to be very general, and thus they are assumed to possess a lot of useful math information (but this is not true), and

(2) they fit into the meta-structure of the assumptions of math, most often the experts seem to follow the laws of algebra in their meta-structures of math…

(neglecting that fact that, in an elementary-way, they are quantitatively inconsistent, and thus (or subsequently) one is describing a meaningless and non-existent set of math patterns, where this is because they fail being mathematical properties at the most elementary level, ie in regard to the elementary properties of reliable measuring) and thus, they (the properties of algebra) can be manipulated by logic (since they seem to follow the laws of algebra), so that these algebraic patterns, which are quantitatively inconsistent patterns, can be put into mathematical proofs, but the patterns which are in the conclusion of the theorem's demonstrated property (or in the proof) are meaningless and without content, since the math properties, which are also a part of the system's assumed math properties are properties of: non-commutative-ness, non-linearity, and indefinably randomness, thus, the quantitative containment structure is quantitatively inconsistent, so that the descriptions of these, so called, "proved properties" have no meaning.

And subsequently a there are a lot of proven math properties which are associated to this quantitatively inconsistent language,

Where this is mostly because abstract quantities which are (locally) non-commutative, non-linear and indefinably random are (or can be easily) related to algebraic constructs, so as to appear to be calculable, as well as appearing to be contained within quantitative sets, and, seemingly, precisely related to other math constructs (but there is no meaning which can be associated to the descriptive construct), so there are the fairy-tales (or the science-fictions) of:

(eg general relativity's program):

For example: energy-density distributions being relatable to curvature, and, subsequently, related to solving for the general metric-function (for the geometry of the space's shape related to the energy-distribution), and, subsequently, using the metric-function to determine the geodesics for the physical system, and thus to determine the "material interactions" and the, subsequent, system-dynamics (for the system's components), so as to identify the "true nature" of the physical system's measurable properties, supposedly, within a stable system, (but such a stable property is outside of what such a descriptive context can actually describe)

But, nonetheless, these ideas are all related to theorems which are rigorously proved, but this is in a descriptive context which is both non-linear, non-commutative, and subsequently, it is quantitatively inconsistent, wherein there is neither any form of reliable measuring units, ie there is no quantitative consistency, nor are there any stable measurable properties, which exist within such a description, but the solar-system is, apparently, very stable (and planar), but there are no valid descriptions of these properties made in a quantitative descriptive context.

That is, this language structure, followed by the intellectual-class, is all nonsense, if they insist on basing the math patterns, which they describe on the quantitatively inconsistent math structures which are (locally) non-commutative, non-linear and indefinably random.

Particle-physics is even worse, but it is the two descriptive contexts of general relativity and particle-physics which completely dominate the physics departments, of the, so called, public universities

(ie the universities of the indoctrinated, or the universities of the gullible, or the university of idiots).

And

The, supposedly, highly competitive social-channels through which our society identifies its most meritorious elites, ie the (so called) superior intellectuals of the society,

Who "prance around" the, so called, "institutions of higher-learning" doing their most notable, and prize winning, nonsense.

[all they really do is to become the icons of a (fake) high-intellectual value within the propaganda system, where it is the illusions created by the propaganda system which are the main images within society, but, which really uphold the, so called, superiority of our arbitrarily determined, so called, highly-accomplished intellectual culture]

The ridiculous intellectual activity is actually, narrow, dogmatic, and a bunch of rubbish, ie useless.

Why is no-one pointing-out this "state of delusion" in the academic, and expert, world, and the uselessness of these (quantitatively inconsistent) (fake)-patterns which are being identified?

Is this about the effectiveness of spying and sending-out agents to interfere and contacting militarized management so as to coerce the people expressing unwanted ideas?

But

It is also about the theater of illusion, created so the public will aspire and compete, in meaningless competitions, whose order and rules are maintained in the "high-valued technical peer-review journals" which represent the funded "interests in mathematics" which is being done in the university math departments all of the descriptive languages of technical math, related to material-based creativity, eg quantum physics, particle-physics, and general relativity, and their derived theories, eg string-theory, are based on:

materialism, and either indefinable randomness or non-linearity (or both) and they are all brought into a descriptive context of non-commutative-ness, so that they are not quantitatively consistent sets, but which are, nonetheless, used for a (failed) model of the quantitative containment of a system's measurable properties.

For example, trying to find the stable set of spectral-orbital properties of the many-(but-few)-body quantum-and-orbital systems

(but in this descriptive structure this is never being achieved, since "in this descriptive structure" stable patterns are impossible to identify),

That is, the dogmatic and required (by peer-review) descriptive context (due to both the math structures and the math methods used) is many-valued, discontinuous, chaotic, and there are no stable patterns in the description,

ie there is no relation to stable patterns which either "it can describe" or stable math patterns "from which the description is built," it only describes fleeting, unstable, vague patterns

This quantitatively inconsistent math structure is used to model: general relativity, quantum physics, particle-physics, and all the other derived speculative physical theories, where vague models of, so called, particle-properties (point-particle collisions put into a containment set whose basis is randomness, are focused upon and "explained" by theorems proved in these derived theories [note that geometric point-collisions and randomness are logically inconsistent math constructs], eg these are sting-theories twister-theories grand unification theories etc but are these "assumed to be particle-properties" (observed in particle-accelerator detection-chambers), in fact, really about identifying collisions between "distinguished points" of stable shapes,

Where these stable shapes might be defined for many different size-scales, and for many different dimensional-levels?

This is a much more interesting model of the, so called, "point-particle collisions," since the distinguished points of these stable shapes are (or can be) related to a quantitatively consistent math construct.

The issue in math is stability and measurability

vs.

Indefinable randomness and quantitative inconsistency (as well as logical inconsistencies)

Or

Definiteness

vs.

Arbitrariness

But definiteness is about intuitively fitting measurable properties into existence in a careful manner.

And

Definiteness is not about forming lists of algebraic assumptions which seemingly can be applied to any arbitrary quantitative pattern, and insisting on arbitrary contexts, and arbitrary interpretations, placed in quantitatively inconsistent and logically inconsistent, but seemingly, precise descriptive contexts.

That is, this is not

science (or materialism)

vs.

religion,

That is, science vs. religion, is a complete mis-representation of everything, ie science should not be based on the idea of materialism, and religion is like string-theory (or general relativity), ie it is all a bunch of carefully stated arbitrary nonsense (perfectly suited for arbitrary use in a propaganda system).

The true sets of opposition are:

stable patterns and definite descriptions

vs.

Arbitrariness and its main relationship to the delusions of a quantitatively inconsistent set-containment context

Or

stable patterns and definite descriptions

vs.

ineffective science and religion

Identifying some important issues in regard to the failures of modern math and science

Though this math context (materialism and partial differential equations) began with various aspects of classical physics, and was based on inertia and force-fields, so that for the solvable descriptive contexts, the descriptions are accurate, and the descriptive context is a good context from which new creative contexts emerge,

eg mechanics, guiding rocket flights through the solar-system, electromagnetism, thermodynamics etc.

However, within this (original) descriptive context there has essentially been one description of a "more than one-body system," namely, the gravitational 2-body system, which is a problem which was identified and solved by Newton, but mysteriously constrained to a plane, in the, so called, 3-space, which material is assumed to define.

But the math and physics which has extended beyond the classical viewpoint has not progressed past this point.

That is, in Newton's gravity and Bohr's and Schrodinger's H-atom (but "in the H-atom of Schrodinger" the radial equation diverges, there is not [in reality] a solution to the H-atom, other than Bohr's discrete orbits, which were perturbed by A Somerfeld) (where again this occurs on a plane, WHY?) only the 2-body problem has been solved, and in general relativity only the 1-body system is solved, but the math community has not been able to progress "at all" past this type of system formulation.

As long as these math models are based on: materialism and (partial) differential equations, where the coordinate structure needs to be consistent with, the (assumed to be contained system) system's shape in a context defined by the idea of materialism, there will continue to be failure.

But (or, that is) (there is an alternative)

the description needs to be quantitatively consistent, if one wants to be able to describe stable patterns (and/or measurable properties) of (the) a physical system whose measurable properties are observed to be stable

These properties of stable-ness of form and an implied reliability of measuring are contained in the classical metric-invariant spaces, (ie the SO, SU, Sp, etc, classical Lie groups) in particular the metric-space of non-positive constant curvature, where the metric-functions can only have constant coefficients, and the context of stable patterns defined by D Coxeter, in regard to the set of discrete hyperbolic shapes which are defined over many-dimensional levels (up to hyperbolic-dimension-10, ie where space-time is a 3-dimensional hyperbolic metric-space). That is, the observed order of the physical world cannot be derived from the idea of materialism and, subsequently, using both the properties of local linear measuring and physical laws (to be used) in order to define a physical system's (partial) differential equation. Such a context is mostly related to perturbing a material-component which is already within a spectral-orbital structure, which is intrinsic to the spatial structure of the material-component's containing metric-space, but this exists in a many-dimensional context, which is organized around stable shapes, ie both metric-space and material-components are stable metric-space shapes, ie discrete hyperbolic shapes.

(alternative viewpoint)

That is, a force-field acting on material, in the material's own (implied) containment context, can only permute material orbital-motions, in regard to a 2-body system context, while (in regard to a new set of ideas about physical description) stable patterns are a result of stable shapes, which are defined over many-dimensional levels, where it is in this new context (of stable shapes defined over many different dimensional levels) in which the principles of general relativity can best be applied within stable (solvable) math contexts.

The point (or the property) which this new context (of stable shapes, which are defined over many-dimensional levels) identifies, in regard to "the sets of stable shapes defined over several dimensional levels," is to both (1) identify stable shapes, or stable math patterns, ie both stable material systems and stable metric-invariant metric-space containment contexts, and (2) to identify a finite stable spectral-set upon which a consistent quantitative structure can be generated (or upon which a consistent quantitative structure can depend) so that within a stable system (where a stable system equals its stable shape) the coordinates and the stable shape of the system are consistent, in the sense of being: locally linear, metric-invariant, and continuously commutative (ie completely parallelizable shapes) ie dimension and system shape are consistent with both one another and with the geometric measures of the system's shape, and in a "one-to-one" and "onto" linear map-structure of local coordinate structure fitted (or mapped) onto the stable shape of either the material system or the metric-space shape's stable orbital properties, (where metric-invariant metric-spaces also allows for reliable measuring on the material components contained within the metric-space)

and where

these stable shapes are related to "rectangular" convex polyhedral fundamental domains, so as to possess distinguishable, separate, discrete spectral-orbital properties (whose spectral-values are associated to the geometric-measures of the face-structure of the rectangular polyhedral simplex) and which are consistent with the shape's geometric measures, ie where this (finite) discrete set of spectral values is the basis for the finite spectral set, upon which the quantitative structure depends (or is derived, or is generated).

Consider that,

It is meaningless for propaganda to state…. that "the particle-tracks of a high-energy "particle"-collision validate the existence of the Higg's particle, and subsequently, all of particle-physics is, thus, verified," if particle-physics can not be used to describe the stability and the stable spectral-structure of a general nuclei.

Or, In other words:

If particle-physics can not be used to describe the stability and the stable spectral-structure of the general (non-radioactive) nuclei then it is meaningless for propaganda to state that "the particle-tracks of a high-energy "particle"-collision validate the existence of the Higg's particle, and subsequently, all of particle-physics is, thus, verified."

How is it known that any of the "tracks" in the detection chambers are a result of a particle's motion. Where some structure is identifying "tracks" in a detection-chamber, and that these tracks are, in fact, really a property of a distinguished-point of a discrete Euclidean shape, and how does one know if the protons, in the so called "point-particle-collisions," which are formed in the particle-accelerators, are not (really) also, really, the distinguished-points of discrete hyperbolic shapes,

Note that the isolated and cooled-electron (of Dehmelt) does not hover as a charged point-charge, but rather forms into dynamic orbits, ie the elementary-particle is better interpreted to be a discrete hyperbolic shape with a distinguished point, than as a point-charge.

The meaning of this announcement (ie that "the particle-tracks of a high-energy "particle"-collision validate the existence of the Higg's particle, and subsequently, all of particle-physics is, thus, verified.") of what is most likely another example of dry-labbing of data, ie scientific fraud or, equivalently, scientific misrepresentation, that is, the only model of "particle-physics" is non-linear, non-commutative and indefinably random, so there are no stable patterns which are being described by this language, yet it is claimed that all particle-collision data from particle-accelerator detection-chambers can only be related to the descriptive context of particle-physics. This is simply a fraud. (furthermore, the descriptive context of particle-physics cannot be used to describe the properties of stable systems, and can only be used in the context of bomb-engineering, the main focus of particle-physics is on particle cross-sections and their relation to rates of reaction in the instrument of the nuclear bomb)

Therefore,

The claim that "the particle-tracks of a high-energy "particle"-collision validate the existence of the Higg's particle, and subsequently, all of particle-physics is, thus, verified." is an announcement which is, essentially, announcing a heavily funded descriptive math-model, which possesses no content, in regard to the properties of the physical world,

Rather it is a statement, which is all about:

stating that, "the research into detonating a gravitational singularity using the cross-section models of particle-physics will continue full-tilt."

And by incorporating the fake-knowledge of particle-physics into the focus of the, so called, scientists, of all the high-valued and uniformly based propaganda totalitarian nations, then the result of this will be that the focus of "the knowledge of physics" will remain on war and destruction, and not on the relation of descriptive measurable knowledge on the true relation which exists between "life and existence."

One sees from the intellectuals (ie those with a voice within the media, and in the meritorious institutions) a group of people supporting a descriptive context, a, supposedly, rational structure, which has been organized and used, so as to no longer have any meaning (incapable of expressing any practically useful content), where this has been true since about 1910, and its, so called, pinnacle of intellectual greatness ie particle-physics, is, in fact, an expression of developing the ultimate instrument of destruction (and this is not high-intellectual-culture, rather it is the 19th century model of chemical reactions, a model of chemical reactions which causes the chemical models of life to be complicated and unfathomable), and with no other redeeming creative attribute.

Thus, the intellectual class is really about representing their own arrogance, and their own self-importance, and their individual-superiority, which has been defined for them within the propaganda-education-indoctrination system, ie they are the social icons of "inequality as a law of nature," eg the top .001% of the normal distribution of the random property, eg of, so called, individual intelligence (this is best described as barbarity, especially, since their high-valued truths cannot describe the fundamental stable properties which are observed for physical systems, ie their supposedly, superior intellectual interests are, in fact, massive failed intellectual endeavors).

While the right is also all about arbitrariness, and their, so called, superior ability to "get things done," and subsequently, they express their own type of arrogance, ie their self-righteousness based on arbitrariness, which is used to justify their barbarity (where their barbarity is justified because they have demonstrated their social-worth, by their ability "to get things done," mostly through their barbarity).

This barbarity, based on arrogance, characterizes the justice-military system, as well as monopolistic businesses.

Note that, the Roman-army built the superior Roman-culture with bricks, while the modern business-army builds product, essentially based on the, seemingly, absolute truths of 19th century science (ie the equivalent of the Roman-army brick-laying).

The intellectuals claim to be rational individuals serving a collective (a consensus) on a rational march to an absolute truth, while

The right claims to be individual people of action, but such a "person of, so called, 'individual' action is only successful if they are serving in an army," which, in turn, collectively supports the few in the investor-class

The "great sin" of the human is to always mis-represent themselves

The intellectual-class gives no review about violence, which has any sense of reality to it,

ie violence is not simply imperialism, rather violence, in a totalitarian state, is constant and unrelenting, it is arbitrary, and it is justified by arrogance, the arrogance of the right, of those serving the ruling-class so as to be granted impunity for the arrogant behaviors, while the arrogance of the intellectual-class is the (an) arrogance based-on the arbitrary intellectual language structures which serve the same business monopolies, for which the continual violence is needed in order to defend and maintain the social position of the ruling-class,

ie the intellectuals and the bullies both collectively serve the same monopolistic businesses, and both arrogantly claim to serve a superior (but arbitrary) belief structure, where these beliefs are given high-social-value by the ruling class, and both of these opposite sides (intellect and action) are controlled by the ruling-class, and both sides are motivated by their own type of arbitrarily (based) arrogance.

(And a belief that social inequality is a law of nature, ie circular emotional pathways used by the investor-class to get society to collectively support the arbitrary high-value defined by the interests of the investor-class, supported by the arrogant intellectuals and by the arrogant violence of the people-of-action).

Chapter 26
Topology

The issues of topology and shape are placed into a context of homology and co-homology, in regard to a topology of a shape, where, in the case considered in this paper, the topology is based on the metric-space's metric-function, and the set of geometric-measures of subspace-regions which are related to differential-form-measures of subspace regions defined on a (compact, ie closed and bounded) shape, where the (different dimensional) subspace structures (separate, independent, linear subspaces, or hyper-spaces, or hyper-planes) within the space are best (or most intuitively) thought about in regard to the face-structure of simplexes, in particular, in regard to rectangular simplexes, or rectangular-ly-related (convex) polyhedrons, the more general simplex structures, are the basis for defining homology which, in turn, define the particular dimensional sub-regions, which (also) define the boundaries of regions of definite-integration, where definite-integration depends on partitioning the region over which one is integrating, so that the small-cubical-partitions (of the local orthogonal coordinates of the shape) have lower-dimensional boundaries (or faces), to which the fundamental theorem of calculus effectively defines the evaluation of an integral, in regard to a definite-integral defined over (on) a region of a function's domain-space, ie the function's defining domain region,

Namely, to be the anti-derivative-function which, in turn, is evaluated (or integrated) over the "domain-region of definition's" boundary, ie summed-over the (small) faces of the domain-region's ([in the limit of the] ever-refined, ie ever-smaller) partitioning-cubes.

The topology of a shape is (quite) often related to the holes which exist within a general topological (manifold) shape (this is true for the topological ideas of homotopy (or deformation groups), homology groups (defined in regard to boundary-maps, in turn, defined on the faces which compose a simplex), and co-homology, defined in regard to the exterior derivative (which is defined on differential-forms) which, in turn, are (differential-forms which are) evaluated on the faces of a simplex (or a region's limit of refined sets of partitioning cubes).

Differential-forms are based on (or logically related to) the metric-function's geometric-measures on the metric-space, and they are locally defined over a set of facial polyhedral-subspaces, (which

561

are, in turn, defined by the local coordinate vectors defined for a simplex and its set of faces, where the different directions [of the faces local coordinates defined at a point] define an alternating-form {or a local differential-form}) and, thus, since the topology of the shape is being determined by a (the) metric-function, the local directions of a face's, where each is related to a length measured by the metric-function (and related to a local geometric-measure of the face's region by the algebraic properties of an alternating-form) the geometric-measures of the differential-forms are (should be) topological invariants.

The holes in a shape...., when associated to a topology, where the topology is defined, in regard to a metric-function of a metric-space;...., are related to the integral-values over subspace regions, or evaluated over boundary-regions, within a metric-space, so that the subspace regions, which possess a non-zero integral over a cyclic (periodic) region (of the given subspace), identify the holes (which possess the dimension of the region) in the metric-space shape.

That is, an integral over a periodic (or cyclic) region in a space (or on a shape) which is non-zero implies that the periodic shaped region is related to (or defines, or identifies) a hole in the space (or in the shape).

A periodic p-form whose definite integral (over a periodic-region) is non-zero, is, in turn, related to a region-bounding (p-1)-face whose integration defined over a (p-1)-form is zero. Or the boundary-map [defined between faces (or cycles) on a simplex] of (or defined on) the p-face's bounding (p-1)-face structure is zero, or the exterior derivative of the p-form is zero. (?)

This is fairly complicated and abstract

But

The main-point is that these ideas, about holes in shapes, are fundamentally related to space's of constant curvature, (where the Thurston-Perelman geometrization theorem identifies these constant curvature spaces as being fundamental) where the constant curvature spaces are metric-spaces with metric-functions of particular "metric-function signatures," which are, in turn, related to particular types of constant curvature shapes. Where the three different types of constant curvature shapes are:

(1) Euclidean space is (can be) related to spheres of constant curvature-(+1) and

(2) Euclidean space is (can be) related to tori of constant curvature-(0), while

(3) in a hyperbolic metric-space the set of discrete hyperbolic shapes of genus-g whose shapes are made of g different toral-components, and which have a constant curvature-(-1)

so that holes in space are either the inside regions of n-spheres, or the set of hole structures defined on the shapes of moded-out rectangular simplexes, so as to form a torus (or doughnut shape) or

the set of hole structure defined on the convex polyhedron defined in relation to a set of g (in number) rectangular simplexes, in turn, attached as a string at "diagonal" vertices of the rectangular simplexes, and so that the vertices are deformed so as to form a convex polyhedron, so that the new shape of a compact convex polyhedron (also moded-out) has the hole structure of g separate toral components which compose the moded-out shape of the polyhedron.

That is, the hole structure of the torus is the main way in which shape is related to the holes of a stable shape.

thus there is the idea that homology and co-homology are better ways in which to discover holes in shapes than by deformations of spheres of various dimensions which are immersed in the shape

The homology of an n-dimensional rectangular simplex (or face-structure which gets moded-out) can be estimated by considering the doughnut shape of a moded-out n-cube and the holes in the shape which result from the moding-out process of identifying opposite faces so that these opposite faces define a periodic hole region whose base measure of the cubical-face is multiplied by 2(pi)r so as to geometrically-measure the hole region defined by the identified opposite faces, etc

All the different dimensional p-faces of the n-cube, or the number of "n chose p" combinations, identify the number of (p+1)-dimensional holes in the n-torus, etc

Multiply this by g for the hole structure (or the number of holes of the different dimensions) of the (-1)-constant-curvature shapes, when such a (-1)-constant-curvature shape has genus-g.

Chapter 27

Questions about quantity and shape

The extent to which either a metric-function (or a pde) can emerge from a single-point in space?

This is about both geometry and the "laws of physics" but

Is this about the existence of higher-orders of differentiations at the given point (or at a point) [ie higher-order numerical representations (or greater precision) of a function's value (but is the function's value given in this (Taylors polynomial related) manner actually valid (or accurate) [does a shape have a true relation to such quantitative precision]?

The derivative must be related to a local neighborhood (in a [sequential] process which defines a limit).

Is the set of quantitative (or algebraic) properties, defined on such a neighborhood, central to determining if quantitative-structures (eg methods of measuring a shape) are valid for the descriptive context which is being defined by a shape (eg curvature and/or non-commutative patterns of measuring, which are based on the combined properties of coordinates, where the properties of spatial-measuring on coordinates are defined by a metric-function {which is a local construct in regard to sets of coordinate-functions, which, in turn, define the shape (being measured)})?

Answer

These constructs do not emerge from a single-point, rather the main issue of physical description [and whether a physical system (or shape) have a true relation to great quantitative precision, {is there a natural limit in regard to a need for great quantitative precision of a system's measured properties?}] is about how stable spectral-orbital shapes relate to the more elementary issues about measuring eg (such as) if the property of commutativity is maintained by the local models of (linear) measuring [in a local neighborhood].

How can sets of measuring methods (or measuring-scales) be attached to shapes (and so that the shape remains stable)?

At what point do curved shapes (or curved measuring constructs) move outside of a quantitative structure?

To what extent does the math (quantitative) property of non-commutativity cause measuring done in the context of measuring a shape's properties (or its form) lose its meaning?

Math is about descriptions of measurable patterns, but the patterns must be stable, ie either quantity and shape,

or

quantity and a set of stable, well defined elementary-events, which define a set of random possibilities (whose occurrence can be counted in a quantitatively consistent context, eg fair samples and well defined events), in which case one wants models about betting on the occurrence of a fair sample of these (fixed) events (where an event is a subset of the elementary-event space.

The interesting math context is quantity and shape

If a shape's coordinates do not identify a quantitatively consistent context for numbers and for measuring then the patterns (defined in such a context) are not reliably measurable, and the patterns cannot be stable, so the description has no (meaningful) content

Thus, the shapes with meaningful content are those which have local measuring properties which are:

1. Linear,

2. metric-invariant, and

3. continuously commutative, except at a finite number of points (or except at one point), and

4. the metric-function has only constant coefficients,

or

5. if the coefficients (of the metric-function) are "functions of the other coordinate variables" then the shape must be based on a "parallel geometry," which turns out to be:

1. lines,

2. Cubes (or rectangular-faced simplexes),

3. cylinders, and

4. "orthogonal-sets of circle-structure shapes with orthogonal coordinates (which allow local parallel transport to be commutative)" (tori, and discrete hyperbolic shapes [where discrete hyperbolic shapes are shapes based on many toral components, which are attached to one another in a linear fashion, but can be folded at the points between the toral-components], note: where a torus is a discrete Euclidean shape).

Physical description is all about putting these shapes into a many-dimensional descriptive context, wherein the observed stable properties of spectral-orbital systems…, which exist at all size scales, and are composed on many-(but-few)-bodies…, can be described to sufficient precision, and the descriptions is relatable to practical creative development. This means using the above list of stable shapes which are placed into a higher-dimensional context and the topologies and interaction structures are consistent with observations and they lead to practical uses or they lead to "practical" (or realistic) exploration of the existence of which life is a part.

The main issues concerning physical description are about the containment of a lower-dimensional component (material-component, or equivalently a stable metric-space-shape) within (or upon) an adjacent one-higher-dimensional stable shape.

This can be "perfect-fits" of shapes which are equivalent to the higher-dimensional shape's faces,

or it can be condensed material, where condensed material is about the relation of existing shapes of the same dimension and of different sizes, eg the atoms and the solar-system both being 3-dimensional shapes, but the interactions which are most significant within this dimensional-subspace (of a shape-dimension-containment (subspace) sequence) are the shapes which are the size of the solar-system, so that the force-field manifests more strongly when identified over a greater size-scale than when identified over the size-scale of atoms (eg a form of dark-matter) but the smaller forces (of this higher-dimensionally defined force-field) cause the van der Waals forces between atoms.

That is, it is from the context of stable systems (the main issue of physical description) that the other aspects of (the physical descriptions, of) local measuring can be related to the descriptions of material-interactions so as to be placed in a meaningful (descriptive) context.

That is, material-interactions are not central to physical description, where this is because they are mostly non-commutative quantitative relationships, and thus they are both unstable and/but they are also second-order (or they are lesser) forces (than are the orbital-geodesic "forces"), where the material-interactions only perturb the stable spectral-orbital structures which are guiding the dynamics of (most) material-components when they are within stable systems, where these stable spectral-orbital structures are mostly (defined for the dynamic material-components) by the shape of space upon (or within) which the material-components are contained

(but now this is occurring in a new context of higher-dimensions, so there exist other [seemingly] hidden relationships, note: the differential-form associated to the electromagnetic-field has a hidden structure to its description [in space-time]).

Chapter 28

propaganda of the experts

The propaganda of the experts

Consider the expressions of the highest echelons of the expert-class, which is now claimed (by the propaganda-education system) to be a part of the intellectually superior genetics, which is now (2014) associated with those in the intellectual-class, eg the physics community

The point of western culture has been that it is a culture which is to be based on lying stealing and murdering (and such behavior's relation to terrorizing the public, so as to control the public) so that this type of the behavior of the ruling-class is sanctioned by the social-institutions called justice-systems, education-systems, and now the propaganda-public-information systems of the media

And

To present "to the conquered many" the illusion of technical progress (and a high-valued civilization; but mostly based on barbarism), based on a fixed form of controllable technology, or a set of engineering principles, which also form a fixed narrow category, which is controllable by the ruling-class, eg in Roman-time this was "the laying of bricks" to build large structures eg aqueducts, big-buildings, sewers, etc

Now it is communication systems, and military weapons, along with the brick-based building of buildings with cement, which the Romans seemed to have also done on a large scale, based on a fixed technical process, apparently, easily communicated between others (yet, still hidden (or remote) from the public capabilities, ie the knowledge is hidden, and the process is expensive) which form the fixed set of technical abilities upon which the high-valued-ness of the civilization is claimed to be based within the propaganda-communication channels of the (Roman) civilization [note that, this identifies the context of a product producing corporation]

Where technical progress is (seems to be) the (main) glue upon which the many stay loyal to the few, in order to be able to master the technical capabilities which are controlled by the few (through investment), and to, subsequently, become an "equal" (or a partnering-cohort) with the few.

Consider the expressions of two of the, so called, genetically superior intellectuals whom have expressed, in popular writing, in books about their ideas about physics, namely, M Gell-Mann "The Quark and the Jaguar," 1994, and L Randall "Warped Passages," 2005.

Gell-man writes a books mostly full of mis-representations about the subject of physics, without careful enough analysis of math or about physical principles. That is, he floats on the set of principles offered to him by the educational-process of formalized axiomatics

His voice is the voice of absolute truth, as expressed to civilization by a "Nobel-prize-winning-superior-intellectual-person."

Everything he writes is an expression, he believes, of an absolute truth, which he gracefully provides to the public about science, as provided by a genetically superior and superior intellectual person, who is one of the very few who is actually capable of discerning truth for the public, at least, that is the story provided to us by the propaganda-system, and apparently, Gell-Mann truly believes that is his rightful social position.

Yet, nearly everything he says is a mis-representation of the subject matter he covers.

For example he expresses Godel's incompleteness theorem (which he does not name) correctly as "the limitations of axiomatic expressions" but then mis-represents this as being an expression about "the undecide-ability of math statements" whereas, since math is about definitive statements in a well defined (language) context, Godel's incompleteness theorem really means that "the assumptions and contexts and interpretations of math and science need to always be revised" by the science-math community, ie it means that everyone is a natural participant in the process of equal free-inquiry which is the true basis for science and math.

Einstein was the last scientist who expressed his ideas at the fundamental elementary level of assumptions, but, apparently, Einstein thought that the quantitative structures of math are absolutes and that a non-linear math context was capable of expressing, or describing, the actual measurable patterns which exist, but when the quantitative structures upon which a description is based are quantitatively inconsistent then the descriptions of the, so called, math patterns cannot possess any meaningful content, and the descriptions are arbitrary and practically useless.

Gell-man provides two very nice sets of expressions

1. He characterizes information-theory by the means in which (or through which) in regard to the way in which information-theory characterizes information, namely, as a relation between randomness and the compressibility of information (which are a part of signals being sent down a communication channel (or through a wire))

2. He deals with particle-physics in 21 pages in his book "The Quark and the Jaguar," of which about 12 pages fairly completely outline particle-physics,

no small feat, since most described properties of particle-physics are all about mis-representations concerning the aim (or logic) of the subject ie mis-representations about the descriptive context, it is a descriptive context which has nothing to do with material-interactions, though that is how it is represented, and it only has to do with the probabilities of collisions in a chaotic random context of a reaction, in an explosion, where the rate-of-reaction (and the energy given-off by the explosion) is determined by the probabilities of component collisions.

Between these 12 pages of Gell-Mann and a succinct statement provided by Y Manin, that "the integral of the wave-function acted upon by derivative-connection of the internal particle-states, provides the perturbed adjustments to the wave-function," and where Gell-Mann identifies (in the 12 pages) the context of the internal particle-state structures, but where it also needs to be noted that though particle-physics is locally, what is called, gauge-invariant, but which is actually a global model,

(ie where "local invariance" means that the interaction equations are based on [or derived from] variations of a covariant scalar value (invariant to the local transformations of the fiber-group)) [or the equations of material-interactions, as modeled by (or as) the particle-collision, and related particle-state transformations of particle-physics, which really only relate to the determination of (or the modeling of) a particle's cross-section (or probability of collision), are invariant to local fiber-group transformations

{but this type of model for material-interactions is only about determining particle-cross-sections, ie a measurable value related to the probabilities of particle-collisions, and it has nothing to do with the geometry of material-interactions}]

That is, it also needs to be noted that neither particle-number, nor energy (of the particle(s) in the particle-energy-level) are conserved in this model of material-interactions, ie there is no model of a global system in this descriptive context, and the wave-function is simply a model of some arbitrary energy-state (or energy-level) in which the particle-structure of interest is being described, in regard to the (other, or external) particles which are colliding with the particle-of-interest, in its arbitrary energy-state (or arbitrary energy-level).

The whole descriptive context is useless, except for the relation it has to particle-cross-sections, and this, in turn, is related to probabilities of collisions in a chaotic transition process between stable states of material where this chaotic transition models an explosion, ie the context of energy is arbitrary and related to a transition (system) modeled by (as) a probability-of-component-collision context.

Furthermore the description

(It) is about a math context which is: non-linear, non-commutative, and indefinably random, so it is meaningless, and without the capacity to have any content, other than the externally imposed content of modeling explosions, ie where one sets-up the conditions which facilitate a fast reaction-rate for the materials composing the bomb (or reaction). Where these explosions are, essentially, detonated by possessing a critical-mass (or possessing a chemical reactive property) where the unstable property of radioactive materials, when it has the property of being a critical mass (or when a small amount of energy is introduced, to amounts of "pure" types of chemical materials, so that it can then initiate and continue a rapid chemical reaction), then it is, subsequently, related to a chaotic transition process which can exist between (certain types of) relatively stable states of material.

That is, this is a chemical reaction rate model of "chemical reaction rates due to probabilities of collisions" (where the collisions need to be at [or more than, or at-least] the right amount of energy) between two types of materials,

But the molecular-component model of chemical properties of material…., which can exist in various types of material systems ie material system which can be in the various states of matter: solid, liquid, and gas; or solids (or condensed matter) and fluids (ie liquids and gasses)…, has not been all that successful at descriptions which lead to the understanding and control of chemical properties and the control of chemical reactions.

Unfortunately, the so called intellectually superior community of (particle) physicists accepts this context as the set of assumptions, which are the only possible ways in which material-interactions can be modeled

And

This is true for both Gell-Mann and Randall.

The rest of the book is delusional, by Gell-Mann, it is mostly about entering the non-commutative, non-linear, and indefinably random descriptive (math) context of "complexity," which is about trying to build a practical model of the world from the axiomatic formalized set of assumptions about existence, which lead to there being no capacity of such a language to be able to describe the properties of the world in a manner which is either a wide-ranging and accurate description, and/ or which is also practically useful in regard to further technical developments.

He also writes about, what he calls, the problem of the mis-information provided by others (to the public) over the inter-net and other communication channels

His solution is that (or seems to be that) only the few who express the absolute truths (for the narrow technical interests of society, ie or absolute truths as defined by [and which are related to the narrow {monopolistic} interests of] the ruling-few) should be allowed to express ideas about subjects through the communication channels

The perfect message for the interests of the ruling-few

Whereas the only solution (in regard to the incessant presentation of mis-information over society's communication channels) is to allow free access as well as equal free-inquiry to be a part of the educational institutions, so that all ideas are expressed, and they are expressed and/or framed so that the assumptions upon which the expressions of ideas are based are made clear, so that they can be easily categorized by a computer and made available for everyone

Instead there is the development of the expert language which builds on top of already expressed so called expert expressions where the assumptions are hidden, and not ever re-evaluated, and the context is mostly nonsense ie based on non-commutative, non-linear and indefinable randomness assumptions which result in the descriptive language to not be capable of possessing any meaningful content, rather, only expressing formal language relations related to a world of illusion

And then there is L Randall

whose main statement is that "she is around the top people, ie the intellectually superior people of all of society (who are in society's, so called, top educational institutions), and she talks to them about particle-physics and string theory, so she is also providing, in her popular book "Warped Passages," the absolute-truths about "what we know about all of existence" and it is about the relations that exist between material and material-interactions, where material-interactions must be modeled as particle-collisions, which are related to internal particle-states, where this relation is made through equations (based on derivative-connections) derived from variations defined on covariant scalar values, as required by the ideas of general relativity, ie the invariance of equations as the basis of physical law, and where, it is assumed that, physical law is (assumed) to only be about the context of material interactions modeled as partial differential equations, which are invariant to local fiber group transformations, where material is reduced to elementary-particles, which interact through particle-collisions,

ie the chaotic transition model of material-interactions, where the model of a system is arbitrary (particles exist in an arbitrary energy-level) and material-interactions are only related to particle-cross-section properties

(where geometric properties can only be introduced through a connection-derivative, which is associated to a metric-space "system containment-set" [but this is not true for particle-physics] the cross-sectional properties of the geometric point-particle collision (which models material-interactions, when in a context of material reducing to elementary-particles) is contained in an internal-particle-state space).

So, in this highly unrealistic context which is unrealistic since it is a descriptive context which is not capable of describing the observed stable properties of the world,

L Randall then describes a geometric context of the internal particle-states which is provided by string theory and

Proceeds to present a geometric model of (?) determining mass for elementary-particles, apparently, consistent with the Higgs mechanism of breaking the symmetry to cause mass to form, but,

"To form mass from what?" Apparently, mass is formed from "the" vacuum.

In a quantitative descriptive structure energy and "the vacuum" (the place or energy-quantity) from which energy can either form or energy can be absorbed so that the quantity of zero or the concept of an empty-set can be equivalent aspects of the existence properties of elementary-particles,

ie particle-physics is, supposedly, based on a math-model (for material interactions), but where the quantities (upon which the descriptive functions are supposed to be built) are not well defined, since particle-physics is a math model which is: non-linear, non-commutative and indefinably random (and thus quantitatively inconsistent), so that (or then) zero can play any role which one wants it to play, (1) it is zero (or the empty-set) or (2) it is infinite-energy-value (ie it is equivalent to the big bang), or (3) it is just the right-number (so as to fit data), then in such a descriptive context a scalar-function can be added to a gauge-invariant (or covariant) scalar so that a variation (on such an covariant scalar action-value) of this scalar results in the derivative-connection set of partial differential equations, which, in turn, can allow one to fit data about particle masses to the descriptive structure of particle-collisions, apparently, so that the zero-position of the vacuum energy for a colliding-particle is shifted to show-up as a non-zero energy-value (associated to a point in space of a particle-collision) so that the particle has non-zero mass, ie mass equals energy. That is, mass comes into being through an elaborate descriptive context placed onto an arbitrary and meaningless inconsistent quantitative containment set structure for a (quantitatively inconsistent) description.

And the data from which this descriptive context (of particle-physics) is built, is interpreted to fit into the descriptive language of particle-physics, so as to easily be able to confirm this absurd and useless descriptive context,

This is exactly analogous to what the epicycles in Ptolemy's descriptions of the planetary-motions were used for, ie the epicycles could be easily used to confirm (by data fitting by using the epicycles) the planet positions in the night-sky, and if the "predicted" motions were not confirmed then new epicycle structures could be added, so as to continue to be able to verify the model (of either Ptolemy or of particle-physics) arbitrary phases (of a probabilistic model) can be added so as to continue to be able to fit the description to the data, ie the epicycles of Ptolemy were phases in a probability model, which was to be used for fitting data [this is also what the U(1) fiber group (or phase group) allows in particle-physics, the particle's themselves are simply a many-dimensional phase-adjustment to the geometric properties of elementary-particle collisions, and the cross-sectional scattering phase adjustments needed to fit the data of the particle-collisions].

But nonetheless Randall tries to use the geometry of the (6-dimensional) space, in string-theory, where material does not exist, so as to develop a non-linear and indefinably random geometric model as the geometric basis for why graviton cross-sections are small with massive-particles, where mass is determined to be consistent with the Higg's mechanism (described above as a scalar function which breaks the symmetry of the vacuum energy for a particle-collision) does result in the property of gravity being weak so that the weak force of gravity is also an attribute of the description of the properties of elementary-particles in particle-physics

It seems to be a descriptive context which is more interesting than particle-physics, which, in turn, is highly algebraic, and, essentially, only related to one point in space, ie the point of particle-collisions, where in Randall's descriptive context the description is about geometry, but it is the geometry hidden from experimental measuring, and/but, in fact, it is a model which is non-linear, non-commutative and indefinably random, so it (Randall's description) is a description about nothing, ie a description with no meaningful content.

Randall does a fairly good job at describing her geometric interests, but it is clear that she has no clear vision of the abstract basis for her work, her descriptions of quantum theory and particle-physics are complex and show clearly that she has no capacity to critically analyze the language (of particle-physics) about which her work is desperately trying to extend

Apparently, she is unaware that "all she is doing" is only related to trying to detonate a gravitational singularity, where the whole of science has now (2014) been transformed into physical descriptions which only depend on the properties at one point in space, the point of the particle-collision, and the cross-sectional properties of any identifiable particle, so as to fit into the descriptive context of a large explosion,

(even though these so called particle-events are unstable, and, thus, one is unable to incorporate these unstable events into a valid probabilistic model, since valid math cannot be developed from a context where counting is not reliable [ie counting unstable random events is not reliable]),

The main point is that, both Randall and Gell-Mann are almost entirely about their positions within the propaganda-education-indoctrination system as the intellectually superior beings, which our social system filters-out of the crowd, so that the absolute truth about the world can be reported to us through these experts, so that we can be assured that there can be no other way in which to run and organize society, so that the public can be sure that our civilization is based on absolute truths, which are discerned by the intellectually superior few

Even though these pathetic models of "the true nature of existence" are exactly the models used in the engineering of nuclear weapons, and these descriptive structures are otherwise completely useless expressions, which only express possible truths of a world which is an illusion, ie only related to a word-world which is, otherwise, an illusion.

These people, such as Gell-Mann and Randall, are uncritical, thoughtless, and self-obsessed with their memorized rules, through which they see "a means by which they can realize their ambitions of being arrogant superior-people." This is, effectively, an obsession with an arbitrary dogma, which is no different from a similar obsession associated to the belief in racism, which is, again, an expression of selfish (or self-obsessed) arrogance

And

In fact, this dogmatic obsession where these people seem to be unaware that their actions are really about trying to help (the gone) E Teller, in turn, help the ruling junta, realize a doomsday weapon, with which to terrorize the world (the US empire has always been centered on the use of psychopathic terror, and a false sense of arrogant superiority)

But

Nonetheless, their social positions and their clearly mis-guided efforts, which represent science as an authoritative dogma, which, in turn, represents a, so called, absolute truth, which is really a viewpoint which is only associated to the making of nuclear weapons, and it is not a viewpoint which is related to either wide-ranging (fairly) accurate descriptions of observed properties, nor is it related to any wide range of practical creative efforts, and thus it is an effort which is only about arrogance and the expression of inequality, as racism is also used within society.

The, so called, intellectually superior people..., as identified by society's institutions, which, in turn, are only allowed to serve the interests of the investor-class..., are the intellectual-bullies who force the society into the unfounded belief that "this is the only way things can be," and in the name of knowledge and creativity (but as only the investor-class is allowed to define the value of these attributes) all other knowledge and creativity is excluded, by means of the efforts of these idiot wage-slaves (who seem to really represent a form of institutionally managed autism, with limited language skills, but who are people who are also competitive and arrogantly self-serving, ie they are also psychopaths) who follow: quantum physics particle-physics and general relativity as well as all the other mis-guided literary-expressions of a delusional math construct which are derived from these, so called, absolute truths, where the math models of these absurd constructs (about all of existence only needs to focus on one point, the poit of the particle-collision in space-time) which are based-on math patterns which are: non-linear, non-commutative and indefinably random and the are quantitatively inconsistent thus the quantitative properties of the base space (where it is assumed the system is contained) is not capable of describing a stable math pattern, ie all is just fleeting change associated to a chaotic transition, and whose claim to "being verified" are data and associated interpretations of this data which do not hold water (ie how can people take this crap seriously)

Ie where are the valid descriptions of all the stable many-(but-few)-body spectral-orbital systems that are observed. If these many stable systems cannot be described by the language of science, then science clearly needs new forms of expressions: new assumptions, contexts, organization of math patterns, and new interpretations of what is seen, etc etc

The entire social system is an expression of a very immature expressions of psychopaths, which is all about destruction of everything else, so as to hold high a few elites, who are the willing autistic-psychopaths, who are an active part of this destructive social process (yet they only see [obsess on] themselves) ie those inclined to authoritative dogmatic absolutes and arbitrary ideas which express their superiority, such as racism.

Part II
(social commentary)

Chapter 29

Social Commentary

(a Rambling discussion)

In the propaganda system one can chose between being an intellectual bully or being a physical bully, where both sides depend for "their sense of being superior" on the set of arbitrary choices made by the investor-class whom both sides, effectively, worship, though the intellectuals actually believe that they are in their positions because of their own independent rationality, but they are really the hangers-on of the rational choices made by the ruling-class, whereas the physical bullies seem to believe that they are expressing their individual freedoms (independence through physical domination, miniature-models of the ruling-elite), but they can only become dominant, if they are provided with (or are allowed) impunity,

And this requires that they form (or join) a military-collective which supports the ruling-class,

eg one aristocrat or another (ie intellectually dominant elitists [based on arbitrary dogma] or violently dominant elitists [again based on some form of arbitrary righteousness])

Which side is the more hypocritical? {Either, Or}

If one is truly independent then the spies and the institutional managers either "turn you around" (turn one into a spy-agent asset) or exclude you from any form of meaningful expression, as well as an exclusion from being able to work in society, (where such exclusion is used as a demonstration by the propaganda-and-academic-based intellectuals of such an independent person's intellectual inadequacies, ie "there is only one way in which things can be done," where in science this is expressed as "society already possessing absolute truths about the nature of the physical laws of the material-world" but these laws cannot be used to solve for the very stable spectral-orbital properties of the fundamental many-(but-few)-body systems [ie they have not made any significant mathematical progress since Newton]) [to re-iterate, the independent intellectuals are proven to be stupid, by the intellectual-class, since "our culture…," the intellectual-bullies claim, "… already possesses absolute truths," eg "the, so called, laws of physics"] (ie thus, exclusion in regard to these (or in regard to any such new) ideas being expressed, demonstrates one's (or such an independent intellectual's) intellectual inadequacies, especially, if one is not being successful in the highly-controlled and systematically-compartmentalized and propaganda-dominated market-place, where the terrorized (and highly suggestible) public only acknowledges established "knowledge," ie no new knowledge is allowed by the consumer in the market-place), and "in the collectivist-system of bullying," the justice system and their satellite-militias, will try to bully such a person, such a person as an independent person, (where the collectivist-system of bullying is built from intensive hidden interference, in the lower-classes, since the lower-classes are less likely to support the interests of the investor-class, and it is, thus, the lower-classes which are more dangerous to the investor's interests, than are the indoctrinated intellectuals {but it is the spying on the intellectuals is expressed in the media, while the hidden manipulative agitation of social violence by the spies within the militias [and within the spy-manufactured hate-groups] goes un-reported, or grossly under-reported, since it is central to the process of militarizing society}),

Eg the stand-off between Bundy's supporting militias and the police, was really a stand-off between (1) the police-personnel and their spies (and some patsies) and (2) the police-personnel and their spies (and some patsies), eg "in the evening" the two sides likely went-out to dinner with one-another so as to compare spy-notes and to plan some other PR event, somewhere else.

Note that: Western culture has not progressed "at all," from the social conditions of the Roman-Empire! (or has it regressed?).

In the propaganda-education system, all knowledge is pre-packaged and properly-categorized (there is no room for new ideas), in regard to the narrow interests of the investor-class, especially, in the way in which the investor-class uses knowledge for their technical investment projects, and this (categorization of ideas) both narrows and forms a highly categorized organization of knowledge, categorized based on its use by monopolistic businesses, is also the way in which the entire society uses knowledge. That is, either there are monopolistic businesses, which one can be a sub-part, or there are institutional ways in which to use knowledge, in regard to creativity, where creativity is best described as practical creativity in which some instrument is built, where this instrument serves the interests of the investor-class, but the word "creativity" most often means art and literature, eg motion-pictures, vague expressions and fantasies, etc,

However, though science is automatically thought about as being only related to practical creative efforts, in fact, it is now "closer to the literary creativity of science-fiction," than being related to technical practical creative development.

The reply, to this criticism, is always that the "modern gadgetry" is all about using modern science to develop electronic capabilities, but this is only vaguely true, since the science used is almost entirely 19th century science, ie solvable controllable classical physics, ie essentially Faraday's electromagnetic laws, and Tesla's circuitry adaptations and innovations and contexts, where subsequently, the TV was modeled mechanically by 1910, and realized electronically between then and 1920, whereas it was the fast switches of the TV, which led to the Boolean algebra analogy with "logic-circuits, based on the circuit structure of switches," and the subsequent development of computers.

Note that, the microchip is simply an optically-imaged circuit-board, built on a semi-conductor crystal, whereas the transistor and semi-conductor-diodes (which are circuit components) were developed using thermal techniques applied to silicon crystals, and then "carefully placing metal conductors into the circuit component" (ie transistor, diode, capacitor, resistor, etc) the only ideas or models which come from actual "quantum properties" are the laboratory-measured crystal discrete energy-levels, which are only vaguely related to quantitative quantum-theory models of electron-components trapped in a box (ie trapped in a crystal).

The question became, "How fast can information (signals, or signal-packets sought [based on logic-circuits] from memory-banks) be processed through logic circuits, and how to organize this process?"

The body of science information has come to be totally controlled by the investor-class, in that, it is they whom determine how science is worded (eg physics is defined as bomb-engineering, ie particle-physics) and categorized and compartmentalized, so as to suit the needs of the investments of the ruling-class.

It is not an objective search for truth done by independent rationally motivated people,

Rather (especially since WW II when the society became completely militarized and regimented),

Note: where the "hit on Kennedy" seems to have been an exercise by the military in regard to their total information control, which their spy-techniques were already providing for the spy-agencies, as the CIA did "the hit," set-up a patsy, and controlled the propaganda about the event, within a completely militarized institutional, so called, justice-system… ie the so called, propaganda-and-justice-system… structure.

Education and the so called "search for knowledge" is narrow, competitive, and far too authoritarian, especially, in regard to the education system being based on "graded competitions," where "the rules of the competition" are based on narrowly defined dogmas, a structure of word-usage which best facilitates the relation of academics to the servicing of specialized instruments.

The claim is, that knowledge has become too complicated and difficult for ordinary people to engage in its development, but this is only "somewhat true" (in a society which mis-represents "what knowledge actually is,"), especially, for the way knowledge is related to certain instruments, where the interest of the ruling-class is for "the slow development of their instruments," so as to maintain their (fixed) monopolistic model of market definition, and market domination, ie knowledge is narrowly categorized and defined so as to be based on the way in which businesses want to use knowledge, eg physics became the bomb engineering knowledge (and engineers) for the military industry, and the knowledge is narrow, and has become slightly complicated and seemingly abstract, eg (in regard to spying) vague ways of establishing the order of spectral-signals from any structure which possesses enough stable order for information to be developed (in turn, is being developed), eg relating electromagnetic signals from a relatively stable quadra-pole charge configuration.

But the actual development of knowledge in an independent and rational manner has been thwarted, since this actually interferes (or could interfere) with the investment interests of the ruling-class, ie if a new knowledge is found and "the existing instruments are found to be un-necessary"

Thus, independent interest in the much more set of fundamental questions of physics, which, if not answered, will result in an un-ending spiral of meaningless science discussion, about wildly speculative models, in regard to particle-physics, general relativity and the relatively wide set of derived theories: string-theory, quantum-loop gravity, grand unification, etc etc all related to bomb engineering, where this is because "the model of particle-physics is only relatable to elementary-particle cross-sections which are (ie speculated [or hoped] to be) used to determine rates of nuclear reactions" and general relativity is only being used to model, in a very inaccurate and speculative manner, a mythological gravitational singularity. The point of bomb engineering is now (2014, and since about 1955) that of trying to detonate a gravitational singularity (thank E Teller for this, ie thank a "police state practitioner" for this, the man who crafted totalitarian-science within the US, whereas one might thank the Pinkertons, or the US justice-system, for imposing both wage-slavery on the public and an imposed religion, of "worshipping the particular uses of the partial knowledge of the material-world, which is of most interest to the investor-class and their military monopolies," also on the public).

However, when general relativity is placed in its proper math context, which has now been identified by Thurston-Perlman geometrization, then general relativity can still enter the discussion at a fundamental and elementary descriptive level, Einstein's main value was that he began his discussions at the level of assumption and context but space-time is not victorious over Euclidean frames they do different things and non-local properties can be distinguished but furthermore in regard to general relativity

Education (or knowledge) vs. creativity

Knowledge is developed often in relation to a creative intent, ie to "use what is known"

But rationality (in regard to descriptive knowledge), [if directed by the simple language of assumption, context, and interpretation], can lead to new creative contexts.

Whereas creativity in society is almost exclusively controlled by investment.

And the education-system is about presenting the "narrow band" of knowledge, which best serves the investor-class, so as to exclude all other expressions of rationality.

In fact, the use of spying and its relation to field-agents is all about finding expressions of new ideas and sending out agents or contacting institutional managers so as to interfere with the expressions of new knowledge.

But the main "set of spies" are the indoctrinated souls who fit into the propaganda-education system, ie spying is managed and controlled around the information of the propaganda-education system (psychological control is manifested around a fixed mental construct of the world provided by the propaganda-education system, and where this main "set of spies" are the indoctrinated souls who actually believe the delusional quantitative models which are supposedly crafted so rigorously on absolute truths, ie the world cannot be any other way.

That is, the icon of a top-intellectual cannot be anything other than a superior human-being, who can discern the "absolute truth" about the world, and discuss it quantitatively, in incomprehensible contexts.

But

What systems are being described? Can the stable spectral properties of a general nuclei be described using the, so called, absolute truths about the world? {No} Have there been any new inventions, which are different from "systems which are simply quantum properties, which are most deeply related to classical 19th century physics, by coupling the quantum properties to a classical system?" {No}

That is, the comprehensive list of new inventions

1. the nuclear bomb, but the nuclear bomb is really modeled after the 19th century model of chemical reactions, and

2. the laser, the other modern inventions are quantum properties, where the quantum properties are determined experimentally, so that the quantum properties of one quantum-system can be coupled to a controllable classical system (usually an electronic circuit), eg spin-resonance imaging is about coupling the electromagnetic properties of spin-systems to a classical circuit.

Though S Jobs, in regard to the investor-class, was fairly broad in the products which he developed, but it was a very narrow expression of the specialized knowledge of electrical engineering (related to [or possessed by] the children of electrical-engineers that lived in his neighborhood) and the complex microchip circuits had already been developed {by using the properties of thermal and related to crystals (whose experimentally determined energy-levels could be used in the context of voltage within classical electric circuits, and thus controlled by the circuit's properties)}, and thus, the micro-chip was ready for new applications, in regard to telephones, TV's, and computers, ie his (Jobs, and to a lesser extent Gates) creativity was very narrowly defined.

Gates created a monopolistic business model while Jobs was (at first) ideal (carefully designing an individual computer) and got a bit kicked-around, but eventually he learned to play the game of "monopolistic domination" as practiced by the so called superior individuals

Similarly the knowledge of math and physics have come to be very narrowly defined in that they focus on quantum physics, particle-physics (or bomb engineering), and general relativity, so that E Teller (and the fission and fusion bomb projects, and the relation of these bombs to the military investor-class) turned physics and math departments (around the country) into departments which focus primarily on detonating a gravitational singularity, in the context of particle-physics descriptions, where particle-physics deals with the chaotic random particle-collision probabilities, in regard to the transition between two relatively stable material states, ie the bomb before detonation and the state of material after the explosion (are the two relatively stable material states)

But this knowledge of particle-physics has very little to do with being able to describe the stability of general: nuclei, atoms, molecules, molecular-shape, crystal energy-levels (the critical temperature at which superconductivity can occur has been exceeded, so the BCS theory of crystals can no longer be viewed as being valid), the properties of living systems, the stability of the solar system, and the (galactic) properties of dark matter.

This list identifies the biggest mysteries in regard to physics, but these mysteries, "which are of a much more foundational matter than detonating a big bomb," are ignored with the statement that "these simple stable systems, built from many-(but-few)-bodies, are too complicated to describe," but these most fundamental physical systems are very stable, and this means that their math models should be: linear, metric-invariant, and continuously locally-commutative, ie solvable and controllable, where such a descriptive context lends itself much more to creative developments than does the chaotic and indefinably random context of quantum physics and particle-physics, and the non-linear (and, subsequently, also chaotic) context of general relativity.

That is, one wants an education system which does not serve only the narrow creative interests of the investor class, but which also develops a wide range of new creative contexts,

ie wider and more interesting range of creativity than did the "micro-chip" when it ushered-in (during the 1970's and 1980's) a narrow range of creativity, controlled by the investor-class, a new knowledge based on new assumptions and new contexts, is a new knowledge which can relate the public, through their new knowledge, to their own new creative products, rather than the narrow set of products to which, "what should laughingly be called public education" is now related.

The narrowness of the creative capabilities of the society is the main reason for its economic collapse, so that one can now see (or should be able to see) [that society's creative abilities primarily focus on military instruments and] that this is necessary for a society which is hierarchical, and that hierarchy is arbitrary, though the propaganda system continually states that "the current society possesses absolute knowledge" and

Thus, "this is the only way in which society can be"

But the absolute knowledge is too narrow and it has failed. Yet the hierarchy and subsequent sense of there being an absolute truth associated to the social hierarchy is the delusion which the propaganda system places into everyone's mind, that this results in such a deeply indoctrinated mind, to insist that general relativity and particle-physics cannot be questioned otherwise one is moving away from the absolute truth which ahs been established in the minds that the failed theories of particle-physic and general relativity are to never be questioned since our culture possesses a superior truth ie the truth which the propaganda system has placed into everyone's (indoctrinated) minds. The authority in truth of the religion of particle-physics cannot be questioned in a similar manner as the religious authority in the age of Copernicus and Galileo could not be questioned

If one questions such an authority then one is expressing barbarism, ie "things must be this way" is upheld by every indoctrinated mind

But this is science, no science is in the paragraphs above, and rather particle-physics is verified in the same way in which the Ptolemaic system of epicycles was also verified that is scientific descriptions are verified by the wide range of systems to which they provide valid descriptions and by the great value they have in regard to practical creative development.

However, in regard to the absolute truths provided to society's indoctrinated minds by the propaganda system, the real truth is that:

The abstract math models of great complexity, which are based on quantitatively inconsistent math structures of non-linearity, non-commutativity, and indefinable randomness and placed in a containment context where the math sets are too-big (so that the models can be defined, in regard to convergences), so that the models are logically inconsistent, and thus, this descriptive language leaves one in a state where the descriptions have no content, and no meaning, unless a controllable system is put into such a context where the system has feedback, and the system is designed to realize a purpose which is externally determined, and the feedback is designed to realize this external purpose, ie there are no internal controlling structures in such an otherwise meaningless "quantitatively and logically inconsistent" descriptive context.

When considering the intellectual left one needs to understand that they are both elitist and indoctrinated though they see themselves as quite capable of critical thought, but in fact they are limited and narrow, being embroiled in the technical aspects of their fellow (respected and anointed) elite-intellectual's expressions....,

ie a circularly repeating intellectual construct of language whose intent is to slowly develop an accepted way of thinking in relation to the objectives of the propaganda-education-system,

ie for the public to stay fixed

eg trade is about investing in markets, psychology is about manipulating society's icon's of high-value, where they (the monitors [or gate-keepers] of the propaganda system) are supposedly anointed by the academic and artistic and intellectual community (supposedly based on markets), but really they are selected by the ruling investor-class, in order to maintain the domination of society by the investor-class.

In this setting..., within which they become fixed and essentially irrelevant..., they compartmentalize and categorize and their analysis becomes irrelevant and distant from reality

The inane artist arrogantly expresses their high-social value (as independent thinkers who the public wants to read, but their market-value is really only the market-value created by the propaganda-system) and, subsequently, they demand that their copyrights protect the market value of their creations, ie of their, believed to be, genius, but their fortunes are only expressions of the domination of the investor-class over society, and "the use of the, so called, artist" by the investor, for the investor's needs in the propaganda-education system, a system of narrow dogmas, and subsequently, the so called, fierce competitions of intellect are defined in regard to the narrowly conceived ways in which things "must be this way" ie "there are no alternatives"

The left stands for intellectual deception

But

The right stands for straight-out deception and lying and thievery and coercion all sanctioned with impunity by the institutions of: law politics science and propaganda, and which serve the interests of the investor-class all about dominant control of language, ie the highest authority, and violence

All composed and knit-together for an arbitrarily defined hierarchy-of-value with the investor-class at the pinnacle

The main focus of this deception is the lower-social-classes where propaganda management spying and agent-interference create an illusion of individual domination by means of individual violence

The left is to make the social-institutions…, which are used to control of language and practical knowledge which is related to both language and practical creativity…, to work so as to continue to support the interests of the investor-class

This means deceptions in regard to abstractness, arbitrary technical language, and complications all related to the practical instruments into which the rich have invested and about which social-institutions are organized

The basic distinction between the two "opposing" sides is supposed to be "collectivism vs. individualism"

But both sides are for collective societies.

One to intellectually support the creative needs, practical and cultural, of the ruling-class while the other is to develop the propaganda needed for the society to collectively support, by violence, the interests of the ruling-class

That is, if the left took-over they would do the same, so called, creatively productive things that the ruling-class does, (perhaps, briefly, with more compassion) because that is the intellectual structure within which they have been indoctrinated

The way out is to have a society of individuals as equal creators and an equal market but small collective structures are (or may be) needed to support the equal nature of individuals and their ability to create

That is, creativity is not simply extending a fixed way of thinking or be engaged in manipulating and deceiving others, nor is it about using a limited viewpoint of science and technology and extending "products" in ways in which the, so called, market can continue to be monopolized

And

People are not basically violent animals as the "publicized religions" express, but rather, people are curious and want to be creative, ie they are really gentle when they see one another as being equal

Again

The way out is by using language at an elementary level of assumptions organization containment context interpretation and to use technical language only when the elementary aspects of the (relatively fixed) language have been developed. This, by the way, is the point of Godel's incompleteness theorem, as well as Socrates seeking for all in society to partake in equal free-inquiry.

The language is destroyed by the propaganda-education system

The left vs right

The science vs. religion

The superior person vs. the common low-value person

The liberal left vs. racism

Education vs. intellectual elitism

Are all mis-representations and based in a language without meaning

This is similar to the way in which math and science have been using a descriptive language based in math which also has very little content (ie it has, essentially, no meaning, ie meaning is introduced from outside of the descriptive context, where the math description provides, virtually, no compression of information, it vaguely describes the environment to which a controllable feedback system must respond, whereas the controllable system is based on a solvable classical description), ie it is a use of a language which is almost incapable of compressing information, ie information stays at a level of indefinable randomness, where indefinable randomness is a math property of probability which is based on elementary-event-spaces where either the elementary-events are unstable, or the elementary-events are not determinable. The source of the problems which the language has is due to (or is a result of) it being used almost always…., especially in regard to what is called physical law, where physical law is based on (partial) differential equations (which are used to model the measurable properties of an assumed to be material system)…, expressed in a context where the basic context is non-commutative, non-linear and indefinably random. It is a context where there is not quantitative consistency for the math structures, it is a descriptive structure which, essentially, tries to reduce physical description to a discontinuously chaotic state of random-collisions of point-particles, which exist essentially, in the context of transitioning between two relatively stable states or an unidentifiable system construct.

This is the dominating descriptive context of the professional math and physicist, and (yet) they cannot describe the observed stable spectral-orbital properties of general physical systems (which exist at all size-scales).

Nor is this descriptive context related to any type of new contexts of (or in regard to) creative technical developments, where internal physical properties of the system are controlled, though feedback can be introduced to classical systems so that they can be steered through the sets of fleeting chaotic environments because the classical system is endowed with an external purpose, eg going somewhere, in turn, related to other external inputs eg global positioning

In the left vs. right scenario of the propaganda-system, it is

The collectivist society run by the intellectual scientists

vs.

Individual freedom, where one's freedom is related to "one being able to violent dominate others"

This is far from true, though it seems to characterize the left, in fact, the right are also collectivists, who support the rich, as well as supporting the violence needed to maintain their arbitrary social hierarchy

The militia-right is not expressing the natural (bigoted) activity of the so called uneducated-class, rather this bigotry is expressed as much by agents and it is really an attack on the lower classes essentially, undertaken by the justice-system, so the lower-classes, will not unify so as to oppose the investor-class,

where this attack is based on propaganda and interference by spies and the justice system, and it is used by the justice system and spy agencies to terrorize the public, by manipulating those in the low-classes who they can prey-upon, but its main purpose is to interfere with the lower-classes, so as to oppose their being able to unify themselves against the upper-classes

Unfortunately, the left does not seem to see that the intellectual structures, about which they are so well educated, are the intellectual structures which best support the investor-class, ie if the intellectual-class actually took-over then in such a utopian rule by the intellectuals, the intellectuals would be doing exactly the same, failed, activities as the investor-class now does. Since that is all that the intellectual-class knows, ie they are idiots, they do not fully comprehend, that it is a propaganda-education system, its not simply Fox-news, where one can barely distinguish between Fox and all the other media including the, so called, alternative media. Though there is some useful information available on the inter-net though it is difficult to find and/or identify.

The science vs. religion, is

The materialist vs. the spiritualist

But this is far from true, (1) the idea of materialism has failed, which should be quite clear since the properties of stable systems cannot be described in the context of materialism

(2) spiritualism is about equality, but religion is all about the propaganda of (for) a social hierarchy, in which, supposedly, the superior moral person is to be the judge the inferior public.

Furthermore, the religious side's leaders are more materialists than is the science side's personnel, since the material world is the context in which the religious leaders support the very materialistic ruling-class. This is Calvinism, and it was also the power structure of the Catholic church. Subsequently, the ruling-class sees to it that the religious leaders are protected and maintained as "spiritual leaders." This type of hierarchical propaganda-like religion is also a very good propaganda structure for the rich, who, effectively, become seen, by the public, to be supported by God.

The natural superior people vs. the inferior people, or superior culture vs. inferior culture, is an idea which is built in a social context, where certain types of behaviors or activities fit into a rigidly structured society and these activities are accepted as "God given" and artificial measures of high-value can be used to measure the value of these activities, ie high-value is arbitrary and quite artificial. It is these well defined activities which have become dysfunctional which lead to the decay of the society.

Intelligence tests measure the rate at which a person can acquire the dominant culture, it is not an actual measure of intelligence, where intelligence would be better defined as a person's ability to discern truth, but "who knows the truth, so as to actually judge this attribute?"

Racism is an arbitrary belief about one type of person being superior to another types of person.

Consider that the high IQ person would be considered to be superior to an ordinary person by the left, but what is being measured by IQ is an arbitrary property related to the dominant culture, so this is a form of racism, ie an arbitrary belief, which has no constructive function in society

The intellectual dogmas of the intellectual-class are the arbitrary narrow dogmas, which define intellectual authority in society, and they are established by the investor-class, since they are the narrow dogmas of the intellect which best serve the ruling-class. That is, these high-valued dogmas are also arbitrary, and they become quite dysfunctional

Arbitrary beliefs which have no valid constructive affect within society, but they do help maintain the selfish interests of the investor-class, and they are the basis for continually repeating the main idea of the propaganda-education system, ie that people are not equal.

In a world where 85 people own the same amount as half of the world population one can also be assured that these 85 people own the controlling state in 99% (or more) of all the monopolistic businesses of the world, and that in this set of 85 it is "likely true" that only 10 (or less) are the main operators.

Thus, there is a fixed way of developing and using knowledge, which supports the interests of these few, and this narrow viewpoint in knowledge, and its basis in selfishness and violence, is the cause of the failure of knowledge and the failure of creativity, and the subsequent destruction of the earth due to the narrow list of resources upon which their creative institutions depend, so as to be poisoning the earth, for the selfish interest of the few, in a social structure of domination and extreme violence and extreme inequality.

What the left…., for lack of any meaningful category, since the propaganda-education system has striped-away any meaning (or use) which language might possess…., misses in their analysis of the society is the stark violence through which the society is run and the way in which propaganda is an arbitrary expression which says anything to support itself

Furthermore, any person who is expressing ideas on the left is chosen to express those ideas as a plan within the propaganda-system. That is, the propaganda-education system has brain-washed and subsequently chosen a certain set of wage-slaves, who express ideas which are a part of the propaganda plan of being able to assert the arbitrary nonsense which is expressed over the propaganda system

One gets book-reports from the left… from the works (old books) which the investor-class has allowed to be published

The main point the left misses is that the, so called, mess of violence and bigotry on the, so called, right, (but this) is an illusion built by the propaganda-system, and is really a propaganda-model of the lower-classes, where the lower-classes have been the most infiltrated and controlled by the spy-and-violent coercion processes, used by the ruling-class, since the lower classes can most easily unite in their opposition to the dictates of the investor-class ie Pinkerton was a cop who knew well the war being waged on the lower-social-classes, and who formed an army for the ruling-class's interests, that operated with impunity in regard to law.

Again, as always, the issue is justice, and the institutions which use violence and lying with impunity.

Was it the spirit of J Caesar who said, "if one wants free-speech then go buy a newspaper?"

Furthermore, the so called expert intellectual-class is operating in a dysfunctional intellectual context, ie the intellectuals have not seen the destruction of descriptive knowledge so as to serve the interests of the ruling-class, where this involves being wage-slaves and the easy to manipulate the idea of high-value in society when the communication systems are controlled and the management serves the ruling-class. This is based on coercion which is lost on (which is not seen by) the intellectual-left, ie those whom are allowed to express ideas over the media.

The main function of the intellectual-class is the use them (and their indoctrinated intellectual structures) by the ruling-class both

(1) for their very limited relation to "service instruments and creative productive development (almost all is military)"

and

(2) in the propaganda system to express the idea that people are not equal (about which the intellectual-left is in total agreement, since after-all, they are high-paid wage-slaves).

The worst problem with the intellectual-left's analysis is that it does not define words and it does not deal with: assumptions contexts interpretations etc in regard to the ideas which it expresses, rather it floats on the undefined context of the propaganda-system.

Education vs. intellectual elitism

The true nature of education is that it is about the experts only being allowed to follow in their expressions of ideas, about authoritative traditions and their "job" is to keep adjusting and complicating fixed sets of instruments

This is mostly about militaristic expansion and on well defined military products which are used to destroy the earth related to the selfish interests of the monopolistic oligarchy of a mind colonized by the oligarchical interests

All control of physical systems are based on the 19th century classical physics models

Whereas

Quantum physics, particle-physics, and general relativity and all the various derived theories eg string-theory many-worlds and quantum loop-gravity etc etc are all dysfunctional

Where the math models which the professional math people are providing for society's practical creative efforts, is, essentially, a dysfunctional model of chaotic and transitory and vaguely discernable and unstable patterns through which either a feedback-controlled classical system navigates (moves) with its purpose determined from the external interests, and not internally controllable, in regard to the chaos whose information is not compressible and useable but only to be transitioned across (or navigated)or

As well as.., being related to bombs whose core idea for "control" is determined by reaction-rates governed by particle-collision probabilities, that is, there is no control rather special types of systems made of particular material-chemical-radioactive-nuclear-types are brought into being by, usually, thermal processes of purifying materials.

The lack of development of life-science chemistry implies that… the model of particle-chemical-component collision rates as the model for chemistry between individual mole-types…. is not a valid model of life-chemistry

The main problem is that the math models of physical systems stay in algebra or pde and they stay with the idea of materialism and they represent their models which possess a local context in which the patterns are non-commutative and non-linear and/or they are placed in a context which is indefinably random

They push their failed ideas to their ridiculous extreme

And not consider new creative contexts based on different elementary language structures

One needs to also consider new elementary language structures for

Society

Law

Politics

Economics

Eg

Is the point of society to either create or to dominate?

A good person to critically consider is H Giroux (HG) the French-Canadian (liberal) intellectual, since he considers a greater breadth of the society in his commentaries, than many other liberal-intellectuals, where this is done in an (his) article in, truth-out, 6-8-14,

"The Spector of Authoritarianism and the Future of the Left: an interview with HGiroux…"

But one sees a lack of definition, ie no clear focus on relevant details.

However, when liberals attack foreign policy they spend most of their time presenting the details and they appear to be making a court-case in regard to the illegality of the state actions

Whereas when HG states that "the commanding institutions are in the hands of corporations financial interests and right-wing bigots whose strangling hold over politics renders politics corrupt and dysfunctional" but

What is the structure of this take-over, of what is supposed to be a democracy based on laws?

How has this happened?

The weak answer is business interests (but this idea can be expanded in regard to the structure of the existing institutions themselves, but this is not done in any detail) apparently accomplished with commercials and slogans equating capitalism with democracy

The more direct and succinct answer is that the law of the US is, claimed to be, based on property rights and minority rules, though this is wrong, rather US law is to be based on equality, there is no one who is allowed to explain this in a civilized setting within the society essentially this is attacked by the (corrupt) justice system.

They (the liberals) point-out education is being attacked, they make the clear statement that education is associated not to critical thought but rather it is related to training, that is (but) the education system is the basis for this business based power structure within the society and this is totally ignored by the intellectuals, where the intellectuals are given so called success in their lives due to their agreeing to the indoctrination of the education system which teaches the skills needed by the wage-slave public to work on creative projects within which the investor-class has invested,

They decry the destruction of public education but they do not critique the already authoritarian educational structures, through which they became the academic winners, so as to get to have a voice on the media they along with the best selling artists are in essentially full support of the copy right laws but copy-right is about domination and theft where essentially there is now no content in any published best-seller other than the content which fits so nicely into the propaganda system that is in their arrogance as great thinkers and best sellers they support copyright law though there is very little if any content in anything which the propaganda-education-indoctrination system publishes especially the fact that there is, essentially, no content in "physical review" the, so called, "top" physics publication of the US.

though the liberal also identifies the fact that education is a part of the military industrial complex, but the way in which the structure of knowledge is required to serve military development of war- and spying instruments is not provided since the indoctrination is so very thorough that the liberal sees the indoctrinated assumptions contexts and interpretations as absolute truths, but/and the main issues in regard to education is both its academic failures, as a body of knowledge, and its control by the investor-class and these issues are all about relevant details yet there is no case to be made

There is talk about spying and surveilence-state but the way in which this is done is not clarified, it is still somewhat similar to the methods "through which L Oswald became a patsy for the CIA" on their "hit" on Kennedy, as so stated by a US jury

This is again clearly stated as barbarism and authoritarian punishing state again the moral basis comes from the Calvinistic motto That "God favors the rich" thus if one is not rich then this is because of the propagandistic sins of laziness stupidity and an inability to apply oneself, but the only application to which one can partake is to help the ruling investor-class or to innovate around the way in which monopolistic businesses and the violent-authoritarian-moralistic justice-system (but justice is supposed to be about law and its administration but the laws become Calvinistically-religious or judgmentally authoritarian in regard to a truth, which is not sufficiently demonstrated, eg the science is based on an invalid math-model, will allow

But this type of analysis and well speculated explanation is given

It identifies authoritarian expressions of knowledge but there are not any alternative or new ideas which are allowed to be expressed, this in the main function of spying and authoritarian managers of the propaganda-system, this and controlling the lower-social classes so that the low classes appear to be all about stupidity and violence, which is very far from being true.

The society's power is derived from extreme violence, and the propaganda system makes a spectacle of violence so that the image provided to society by the propaganda system is that power is derived from violence and the way to be powerful is through individual domination gained through violence ie the image of crime-families.

Self interest and celebrating individuals this is the few (mostly in the intellectual class or the so called artistic-class, who are the great supporters of their own individual superiority and that their copyrights are necessary without analyzing the system to realize that their celebrity is that they are expressing the shallow ideas which the propaganda-system wants expressed, ie they are rich because they are an appendage to the propaganda-system and they are not expressing any ideas which have any substance other than being related to the system of social oppression which characterizes the propaganda-media-system.) this results in the selfish unchecked celebrity of individuals, where the high-social-value of society is completely being controlled by the investor-class, the celebrities are actually pawns in the game of societal-rule.

Hedges points out that the radical right is cultivated by the ruling-class for its purposes, but Hedges does not make the case that the spy industries main sphere of both social influence and social concern is exactly the lower-social-classes, since the intellectual class is already thoroughly indoctrinated in the useless knowledge which supports the ruling-class

The propaganda-system (including the so called alternatives,) is exactly a-tuned to the interests of the US military machine (where spying and managing makes sure to exclude any interesting idea which challenges but which is also superior to the accepted authoritative dogmas)

But

This is not the story one hears, rather the story is that the left is the side which needs to be watched and controlled or intimidated

But

It is mostly the lower-social-classes which get the attention from the surveillance-state and the result is that terrorism emerges out of the lower-classes which have been attacked and manipulated by propaganda managers the justice-system, spies and their agents, many key figures are likely themselves agents

The liberal class also mentions the economic and markets both being dominated but it does not give a criticism of the so called economic laws, where there are market players so big that they can rig markets, eg rig the Gold-market nor do they linger on the non-existence of markets in a monopolistic based society

While

The left or intellectual class are mostly self-regulating, based not as much on race, but rather based-on social-class and social status where an example in regard to Chomsky* is given below. Where such social control through the propaganda system is facilitated by the fact that the investments of the investor-class defined high-social-value, especially, in a society composed of wage-slaves.

These things are true, based on Snowdon's revelations, so anything one might say in regard to this state of affairs is the more accurate description of the state of existence.

But the left provides the CIA's manual for revolutions, rather the revolution is about knowledge and creativity and asserting that US law be based on equality (which is the true state of US law, but which the judges and the early Masons based law on violence and theft.

There was W Penn and there was R Williams, and the genius of the native people's cultures, whereas Calvinism (Puritanism) is the worship that "god favors those who possess lots of money" ie it is a religion based on worshipping the material-world)

That is, stories based on accurate models are more accurate than the propaganda system, in regard to its set of arbitrary stories, which are very seldom true (or almost always false).

That is "where stories are needed" then use stories, but where there is out-right academic failure, then there needs to be detailed attacks (criticisms) about the intellectual models being used, but this is not happening, though it exists for the inconsequential detailed attacks on foreign policy by the progressive side, which is given voice on the media

The criticism of the intellectual class is virtually non-existent if the structure of the language which the intellectuals are using is not discussed and instead the criticism is that the intellectual is not thinking sufficiently deeply about the responsibility to the public clearly they have no sense of responsibility to the public since they are engaged in actions which have been given high-social-value, and are related to difficult to memorize intellectual models.

There is a decrying of authoritarianism, nonetheless, the point at which authority becomes arbitrarily dominant in an intellectual construct is not being identified, but this is the most important point to identify.

The cynical evaluations of human-worth given-by the intellectual-class (apparently, based on their arrogance of being chosen to be spokesmen for the intellectual-class, they seem to be chosen for their arrogant personalities) who parade with their victory march of modern intellectual-superiority, but where they march behind a failed partial-knowledge which they have apparently, been duped into believing is so high and mighty and apparently they see it as being on the edge of being an absolute truth, as these failed models are treated by society as, essentially, being absolute truths

The intellectual-class are a bunch of chumps, who have been brow-beaten into their apparent state of worship of the top intellectuals of their class, and who, in turn, brow-beat the public with the endless baloney of an extremely failed state of knowledge.

Chomsky* is the worst to consider, since he is narrow and supports elitism, apparently, because he, himself, is a, so called, practitioner and adherent of science, he proclaims he uses "the, seemingly, enigmatic math structures when he studies linguistics," but he has no evaluation (or criticism) of those methods,

A good example (provided by Hedges in American Socrates, truth-out, 6-15-14) of the arrogance of the intellectual class, is Chomsky's analysis of the biologist E Mayr's claim that intelligent species are likely to rapidly go extinct, thus there are no intelligent extra-terrestrials, whereas extra-terrestrials would get here by using properties of existence which are many-dimensional, whereas Chomsky is a dedicated materialist, in-that he adheres to the material based idea of Darwinian evolution of the species

Chomsky uses the issue of global warming, where this he claims this will "destroy the conditions needed for a decent survival" but there may be worse things, such as environmental poisons, and the fact that our civilization has based there actions on partial truths about radioactivity, ie there is no valid model of the stable general nucleus thus our understanding of nuclear reactions such as fusion reactions are at only a rudimentary stage of understanding they are very limited partial-truths but subsequently in the desire to use "nuclear explosions for the military" our society has amassed tons of relatively-pure and very toxic radioactive substances, and there is certain to be a point where these extremely toxic radioactive substances are no longer to be stored in a safe manner, (since the model of civilization we use is based on barbarity and violence, and out intellectual models are inaccurate and failing, since the social-intellectual structure is based on arbitrariness, ie science is made dogmatic (but based on partial-truths), and subsequently controlled by a few who in effect destroy the intellectual efforts by others to continue and "develop the forms of descriptive knowledge which is accurate and useful, and subsequently, might place 'knowledge about existence' in a context, which is safe, accurate, and practically creatively useful"), and subsequently these extremely toxic radioactive substances will result in the destruction of a great amount of life.

That is, the intellectual arrogance of Chomsky results in his thought processes being mostly mis-leading. He is a linguist, yet he does not rightfully consider the issues which Godel's incompleteness theorem imply, namely, that people should not become arrogant about descriptive truths based on precise languages, such language always fail.

The true intellectual state is that of "equal free inquiry," not arrogance based on dogmas, in turn, based on partial-truths. I have never seen Chomsky ever proclaim that knowledge is best developed based on Godel's incompleteness theorem and the main message of Socrates, that of "equal free inquiry"

Rather

He seems to stand by the arrogant claims of the intellectual-class, where these intellectuals support an arbitrary hierarchical society based on violence, though he "way understates" any skepticism, by saying "if Mayr is right then" we are about to be destroyed environmentally and economically

Well this analysis is absurd, and it is absurd because of the same blindness of the intellectual-class to the arrogance of themselves, the elite intellectuals, without identifying the fact that the intellectual-elites exactly fit into the social hierarchy so as to make the social-hierarchy function, and that hierarchy is failing because the intellectual-class has failed, and, in their arrogance, they are incapable of reflection.

They still seem to believe the propaganda-hype, namely, that they are supporting an intellectual effort which is essentially based on absolute truths.

Unfortunately, Chomsky stands with the intellectual institutions which support our arbitrary hierarchical society

Furthermore, Chomsky's prescriptions are quite wrong. We are already in a collective society, similar to how the Roman society collectively supported the Roman-Emperor, ie a collective hierarchical society.

(but the hypocrite)T Jefferson wrote the correct words, which oppose the western-culture's demand for social collectives, which, in turn, support arbitrary social hierarchies, namely, "we are equals."

Law is to be based on equality.

Law is not to be based on property rights.

That is, we are equal individual creators, but where collective action is best identified in regard to trade but where the purpose of society is to develop knowledge and engage in practical creative efforts, but for descriptive knowledge this is best done individually, by means of equal free inquiry

But

The society also collectively supports the people and each individual's equal individual creative efforts, which can be done in a selfless manner.

Furthermore, the creative efforts need to be adaptable so as to be able to change based on threats to the environment or adapt by changing to any threat identified by means of the precautionary principle.

It is selfish efforts (to achieve distinction as being superior) which the law should most oppose.

Chomsky's example, where in regard to the proclaimed environmental collapse, which is being predicted, he pit's the, so called, backward indigenous populations, who oppose the social conditions leading to such a collapse, vs. the most advanced cultures, and, so called, educated nations, which possess the absolute truths upon which the western intellectual class is built, and who are racing towards the (environmentally) destructive possibilities, though he seems to support the indigenous people in their efforts, he does not assess the issues of intellect in any meaningful way, the west is the leader of violence, stemming from the law of property-rights, and the violent western culture isolates and protects the intellectual-class (they are needed by the bankers), so that they arrogantly dismiss the culture of the indigenous people as backward.

And

Chomsky seems to be blind to the fact that it is his own intellectual truths, in regard to linguistics (eg using the failed indefinably random math techniques to study the properties of language, if there was intellectual integrity in the academic system, it would have put indefinable randomness into question long-ago),

ie the way in which he uses expert language, is derived (also) from an intellectual structure which is arbitrary, and leading to destruction (for both the environment and knowledge). It is a dogma which excludes the voices of the, so called, (backward) indigenous people, or even an expert trained voice, which is calling for true alternative ways to build expert languages, in order to solve the currently unsolvable problems (which are truly the most fundamental problems facing physics and math), eg Why are the fundamental physical systems composed of many-(but-few)-bodies so very stable?

By Chomsky using the trick of reflection, in regard comparing the assumed to be backward cultures with the intellectual advances of western civilization (though this, in itself, is an expression of arrogance, since C Castaneda has shown the ancient Mexican culture to be a far more sophisticated culture than the west), this results in a deflection of intellectual responsibility of the intellectual class, the partial knowledge which is at the root of the social destruction, if the intellectual-class had any integrity, they should have been questioning (back in the 1950's) Teller's arrogant dogmatism (exclusively associated to making nuclear weapons), and the spying should have been "nipped in the bud," in the 1850's (in regard to the way in which the ruling-class was allowed to use the Pinkertons).

But where there should be some criticism of failed math methods, since modern technology has not gone beyond the telephone, the radio, the TV, where with the fast switches developed for TV then this could lead to computers being designed around the fast switching developed for the TV's, that is "the math methods Chomsky brags about using (eg indefinable randomness)" are clearly failed math methods.

The science of the 19th century has become the analog of the bricks which were used by the Roman solders to build the infra-structure of western civilization, during the time of the Roman-Empire (which ended in Byzantine around 1450 AD, about the time the banker-families were choosing the popes, in the (then) Holy-Roman-Empire, or the western part of the (by then sacked) Roman-Empire).

That is, Chomsky is clearly a victim of the propaganda-system, in regard to his personal concern about his "intellectual status" in an authoritarian academic institution, and this weakness (this vulnerability) makes him very ineffective. (Chomsky has been a clear voice in regard to being educated about, the failing of western foreign policy, but by 2000 there were alternative expert narratives which demonstrated the failings of the physics and math communities, but these expressions are excluded from the intellectual-class due to its own arrogance and dogmatic basis)…

To his credit Chomsky points-out that the education system is all about indoctrination, although there is a strong relation between propaganda and education, which he seems to not express, yet he points-out that the education system is still terrorized by the McCarthyism originating in the 1950's.

But he misses that the US is the most successful nation to be able to terrorize its own population, apparently, based on the illusion within the public (formed by the propaganda-system), that "we (Americans) are lucky to live in the US society," but furthermore, this terrorism emanating fundamentally from the justice-system, which has been put in-sync with the propaganda-system ie the sense of terror within the public emerges from both the policing and the propaganda, (but the "allowed to be published" intellectuals never make such an obvious connection), where the failure of modern math models is the most fundamental problem facing society, namely, the failure of its intellectual models in regard to creative and technical development (the technical issues will be dealt with a bit later) which has led to the natural lack of growth in society, in regard to technical creativity within the society, where this lack of practical technical growth is natural in a totalitarian social system, which supports, in a collective manner, the few in the investor-class, and whereas the investor-class does not want knowledge to be developed, so that the investments they have made into the instruments upon which the economic and social power of the investor-class depends (are not put at risk), and where they want these instruments to be gradually developed, so that these instruments and their associated social strategies for monopolistic domination continue to work for them, thus, they do not want new knowledge, nor new creative contexts, which might make obsolete their instruments, and which would result in new competitions in their highly controlled and monopoly dominated markets

H Giroux is correct that "the popular culture produces a subject willing to become complicit with their own oppression".

But Giroux also supports the same intellectual basis, which has failed, but apparently he wants the, so called, equal opportunity for the public to participate, through education, in the narrow knowledge and market structures which have been crafted by the investor-class to support the investor class

This also seems to be the position of Elizabeth Warren (2014), who seems to support the current structure of education-corporate development but this path is the path to militarism

This is called Keynesianism economics.

Which, apparently, is the only path of knowledge, ie narrow and related to exploitation destruction and great violence, which fits with the idea of capitalism

Individualism related to knowledge and creativity and related to a truly free market but such a free market must relate its successful products to cooperative collectives so that domination by means of market domination is not to be allowed

Products can always change because the human being is knowledgeable and creative and can easily adapt to changes

But capitalism is the evolution of the Roman-Empire model of western civilization, where stable social contexts, eg related to the Roman water systems sewer systems and roads as well as big building, which together have led to a set of products which are stably related to the function of a town and subsequent markets and investment where the top investors now take the position of the Roman-Emperor

That is apparently such seemingly stable town structures really depend on an extreme level of violence associated to the Roman model of domination through violence

By 1400, in Europe, the banking, Medici, family was appointing the pope, where the church was the propaganda-education-arm of a hierarchical empire

That is, one finds in Giroux's social criticisms the same failings of the intellectual-class and that he is, seemingly, unaware of... in particular... concerning the education system, that it is narrow and failing and that the basic assumptions and definitions of the discussion are not examined in careful enough cause and effect in regard to how language is used

Apparently Giroux is a collectivist who wants the intellectual-class to be in control, but as a deep insider in the technical intellectual-class, this is also implicit in Chomsky's viewpoint of society

Thus there is a very false dichotomy formed around left and right, liberal and conservative, collectivist vs. individualist

Where this division is quite false since the right is the main group of collectivists who are conservative and are expressing the old European (or western cultural) tradition of a collective society which supports the roman-emperor, or in now a days a collective society supporting the investor-class, ie the monopolistic banks and corporations, while the absurdity of the whole discussion centers on the liberals, who support their own intellectual class as the "correct elitists" who should be running a collective society, but apparently the liberals do not see that their own intellectual structures are exactly the intellectual structures which best support the investor-class, since through propaganda wage-slavery a result of illegal violent support of the investor-class by the justice system, spying, controlling-management, and thus controlling education, thus the education system's curriculum is comprised of the vocational technical categories which are needed by the investor-class so as to adjust their instruments and run their institutions and run their propaganda structures through a system of elitism and high-salaried workers, the so called "real world" of society, but a world picture which is the result of corruption of the governing and legal institutions of the society where the idea presented is that it was the investment in the colonial corporations which built America far from true for the W Penn's Quakers, or for R Williams, but the Puritans are Calvinists and are agreeable to this idea as are all the successful mercantile enterprises of the North Americas

That is both sides support exactly the same set of social properties which have led to the failure of society in its elitist totalitarian state of existence

The right best expresses the idea of individual freedom, which is quite divorced from equality, thus it is the great enemy of the lower social-classes, ie the 90% to 95% of the population, but it is the expressions of the militias and of the Indians (on the Indian reservations) so complete is the propaganda coverage of the society

In western thought (or action) the only society which stood-up to the oligarchy was the original US, American declaration of Independence which proclaimed that US law was to be based on the idea of equality for both equal free inquiry and for equal free-markets where everyone was to have a 300 acre farm so as to not be dependent on the society but this was being expressed by white European land-owners in the US who were rich educated racist and believed in the superiority of their investment based culture, which was then failing but the new world was found by the set of violent and elitist (superior) Europeans who then did as they had learned in Europe they lied stole and murdered to exploit the world for their selfish gains

Where both the violence and the intellectual superiority are still being expressed by the right and left respectively.

So why have the intellectual structures of the west now failed since they are based on math structures which are non-commutative, non-linear, and indefinably random, where indefinably random means that it is a probability-description which is based on elementary-events which are unstable-events, eg elementary-particles, and/or events which are not determinable, based on the descriptive context, eg quantum physic's inability to describe the stable spectral properties of a general atom, but the spectral properties of general atoms are stable and this means that these are controllable descriptive contexts, and cannot be determined (described) from a quantitatively-inconsistent descriptive context, where the context of non-commutative and non-linear and indefinably random is quantitatively inconsistent descriptive construct and cannot describe stable patterns rather it is a context which has no meaning and no content it describes fleeting slightly distinguishable varying relative inter-relationships which are transitory, this is the same context as exists when a nuclear bomb explodes due to nuclear reactions and the resultant random collisions of the components which have been broken apart by the reaction where the original state before detonation was relatively stable and the end-state of the remaining various elements is also relatively stable, ie a chaotic transition between two relatively stable states the stableness of the original and end-states cannot be described by this same chaotic quantitatively inconsistent structure, as is now being attempted by the intellectual class, actually they are really trying to detonate a (mythical) gravitational singularity, but they apparently are actually not very smart and they seem to be unaware of this

The molecular and evolutionary model of life are also based on such a quantitatively inconsistent descriptive basis, where modeling chemistry simply in terms of particle-collisions between specific types of molecules, might not be a valid model of a living system's actual operating mechanisms, and then trying to relate this chemical model to some extremely complicated chemical-causal-chain of specific molecules and their reactions back to the DNA molecule's properties might actually identify an absurdity of complicated purposefulness, which (where) mental capabilities cannot be so purposeful and unified, in their quantitatively inconsistent and indefinably random descriptive context.

That is the investor-class has turned the descriptive context of bomb-engineering into university physics department curriculum, and thus the cultural knowledge is being destroyed for monopolistic and intellectual domineering interests, ie expressions of psycho-pathy and domination are what get expressed, not valid ideas.

The attack on the public by what appears to be militias and individual freedom defined by violence is not a natural expression of the public, but rather it is an attack on the lower class by agents and spies of the justice-system, so that the public cannot get unified in their opposition to the investor-class. This is all orchestrated by: propaganda, social management, spying, and militarization (managers taking Ques from their further, but hidden, superiors), so as to both divide the lower social classes, and to have a way to manage a form of terrorism, apparently, emanating form the lower-social-classes, and which can be used against the public so as to terrorize the public.

It is collectivism (and collective thought) which causes bigotry and arbitrary dogma…,

(note, where both left and right are collectivists, and where both essentially support the investor-class, one supports the violence needed for inequality and the other supports the elitist intellectual structures put in place by the investor-class)), not individual independence, as simply envisioned in the Declaration of Independence and the Bill of Rights, where individual freedom is about equal free-inquiry, and the relation of new knowledge to new practically creative contexts,

ie equality to exist, communicate, and build in a context of selflessness but willing to save or protect resources of the earth when necessary, following the precautionary principle. And

Putting creations, if one so wants, into an equal and truly free market, ie propaganda must have great limits and monopolies must become collectivized, ie the domination of society by individuals or by small groups is quite destructive due to its relation to narrowness, ie hierarchical societies are regimented societies

Descriptive knowledge, in particular measurable descriptive structures, and, in particular, those measurable descriptions where measuring is reliable and the patterns being used and the patterns being described are stable, in this descriptive context it is necessary that the language not become fixed and narrow and dogmatic and overly authoritative, that is, the fundamental simple basis for a precise language used as a basis for useful knowledge related to practical creativity, be continually challenged, and questioned, so that always new: assumptions, contexts, definitions, ways of organizing math patterns, new interpretations, and new ways of considering the containment-set of a measurable description, always be re-considered and expressed and its relation to a creative intent be also expressed the context of the new creative thing or process the practical creative context may not be a material context.

The same type of elementary discussion about assumptions contexts interpretations etc, also needs to be considered when it comes to social institutions, in regard to, say, law (should it be based on property rights and minority rule, or should it be based on equality), and trade (should it be based on monopolistic domination built around propaganda, or should an equal and free-market exist), and propaganda-education systems should education be about equal free-inquiry where the intent is to engage in practical creative building? or should knowledge primarily play a role of defining intellectual-domination based on authoritative dogmas within the propaganda-education system? and "if one wants a democracy?" and "what does that mean?" Are there some principles which a democracy cannot change? That is, are their certain human traits which one wants one's society to focus-on and develop, such as people's capacity to develop knowledge and to relate it selflessly to an intention for practical creativity done in a selfless manner, should the idea of materialism be permanently attached to science and separated from the ideas of religion? Or can knowledge have a wide range of possible relations to existence, including a many-dimensional containment-set structure for existence which contains the material-world as a proper subset? ie can science and religion be unified?

Chapter 30

Monopoly vs. equal creators

The current social structure is both

the usual western structure of oligarchy (only a few can define high-value) whose propaganda is centered around technical engineering prowess, ie organized around certain instruments, which define a narrow structure for social regimentation, as well as for knowledge and the limited range of creativity, and but where the structure of oligarchy has been based on both the industrial revolution and trade based on the interchange of money, where, in turn, product-production and trade is (has been) organized around investment by the few (very rich), whom invest into certain narrowly defined sets of instruments, which are, in turn, used to identify the narrow monopolistic range of knowledge and related creativity, which is the narrow range of knowledge which is allowed… for society to consider… by the investor-class, so that nearly all of society's efforts serves the oligarchy eg the factories along with developing a certain narrow range of instruments, which form the basis for the technical aspects of a so called public education (limited in its [knowledge's] range by the interests of the oligarchy) to be used for oligarchic social control, and this type of social organization has defined western culture, organized as an oligarchy, where this oligarchy, it is claimed, define the true high-value of human capacity, where this social structure is managed or administered…., eg management; and spying which are in close relation to the technical experts whom serve the narrowly defined interests of the oligarchy…, by means of great violence and a structure of law based on property rights and minority rule, but now

There are new sets of instruments, which allow for great ease of communication, by typing and up-loading to web-sites, and accessibility to the communication system, where the ideas being expressed can be easily identified and classified, and it can easily be determined if the ideas being expressed are new ideas, or if they are the development of old ideas, where the social model of technical education has been based on the narrow development of a narrowly defined set of specific type of instruments

But this narrow definition (determination by the investor class) of "what ideas to develop," in the technical sense, is not necessary, since there are now cheap instruments which can be used to develop a wide range of ideas

That is,

Why should the creative productive efforts only be used to enhance the rich, and to enhance the military-justice-system complex which supports the investor-class's monopolistic control over society?

That is

The information system based on identifying strings-of-symbols, ie the same way in which the library identifies literature, can be used to also identify the structure of math-models of existence

Rather than only basing precise descriptions on "materialism and partial differential equations {associated to either randomness or geometry}" there can be other ways in which to develop descriptive languages, ie when a language is incapable of describing the observed patterns of the world then, as Godel's incompleteness theorem suggests, the assumptions, upon which the (clearly seen to have failed) precise language is based, need to be questioned and changed in fundamentally new ways

ie new assumptions, interpretations, organizations, new contexts, and new ways in which to emphasize the set of properties (eg axioms) upon which certain aspects of a precise language need to be put-forth

And

This needs to be done on an information system which is accessible and open

So that new quantitative models can be considered and their relationship to new creative contexts can be identified, so that the productive instruments of society, eg affordable book publishing and affordable 3-dimensional printing of practical models of new inventions can be put-forth, new precise descriptive contexts which are related to new interpretive aspects of a precise language….,

ie a language where measuring is reliable and the patterns of the precise description are stable,

…, can be quickly applied to the new ways of thinking and to new creative intentions.

Instead of

Putting an edifice of intellectual inertia in place, where there are institutional ways of seeing the world and the efforts to improve a narrow limited idea about what types of instruments are to be worked upon by the institutions which serve the interests of the oligarchy.

That is,

The new way in which to develop both new ideas and new contexts for inventiveness is through the development of a web-site where ideas are classified based on the set of math assumptions upon which they are based and in turn relating the new contexts to direct means of developing a new context, ie a new viewpoint in regard to the relationship of ideas and practical processes.

That is,

One does not want the new way of doing things to be managed by people who are already indoctrinated to function within the current system, ie those who are considered to be meritorious, their so called merit has proven itself to be mostly related to failure even though they get a very large boost from the propaganda-education-indoctrination system, a system of mental inertia.

That is

If existence is given a many-dimensional model then "what types of 3-dimensional shapes or material-configurations can couple to the higher-dimensional shapes?" and "what are the natures of the higher-dimensional shapes, which, in turn, are related to the 3-dimensional material world, which, in turn, have a relation to higher-dimensional geometric properties?"

For example,

The oscillating shape, of dimension-3, is related directly to energy generation, but its oscillating geometry is in a 4-dimensional space, but clearly the 3-shapes which we experience have a relation to these 3-shapes and thus also to 4-space.

Some simple examples of oscillating 3-shapes which are contained as geometries in 4-space are:

1. radioactivity, as well as

2. life-forms such as

2a. plants or

2b. Bacteria

Etc.

That is the new models require new viewpoints about relationships to shapes and containment spaces

Another example is that, the life-form of an oscillating 3-shape may be quite large, eg as big as the solar-system (or bigger), but such a shape would have a relationship to the shape of the solar system, while the entire solar-system, including a model of the sun, may require a 5-dimensional shape which would be contained in a 6-space, and this would require new ways of considering coupling, in a practical manner, to shapes, which in turn, where such a coupling geometry may be relatable to accessing cheap clean energy sources for earth. That is, such high-dimensional geometry of the solar-system and/or the sun can be related to the geometry of the condensed material of the earth in the earth's solar-system orbit and this means that the geometry of the earth, eg its mysterious internal structure, might be easily coupled to the oscillating energy-generating properties of the sun, eg some idea which say Tesla might have been trying to realize in his ideas about antennas and the formation of electromagnetic-fields. Such (magnetic) fields would be related to the toral-shapes or the stable circle-space shapes in 4-dimensions.

That is,

The old, so called, experts or the old set of people which composed the, so called, meritocracy-of-intellect, have failed both at imagination and they have failed at analyzing the condition of their own descriptive language,

And thus

It is not such an indoctrinated set of people who should be either the managers or the set of consultants to which the new ways of revolutionizing the way in which creativity and knowledge are both realized and used in society through information systems which easily couple to descriptive knowledge which might be possessed or be considered by anyone

Remember, S Jobs main contribution was his willingness to keep considering new ideas upon which to develop, rather than imposing a rigid dogmatic viewpoint about what type of knowledge should be narrowly focused-on so as to slowly improve the instruments upon which his monopoly was to be based an economically expanded rather than expanding ideas and new creative contexts, ie the monopolists want fewer ideas they want fixed dogmas.

Though they (the old experts, of the monopolistic social context) can contribute to such new endeavors, but there is little reason to hold their opinions in any (high) regard, ie in regard to it (their, so called, expert opinions) having any (real) value.

Chapter 31

A world where there are no
patterns for one's language

This is the typical context of social interactions

The point of material-based societies, ie societies based on lying stealing and murdering, is about the social organization, (organization of violence and communication) by the psychopathic leadership of the society

Language can be constructed so as to appear to possess some stable properties, that is, a particular narrow viewpoint can be expressed, eg related to a social process (or to production or building), and then this process is made stable by its continual repetition over the propaganda-education-indoctrination system of the society or through its relatively stable social processes ie its continual repetitious use within society

So that the "product" of the particular actions can be given value, so that this value can be associated to a normal curve, ie related to samples of averages in regard to some vaguely distinguishable feature of the action (or process) or product of an action.

The normal curve can then be used to distinguish high-value within society

In technical (or precise) language there are quantitative structures related to stable patterns and there are quantitative structures related to non-commutative, non-linear, and indefinably random math constructs, so these math constructs are neither quantitatively consistent nor are they in any way relatable to stable (math) patterns

Axiomatic formalization is "all about" making the later, the norm of high-valued technical discussions in academics and in the businesses which are related to adjusting complicated instruments (eg military and communication instruments)

And it is the math structure which characterizes quantum physics, particle-physics, general relativity, and all derived (fake)-math constructs, eg string-theory etc

Particle-physics is about the business of bomb-engineering

It is essentially based on "the trails" of apparent particles, observed in cloud chambers (or bubble-chambers, or now in detection chambers where apparently the asymptotic particle-paths are detected at the chambers confining walls and then the so called particle-paths are re-constructed based on the energy of the collision impact)

It is entirely an empirical model which came out of bomb-engineering

The only empirical property of significance of particle-properties is the probability of a particle-collision

While the model of particle-physics is based on

(1) the existence of distinguishable material-particle-components: electrons, protons, neutrons, and the difficult to detect (if it really is detected) neutrino, and then

(2) the classification of these particles into leptons and hadrons, ie hadrons are particles of the nucleus,

where it can be assumed that the nuclei and the atom have stable material-component properties associated to stable spectral properties the light spectrum of the atom and the x-ray spectrum of the nucleus so that these distinguishable spectral properties are associated to these system's numbers of components of the different types

Since systems with the same number of component-types always possess the same stable spectral properties one can assume that these system form in a controlled (or quantitatively consistent) context, or, alternatively, they form so as to be a part of a consistent quantitative structure (based on resonating with a finite spectral-set in a high-dimensional containment space, where the containment-space contains the stable finite spectral set, upon which a consistent quantitative structure is built).

(3) but the classification is re-done into sets of quarks and leptons:

three leptons placed in 3 independent families, where the families (of the same particle-types) are supposed to be dependent on the energy-level of a material-interactions, and where the interaction is modeled as a point-particle-collision

And three quarks, also

Placed into 3 families similarly distinguished, by energy-levels, as are the leptons and their 3-families.

But the leptons are too small to be easily distinguished by particle-collisions in cloud-chambers and the quarks have the properties of being un-observable

Where this math model is based on non-commutative and non-linear math structures defined in a context of indefinable randomness, ie it is a description without any meaningful content, rather the entire descriptive context is only based on empiricism (tracks of particles in cloud chambers) and the relation of this empiricism to the particle-collision model of (material-interactions (?)) bomb-engineering related to the (of) probabilities of particle-collisions

Note that the only particle in this descriptive context which has empirical mass big enough to be observed in a particle-accelerator is the massive field-particles of the weak-force, where the hadrons are modeled as the unobservable quarks, in reality, the hadron, and its associated (usually) stable nucleus, are complicated bounded stable shapes, as are the leptons as the stable orbital patterns observed from Dehmelt's isolated electron has shown, and the model of the neutrino is that of an unbounded stable shape, which may, in fact, be bounded by the metric-space within which it (and/or the stable atom of which it is a part) is contained.

It is claimed that the Higg's particle (the particle which is supposed to confer mass onto elementary-particles) is claimed to be detected as "going between" pairs of the massive field-particles (of the weak-force)

That is, particle-collision data fits into the empirical particle-physics model, just as planetary-motion data fit into the empirical "epicycle model" used to interpret data, in regard to Ptolemy's model of planetary-motions in the heavens.

Note that particle-physics has not been able to express a model for the stable nucleus, since there is nothing in its math model which allows its descriptions to possess any meaningful content, Rather (in particle-physics)quarks are put-together to describe a proton which is supposed to be unstable, but the data shows that the proton is stable,

So (now) theoretical models, which have no meaning, will be identified so as to design experiments so that proton instability will be detected based on the interpretation of newly designed experimental data interpreted within the meaningless theoretical (but really only an empirical) model, as arbitrary data from particle-accelerators is interpreted to support the meaningless claims (about the confirmation [or attachment] of mass to elementary-particles, by a scalar-field-particle, ie the inertia of an elementary-particle is about colliding with a massive field-particle, ie the Higg's particle), about a meaningless model of "particle-collision based" material-interactions

Which is only related in a "practical" way to bomb engineering, where the model of a bomb explosion is the same as the model of a chemical reaction, in turn, based on molecular collision-rates, where this chemical model was determined by the 19th century chemists (and seems to have no ability to guide the understanding of controlling chemistry, oir the inter-relationships between molecules, especially, in regard to the chemistry of living systems, eg drugs are found in plants they are not made "based simply on the knowledge" of chemistry).

But the size of atoms the size of their stable spectra and the size of a nucleus and the size of its stable x-ray spectra as well as the, mass=energy, and then energy= h x frequency, (light wave-length) size of the massive filed-particles, can all be used to identify the finite spectral-set upon which (or to which) a stable model of existence can be related, a new model based on a finite spectral-set of a high-dimensional space partitioned into (a finite number of) stable shapes. Where the massive field-particles are really stable shapes (with distinguished points, so it appears to be a point-particle) into which a small discrete hyperbolic shape may be broken

Though the right (the oil-lobby), with a small amount of justification, questions models based-on probability, which may or may-not be well defined, (however) no-one questions the math models based on indefinable randomness, where probability is not based on a well defined elementary event space, as well as math models based on non-commutative math properties, as well as math models based on non-linear models, which dominate our quantum models and the particle-physics models as well as the general relativity models of existence so that these theories possess no meaningful content and thus are not relatable to either accurate descriptions or to a context where new creativity is fostered by the descriptive context, since the math contexts of: non-commutative, and non-linear, and the indefinably random have no meaning.

This is an important aspect of the failure of the society to produce new creative contexts so that new range of creative efforts can come into being

The point of a personal communication system which is tied to the entire culture is about developing new sets of descriptive languages so as to facilitate new ways in which to create and produce new product (simply apply a need for new sets of assumptions for a math structure, and distinguish descriptions based on new ways in which to use and organize a precise language), not to communicate a bunch of babble and self-centered baloney, though this is also fine, if it is not spied-upon.

Wikipedia is about the dogmas which the bankers want expressed in academic society, this is OK,

but, these dogmas need to be reviewed and criticized in a deep manner, especially, if they are not related to accurate description nor related to developing new practical creative contexts (whereas most of the academic patterns which are based in formalized axiomatic [ie whereas nearly all academic concerns are so-based] and thus are unrelated to any practical aspect of the world we experience (or only related to some chaotic context of the world we experience).

Chapter 32

The plantation

The plantation

Are the perturbations of the otherwise circular-orbits of charges (or other material), defined on discrete hyperbolic shapes, only possible for condensed material, which are in an orbital confinement within the material's containing metric-space shape, where condensed material is the size and shape material which does not fit into the spectral-orbital faces in a very close approximate fit to the geometric measures of the spectral-orbital related face (or the shape's natural spectral-measures)?

The issue which is missed, by those who work on the plantation, and who work for the plantation-owners

The issue is that the so called public institutions are all about class warfare, the institutions are doing things which are quite wrong and unlawful for example the US institutions are exterminating its unwanted people but where this system is hidden by the, so called, complexities of law and in the propaganda of distraction, and of course the very hidden spy-networks and associated institution managers, but the wrong-ness is not dealt with by the institutions, but rather calmness is called-for and is re-instated in the local social place of the wrong-doing and the social doings in that region are claimed to have moved-on, ie the institutions are allowed to engage in class-warfare in an unlawful, or wrong, manner without consequence. It is not a country (or society) which is based on law, rather it is based on social-class (and social-value, as defined by the ruling-elite).

This is the main issue of our society

I. The society is corrupt, and the, so called, leading and educated of a, so called, top 10% (hired by the propaganda-education-indoctrination system) all support the corruption

That is, there is absolutely no individual freedom, either for general intellectuals, or for people of action (ie no individual freedom for those who are not serving monopolistic corporate interests), it is all a collective-society, which supports one set of guiding principles, namely, those (relative) expressions (only about a material world, or about a hierarchical personality-cult) which support the interests, which, in turn, support the ruling-class, ie Calvinism in the extreme.

There is enforced (by the corrupt justice system) a required belief- (for those in the public) -in the religious beliefs of the ruling-class, which get expression in the propaganda-education-indoctrination system.

And the political propagandists who are supposed to determine how the society is organized, supposedly, for the public interest, but the public interest is cruelly defined (by the propagandist-politicians) as the interests of the ruling-class.

That is, the politicians are simply the most self-serving and selfish people in society, who are a part of the propaganda-system, which is owned and controlled by the ruling-class.

This is the essence of the (an again) needed "abortion-rights movement" to fight the enforcement of the religious beliefs of the ruling-class, on the public.

This is the essential context which defines the deep corruption of the justice system.

This is about human rights…, in a dysfunctional society, in a society which is constructed based on the violent and terrorized public whose personal expressions are not allowed by the dominant ruling-class…, vs. the selfish interests of the ruling-class, which the US institutions of justice and politics and the communication systems, through which expressions of ideas are to be filtered (ie through the propaganda-education-indoctrination system) so as to fit into the interests of the ruling-class, are dedicated and organized to support.

The justice system is a counter-insurgency operation against the public where the spying is all about exactly supporting this operation

This has happened, not in a democracy, but in a police-state governed through military acts and propaganda and through hidden spying operations which are an integral part of the, so called, justice system.

This is against the law

The country needs a new continental congress so as to do away with the arbitrary stand by those governing the US institutions which are entirely allegiant to the few in the ruling-class (the paymasters to the wage-slaves), to which these institutions (full of selfish and domineering people) have now come to serve (ie to serve the ruling-class)

II. The other main issue is that the math and science high-intellectual-value of the culture is mostly a bunch of baloney.

Copernicus supplied the answer to "how the motions of the planets are to be understood" but still today (2014) "there is no rational explanation [based on what is now called the (so called) laws of physics] as to why the planets have that organization along with their associated motions and to also remain stable."

The point of the Thurston-Perlman geometrization proven-theorem is that there is simplicity in stable-math patterns, and it was shown by Perlman that

When math structures are quantitatively inconsistent, ie sets of operators and (partial) differential equations which define (physical) systems (or of math structures) which are non-commutative (not single-valued), non-linear (no valid definition of a measuring scale), and the associated context of indefinable randomness (unstable random events and incalculable random events) which are associated to either geometry, or to harmonic eigenvalue structures, of (physical) systems (or to abstract math structures, eg economics, psychology etc), then the (physical) systems contained in the (invalid) quantitative structures of the base-space (eg coordinate metric-space), are both unstable and indefinable, and either disappear or disintegrate to simpler stable forms, which are identified in the geometrization theorem.---- where these types of nonsensical quantitatively inconsistent systems are examined by mathematicians "all of the time"---

wherein the system is, supposedly, contained,

ie when these systems are within such a containment space,

(ie when the operator models of systems are non-commutative non-linear, and/or indefinably random)

Virtually, everything which is modeled by a general context of math,---- and its relation to sets of operators or (partial) differential equations, is quantitatively inconsistent (ie non-commutative, non-linear, and/or indefinably random) and thus it is invalid, and not to be trusted, in regard to the so called properties, which it is trying to describe so that the constant talk about defining "metrics" ….

(supposedly measurable qualities, in regard to seemingly distinguishable properties, but these properties, ie usually random events, are unstable and are unrelated to quantitative structures (or related to coordinate domain spaces in quantitatively consistent ways)), so as to claim to be able to measure the properties of the, so called, systems, which one is modeling in a quantitatively inconsistent manner, and thus, (after making the measures, in regard to unstable patterns) come to conclusions about whether there are measurable affects which might define the, so called, "success of the system," determined by these phony claims of observed measurable properties ----, are virtually all about meaningless descriptive contexts, which are devoid of any content.

It is all baloney wrapped in a context of a phony claim, that these are, so called, precise descriptions which can be measured (but within a quantitatively inconsistent context), and thus the descriptions have a claim to being called scientific (but this is not true).

Rather it is virtually all fraud.

The set of relatively complicated math models, eg the very stable many-(but-few)-body spectral-orbital systems, which can be described in a quantitatively consistent, and thus valid, manner is very limited.

This statement

Along with the Godel incompleteness theorem which states that "precise languages can only describe a limited range of patterns, so the language's basis in assumption, context, interpretation, organization etc need to constantly be reviewed and changed so as to be able to describe the set of patterns which are being observed

Together

Require that math descriptions (math models) be simple and be constantly criticized and reviewed and new elementray viewpoint continuously be considered

This is something which an inter-net and computers can be used in regard to communication and organization of expressions based on their elementary assumptions

Rather than using the communication systems as a basis for a police state to support the arbitrary interests of the few in the ruling class

The issue for science has been, "why do we observe the particular properties of systems, which when these systems have a certain number of distinguishable components, they always possess a precise set of very stable spectral-orbital properties?"

Almost all of the intellectual categories of our civilization are arbitrary, and wherein most descriptive knowledge in each category has reached its limits in regard to both its narrow specialized use in civilization and the validity of its descriptive context.

That is,

The intellectual-class, and their arrogant claims of "consensus and truth," are upholding an arbitrary and mostly failed context of thought,

namely, that quantitative structures are still valid when the set of operators and (partial) differential equations (or physical laws) which are used to describe (or model) physical systems whose set-containment is dictated by the arbitrary claim that materialism is an absolute truth, and which are modeled by: non-linear, and/or non-commutative, and/or indefinably random operators or contexts of macro-geometry or micro-reduction to random particle-events in space

Rather

In such a descriptive context there are no quantitatively consistent patterns, and thus the descriptions are without any meaningful content

According to the so called "consensus of scientific truth" our culture still "does not" possess a rational measurable description as to why Copernicus's model of the solar-system "is 'the way it is.'"

Or more generally,

Why are the many-(but-few)-body spectral-orbital systems, which exist at all size scales, stable?

Furthermore,

These stable micro-spectral systems are uniformly the same, in regard to their spectral properties, which means that these systems formed in a very controlled context, ie they form into stable geometries.

The people of action, as well as everyone else is also deluded by propaganda and (also) arbitrary expressions, and in their behavioral reflection of the ruling-class, which is the dominant social construct in our western civilization. That is, the public expresses ideas in a context which has an abnormally strong adherence to desiring for those expressions to be dominant over all other such expressions, ie a totalitarian absolute truth enforced by a Roman Emperor, or by Machiavelli's "Prince," or by the "godfather" of the thugs, or by the Nazi Furor, but based in a collective social order, with only one dominating set of social and intellectual guidelines.

Both sides are reflections of the domineering behaviors of the ruling-class, and they both express deep passion for the arbitrariness of their mostly meaningless expressions, especially, when measured in regard to the true heritage of being human: individual-knowledge and individual-creativity and not living only in a material-world

Instead of the empty clash to be top cowboy of a one-horse town, or arbitrary expressions of domination within a fixed narrow and limited category (a category of action which has been exhausted).

The world is a mystery and its described properties have the limitations of the descriptive language, ie language must always be changed, especially, if it loses its accuracy and/or it is no longer useful for practical creative development.

Thus, many religions seek to see the world as "it really is" the person seeks to have (or has had) experiences an existence beyond the experience of the material-world, but there are not word categories which can accurately describe the experience

The point of civilization is to place things into limited categories, so as to identify the value of life in regard to that which matters to the ruling-class's arbitrary categorization of usefulness, and their subsequent use social forces to push life into this context (the activities of the public are only to be those activities which help the ruling-class)

However,

The only way in which the world (ore western-cultured world) is different from its interpretation within civilization ie as judged for its usefulness by the emperor, is that the ideal-world, which the world really is, differs from the knowledge of civilization where the knowledge of civilization is defined only in regard to... (the property of existence only being)... the material-world, and its, subsequent, definition in terms of dimensional containment

To experience the ideal world which is different from the material-world is to perceive the "world as it really is" that is, the basic tenants of religion are opposed to civilization as defined by the ruling-class of the civilization

Thus, the emperor lays-claim to a religion: a religious hierarchy serving the Emperor, but where since the banking-class rose out of the regimentation of the Roman-Empire, this is (has been) Calvinism: The rich are the superior rulers, and the world is the material-world, where we (in society) strive to be rich, and only "the truly good" are rewarded with riches. This is an arbitrary world whose references to value are determined by the (selfish) rich, ruling-class, and (of course) value is defined within (or by) the material world.

The Romans could lay bricks and mold cement and do big engineering projects, so as to entice people into their civilization, but only after the Romans brutalized them whereas the current civilized world, can build complicated instruments from the "geometric" descriptions of controllable systems, from which things can be built, for solvable classical systems and

Described in the context of discreteness, for the quantum descriptions, but the controlled systems are only relatable to a geometrically-measurable context, where the patterns are stable and the measuring is reliable, whereas the quantum description is random, but further, it is indefinably random, so it is arbitrary, except for the precisely measured discrete energy-levels of (usually micro-) system, but these precisely measurable properties of general quantum systems are not describable [based on using the, so called, laws of physics today (2014)].

That is, only solvable geometry is both accurate, when in a context where measuring is reliable, and of practical use in regard to creative development.

That is, the technology of today is similar to brick-laying in that it depends on 19th century knowledge of material, but knowledge guided by business interests adheres to the practical material-world

Where the instruments are mostly military and communication-systems ie instruments which have been used to control the public, and protect the wealthy

That is, the world is a mystery, and the descriptions of uniformly stable systems (especially, micro-systems)…

(ie an identifiable property, as well as a uniformly detected property, such as the number of components composing a system, determines the measurable spectral properties of the general micro-systems, where such uniformly stable systems had to have formed in a controlled context),

And this means that all physical description is stil, Are still, in the category of a well defined mystery.

But the categorical context of the world, (which) the ruling-class wants fixed, so that the development of further knowledge…, which is different from solvable classical (geometric) systems…., is not desired as a part of the category of (proper {or peer-reviewed}) knowledge (this is about the relation of the, so called, patents to the legal control of knowledge within society), especially, since the category of "the number of component-particles [particles is the material-related interpretation of the allowed (or used by the ruling-class) descriptive context of micro-systems] which are composing a micro-system," is being studied (by the big monopolistic military businesses) in regard to the relation of these components to an explosion, or a reaction-rate, where a reaction rate in a thermal-statistical system (ie a confined system) can be determined in the context of particle-collisions in the context of particle-cross-sections.

The relation between geometry and quantitative inconsistency (essentially, there is no valid relationship),

And this is because when there is quantitative inconsistency then the descriptive language (which is quantitatively inconsistent) is arbitrary and possesses no meaningful content,

Though there might be a descriptive context related to a "practical" use, ie a use by some big monopolistic business, eg particle cross-sections are related to rates of reactions in bomb explosions, and thus, the measurable particle cross-sections are related to the big monopolistic businesses of the military. This is the content of particle-accelerators and particle-physics, which is a quantitatively inconsistent language, and has no meaningful content, rather it is a ruse (a deception) used to focus efforts of the wage-slave bomb-engineers (or also called university physicists) on bomb-engineering

It might be noted that as a physical system moves away from a linear, solvable descriptive context (ie the set of operators or the partial differential equation which identify the physical system) the classical non-linear (partial) differential equation is still being defined in a geometric context, so the critical points of the differential equation (and their associated limit-cycles) are good (or OK) geometric approximations of the system's behavior "to first order!" so this means the reliability of feedback systems which navigate the limit-cycle descriptive context need many feedback mechanisms in order for the feedback system to have some reliability of its own (beyond first-order) in the mostly meaningless non-linear context of a geometric differential equations which can only possess a limited quantitative relation to their (own) critical points, however, the system being non-linear is quantitatively inconsistent, and is thus chaotic (eg discontinuous), so the first-order level of the description is not "a given" (ie it is not certain by any means) and thus, it (the first-order geometric-context of the partial differential equation) needs to be monitored, ie the geometry of the partial differential equation which, supposedly, defines the physical system is not necessarily valid.

That is, a map based on stable math patterns (and in a context where measuring is reliable) which can be used to identify the basic math patterns in higher-dimensions, which can be stable...., and thus, real, and/or thus, the types of patterns which one might perceive when seeing "the world as it really is"… can be a very useful expression fro both religion and for science and thus for practically useful new creative contexts.

Number of particles in

It is all about a carefully choreographed set of distractions, and other ways to divide, so as to use force, so as to, effectively, have the ruling-class remain dominant (both sides of the distraction are being bank-rolled by the investor-class, and only certain ideas are brought to the public by the propaganda-education-indoctrination communication-spy system, ie the military expenses which are all being used to both help the elite realize their interests and protect the elite.

Eg talking about a historic perspective of, say, T Jefferson, when it is only the exact words which are relevant

Eg

The Declaration of Independence proclaims that "US law is to be based on equality"

The nature of T Jefferson (a great hypocrite) or the nature of the ruling-class (ie slave-owners) is not relevant

Only the principle that "law be based on equality" is relevant

This was the first real break from the Roman viewpoint about society which was expressed by a European colony, and where the Roma viewpoint characterizes the western-European viewpoint, which is a viewpoint of arrogance and violent barbarity, and subsequent regimentation of society by the banks, by wage-slavery, and by the propaganda system.

In other words

The issue is the collective society which supports a very narrow hierarchy which define social-value in an arbitrary manner, and which control society based on wage-slavery associated to narrow markets which are based on particular types of resources and particular types of technical instruments and how these instruments are used and organized in a social structure of control by the few within society, ie law being based on property-rights and minority-rule, ie the same set of laws which were the basis for the Roman-Empire, where the Roman-Empire was a collective society which upheld the Emperor's power,

and the failed rules of society which are protecting social structures, which, in turn, are organized around propaganda, ie the propaganda-education-indoctrination system wherein only certain voices (or belief expression modes of communicating) are allowed to be expressed and this is all about issues of domination

Thus there is the dominant bomb-building intellectual-class (a failed and arbitrary viewpoint, which is placed into the propaganda-system as an (almost) absolute truth) who seem to mostly be institutionally managed autistic types, and the seemingly more arbitrary physically domineering types of people who want to base their domination on arbitrary moral grounds (ie based on what they "say," ie a viewpoint quite similar to the actual viewpoint of the very few in the ruling elite, and also quite analogous to the power of the Roman-Emperor, but now it is the elite bankers.) ie people come to be reflections of their culture.

Believe in the self, or gather together for power within a group, ie a group of dogmatists

The issue in Gaza is that the Gazans have been illegally imprisoned and punished by another country

The issue for the rest of the world is that the public has been illegally imprisoned and punished by an oligarchy

At the core of western thought the Judeo-Christian-Islam viewpoint is the claim that 'we are all equal creators" so why not organize a society to allow the expression of this idea?

What opposes this idea is both arbitrary collective-moral-ism and the capitalist-socialist viewpoint of collectivism where wage slave serving the elite make sense

Eg The issue in the US is that the public has been taken over by a fascist police-state, which continues to act in a criminal manner spying, lying, fabricating evidence, manipulating the media, where it hides its criminality by claiming state-secrets, ie the state secret of being at war with the publicwhere fascism is a government-and-justice system working for the corporations, or working for those entities which are extremely wealthy.

The only individuals which "count" (the only individuals whom possess freedom and they unanimously express their so called freedom as their right to dominate the public and the world's resources, they express selfishness and they claim they have gifted everything to society, ie they then use their freedom to lie and deceive and steal and murder) are those with great amounts of wealth so that the national interest is to protect these few and to serve their needs (to be dominant), where this is clearly a criminal act according to the legal documents which defined our nation's beginning, in fact one of the main crimes is to destroy the nation in order to serve a world-government but which is really a small set of very wealthy people.

This basing government on serving a personality-cult which worships the very wealthiest people on earth, is actually an exaggerated form of the Calvinism-religion, which is the, apparent, belief structure both of the government-and-justice system, and it is a (primitive) religion which the government-and-justice system has imposed on society, a religion which worships materialism and wealth

The courts use information in their courts which is obtained illegally, they lie about its use in their courts, and then they claim national security to protect their criminal acts, nonetheless they do not do anything to the ruling investor-class, that is, they criminally to not enforce the law, but when they claim to enforce the law they do so in a criminal manner

Clearly this system is rotten to the core and needs to be replaced

A new continental congress

The declaration needs to be re-instated

And the Bill of Rights needs to be enforced but the constitution quickly led to irresponsibility, control by propaganda, and criminality

We need town meetings

Based on good use of communication systems

Ie language categorized based-on assumptions and

it is a belief system which is imposed by means of extreme violence, and a highly controlled propaganda-education-indoctrination system, ie all language and thought is controlled based on propaganda and an extreme level of scrutiny and a system of agents and managers all who are a part of the institutional control, though they are only about 5% (or less) of the population, while the rulers seemed to be basically 85 individuals world-wide (with their controlling stakes in conglomerates) and of these there are likely it is "5 to 10" who are the dominant controlling individuals.

Caesar would be quite proud of the direction of western culture, where science is only about war, and meaning is only about the arbitrariness of propaganda, and the violence by which this arbitrariness is imposed, and the adhering forces (the technical and social forces which are used in the propaganda system to justify hold the empire together) are the technical illusions of development, eg laying bricks to make an illusion of a civilization (the bricks are now 19th century science, primarily used for war and communication control).

The corporations and the justice-system engage in crimes, and they are never brought to account by the governing-and-justice system, and are allowed to continue to illegally oppress the public, through both the unaccountable institutions and through extreme violence and spying and its associated managing institutional reach, so as to cover-up the overwhelming acts of: lying, stealing, and murdering, which the system, itself, commits with impunity.

The political processes are a shame, which are controlled by the propaganda-system, and the corrupt governing-and-justice systems,where the, so called, commander and chief (the president) [is not the leader of the public militia, armed as it (the public) sees fit (ie the second amendment)], rather

He is the leader of the military-industrial-complex where this army is armed based on both how the society uses science and based-on the workings of the propaganda-spy-system of public control, (the commander and chief) is the head of the police-state of the "corporate captured" government-and-justice system based police state, where the government is working for the corporations, and where the corporations control the propaganda-education-indoctrination system.

The country's army is both the spies, the police, and the foreign invading armies and equipment.

The entire justice system is completely illegal, and as rotten and corrupt as an institution can be, and institution which upholds the idea of materialism and material-wealth, which is what it also, effectively, worships, where the operatives are propagandized-robots, who have been selected (by the police-sate) for their cruelty and their being easily manipulated, ie their thoughtless-ness.

It is this religion of Calvinism which Eisenhower instituted after WW II in the context of the police-state enacted through Truman

Though Calvinism has always been the way of the capital-class, ie the investor-class, eg G Washington and unfortunately also B Franklin, (Franklin is a bigger hypocrite than Jefferson since he espoused the virtue of being a Calvinist, but his entire life was enabled by the more equal society of Pennsylvania), Where T Jefferson actually stated the basis in law, (law is to be based on equality, as opposed to Roman-law based on property-rights and minority-rule) where this proclamation "law is to be based on equality," should have separated America from the mentality of the Roman-empire and of the European culture, but the colonies were far too violent, the settlers exterminated the native peoples whose culture was better related to a social law of equality, ie the European settlers should have adhered to the cultures of the native peoples.

The education-indoctrination system the social context of wage-slavery and a management hierarchy of a limited set of institutions (also controlled by the investment of the investor-class) allows a narrow and arbitrary but very complex viewpoint and context for interpretation of information to also be repeated incessantly by the, so called, educated population

Where the viewpoint constricts the sciences and their relation to the creativity within society, and the so called objective viewpoint of social events, ie news, gives a story of an implicit subordination of the public and its institutions to the arbitrary and violent social structure defined by the ruling-class, where the subordination is maintained through violence and indoctrination

Ie the need to express new thoughts

ie Marx and capitalism are both about "a collective society which upholds an elite few,"

What are now called "Markets" and "profits," are simply expressions of ways to organize society based on inequality,

ie this is all about propaganda,

Whereas

Jefferson's ideas were different (ie base law on equality), as well as the ideas from the cultures of the native peoples (were also different)

"We are all equal creators" is a good rendition of the golden-rule

Idealism and materialism are equivalents whereas materialism has shown itself to fail as a basis for the development of new practically creative contexts

Materialism does not need to be the basis for science, and The relation of science to sets of military instruments and communication instruments, are not science's main (or only) use in regard to the expressions of sets of measurable descriptions which can be related to new practical creative new contexts

What exists is too rotten and corrupt to be allowed to continue (to exist)

We the people

Need to Start over, with a new continental congress so that, law is based on equality, and the law is upheld (get good people as managers, and continually review their actions and decisions, so that the law is maintained)

Markets built from each person being an equal creator

So that they can "gift" their creative efforts to a (truly free) market

It is the earth which sustains us

Use the spy-ware, so as to use it against the (deposed) ruling-class and its managers, operatives, and spies

Chapter 33

Collapse

One can consider that the US military is trying to find ways to de-stabilize the US, but it has not succeeded, since (apparently because) the US population has been so thoroughly terrorized by their justice system and centrally planned, and narrowly defined monopolistic economy of central-planning based on selfish interests, that they (the public) have only a most narrow viewpoint about any possibility different from "what is" the expression attributed to M Thatcher in the 1980's.

So that all the actions of the public, who have been coerced into being wage-slaves, must produce value to the investor-class, so that this (production of value for the investor-class) identifies a person's value within society,

Where it needs to be noted that when Marx was talking about capitalism, ie when he was discussing a collectivist society, the collectivist society which he was considering was the collectivist-social basis for capitalism, and he is basically interpreted (by the indoctrinated intellectual-class) to mean (or by the so called radical (or indoctrinated and rewarded as being superior in their competitive abilities) intellectual-class) that "the intellectual-class should run the collective society, instead of the investor-class," but both sides (intellectuals and investors) are equally ignorant [where the (elite) intellectual-class really simply parrot the failed intellectual structures which have been defined by the investor-class].

The intellectuals seek to over-take those (in the investor-class) who brought them into existence, and the investor-class rewarded them as being superior intellects,

ie the intellectual-class truly believe that there is no other way than the way shown to them by the investor-class.

Whereas the other (or regular, or less rewarded in regard to society's highest-values) wage-slaves are manipulated by the spy-intelligence system so as to represent coercive militias, which are racist, and violent, so as to terrorize the public, these heavily manipulated will often be small business mangers or possess important (or high positions) in a church, eg tea-partiers or members of the spy-class.

That is, the intellectual-class, essentially, stand for knowledge, (most often descriptive quantitative models of existence (or experience), as interpreted by the investor-class) which is authoritative, and it is presented (to them by the investor-class) as the root to (of) social power, ie it is presented through the media as being an expression of an "absolute truth"

While

The rest of society is embroiled with the "spies and infiltrators" who oppress the public so as to conform to the interests of the ruling-class, ie the paymasters and investors

That is, if the society was led by the intellectual-class, then these intellectuals would be doing exactly the same types of things which the investor-class is also doing, because that is the only things which the intellectuals have stood for, namely, the set of ideas which best support the investor-class

Whereas, the other types of thinking, which have been excluded by the investor-class, but more importantly the other (alternative) ideas have been primarily excluded with the help of the intellectual-class, whose ideas are consistent with exactly the desired actions of the investor-class.

This is the result of the propaganda-education-indoctrination system, which rewards those who learn, and do, "exactly what the ruling-class want the intellectuals to learn and do"

That is, the main issue with education for the investor-class is that they want the instruments which they depend upon to be adjusted and developed or improved to better help serve the interests of the ruling-class, thus education is always about staying in the same fixed context and extending the development in the directions which may help develop the small set of instruments which best serve the ruling-class's narrow and selfish interests. Thus prizes are used to define high-technical-value and to define the superior intellectuals and they are always about extending the already established context of thought, ie extending the context defined by formalized axiomatic and the set of ludicrous set of patterns to which this failed set of axioms and traditional contexts has led ie it has led to an illusionary-world rather than deeper understanding and expanded technical creative developments, ie this is because formalized axiomatic is not capable of describing the observed patterns of the (material) world, rather this traditional axiomatic structure leads to a few subtle developments in the detonation of nuclear weapons and breaking encryptions and using information (or communication) systems for spying, and interfering with the public, and very-little else.

What the intellectual-class learns, within and through their competitive filtering-out process, of the, so called, educational system (or educational process, but which is really an indoctrination process) [in some of (what is considered to be) its highest forms (according to the ruling-class)], ie the quantitative models from which technology is built designed and developed, ie

What is learned, is about that which is derived from a (quantitative) system of language, and which is the basis for the instruments, which, in turn, form the basis for the technical power of the ruling-class, but which, in turn, identifies a technical quantitative language which possesses obvious limitations, as to the types of patterns, which any such (or which one) particular organization of, the quantitative structures (in regard to its axiomatic formalism) can describe.

If the quantitative language proceeds after both accurate descriptions of a wide range of systems and the developed descriptive context results in ever wider new practically creative contexts,

Then...

This limited set of expressions, of a quantitative truth, need to be both quantitatively consistent, and measurably reliable, and it is a language which needs to have solid relationships with very stable sets of (math) patterns, so as to (or if it is to) have any value as both

(1) an accurate description over a wide range of systems (or over a wide range of applications), and so that

(2) the descriptions provide a context of both use and control over the observed set of stable patterns which are observed, in regard to the measured properties of existence.

That is,

What (for science and math) should be a method of inquiry,

ie equal free-inquiry (as prescribed by Socrates), to lead to practically useful knowledge in regard to ranges of practical applicability of new and ever more widely applicable descriptive contexts,

But

Science and math has been turned into a quest, by the so called superiorly-genetically endowed intellectuals, those who are, supposedly, capable of discerning truth for the rest of society, those who win the prizes for intellectual prowess, but (and) where that truth is (ie is really represented as) an absolute truth [but Godel's incompleteness theorem showed that such an idea as an absolute truth is a bunch of nonsense]

But nonetheless this axiomatic representation of an absolute truth is derived from a fixed form of axiomatic formalization, which is defined by the academic institutions, which, in turn, serve the interests "in education" of the ruling-class, and this axiomatic formalizations was instituted, or initiated, by D Hilbert (around 1910-1920), and is essentially represented in the formalized published program of Bourbaki, ie volumes of formalized axiomatic based math.

But this formalized set of descriptive contexts, which is based on particular types of interpretations of observed patterns, and based on certain ways in which to organize the patterns of a, supposedly, quantitative language, and ultimately dependent on the set of axioms, which the descriptive discipline claims to follow, has already been shown to fail to be able to describe the observed patterns of existence, and thus its elementary structures of assumptions and pattern-organizations need to be changed, if one wants a "true development" of a an accurate and practically useful quantitative-based descriptive language.

But this is not what is happening inside the so called public universities,

Rather, the (known to be failed) formalized axiomatic continues in its traditional and improperly authoritative forms

Namely, it is primarily based on the axioms of algebra, eg the real-numbers and complex-numbers, which are the number structures of the coordinate containment sets for models of physical systems, and the relation to (of) these coordinates to both

(1) metric-function models of measuring, within a coordinate system, and

(2) the set of group (or matrix) measuring-invariant, (or what has, errantly, come to be the equation-form-invariant), transformations, which are supposedly related to local coordinate properties in regard to models of local measuring of the system's properties, where the system is assumed to be, contained-in a coordinate set,

As well as

(3) being related to the set of the axioms about local numerical relations, ie continuity and defining derivatives,

As well as

(4) the existence of local linear models of measuring in regard to the set of measurable world-properties of:

position,

time,

motion,

energy,

inertia,

charge,

temperatures and

numbers of particles within a closed (for) collective sets of thermal systems

But the equation-form-invariant properties of local coordinate transformations,

ie the types of algebraically allowed local coordinate relations, which are supposed to allow for a consistently measurable structure

are about equations which define material-interactions, but these equations have failed to identify, within a quantitatively consistent descriptive context, the reason for the observed very stable spectral-orbital properties of the very prevalent many-(but-few)-body systems,

That is, the equation based descriptions of material interactions have only been able to describe the properties of the two-body system, which possesses a great deal of other symmetries

eg spherically symmetric interaction-force-fields and essentially all interacting-materials are placed onto a 2-plane (for no good reason) although the spherical-harmonic of the H-atom seem to allow for angular-momentum variations of shapes in 3-space

And the non-linear descriptive structure, which is based on the invariant-forms-of-equations, only apply to a one-body system, which possesses an arbitrary property of spherical symmetry which is also applied to it's a descriptive set of assumptions (or contexts)

Nonetheless, it is the set of non-linear and "non-commutative sets of group operator-actions" to which the, so called, superior intellectual-class adheres, in their uncritical beliefs, or rather their deep worship of the high-value which the ruling-class defines for themeither on the, assumed, set of measured properties (or functions), defined (usually locally) on a set of system-containing coordinates (operators applied to the local coordinate properties) or applied to system-containing harmonic function-spaces,as well as a set of "indefinably random" constructs, where "indefinably random" means either basing probability on sets of events which are unstable, and thus counting such events cannot be properly contained in mathematics, ie the counting of unstable events is numerically unreliable, or defining a supposedly probabilistic context, in which "the assumed to exist" (or already observed to exist) set of events (set of spectral-values) cannot be calculated, since the math structures are non-commutative, and/or the math constructs are improperly defined, ie the assumption that a partial differential equation (or local models of measurability) applied to a function-space (or a single-function) is the only valid descriptive context through which the measurable properties of the world can be described in a valid quantitative manner, which the math and science community feverishly, and obsessively, consider so as to, in their haste and arrogance,

So as to exclude any other sets of:

Assumptions, Contexts, ways of organizing a descriptive language, or the types of interpretations to which the observed patterns can be relate instead

Only consider the descriptive contexts which are related to the set of axiomatic formalization which, in turn, is associated to particular assumptions about physical descriptive languages are allowed

In regard to the relationship which these fixed formalized axiomatic structures can have to a descriptive context so that it is only in this context that the absolute truths of the (assumed to be) genetically-superior intellectual-class can be identified and their value to the investor-class judged

That is, the intellectual-class the highly touted superior intellectuals are, in fact, a bunch of chumps, who uncritically follow, serve, and, apparently, worship the descriptive contexts which the ruling-class provides and/or imposes upon their intellectual efforts

Consider the AMS Notices journal October 2014.

In this journal there are both the advertising for research into financial models, the advertising for grants related to (and determined by) a rich mathematician, and the announcements of the Fields-prize winners, ie the Nobel-prizes for mathematics, ie the exultation of the superior-intellectuals in the intellectual-class.

These categories of concern for the math community as described by the AMS Notices journal are all about math structures which are: non-linear, non-commutative and indefinably random

There is the statement that the non-linear pde's are some of the most prevalent types of pde's so they need to be "the most carefully considered" types of math constructs and the statement that the most important natural phenomenon are governed by non-linear pde's

One can also extrapolate to another command, imposed upon the math community, that because non-commutativity is so prevalent, then it also needs to be considered very carefully.

But the importance of non-linear pde's, as well as sets of non-commutative operators, in regard to such types of related equations being fundamental to governing natural phenomenon, is a very big assumption which needs to be looked-at in a critical manner

The better assumption is that because pde's and the use of operator structures have not been of any practical use in regard to describing (formulating and solving) most stable physical systems, where it also needs to be noted that stability implies both linearity and controllability, then these non-linear and non-commutative models of physical systems do not govern physical phenomenon, and subsequently, one needs to consider other descriptive contexts and new sets of assumptions and interpretations of observed patterns.

That is, one can conclude that the investor-class wants to maintain the nincompoop (ie weak-minded) basis for the so called intellectually superior intellectual-class in a condition of babbling about irrelevancies in regard to discussions about math,

And, subsequently, the social forces, associated to such a set of well-defined superior intellectuals, will cause the improperly defined math-authorities to continue to follow their irrelevant, and mathematically inconsistent, descriptive contexts,

Apparently, in the name of their propaganda-defined role-identities of pursuing their idea of an absolute truth which can only be attained by the set of (apparently, assumed to be), the genetically superior, and the intellectually superior few

Thus, the formalized math community are, for no good reason, committed to trying to describe sets of stable patterns which are assumed to exist within a quantitatively inconsistent set of containment set structures, but which are still related to some form of instrumentation, which, in turn, serves the purposes of the ruling-class (where the ruling-class does not want technical development since it puts their investments at greater risk),

eg particle-physics is dedicated to detonating a mythological gravitational singularity.

So (because the math community insists upon using non-linear and non-commutative (and indefinably random) math patterns as the basis for its descriptive constructs, then this means that, the quantitative properties (contained in either the coordinate spaces or in the function-spaces) of the [so called] patterns which these mathematicians are trying to describe are "patterns" which will be: (1) improperly scaled, and (2) exist in a discontinuous context, and (3) the functions, eg coordinate functions, are multi-valued,

This means that when one tries to measure the system's measurable properties, (ie properties which have been determined by a system's formulated set of (what, according to the idea of physical law..., are supposed to be the) system defining set of equations (for the physical system), and a, subsequent, calculated set of solutions; which is the purpose of a quantitative description), then both the scale properties and number-properties of one's measuring device, eg ruler, are not consistent with the number-system in which the properties of the calculated solution are represented.

Thus, there cannot be any identification of patterns (which exist) within the containment set, upon which the partial differential equation math structures are being defined, in any of the following sets of math interests, which are either non-linear, or non-commutative, or indefinably random, which are being modeled within the context of formalized axiomatic math structures, and which are (almost always) applied to:

1. dynamics

2. stochastic processes

3. non-linear-geometric number-theory and

4. the considerations of 2-dimensional discrete hyperbolic shapes (or Riemann surfaces), but (or that is) these math patterns (discrete hyperbolic shapes) are not considered in their simplest math forms, which allow quantitative consistency, but rather the categories which the mathematician must pay attention to (so as to be consistent with the basic concerns of the math community of authoritative experts) are about abstractly developing these math-objects into more complicated constructs, where the descriptive structures (usually algebraic structures) are non-linear and non-commutative

That is, when one hears that there is a simulated computer-model of a formulated and calculated system so that the system is modeled in regard to either ordinary differential equations (OD) or by means of partial differential equations (pde) and the local properties of this model are non-linear or non-commutative, or indefinably random then the quantitative structures associated to such a calculated and computer-simulated system-model are unreliable and the quantitative effort is not capable of actually identifying a pattern which can be relied upon as a valid quantitative description of such a system.

And

This set of locally non-linear, non-commutative and indefinably random types of (physical) system-models by differential equations is the central focus of the so called intellectually-superior math and physics communities.

This is not an anti-science or an anti-math statement rather it is a statement about trying to save math and science, so that

The, supposedly, carefully considered proclamations of science and math can possess some validity rather than being all about nonsense, and it is this continual expression of nonsense, which has destroyed the value of the expressions of math and science communities where this destruction has been a result of (or done by) the manipulation and management of what has become an expression of arrogance and obsession with being intellectually dominant (as opposed to an expression of truth) for the participants in the math and science community, who, nonetheless (despite their, essentially [or virtually], only expressing nonsense), they are still serving (through their expressions of nonsense) the narrow instrumentation needs of the ruling-class, ie physics is still almost entirely only about bomb-engineering, and they are providing the social-role which is a key part of the propaganda-education-indoctrination system which wants to convey the impression to the public that the way things are is the only way things could be, since the, so called, top intellectuals have all been tricked and manipulated, and carefully managed, administratively (in our highly militarized, police-state, society), into supporting all this nonsense.

That is, the, so called, genetically superior intellectual-class always try to push the math patterns to a point where there is both

(1) quantitative inconsistency, and

(2) the construct has only an "(abstract) delusional language relation" to a, so called, description,

ie a description which does not possess any form of either a useful pattern, or of a stable pattern, except that the descriptive language be self-referential to the math language which describes primarily delusional patterns, which are only related to the language of abstract, quantitatively inconsistent, sets of math structures, not related to experience or to measurable properties.

This is a result of the process, begun by D Hilbert around 1910, of axiomatic formalization, and which got more entrenched through the writing of Bourbaki volumes of formalized axiomatic math mostly motivated by algebra but also relying on the notion of convergence, where the real-numbers have been made into a set which is "too big" by using the "axiom of choice," so as to represent an infinite number-identification-process as being a finite process, and which is associated to the construction of the real-numbers, so as to build the real-numbers into being (a) far "too big" of a set, resulting in logical inconsistencies.

ie why think about the unfathomable, where the language has no relation to either accurate descriptions of a wide range of observed patterns, or to reliable relationships to a wide range of measurable properties or to a sense of a practical usefulness of the descriptive context to a wide range of practical utility.

But, nonetheless,

This is what is being done by the so called intellectually superior few (ie the autistic-types with rudimentary language skills, but those chosen [by the institutional mangers] are those who obsess on being dominant, ie the psychopathic equivalents of those few in the ruling-class)

And the extra condition of the rich (so called, top) mathematician who worked on encryption, and, apparently, got rich by spying, and (perhaps, or most likely) doing things like; getting information a split-second before the regular communication channels provided the same information, and then using buying and selling types of algorithms to take advantage of this early information,

ie most likely he got rich (based on the traditional way in which the investor-class get rich) by stealing.

How the, so called, intellectual-class deals with the very important quantitative descriptive structures when under the guidance of the ruling-class (investor-class) through the propaganda-education indoctrination system:

There is a problem with trying to impose a quantitatively inconsistent algebraic context, in regard to the issue of modeling (physical) systems,

In what is supposed to be, a local linear measuring construct, which one wants to impose on a function (or number-system) but where the function is, supposedly, defined in regard to a quantitatively consistent local coordinate context, and most often, also, metrically-invariant local coordinate context,

Thus, the goal is to try to formulate a causal construct of material-interactions, whose quantitative descriptions are assumed to be based on the definition of certain sets of partial differential equations,

Where this is supposed to provide (or be) the reason as to why a physical system enters into a stable context, but there are the problems with this descriptive construct, as mentioned above, of forming a quantitatively inconsistent, algebraic context for the sets of measuring operators, where this is due to these operators being non-linear and placed into non-commutative descriptive contexts (also see below), then there are the problems with the ideas which enter the descriptive context in regard to the dual sets of descriptive contexts defined by the Fourier transform, when the description is modeled to be a harmonic function-space, ie a probability construct, but again (the experts) not being able to form commutative sets of operators in the context of "harmonic function-space system containment" models of physical systems through which the spectral properties of the material systems one is supposed to be able to identify where this inability to determine a set of commutative operators for this (quantum) math structure (associated to function-space models of a physical system) leads to indefinable randomness (ie the inability to calculate a stable system's observed spectral-set) and

The fundamental issue, as to whether, unstable, and vaguely defined (or vaguely distinguishable) random events, can form a basis for a valid probability description?

The clear answer is, No!

Apparently, the goal of the intellectuals (in the formalized axiomatic math community) has been to:

(1) slowly develop certain types of instruments, and

(2) describe every detail of existence, as if it is a quantitative construct, so that it is assumed that this is to take place within a quantitatively consistent set, (but, nonetheless, non-commutative and non-linear operator contexts are used, as well as the description of indefinably random contexts are continually attempted) and this goal (to describe all details of existence) results from the observed properties of vaguely distinguishable (and unstable) patterns, in what might be assumed to be a measurable context ie the flow of existing events and an assumption that all descriptions even

non-commutative patterns can be related in a quantitatively consistent manner to an inert set of system-containment math structures (ie coordinates or function-spaces) which will maintain their relation to quantitative consistency regardless of the types of operators which are being used to try to identify measurable properties.

The alternative set of assumptions, which the intellectual class actively excludes

Consider a new assumption, in which, the reason that there are stable patterns for some material systems, also has to do with the definition of material, but where (this definition of material) is also equivalent to the definition of the material-system's containment set of coordinates, that is material and a metric-space and the stable material systems are all forms of the stable shapes which are naturally associated to a metric-space and its discrete isometry subgroups, and these stable metric-space shapes are models of both material and stable spectral-orbital material systems, where both are stable metric-space shapes, but the material objects are separated from their metric-space containment sets by at least 1-dimensional level, where this pattern of a descriptive context exists within a many-dimensional model of the stable energy-shapes which are naturally defined by mathematical patterns(either isometry subgroups, or when the pairs of opposite states of material (or of the metric-space's material set of "two metric-space properties," ie a physical property and its opposite, or local inverse) then these stable shapes would be related to the stable discrete unitary subgroups)

Relation of this new context to the usual context

That is, when the description is based on local measures of physical properties, so that material is an extra added measurable attribute, as is the case in regard to the current (2014) set of assumptions of physics, and they are patterns to which any new alternative descriptive structure needs to be consistent, then there needs to be both quantitative consistency (linear metric-invariant and continuously commutative and parallelizable, [related to natural-coordinates for the shapes considered in the description]) in the sets of operator types, which a model of local measuring allows, so as to remain quantitatively consistent, as well as a definition of the material, other than it being simply a "measurable quantity,"

ie the idea of material directly related to an operator construct (as the classical idea of inertia is related to "force divided-by-acceleration" or to a geometrically based gravitational force-field) or (both charge and inertia, as related to charged material's geometrically-based field-properties).

Where these conditions are satisfied by the stable shapes of the discrete isometry shapes

(ie back to the alternative context)

That is, the metric-space constructs of the discrete isometry subgroups, or equivalently, both the set of discrete hyperbolic shapes and/or the set of discrete Euclidean shapes together allow for the constructs of energy-shapes or energy-lattices and/or momentum relationships.

That is, material is something which might (or should) be able to either fit naturally into stable shapes in regard to the next higher-dimensional discrete (hyperbolic) shape or to condense into a form of material, which can be in orbits, which are defined by the stable shape of the metric-space, within which the condensed material is contained.

Social forces which influence intellectual efforts

There is a further issue, which is all about social constructs, and it is related to the existence of social classes and the subsequent viewpoint as being privileged by being in such a social-class, this is a copy of the higher-social-classes, which have been defined by the investor-class, and

This formation of an intellectual-class was done by means of the investor-class also defining, through the media (which they own), the idea about what has value in society, but once in a social-class, of high-social-value then this implies for an individual "social dominion," and the delusional belief in one's superiority, and in one's, so called, superior outlook (so called, by the media).

The problems with the current descriptive context

Thus defining material interactions by means of operators, in which material is an extra measurable entity, as is the case of the descriptions of classical physics,

But in contrast, in particle-physics, a definition for material is lost in a non-geometric (exclusively algebraic) operator structure of particle-physics, wherein material is defined as unstable events, which are associated to an unrealistic, vaguely, geometric model of a particle-collision, which is defined for the main purpose of defining an elementary-particle's cross-section, in a probabilistic based description ie in the probabilistic context of harmonic-functions and dual sets of physical properties (or physical operators), and does not possess any relation to a definition of a geometrically determinable material-interaction, which, supposedly, can cause a system to form into a stable structure.

[Or, and only has a remote (if any) relation to a definition of a material-interaction, which, in turn, can cause a system to form into a stable structure.]

Or outside of particle-physics, where material, or a material-system, is well-defined(based upon)

A description of a physical system still falters, since the description is based uponOr represented in the context of any type of non-linear, or locally non-commutative, operator relation

In which the local construction of derivative and integral operators, by means of limits (and convergences within a continuum), is assumed to allow, in regard to: the derivative, and derivative-connection, and the integral

And their, subsequent, formation into pde's

But, in this case nothing (neither material nor a [stable] material-system) gets defined in any type of quantitatively consistent way.

Apparently, there is a belief that because one does follow the math axioms of both algebra and limit-relations then the quantitative sets upon which these math constructs are made will necessarily remain quantitatively consistent, but this is not true. Stability must emerge as a math property from very simple math constructs: linear, metric-invariant, continuously commutative, and defined upon parallelizable shapes.

Nonetheless, these math properties are virtually excluded from the descriptions which are considered by the intellectually superior-few ie society's dominant intellectual-class

Whereas these simple quantitatively consistent math properties are the math properties which characterize the shapes identified by Thurston-Perlman in geometrization the stable shapes of the continuously orthogonal circle-space shapes, basically, tori the discrete Euclidean shapes and strings of toral-components the discrete hyperbolic shapes

Yet, stable systems are the norm, not the exception.

Nonetheless, one wonders why the community of, so called, expert mathematicians tries so hard to exclude new math constructs, and new contexts, based on the set of stable continuously orthogonal circle-space shapes, where there is quantitative consistency as the basis for an expanded descriptive context.

Reviewing the alternative set of assumptions and its descriptive context

That is,

1. the definition of material,

2. the set of stable spectral material systems, with charges fitting into stable shapes, so as to occupy spectral-orbital structures of the system's stable shape, as well as

3. the sets of stable orbital structures, for condensed material contained within metric-spaces are all a result of the internal properties of metric-spaces, and their natural relation to a set of stable shapes, where this intrinsic property of stable metric-space shape is (can be) constructed (so as) to be related to either stable shapes of energy or continuous shapes of interactions, and the strong dependence of descriptions of interactions upon identifying spatial positions

Which are what is necessary so as to allow the three above types of properties to be described in a mathematically consistent manner, And it (this descriptive context) is a quantitatively consistent math structure and depends on a (the) descriptive context to be many-dimensional

But

Where the main attribute (or property) needed for a wide range of physical descriptions is to determine (or find) the "complete finite set" of stable spectral-orbital values, which can be defined for, and associated with, the set of metric-space shapes which need to be defined for all the different subspaces of all the different dimensional levels which can be defined for a hyperbolic 11-dimensional metric-space, where eleven this is the first dimension (ie hyperbolic dimension-11) where the hyperbolic metric-space does not possess a natural stable discrete hyperbolic shape (as was shown by D Coxeter), where this finite spectral-set depends on there being only one maximal shape defined for each such subspace so that there exist natural sequences of dimensional-size set-containment schemes for shapes contained within (some) higher-dimensional shapes, for this finite set

Basically, the formation of stable systems comes about due to either collisions take place or multiple-sets of lattices are simultaneously a part of some given system (eg a living-system) and when the energy and relative positions are "right" in these two contexts then a new shape of lattice can form due to it being in resonance with some subset of the finite spectral-set, which is defined for (over) the entire 11-dimensional containment-set.

There are new ways in which to interpret the idea of some of the smallest fundamental material-components eg protons, electrons, neutrons, and neutrinos, respectively, which enter into the shape-construct of a material-system.

The math properties of the stable shapes used by the new descriptive context

This descriptive context, in regard to the descriptions of these shapes by pde associated to differential-forms, in turn, results in: linear, metric-invariant, and continuously commutative, and parallelizable, natural local coordinate structures, which means that these math shapes are solvable, and they are extremely simple shapes, and they are defined in a quantitatively consistent context, so that they can be put into a descriptive context in which material-interactions can be defined for all the various dimensional levels within which these shapes are defined, so that these interaction mostly define unstable quantitative contexts, but the (associated) set of second-order interaction pde's, ie parabolic hyperbolic or elliptic, can be used to account for the physically observed properties of systems, including understanding the stable spectral-orbital properties of the many-(but-few)-body systems, which occur at all size scales and in all the different dimensional levels and this is possible because the genus (the holes in the shapes) are newly interpreted as the basis for the stable many-(but-few)-body spectral-orbital structures for physical systems

There exists (in the literature written by m concoyle) a rather complete description of a new math context, which is a quantitatively consistent descriptive context and it (the description) solves for the stable many-(but-few)-body spectral-orbital system which exists for all size scales, and in all dimensions where a descriptive context can be defined.

Perhaps a review of the basic assumptions of the four (or five) different categories of physic's descriptive contexts and then relating these different contexts to the types of quantitative inconsistency operator structures rather quickly enter into the discussion

Probability:

This is about determining the frequency of event-outcomes (or probabilities of outcomes of random event experiments) for a finite set of stable random events or well defined both distinguishable and stable random events, or the set of calculable observable and stable events

Classical:

Is about physical descriptions, both accurate and useful descriptive contexts where the rule for the description applies to a wide range of systems which can be related to local measuring of the system's measurable properties where the description is associated to spatial positions motions and changes in motions of an object, ie local models of measuring properties associated to both spatial positions of a material object and its geometric relation to distant material as well as a well defined inertial relation of the object, ie the relation of an object's changes in motion due to the object being influenced by force-fields in turn dependent upon the material-distribution of the material which surrounds the motion-changing object

General relativity:

The idea that arbitrary coordinate frames of reference (or diffeomorphically [ie invertible at a point] related) are equivalent in regard to describing the locally measurable properties of physical systems, eg spatial-positions and motions etc, so there is a need for covariant (ie tensor, or local coordinate transformation invariant) equations used to describe the material object's properties

And

Since inertial-mass is equivalent to gravitational mass and mass is equal to energy then the geodesics of the system's natural coordinates, ie the coordinates in which the local metric-function has independent diagonal directions, determine the material-object's motions, to find this information one needs both the form of the local metric-function in a neighborhood about the particular point which defines the object's spatial position and the parallel direction equations which can be determined by using an inner-product where the metric-function is also an inner-product so as to determine the local geodesics

But this is all non-linear so no actual patterns can ever be determined except for the one body system which is assumed to possess spherical symmetry.

Quantum physics:

Measuring is equivalent to associated sets of operators (which are related to classical properties of measuring) and the assumption of the containment of the system's spectral properties as well as probability distribution (densities) related to "the probabilities of a random spectral-event at a small region about any particular point in space and time" are contained in a harmonic (all the functions define [repeating] periods) function-space, where harmonic implies being related to probability distribution functions and the set of dual relationships to other function-space which is identified by means of the Fourier transform between the (dual) function spaces where each of the harmonic functions possess a relation to a wave-function where the properties of the wave-function depend on two terms, (kx-ft), so that each term is a multiple of variables so that the subsequent product defines the property of "action," This is both the context of the Fourier transforms as well as the context for pairs of non-commuting operators which define action and which are dual operators to one another so that such pairs of measurable properties define the uncertainty principle of quantum physics.

A set of measurable properties, or equivalently the system's set of (commuting) operators, are needed to diagonalize the function space, so as to determine the spectral-set of functions (for the harmonic function space), as well as the associated set of the system's (defining) spectral values, for any given quantum system.

But, the set of commuting operators can never be found for a general quantum system, most particularly for the stable many-(but-few)-body spectral-orbital systems found at all size scales, in particular for the prevalent small spectral quantum systems of this stable many-(but-few)-body type.

Particle-physics:

In the U(1) x SU(2) x SU(3) fiber group of particle-physics only, U(1), associated to the derivative-connection for the electromagnetic-field, ie referred to as an Abelian connection, has a geometric relation between the distinguished point of a particle-physics system and distant material geometry (namely the geometric relations of classical E&M theory)

Otherwise

The descriptions of particle-physics are only about particle cross-sections, ie probabilities of particle-collisions, but for field-particles which essentially possess no geometric structures, eg flow-densities or flow-directions, for an erroneously model of a distinguished-point-particle and a point-particle-collision which supposedly models a material interaction. The collision structures of elementary-particle collisions, ie the particle-scattering information (or scattering-probabilities) upon which the determination of an elementary-particle's cross-section depends, are related to a set of unstable elementary-particles, so it is a descriptive structure which describes very little which can be related to stable patterns of physics, rather it is a model which can be used to determine rates of reactions in a chaotic context of many colliding components in a context of transitioning between two relatively stable states of material ie it describes the energy of an explosive transition context.

This structure of internal particle-states associated to the SU(2) x SU(3) part of the fiber-group is vaguely supposed to be used to perturb the wave-function's spectral-energy-values, but such a construct is really outside of what mathematics can actually do, since it is a non-linear, non-commutative and indefinably random math context so that there is not actually any content in this descriptive language other than its relation to the instruments of nuclear bombs.

Each descriptive context has obvious limitations in regard to it actually describing by formulating and solving a system's equations, any pattern which is stable, except classical descriptions

Most of the classical system's described by applying the physical law to determine a system's set of equations end-up being non-linear equations

Yet, classical physics, in its solvable context, has led to useful information, so as to be able to put-together and control classical systems which are solvable. This set of solvable classical systems has been the attribute of physics which has led to greatest technical development derived from science, and to new practically creative contexts, and is the descriptive attribute which best characterizes science and math

That is,

Following data is not science, Ptolemy was able to follow data which he did with the epicycle model of the planetary-motions, and now (for the last 70 years, where now is 2014) it is this "following of data" which is what characterizes particle-physics as well as quantum physics and general relativity.

The following of data by the physics and math communities due to the non-linear, non-commutative, and indefinably random context of the current (2014) types of math models for physical systems has not led to anything, except to proclaim, that virtually all simple stable physical systems are far too complicated to be able to describe. But this is just baloney. Other assumptions need to be considered, and other ways in which to construct math based languages need to be considered, where this takes place at an elementary level of discourse. The expert class needs to yield to other alternative ideas, due to their failures

This is a bizarre society

The leaders (the investor-class) fund the very terrorists which they then go to war against

This is all within a highly militarized and highly propagandized society, in which

there is a development of an intellectual-class, who spend all of their time doing utterly useless investigations into quantitatively inconsistent math patterns and math models of measurable descriptions of existence but where their models of measuring are invalid,andeven though the math language upon which they base their quantitatively descriptive models of existence cannot describe the most common and wide spread and the most fundamental type of physical system the stable many-(but-few)-body spectral-orbital systems which exist at a wide-range of size-scales they insist on their descriptive context as being nearly the absolute truth about the nature of existence, namely, that existence is based on materialism and material systems have the properties that they do have because their interaction constructs are based-on invariant, (either form-invariant [diffeomorphisms], or metric-function-invariant [isometries]), ordinary and/or partial differential equations, {which either directly model local measuring of physical properties for material (or energy) within a context of shape (system-containing coordinates), or there are (usually locally defined) sets of operators which represent the set of physical properties of a system, in turn, modeled as a harmonic function-space which carries the implications of both a distribution of randomness for material reduced to random point-particle-spectral events and the fixed spectral structures of what is usually a small (quantum) material system}

Then the quantum description is improperly identified to need a model of material interactions based on both material reducing to point-particles [but virtually all of these point-particles are unstable] and these point-particles colliding in the context of point-particle (classical) collisions, which the uncertainty principle of quantum physics does not allow, but where these particle-collisions are further modeled as SU(2) x SU(3) derivative-connections, which, apparently (?), fit the set of elementary-particle scattering properties (well, ?), but the only property of interest…., in these particle-collision, so called, models of material-interactions, is the collision-probability cross-sections of these unstable elementary-particles (the only property which these particle-collision scattering models [of the connection-derivatives] can identify is the collision-probability cross-sections of the set of unstable elementary-particles, which only emerge out of high-energy particle-collisions which occur in the detection-chambers (eg cloud-chambers) of particle-accelerators)

But

None of these partial differential equation models of material interactions can describe the properties of the very stable many-(but-few)-body spectral-orbital systems, from: nuclei, to general atoms, to molecules, to crystals, to the solar-system, to the stars in galaxies.

Nonetheless they (the intellectual-class, ie those who preside over the education-indoctriantion part of the propaganda-system) continue to insist that this descriptive context is essentially an absolute truth since they claim to correlate this descriptive language with a few measurable properties, but where they ignore the vast array of properties of very stable systems which are outside of their descriptive context out side of their set of: assumptions, contexts, ways of interpreting the observed patterns, the ways in which they have organized their quantitative descriptive patterns of language, etc etc at an elementary level of language, they need to re-build their failed efforts

But

Instead the social-class-system (both satisfying the interests of the ruling investor-class) and maintaining the role of the intellectual class in the propaganda-education-indoctrination system is much more important than any realistic reference in regard to trying to get a measurable description to describe in a "form of description" which is both an accurate and practically useful truth, in regard to the wide range of prevalent and fundamental systems which such a descriptive language should be trying to describe.

This is a society which is all about arbitrary high-value determined by violence and within such a coercive society power is organized around banking the military and creative production

Thus, the ruling-class controls creative production through a monopolistic business institutions

That is, trade is fit into a system of:

coercive control, and so that, the creative efforts of the public, which is so oppressed, must fit into the narrowly defined constructs, which revolve around certain particular instruments about which the ruling-class has organized their social-power, ie instruments of coercion (bombs) and instruments of information

And this is why the axiomatic formalization, best associated to D Hilbert (around 1910), has been so useful to the ruling-class, since it has the intellectuals babbling about inanities and complications, which, in turn, allows the intent (of the ruling-class) on the slow development of only certain instruments, about which they have organized their social power, to proceed in a limited and hierarchical way, or highly controlled and based on elitism

The intellectual-class has been educated and indoctrinated to suit the needs to the ruling-class

Thus their viewpoint about creative development are all centered around the same ideas which the bankers have about creative development, but this system of knowledge (ie the categories of study and the structures of the carefully used languages most centrally the language of math and science) and its relation to creative productivity has shown itself to be a failure

Now capitalism is a collectivist society, and it is based on the violent control of property, so that resources, found on property, are used in certain narrow ways, so as to serve the needs of the lying, thieving, murderous ruling-class, ie those who violently control the property. But capitalists incorrectly claim that the free market allows the most creative individuals to acquire the most freedom and power. This is incorrect since knowledge and creativity are suppressed by the bankers, and instead, only individuals who can serve the needs to the investor-class are allowed to prosper, those whose efforts fit into the dominating monopolistic "market" structures

Whereas socialism is claimed to be all about collectivism, which places the intellectual-class into the leadership roles and it is over the same structures which the bankers put in place, ie it is collectivist and it is also monopolistic and narrow

The property of humans which societies should most try to develop is the interest that people have in knowledge and practical creativity and this is best done so that each person is seen under the law as an equal creator

Then the narrow monopolies and the coercive rule (over society), which form an unnecessarily narrowly collective society, would be controlled in order to develop practical creative development based on equal free-inquiry

One needs to allow equal free-inquiry, and an associated set (type) of practical creative efforts, for which the entire society is to be in support, so that these creative efforts can be traded in an equal free-market, or given the world as a gift, but in order to maintain both an equal free-market and to see each individual as an equal creator and an equal free-inquirer then dominant social patterns cannot be allowed to be a part of society and to subsequently dominate

Chapter 34

Social choices

The social structure which exists needs to be dismantled

It is a miracle to be alive, but does the public want to settle for the crumbs fallen to the floor from the table of the (oppressive) ruling-class, when, in fact, the ruling-class's arbitrary viewpoint, about how the culture should be organized, has resulted in the poisoning of the earth, and a non-stop violent attack on the public, and a destruction of rational structures which could be of help to society, a society which needs to be about creativity, and its wide ranging set of knowledge, which is to be can be of use, to a wide range of practical creative endeavors

Two choices

I would like to present two choices, in regard to how to organize society which can best serve the human attributes of the individuals (of which society is composed), so as to be based on what might be true about human nature

1.One is based on humans naturally seek knowledge which they seek so as to be creative in the context about which knowledge is gathered

If this is the true basis of most (or the best) human behavior, then society should be based on equality so that each person is an equal individual creator who can express his interest in equal free-inquiry so as to gain knowledge which is in the (correct) context within which the individual seeks to be creative

In this society law would be based on equality and an equal right to free-inquiry and freedom to possess one's own beliefs and to express those beliefs mostly as an expression of the person's acquiring of useful knowledge while one's creativity would be a gift to the world as an expression of gratitude for the earth which has brought the individual forth and nourished them

This would be a society based on individuals, and their individual freedoms to know and create, where the material-products created would require some forms of collectivist social behaviors, which have well defined limits on their expansive powers, where the collectivist behaviors would be associated to products or processes which become (arbitrarily) valued by society, ie traded in large quantities, where these limitations (on collectivist behaviors, or collectivist groups) are required in order to maintain individual freedoms of individuals as equal creators, so that the creativity of all the individuals in society becomes the basis for a trading market which has a wide array of examples of selfless creative efforts

That is the society needs to promote the common welfare so that individuals survive and can express themselves to the community and to enable them to create what it is they want to create

As opposed to the collectivist behavior of a capitalist society where domination narrowness and destruction becomes the effect of the collective society

2. The other is based on arbitrariness, hierarchy, and violence, (this is the same viewpoint as the viewpoint of the Roman-Empire) where an obvious starting place in this viewpoint is the viewpoint in which society is now (2014) being organized, and it has three pillars of arbitrariness imposed and maintained through violence and institutional terror.

I. Violence best characterizes the way in which society is organized. Violence is used to impose arbitrary social structures

II. A Propaganda-education-indoctrination system controls thought and language and markets, it categorizes and limit's the range of both creative activity and expressions of ideas, by only identifying a small set of categories related to "allowed behaviors within the collectivist society" which was built from the investment interests of the investor-class, it determines the relation between knowledge and creativity, which serves the collectivist (and arbitrary) social hierarchy which is run by the investor-class, where the collectivist actions of society are only allowed to serve the interests of the ruling-class

III. The money-system, which is the means through which "value" is expressed within society, for material and for processes (to be done). This value is arbitrary, so the nature of money is also arbitrary, but it is used to organize social power structures within society, where these social transaction processes are used to funnel social power back to the investor-class.

Production, markets, efficiency, are arbitrary expressions which depend both on fixed ways of the, so called, life-styles of the public, within a propaganda system which controls knowledge and creative categories

Note that the so called, intellectual-class is, usually, the strongest proponent of the very limited sets of knowledge categories which are associated to the very few creative contexts of society, where these creative contexts are associated to the very limited range of a failed form of absolute knowledge, which defines the intellectual-class (who are authoritarian and dogmatic)

I. a hierarchy of people skilled at violent actions against others, similar to what one might imagine the law of the jungle would have been like for the dinosaurs, where in this context, value would naturally be property rights, and the violence associated to defending one's property, where in such a defense one is expressing their desire to own property by defending it, as theirs (an expression which goes against the personal love for mother-earth).

In this type of society, law would be based on property rights and minority rule, and the society would be designed to be a collectivist society which violently supports the interests of those who most successfully acquire (steal) property, and where creativity is based on practical development of material-value

There is already an implicit context of value, but it is arbitrary, nonetheless, where value is domination (by a few, through violence) of the property of a material-world,

ie the earth, which provides life and gives life the nourishment it needs

There is also an implicit assumption, of there being absolute truths, which are defined in regard to a collectivist social hierarchy, so as to negate the symbolism of things being arbitrary, rather the propaganda system expresses the idea that these things (elite hierarchies) have to be this way, because they can be determined from a set of absolute (authoritarian) truths, which the superior intellectuals of the world were able to discern from within the propaganda-education-indoctrination system

ie everything is a relative expression of arbitrary value, which the collective society is forced to support due to it being coerced by the elite, top of society.

Instead the investor-class need (wants an appearance of there being) a basis in truth for a, so called, elite hierarchy

So this is the basis for absolute authoritative truths which the elite-intellectuals of the society express.

These truths have been:

Materialism

Religious based moral-ism

The need to Control of language and thought

Claiming that coin-based trade is the basis for the superior of the investor-class… in a capitalistic society… to be determined, through a, supposed, free-market, but it is really a highly controlled and very-narrowly defined market, which is based on domineering monopolistic big businesses and their related propaganda [but the story told is that "there is a free market" from which the public freely chooses (and buys) the products it wants, and these products are provided to the public by the most intelligent people in society, so these superior people (who make the products which the public desires) deserve to run society] (but this is a model based on indefinable randomness, ie it is an arbitrary expression, ie there is nothing free about markets, and the social context is the main determining factor for the set of desired products by the public, and not simply the set of products

Whereas, in reality the products are narrowly defined and controlled by monopolies, so as to fit into social, and societal, contexts [or to fit into life-styles])

The micro-chip was developed and then it was systematically exploited by monopolistic economic forces, ie by the investor-class.

That is, there were many programmers, and a few who tried to manage technical development, but both the investor-class and IBM were still a main players in this story, ie programmers and program-marketing either emerged from IBM, or sought the help of IBM (ie IBM was the monopolistic business which was already based on computer programming [on mainframe computers]).

Markets and products, and subsequent sales, are about social contexts, which determine a limited viewpoint about products, and the sales of products also depend upon luck, but now they are mostly determined by an all-controlling propaganda-education-indoctrination system, controlling all aspects of the lives and thoughts of the public.

That is, this is all about the illusion of there being absolutes upon which one can base one's judgment of value, both truth-value and social-value:

the intellectual-class upholds the illusion of absolute truths, while the police-state (or people of action) upholds the illusions of social-value [moral-ism, fixed life-styles, the illusion of a free-market based business model, ie to buy goods and, subsequently, build businesses (but, only monopolies or those who can partner with monopolies can survive in highly controlled markets)]

But

It is all, far too narrow, and the absolute truths (of the intellectual-class) have failed, yet the experts and the moralists insist on upholding the visions and doctrines in which they believe, ie the beliefs which they learned from the propaganda-education-indoctrination system (of control of language and thought), ie they are the same beliefs which uphold the investor-class

II. Religion, and propaganda

This has been implemented through Calvinism, and an authoritarian material-based science, where Calvinism worships the material-world and it assumes that God shows the world the best people by making the best-people rich thus law is based on thugs, who are highly judgmental, asserting arbitrary things, ie and lying, so that the justice institutions supports these thugs, since they are supposedly protecting the wealthy, and thus the thugs stand for righteousness, as seen through the, so called, eyes of the law

This context of the religion of Calvinism is the implied context within which the justice system operates, and which is a context which is seemingly hidden from the public, ie it is a context which is never discussed, but

It is the context of Calvinism is used by the personnel of the justice system, so as to identify themselves as a set of morally-superior people, vastly superior to the public, and thus it is used to justify their inhuman and illegal actions directed against the public (this is primarily and administrative issue, and a managing personnel issue, that is, the spy structure, upon which the courts are now based, leads directly to the violent militias and racists, which are so closely controlled and manipulated by the spy-justice-military system, ie the hidden army of the police-state).

On the other hand

The science (of the arbitrary society held together by violence) is to be based on materialism,where this has led to weapons research, and to the development of communication and information-processing systems, where the communication systems are used for spying, thus the justice systems is a totally rigged-game about controlling all information about the public, where this control is gained through spying, thug-gery, and both secret and overt attacks on the public (at the request of the ruling investor-class), ie the justice-military system is the personal armies of the rich; exactly as in the days of the aristocratic Romans, eg at the time of Caesar.

But

In fact, the trading of a narrow set of products and oppressing the public is all a bunch of arbitrariness, so that society, eventually, decays, where

this decay has been a result of (or it is based on) both its own logic, and due to the focus on domination and violence upon which its arbitrary social structures are held together

In the first scenario (concerning organizing society so as to have society consistent with the nature of human beings) the context is about "how one seeks knowledge," and "what one wants to create,"

ie the context of a creation within each individual.

That is, in regard to descriptive truths, the range of useful description is limited by the assumptions contexts, and the ways in which one organizes the language, thus, equal free-inquiry is a process which is trying to always be related to a truth, which, in turn, is also related to "practical" creativity, and new creative contexts, where both wonder and creativity is at the core of the very nature of being human

In a society based on an arbitrary social hierarchy, which is realized and maintained through violence,

The violence directed at the public comes from the justice-military system, which is engaged in class warfare, so as to serve the interests of the ruling-class,

where this takes place in an imposed form of the Calvinistic religion, which defines all justice system actions, and it is both the religion and world-view upon which the justice police-state system is based, and

It (the justice system) imposes Calvinistic religion on the public, who are wage-slaves, so as to define the, so called, righteous nature of class-warfare, of the ruling-class against the poor public, and which is a conflict which is headquartered, by the ruling-class (and their political minions), within the justice-military-state social structure over which the ruling-class rules

II. Other aspects of the propaganda-education-indoctrination system

In today's collectivist capitalist system the propaganda-education-indoctrination system is populated by the, so called, intellectual-elites, who are well indoctrinated in the, so called, knowledge, which is defined by the interests of the investor-class. (based on its context, this knowledge, held by the intellectual-class and which defines the intellectual-class, is considered to be an absolute knowledge, ie things which can be described by science cannot be described in any other manner. But this is far from being true, in fact, it is quite false.)

This means that all published material…, except self-published material…, which is allowed to be considered by the intellectual-class, either in the media or in the academic surroundings, is all fundamentally flawed.

It is based on an authoritative form of knowledge which is failing to be able to describe, and subsequently use in new practical ways, the observed properties of the world

This is mostly due to the fact that the discussion is not based on assumptions and criticisms of existing expressions, especially, if the existing technical expressions are not capable of describing the observed patterns, eg of stable many-(but-few)-body spectral-orbital systems, which no such expressions in modern physics or modern math are capable of describing these fundamental systems,

All participants in the intellectual-class believe in both inequality (especially, the inequality of intellectual capability, ie those in the intellectual-class are the superior intellects of society) and subsequently they believe in social hierarchy where this is (circularly) related (in logic) to the social construct of an absolute authority, which, in turn, is associated to the intellectual-class, ie they believe in the particular form of formalized axiomatic which defines an absolute authoritative context, upon which is based the (academic) competition through which they came to be determined to be through examinations to be, "the superior intellects of society," and, thus, validating their belief in their own intellectual superiority over the public,

But

This is really a belief in arbitrariness (as can be demonstrated by means of the actual implications of Godel's incompleteness theorem), where this belief in an arbitrary authority is exactly equivalent to the arbitrary system of beliefs related to racism

Without equal free-inquiry, there will be a belief that an authoritarian culture is an exceptional culture, thus justifying imperial expansion

Instead (of equal free-inquiry, and elementary discussions about assumptions contexts etc etc) then academic intellectualism is about developing, in a very narrow context, namely, slowly developing the instruments which already exist, and which the indoctrinated intellectual-class maintain, (eg the nuclear bomb) and

It (this intellectual narrowness) is (has come to be) based on formalized axiomatic, whose, apparent, intent seems to be to be able to tailor-make knowledge to fit any purpose which might be of interest to the investor-class, (but according to Godel's incompleteness theorem, it is an impossibility for a language...., which is based on a monolithic [and absolute] set of: assumptions, contexts, interpretations, and a single way in which to organize the descriptive patterns for the authoritative and, and which is assumed to be all encompassing language, where there is no room in the authoritarian discussion for criticisms of the dogmatic doctrine, which is implied in the fixed context of formalized axiomatics)but rather than (axiomatic formalizations of math and physics) making the language versatile and widely applicable, instead it makes the language dysfunctional, where this has come about by trying to extend the descriptive context defined by a partial differential equation (or pde)...., which is used to make locally measurable models of material-interactions...., by considering a mathematically wider pde context, than the original context of the classical physics language, which was a context of being: linear, metric-invariant, and solvable where in this wider context of pde it is assumed that

(1) the pde is non-linear, based on some general metric-function, and in

(2) the contexts where the sets of operators related to the definition of systems of pde are non-commutative...., this is usually in the context where shape is not the focus of the descriptive properties but rather the containment set of the pattern one is trying to describe is a function-space where, in turn, this is quite often related to issues of probability..., and

(3) the descriptive context of indefinable randomness, where the, supposedly, discernable patterns, are, in fact, patterns which are unstable and/or where the elementary events are unstable, this is an improperly defined probabilistic context (so the probabilities which such an improperly defined context tries to define are not valid) and where the set of observed patterns are not capable of being identified by attempts made through an invalid (or quantitatively inconsistent) set of formulations of "system material-interaction models" based on pde, and subsequent calculations, in regard to solving for these pde's descriptive contexts, such as contexts when there is an assumption about modeling a particular type of system as a function-space, ie where it is assumed that the physical system is not a shape which is contained in a coordinate-set, but some type of shapeless physical system, but these shapeless models have not been shown to possess any reliable and stable properties (such as the very stable properties associated with a stable many-(but-few)-body spectral-orbital system), but in the, quantitatively inconsistent, yet, nonetheless, still assumed to

be describable, contexts (of invalid pde [physical] system-models), the quantitative structures, ie the system containing quantitative sets, upon which the description depends…, eg either in coordinate spaces (with coordinate functions) or in function-values or for function in function-spaces, etc, these quantitative sets (ie coordinates and/or function-values)…., are quantitatively inconsistent, where this is because the either non-linear, or non-commutative, and/or indefinably random operators (or operator-contexts) used in such a description, actually, cause the resulting quantitative structures of containment…, (or description) such as the base-space coordinate structures or the set of function-values…, to no longer have:

(1) reliable measuring-scales, and

(2) the measuring-scales become dis-continuous, and

(3) the (function-dependent) quantities become many-valued,

Thus, the number-systems upon which the description depend are not reliable number-systems, so that, the description cannot carry…, within "such a failed and unreliable quantitative structure"…, any reliable (or stable) patterns, ie the description cannot contain any meaningful content.

That is, the context upon which formalized axiomatic was formulated, namely, using pde to model the fundamental (material-interactive) aspects of a (physical) system which one is trying to descriptive in a quantitative context, is a failed descriptive context.

Note: pde's are used to, supposedly, determine (by math-solution procedures) the measurable properties of a (stable) physical system, where pde's are supposedly, determined by physical law. That is, the pde's used to model the material-interaction properties of material systems (or of what are supposed to be models of material-interactions which turn into stable systems) are models based upon the idea of local measuring (or sets of operators which represent a system's measurable set of physical properties), which, supposedly, identify (in the set-containment context assumed by the pde) the actual set of inter-relationships of the particular quantitative types, which, in turn, cause the system to possess the measurable properties which, in turn, are observed.

Unfortunately it is a descriptive context which only works for a few cases, eg the 2-body system.

Thus, one needs to conjecture other ideas, eg it is not "the pde model" which is valid, rather [in the descriptive context of the pde] it (instead) is the principle fiber bundle of the system's isometry (or unitary) fiber group and its subsequent base-space of a (resulting) stable metric-space and their associated set of stable shapes (resulting from the [covering group, discrete isometry subgroup, and conjugation-class] properties of the fiber group), {which can naturally be associated to a metric-space of (what need to be the stable) non-positive constant curvature metric-spaces whose metric-function has constant coefficients and is continuously commutative (so as to defined the shape's natural coordinates) or whose local (natural) coordinate parallel transport structure is locally commutative [and so that (in such a shape which possesses holes in its shape) where the holes in these metric-space shapes are related to either different energy-levels (in which a set of lower-dimension material shapes, which occupies the spectral-faces, fit perfectly into the spectral-orbits of the stable shape), or to different stable orbits for the condensed material which occupies the orbital structures (and which can be related to intricate lattice patterns in both fluid and crystalline materials [which is condensed material])]} furthermore., and the context is many-dimensional.

There are alternatives (but the typical person has been brow-beat so badly by the violence and intellectual bullying, which are all defined in an arbitrary context, that a person is afraid to challenge intellectual authority, but it (that narrow, and uncurious, authority) is a narrow authority, which only serves the (creative) interests of the investor-class) (thus, it is an intellectual structure which really should be contested, and it should be contested by everyone at an elementary level of language ie at the level of assumption, context, interpretation, and the way in which the descriptive language is organized etc)

There are alternatives (things do not "have to be done this way")

Instead (in regard to math and physics, why not) try using E Noether's symmetries so as to associate physical properties directly to (metric-invariant) metric-spaces, of non-positive constant curvatures, and the discrete isometry (and/or unitary) subgroups of the metric-invariant (or Hermitian-invariant) fiber group of local coordinate transformations, which is associated to the metric-space, be used to define the stable shapes, upon which stable patterns (shapes) of (physical) systems (as well as the metric-spaces themselves) can be modeled, so that various different metric-space sets, which are a part of the set of properties associated to a stable physical system, all together, identify properties (for the system) based on stable shapes,

ie stable shapes which identify the stable patterns which are associated to a stable system's measurable physical properties, in turn, associated to the metric-space types, there are the main different metric-space types of:

(1) the stable energy-shapes of a hyperbolic metric-space, and there are

(2) the continuous mathematically consistent contexts of inertia and spatial positions of Euclidean metric-space shapes,

And then also the new description is:

based on new ways of defining the local processes of measuring change (which can occur), and where the changes are discrete, based on two (opposite) metric-space states (which are locally the inverses of one another (ie the local inverse structure identifies an equivalent opposite metric-space structure)), and

The local descriptions of (system) changes are also associated to stable shapes (or stable patterns) through both non-local constructs (based on relatively stable discrete Euclidean shapes, which can define a brief action-at-a-distance geometric relationship), and through processes of interactions which relate the changing-components to resonances... of the systems which are interacting... to a finite spectral-set,

Where, in turn, this finite spectral set (which is defined for an over-all high-dimension containment hyperbolic metric-space) is defined on a high-dimensional metric-space containment set, where the high-dimensional metric-space containment set is, in turn, "partitioned" into stable shapes,

ie the set of stable patterns upon which the description is based, ie both metric-spaces and material-components are both modeled to be stable metric-space shapes (but defined on different dimensional levels, or for the same dimensional level they are defined for the different subspaces), furthermore, the shapes (used in the "partition") cause the higher-dimensional containing metric-spaces to "not be related to their lower-dimensional subspaces through the math property of continuity," rather there are discrete differences between dimensional levels (and between subspace-shapes of the same dimension) which topologically separate the different subspaces (of the high-dimensional containment set) [ie topologically separate the different subspaces by means of various topological properties, defined between the shapes of different dimensions, topological properties of being open-closed and/or being closed, depending on the dimensional level within which the topology (ie properties related to continuity) is defined],

(ie a high-dimensional metric-space containment set, which is "partitioned" into stable shapes, is "partitioned" by a method of "partition" which is based on the properties of: dimension, size of a shape, and a set-containment structure-tree, and a rule about stable discrete isometry (hyperbolic) shapes for each subspace [where the rule is: only one biggest stable shape for each subspace] which requires that a finite spectral-set be defined for the entire high-dimensional containment set, but where resonance can occur between any shape (contained in any subspace) and the entire finite spectral-set)

Furthermore, transitions between two-different stable systems, which is considered to be a part of the properties of physical systems, can occur (1) in a context of many different (and stable) sets of lattices (ie stable lattices related to condensed material), (2) during some special interactive contexts, and (3) during collisions between different stable material-components, in particular, when the energy-of-the-collision is within the correct range.

That is, systems do not enter into new stable system-types based on detailed locally measurable relations, described through pde, which, it has been assumed, allow system formation during material-interactions, eg as Newton was able to describe the construct of a stable system for two-bodies in regard to gravitational interactions, (but no-one has extended this description, based on pde, to the very stable many-(but-few)-body spectral-orbital systems, which are so very prevalent within our existence.

Instead, the new descriptive context, is more about stable metric-space shapes, and their relation to being in resonance with a finite spectral-set, which, in turn, is associated to a high-dimension containment context.

It might be noted that discrete structures become central to the descriptive context in this new descriptive context based on stable shapes and resonances of these stable shapes with a finite spectral-set defined for an 11-dimensional hyperbolic metric-space, which has been "partitioned" into shapes (so as to define the finite spectral-set)

But where

The discrete properties emanate from a fiber group of both each different metric-space and for the 11-dimensional hyperbolic metric-space,

Where these new types (or new interpretations) of discrete properties (which emanate from a fiber group) are:

1. The double-cover simply-connected spin-group, whose spin-rotation period defines a discrete time interval, and material-interactions are to be defined as discrete local operators, which are related to both (1) a local inverse structure and (2) to a toral geometry which is redefined at each different time interval, by the geometry of a toral-shape connecting (touching in a tangent manner) the two interacting materials, so that the tangent geometry of the discrete Euclidean shape (which defines the local geometry of the interaction) is re-defined by an action-at-a-distance toral shape, ie re-defined for each discrete time-interval.

2. The discrete isometry (hyperbolic and Euclidean) subgroups…, [and their two-metric-state complex-number coordinate system's relation to the discrete unitary subgroups],… which define the set of stable shapes,

3. The discrete set of conjugation (fiber-group) classes, which, in turn, define the discrete set of Weyl-angle folds, which, in turn, can be defined on (or between) the sets of toral-components, where the toral-components, in turn, define the toral-component shapes of the set of very stable discrete hyperbolic (or unitary) shapes.

4. The shapes define topologically separate sets of metric-spaces, where this definition is between the different shapes of the subspaces of the same dimension, and/or between the different dimensional levels, where in the higher-dimensional metric-spaces the lower-dimensional metric-space shapes appear as closed shapes with boundaries.

And (outside of the isometry or unitary fiber group)

This set of discrete relations (identified in 1-4 above) are defined at particular discrete points or boundaries, and this allows each separate shape to have its size be defined by a discrete multiplicative constant, which is (can be) defined between either discrete structures or discrete processes

5. The discrete multiplication by a constant….

(so as to identify a change in relative size between the various discrete structures, which are identified in this model of existence) can be defined between either discrete structures or discrete processes.

Note that,

The material-interactions are still defined between material-components, but they are

either mostly irrelevant, ie non-linear, relations, which can be related to feedback systems, or they define a perturbing construct, which is attached to the orbital structures of (usually condensed) material, where this condensed material is in the orbit structures, which, in turn, are defined by the geodesics of a stable discrete hyperbolic shape of the metric-space within which the condensed material is contained, but where the material-interaction structures can perturb these geodesic orbits.

That is, pde play a relatively minor role in the descriptions of stable material properties.

Back to II.

The way in which the investor-class uses math and science, is that, science and math are used to develop only certain narrowly defined instruments of interest to the investor-class, namely, for developing spying instruments and military weapons,

Where this is done within a moralistic social context, where violence has the main determining basis for all judgment, that is, the way in which all discussion is framed, it is all about absolutes, and a necessary requirement to "not question authority," ie the authority of those in the socially superior intellectual-class, so that this fits into the justice-military police-state administration of society, and determines the control of the actions of its people.

That is, science and math have failed, and there is no sense (by the intellectual-class themselves) that this has happened, where this is due to the total effectiveness of the propaganda-education-indoctrination system

More than anything else this is an expression of Calvinism and the religious belief in the society's elites, the investor-class and their servile intellectual-class but a person in the intellectual class is considered quite superior to anyone in the public, ie they have a higher-rung of the Calvinist social-hierarchy of God-given value, which is the psychological model through which the justice-military police-state attacks the public

III. Another pillar of society which again is arbitrary (ie money)

The monetary system is really an expression of value, but value in an arbitrary system is arbitrary. That is, money (or arbitrary-value) is more of an expression of the ruling-class's domination over the wage-slave public than it is anything else.

Furthermore, this is essentially the idea of the modern theory of economic markets, namely, arbitrary markets developed by the propaganda-education-indoctrination system. Built from the ability to force into certain limited categories of one's living existence and the ability to continually repeat a context which fits into the limited categories of one's living existence. A very circular vision for entrapping the minds of the public, ie a waste of people for the selfish interests of the few. Circular thought, or a trivial tautology forcing a mind into obsessions.

What to do, in order to re-gain a society where law is based on equality, and promoting the common welfare

The main point is that the system of Calvinism has failed, ie it has been proven to be not righteous (but rather evil, it expresses selfishness in the context of materialism), ie anyone from the public is better than any of the current officials which still uphold this Calvinist social system of arbitrary violence and arbitrary intellectualism (bankers, oilmen, judges, administrators, the militarization of society within all institutions, the infiltration; interference-with; and spying on all social classes, and the, so called, top intellectuals who uphold through a limited rendition of knowledge the productive interests of the investor-class)

To de-stabilize an already failed system, one simply needs to continually repeat that the system is arbitrary and failed, and to ask questions which focus on what is not being accomplished (or described) by the expert-class,

And, furthermore,

that the public is troubled by their collectivist social structure of capitalism (and its associated and implied religion of Calvinism), which is only being used to support the ruling-class's interests, but which is incorrectly being attributed with the economic rules through which our society is, supposedly, organized (capitalism) being based-on individual freedoms where it is claimed that one can build one's own empire through intelligent use of the, so called, free-market, but, in fact, our society is really organized around violence and hidden police-state activities and It (the extremely limited social structures through which the society is organized) is not promoting the common welfare

And

The oppression should be opposed

This can be done by a new continental congress so that law comes to properly, in regard to US historic documents, based on equality and not the laws upon which violent empires are based, namely, property rights and minority rule

Instead the public needs individual freedom within a context of equality, eg equal free-inquiry, individual freedom to know and to create, in a context of selflessness, the human being is built to know and create and these are the activities which give purpose to the human life, and not simply acting to support the selfish interests of the ruling few in a collectivist society, but where around the process of trading one's created ideas which may be produced in-mass, since our society possesses the knowledge to produce products at a larges scale, eg food production could be a large scale effort, there are (or can be) collective groups, but

It is the idea of individual creative persons which is to be the point of what needs to be protected, not a bunch of collectivist drivel…., about either collectivist capitalist efficiencies or that all of the limited set of categories (or knowledge and creativity) which the bankers have imposed on society need to be made collective, drivel…., from the bankers or the intellectual-class where the intellectual-class only knows what the bankers have required them to know, the individual freedoms of the public need to be protected from the core structures (ie the investment in limited categories and the associated form of limited and absolute authoritative knowledge which is espoused by the intellectual-class), of the already failed state, which only serves the ruling-class,

But

The new society must always embrace and support the pre-cautionary principle, in regard to a collectivist action's (usually related to the production of certain products) possible harm to either the environment, due to the poison, related to the product's production, which is entering the environment, or to society, and its individuals, to which collectivist actions need to respect, or adhere, in regard to curtailing their (with there being a good reason to associate the collectivist behavior with possible) harmful collectivist actions

That is, created products are subordinate to the creativity of the people, that is, if a created product may be doing harm to people or to the earth then, by the pre-cautionary principle, the people need to create new products or change their life-styles

The weakness of the current system is its propaganda-education-indoctrination system, where science religion and the complete illegal and immoral sets of actions done by the ruling-class and their minions are all failed social constructs, while the justice-police-state-military construct is at the core of the actions of the ruling-class. It is from these institutions that the violence and coercion needed to maintain the arbitrary social structures are brought forth and activated against the public

The key is to insistent on talking about issues, such as the basis for law in the US, or the failures of science, the invalidity of indefinable randomness (which is the basis for biological evolution, most of psychology, particle-physics, finance, sociology, or the failure of religion to provide quantitative alternative models, ie religion is used solely for the purpose of defining a social hierarchy, which is what institutions do but it is the opposite of the main message of religion which is the seeking of equality so as to search based on individual freedom to know and create and to act in a selfless manner (remain a child in wonder about the world) etc etc) or the new religious possibilities associated to new ideas about science-descriptions etc etc, where these are ideas (ie the challenging of ideas) about which the propaganda-system, which is controlled by the ruling-class, will never speak

A revolution is natural since

1. There is no justice-system in the US, there is only a police-state a network of spies and managers and administrators and judges, and it is, clearly, deeply connected to the media, and whereas all of these people know about this whole police-state system, and are willingly participate in it, and, in turn, cover it up. This system is only concerned with protecting the paymasters

That is, this system has imposed the religion of Calvinism of the entire population of wage-slaves, where Calvinism defines the arbitrary self-righteousness associated to the way in which the entire system functions, ie it is both an illegal police-state and it is an illegal imposition of religion upon the public.

2. In the police-state there are on the one side the ordinary people, who are harassed and manipulated to the police-state agents, and manipulated into militias with there also being many L H Oswald types also being manipulated for the domestic terrorism done by the secret state so as to keep the public ever fearful and who serve the oppressive other side the very few, who apparently, are highly valued by the police-state staff, in the name of arbitrary high-value, and absolute forms of "knowledge" which define superiority, eg racism and authoritarian science.

3. The government, or what is really simply a group of propagandists, has destroyed the US, there is no law there is no recourse through political institutions, where this is (has been) allowed by the total capture of the justice-system by the military-state, and the total capture of the political system by the propaganda-system.

4. The intellectual-class are the totally captured intellectuals by the propaganda-education-indoctrination system, which has deep ties to the construct best attributed to D Hilbert of axiomatic formalization where the natural interest of the investor-class in knowledge has always been to have the intellectuals take-care of and maintain and slowly develop the instruments of importance to the social power of the ruling-class and this has always been a function of the US overly authoritarian and dogmatic educational system yet many teachers actually teach so as to get people to think and be critical, though this might only be for a rather specialized set of issues, it seems to be enough to cause the ruling-class to attack the education system so as to oppose any free-thought within the public

The internet also allowed for better, or more honest, information transmission about world-events and about specialized topics which the public is capable of thinking about and making decisions, such as the poisoning of the environment, and its relation to both health and survival

And

Snowdon might well be a selected individual whose (hidden) mission was actually to destroy the internet, yet the report (of over-whelming significance) which Snowdon provided is the report about "the police-state, and the active relation between spying and agents who in a coordinated effort attack the public, so as to disrupt the natural flow of human events within the public," an outright criminal annihilation of any form of US law, which completely justifies the nullification of the current social institutions (law, governing, economics, propaganda, education), and instituting a new continental congress,

But thishas not been the point about which information-providing structures have continually dealt, but it should be the point about which information-providing structures should continually be dwelling upon.

Where if this was done, then the criminal behaviors of both the court and the political personalities and associated economic institutions, would come to be clear, though the (bribe-taking, and revolving-door) criminality of the politicians (and corporations) is already make this clear [but it is called legal], and the courts are too corrupt to stop such criminality, ie the justice system is a part of these operations, thus, (or so one should clearly see that) the justice system is whole-scale involved in criminality (it is all deeply corrupt, in the name of their religion, to protect and serve the stinking-rich, whom the justice system defines as the only people of virtue).

Nonetheless

There is yet another deeply troubling social construct which is quite similar to the imposition by the justice-system of the religion of Calvinism on the public, so that one is entering the halls of a Calvinistic institution when one enters a courtroom, where Calvinism is the worship of the material-world and a person has virtue, and is above the law, if they possess a lot of money, where the law is to impose the requirement on an individual that a person possess value as defined by how they serve the investor-class ie the people of virtue as defined by the imposed Calvinistic "religion" rather it is a point of arbitrary self-righteous moral-ism through which the police-state operates with impunity

And that is the construct of the intellectual-class, where these are the winners of the academic competitions which are defined in earnest in the 10th-grade and above part of the education system, where in this context the authoritative dogmas upon which the academic contests (which each person tries to win) are based, is the set of intellectual constructs which best serve the creative and/or productive interests of the investor-class. The so called, winners of these contests are the thoroughly indoctrinated set of people who can (are allowed to) enter into the intellectual-class and the most technically inclined have entered into the thought space of formalized axiomatic whose assumptions is that the set of axioms contexts interpretations and ways in which language is put-together are perfect ie there cannot be any other way in which to do things, ie "there are no alternatives" to this set of axioms and contexts upon which technical descriptive languages must be based

Unfortunately, this was dis-proven by Godel's incompleteness theorem, yet this is actively mis-represented

if the constructs associated to axiomatic formalization are to be believed, then interactions (or all descriptions in which the context of measuring is assumed to be continuous [or probabilistic]) must be [can be] defined in the context of pde, but the vast majority (where one can, effectively, say, all) of the descriptive contexts associated to this context are non-linear, non-commutative and indefinably random and thus quantitatively inconsistent and thus there are no valid descriptive constructs coming from this descriptive context

But

For the thieves of the ruling-class, this is fine, since their power is derived from whatever technical instruments or technical capacities which are present within society, so new development actually puts their investments at risk

In a bomb, the chaotic context of quantitative inconsistency is the norm of an explosion, and detonation is about controlling statistics of collision rates, so making virtually all of academic physics to be solely concerned with bomb engineering is fine with the ruling-class

Biology has come to be about molecular biology with focus on DNA where in chemistry the context is chemical components and reactions mostly related to reaction-rates again associated to the statistics of collisions, but in biology there is a clear, deep, relation between reactions and catalysts, where the current (2014) model of catalysts is almost non-existent since the model is about quantum physics, which is nearly always non-commutative, ie quantitatively inconsistent, and thus, it is a description which is without any meaningful content

That is, chemistry is still a book of cooking recipes, ie and it is mostly a book about the recipes of crude-oil.

Furthermore, the context of biological evolution is an indefinably random context so its descriptive content is basically arbitrary

Furthermore, histories of biological change are at some times (interval time periods) fast and at other times, essentially, unchanging, but

DNA is not a valid model of life, C Venter, if what he claims is actually true (after-all he is a capitalist-scientist, and capitalism is all about lying and stealing), had to put his..., proclaimed to-be, lab-made..., DNA molecule into a cell, and let it be engaged in some cell-process (a process outside of the current (2014) viewpoint about science) before the molecule had the attributes of being able to function in a life-form, {that is, there are two aspects to life, molecules related to inert forms of material and the life-giving properties which emanate out of a cell apparently unrelated to, but closely connected-to, the inert material structures.} and/but

new math relationships, associated to the nature of existence (and related to what is discussed above, namely, in the above context a new model of life would be: an odd-dimensional discrete hyperbolic shape with an odd-number of toral-components), allow for a better model of a living-system, in regard to the energy and organization and a mechanism for an internal control within a living-system.

Electronics is either linear and solvable, and thus controllable, or if there is a non-linear circuit component, then the components can be driven to a state where the properties of the component's functioning are constant, and thus the component properties can be used in a systematic manner in a circuit.

The circuits which seem to be of the greatest use, in a context of very complicated circuit design, are the optically etched circuit-boards, defined on a semi-conductor stratum, and are quite often complicated switching circuit structures used in computers.

This is the one place where instrument development and knowledge might have a valid relationship, but this is always placed (by the propaganda-system) into a context of quantum computing, but quantum systems are the systems which exist in a state of being indefinably random, [and outside of the set of stable shapes, which really define the stable spectral properties of the, so called, quantum systems] and thus, there are no ways in which to ensure a stable context for an indefinably random quantum context, but which is needed for the probability structures associated to the model of quantum computing,

ie the random context may or may not hold for any type of a defined computing process, any large-scale indefinably random physical structures would be difficult to hold and to use, need vast arrays of similar systems all engaged in the same computing process in hopes that one of the set-ups will hold together to identify a quantum computing relation, but the context of coupling between separate systems seems to be outside of the quantum descriptions, so input and output structures are difficult to define

Nonetheless, the entire set of the intellectual-class insist on the useless and arbitrary context of pde's which are non-linear, non-commutative, and in a context of being indefinably random

That is, why are there not many, many criticisms of the useless techniques of quantitatively inconsistent math models and with many, many alternative contexts being identified?

Because the intellectual-class are deeply indoctrinated into a technical descriptive context which makes no sense

That is, the intellectual-class only knows the things which the investor-class wants its intellectual-class to know and espouse

Thus, the useless, alternative media (news) TV shows or computer web-sites are all about providing the propaganda of the intellectual-class, but that is not the set of viewpoints which can effect change,

And

The math physics and biology academic departments need great changes in their structures of inquiry so it is best to base those changes on expressions about patterns which are a part of the academic community than to base those changes on the beliefs of reactionaries, which is what is happening by allowing the propaganda-education system to dominate, and guide, all aspects of how people think and use language

That is, the so called experts or intellectuals which dominate the alternative media do not possess any ideas which are remarkably different from the ideas, which the investor-class wants these intellectuals to possess (or the investor-class want the intellectuals, whom are only allowed to serve the ruling-class, to possess)

That is, the, so called, dissidents of the intellectual-class, whom remain loyal to the idea of science being authoritative, are the ineffectual basis for a, so called, alternative voice to the main-stream media, and though there is some meaningless lip-service paid to saying that they are interested in alternative ideas, nonetheless, the main message of these, so called, alternative voices of the intellectual-class is that the science which is funded by the banker's interests is essentially determining an absolute truth, for which there are no alternatives, ie inequality is a law of existence. This is of course extreme nonsense, not only are there other ways to describe physical systems than to base such a description on the explosions of nuclear bombs, but rather the other ways of using language are the only ways to get to a useful truth, and the idea of inequality is all about defining value according to some existing condition (or discipline) and comparing within such a narrow context.

And

In this respect the intellectual-class is essentially demonstrating the same narrow and unjustifiable authoritarian ideas as do the usually racists, and religious zealot based militias..., which are perhaps the most manipulated part of the population (who are really the general public, but who have been re-classified as "the people of action" who represent violent self-righteous and arbitrary domination, ie the manipulated image in the media of the uneducated, working-class person of the general public), ie manipulated by the spying agents of the secret police..., who serve the secret police in their duty to terrorize the public, for the interests of the ruling-class both sides (intellectuals and militias) are controlled by the agents and administrators of the US police-state society, where the leaders of the two groups are both manipulated and chosen, so as to be the psychopathic personalities who want an absolute demonstration of their individual superiority within the collectivist social construct (or within a collective intellectual-community), which seems to characterize psychopaths (ie those who seek to be dominant over others, in an externally directed context which is associated with their own awareness)

The so called, intellectual-class do not sufficiently identify the institutions and their most significant actions which need to be challenged, the academics support an authoritative absolute truth which is to be controlled and taken-care-of by the intellectually superior few, they oppose the violence of foreign wars but they do not identify the true nature of the US police-state, and the institution which purveys the violence of the west is the justice-military-police-state which supports the ruling-class, they do not simply identify that the governing structure of the US is based entirely on propaganda and instead try to put blame on the inactive citizens, and they engage in meaningless dialog concerning the poisoning and interference with the atmosphere which puts the authoritative intellectuals against the trouble-causing (media) minions of the oil-companies

The country needs a new continental congress, so as to re-establish US law as being based on equality, and there is to be democracy, but everyone's welfare (to survive, and create, and to develop knowledge) needs to be attended-to in a general manner, where this can be done through the internet, where spying is to be reserved for those who were in the top-positions of the former police-state

And

Law is also to be in opposition to people's selfish actions

The new society is all about individual creativity, and the development of knowledge as a result of individual curiosity, related to new creative contexts, where these contexts are practical and not simply dreamy literary expressions, as physics has now (2014) become a community of literary critics who discuss a formalized axiomatic language structure which is only related to an illusionary world (ie the dreamy constructs of a particle-collision which exist in a context of randomness)

Where all of the criticisms of society's institutions and the representation of how they function (arbitrarily and violently, so as to be based on the domination which, supposedly, emanates naturally from the principle of inequality, and run by domineering personality-types, ie psychopaths [it is, still, all the same as the Roman-Empire]), are diluted, and made un-noticeable, by the, so called, alternative media, which supports the worship of the, so called, highly-superior (dissident) intellectual-class, and sales of the books of the, so called, highly-superior dissident intellectual-class, ie they support a market controlled by controlling the language and the thought of the society, where this control is effected through the propaganda-education-indoctrination system, along with the help of the spies and interfering agents of the police-state, ie who work from (or for) the justice-military system. All done in the name of inequality, and the, so called, high-value which the superior-few define for all of society.

Chapter 35

Science criticism

The journalistic web-site Truthdig, writes, something like, journalism is about publishing something (those ideas) which someone else does not want published, and everything else is propaganda, and this is attributed to G Orwell. or social and intellectual states which other people the publishing industry do not want published, but the web-site Truthdig does not publish any commentary about science, which the propaganda-publishers for the science-industrial-military-complex do not want published, there is no strong criticism of the academic sciences, despite their failure to accomplish anything of significance beyond what their empirical investigations can provide, rather the propagandists only fawn over the great intellectual capabilities for the science authorities, and express a reverence for their great knowledge, which, in fact, represents quite a pathetic form of empirical knowledge, ie there are no systematic descriptive contexts which work for describing the observed stable spectral properties of general quantum systems from nuclei to atoms to molecules to crystals, ie no valid descriptions based on physical law of quantum systems which possess stable spectral properties, as there are for the solvable systems in classical physics, and the propaganda-based social-class conflicts, which have been manufactured, by the propaganda-system, to exist between the supposed religious "intelligent-design crowd" (where, apparently, god is the intelligent designer, but there is no model of this, eg a man "existing where?"),

and (vs.)

the tough-minded scientists (whom represent evolution, which is a math model based on indefinable randomness, as well as the anthropic principle that the masses of elementary-particles [and the values of physical constants] are such as to allow life to exist), but where the physics community is actually putting-forth "the physics idea of intelligent-design,"

[that is, the abstract models which they use, which only relate practically to bomb-engineering, and have nothing to do with valid precise descriptions of observed stable spectral properties of quantum systems (or of the stable solar-system) can be used to model a big bang and interpret the so called initial gravitational singularity (which is a continuous model or interpretation within general relativity) in the context of particle-physics and the discontinuity based inflationary model so as to be able {so the speculative claim goes} to predict the development of stars and galaxies (?) and the existing elements etc etc, but they cannot predict the more mundane stable spectral properties of a general nuclei, where such a failure, concerning the observed spectra of a nucleus, is a much more serious matter, a more serious criticism, if one wants to really be considered a, so called, hard-nosed physicist]

(Indeed this is really a debate for mashed-potato heads) so the social-class issue is about "whose vision of intelligent design is superior?" (as should be expected for a civilization built on arbitrary value imposed by great violence by the ruling-class) but both stories are baloney, the only "practical" thing to which today's (2014) abstract theoretical mathematical-physics is truly related, at least as its main focus is related to the context of the detonation of a gravitational singularity by using the properties of particle-physics, where this is the main concern of string-theory and the other nonsense, eg quantum-gravity (derived from particle-physics and general relativity, where the real function of these theories in the propaganda-system is that of being "the icons of the highest intellectual capacity of human beings," which is, of course, a great amount of baloney, since neither of these descriptive constructs can answer any of the fundamental questions about stability of observed patterns: in nuclei or in the solar-system etc, ie these theories are marginal speculative ideas, but which fit into the simple ideas of particle-collision modeled nuclear reaction-rates, and the clear concern is the relation of these high-energy particle-collisions to gravitational singular-points, these singular-point particle-collision models are of interest since they are the same type of models upon which high-energy thermonuclear weapons are also based, the only success that the physics professionals have had in the last 100 years is building "the bomb," and that success was a result of there being critical masses for nuclear material).

Clearly, this is all being driven by the commercial interests of those who have invested in bomb-production, since it ("the bomb") is an important component in the psychology and use of violence which holds the arbitrariness of this social collective together a social collective which is all about supporting the ruling-class, so that the society has been coerced into this social context (ie Little has changed in western culture since the Roman-Empire).

But the track-record of theoretical physics (ie quantum physics, particle-physics, and general relativity) is a record of continual failures, such as, where are the valid descriptions based on the, so called, "laws of physics" concerning the stable spectral properties which are observed for a wide range of general quantum systems of: nuclei, atoms, molecules, and crystals, these stable spectral properties of these systems cannot be calculated,

Instead the claim is made that these stable systems are too complicated to describe but their stableness implies that these general systems are both solvable and controllable, thus the laws of physics have been shown to be failed.

Furthermore, "where is the clean cheap energy of controlled fusion-energy, which was promised to be in existence by the end of the 1950's?" leaves one to believe that the success of such an attempt is not to be expected (though this type of failure is a good thing, since it proves that theoretical physics is a failure, but this is covered-up by the propaganda-system which has penetrated so deep into the psyche of the public so that the so called well-educated people are to be sure to come to the defense of theoretical physics and to revere the intellectuals which compose this failed set of intellectuals, where this take over of education has been done by it having penetrated the public education system, but one can see this as a success story for propaganda and placing institutional managers who are easy to control by the ruling class in charge of educational institutions)

Furthermore, it is likely that the community of experts do not realize the fact that (or the way in which) the military-industry has pulled the wool-over their eyes, but their models carry no content (the proof is that they have consistently failed).

This is due to both the education system essentially based on a contest for what is essentially articulated as god-given intelligence,

Furthermore, education is divided into categories, wherein each category one memorizes properties which are not related to a consistent message, in regard to the relations which might exist between distinct sets of (unrelated) properties about which one is expected to have memorized, eg math is divided into: analysis, algebra, topology, and complex-numbers (concerning these same subjects). Yet, in regard to most of the patterns, which are studied in math, the property of quantitative consistency is never considered as being important. Thus, there is complicated discussion about irrelevant math patterns, where this is the educational context of memorizing complicated math patterns which are mostly useless, it is a contest about memorization and obsession over these irrelevant patterns, and then making claims which may be true in the world of word agreements, ie a world built out of illusions, but which have little or no relation to existence. Thus, there is particle-physics, general relativity, grand unification, string-theory, quantum-gravity etc etc all useless expressions of complicated patterns which lead to nothing but further literary efforts in the mental land of illusions.

That is, what education is really based on, is that it is based on the measured rates of the pupils acquiring the dominant culture, ie it is primarily an expression of social-class, and the way in which the personnel (who can acquire the dominant culture) are chosen; they are autistic, and both obedient and desirous of being intellectually dominant, and carefully managed by both institutional mangers and the propaganda-system, as well as the relation of the education-system to the investor-class, a relation which determines how the education-system is organized (essentially, as vocational schools to serve the banker's interests) where the dogmas, which the investor-class wants to be used by its worker-staff, ie the authoritative scientists, so as to incrementally develop the instruments which are the basis for creative production into which the investors have invested, and which is also the dogmatic and overly authoritative and very narrow basis for an authoritative intellectual competition within the education-system wherein intelligence is defined as "the rate at which someone from the lower-social-classes can acquire the dominant culture" (where this definition is mostly instituted through the propaganda-system and institutionalized by the managers and administrators), but where intelligence of this type (ie obedient to authority), clearly has no relation to a person "being able to discern truth" where this shows "the obvious truth," namely, that the expert practitioners…., ie the winners of the academic contests, those who slowly and incrementally develop the instruments of production for the ruling-class which, in turn, are used for realizing the interests (the goals) of the investor-class…., all have had "the wool" pulled-over their eyes, and "they are busy as could be" single-mindedly developing a "doomsday military bomb," ie a bomb whose military-industrial intent is to blow-apart the solar-system, and their intellectual efforts have nothing to do with anything else in the world, furthermore their math methods are also failures so this is really an effort based (only) on empiricism.

But, that is why the, so called, top-intellects are chosen so carefully, and managed so carefully, that is, the, so called, leaders of the intellectual-class are chosen to also be psychopaths as are the leaders of the investor-class, and as were the Roman-emperors.

Note that the Roman-Emperor Constantine effectively wrote the bible, and it is the essential propaganda-manual for empire. It is literature which identifies an object of arbitrary high-value, which is to be worshipped by the public. It identifies an intellectual and moral hierarchy, to which the public must be subordinate.

(it seems that 300 years after Constantine, Islam re-created this same type of a social-propaganda construct of a social hierarchy)

The point that should be realized, is that copies, by other peoples, of same social structures, are seen as the greatest evil (or threat) to the rulers of an existing social hierarchy: ie Christianity vs. Islam or the western capitalism vs. communism, etc

It was (the [unfortunately a] hypocrite) T Jefferson who wrote the definitive (and very simple) social program, which broke from the western hierarchical society, where law is to be based on equality, and everyone should be involved in farming to some amount, so as to be self-sufficient, but everyone should also be seen as an equal creator, with a wide range of knowledge..., as developed through the Socratic method of equal free-inquiry..., upon which creative intent can be realized, so that the society helps in the production of each of the independent ideas, as best the community can help, and the best-sellers are turned into cooperative efforts, so that the trade-market can remain equal and free, (ie not dominated by selfish concerns) and the way in which material-based production affects the ecology and the earth is to be considered, so that the "precautionary principle" governs decisions about material production's relation to society, so that the trade-market and society can remain harmonious with the earth, [the society needs to team with the world to support its people as equally creative individuals,

Why should those with militias and property, claim to have social-value, and everyone else must work for them?

Answer: No reason.

This narrowness of focus (or interest) limits knowledge and creativity,

where knowledge and creativity are the core properties of human-life], that is, and not have selfish moneyed interests possess, essentially, an absolute dominant control over production, its methods, and its relation to both society and the earth.

That is, the new society of the US (envisioned by Jefferson) was to be based on the individual and equal free-inquiry, and not be based on a collective-society which, in turn, supports the ruling-class, ie the model of the Roman-Empire (the western collective society) (where Calvinistic-Puritanism is based on the collective-society supporting the ruling-class, and Calvinism was more-or-less the definitive Protestant viewpoint, though the Evangelicals were, originally, a very free set of ideas)

The whole idea of western society is that it is circular

It is a social system which maintains and perpetuates itself

It perpetuates social-class,

ie those who are familiar with the methods of affecting control over social institutions

Which is a process which is mostly about being aware of the use of language ie propaganda, and subsequent, spying and coercive agents, or coercive militias, who work for the upper-class, for the selfish purpose of the upper-class, the Roman-aristocracy had their own personal militias and which was the basis for their high social positions (the west has always been brutal)

(eg the scouts of the Indian society who helped the military defeat the native peoples, paying spies and agents)Social-class is a social hierarchy based on arbitrary value(but, for production, ie the basis for social power derived from a market, there needs to be a relation between propaganda and technology)

The (cultural) knowledge which supports the interests of the upper-social-classes is called high-valued knowledge and it becomes the dogmas of the education-system, within which an intellectual contest of acquiring the dominant culture's knowledge-of-interest is played-out

That is, knowledge which the ruling-class sees as the knowledge which best supports their interests is the knowledge which has high-value within the propaganda-education system, and it essentially becomes dogma, and the basis of the singular authority, which is claimed must dominate all discussion which is reduced to a technical question

"The educated" those who either win the educational contest, or who are socially connected, become the intellectual army which supports the ruling-class's investment interests

And

Those ideas which do not support the interests of the ruling-class are marginalized and ignored, since the educated people see any questioning of their intellectual dogmas as originating from "the third-world ignoramuses" or from the "screw-loose dim-wit." This is "the gift of public education" which has been managed by the "loyal to the ruling-class" managers.

Though a lot of effort is given, by the ruling-class, into forming groups which espouse certain types of "ignorant" claims, so that there is always an "us vs. them" social context, which are clashes which are very much about word-usage, but not reducing the discussion to sets of assumptions and interpretations etc, so the discussion is simply about emotions and the social power of the groups (again an idea of being collectives, not equal individual creators), ie they are groups babbling incantations which support some aspect of the collective society, which, in turn, upholds the ruling-class.

Note: The intellectual-class, which so vehemently believes-in and supports the dogmas, which, in turn, relate to the production and instruments which are used by the ruling-class to maintain the social hierarchy, which the ruling-class controls, where this deep belief in dogmas is an expression of the careful management and the effectiveness of the propaganda-techniques the ruling-class uses (though, as much as anything, this is about constant and continual repetition as the main propaganda technique, but it is also about preying-upon (or taking advantage of, and using in a manipulative way) domineering personality-types) but

The authoritative dogmas of the intellectual-class are narrowly focused and becoming delusional (careful proofs of claims which are not relatable to the material world) and this is resulting in the failure of the expert technical capability to expand the creative contexts of society, and subsequently the main focus of the intellectual dogmas are all that can be realized, thus leading to continual development of instruments of oppression, and domination, and instruments which can do great violence and harm, are what is solely being pursued by both the intellectual-class and it is how the economic forces which are available to the ruling-class are being used, so both the technical experts are expressing very limited sets of technical ideas, ie they are becoming arbitrary, limited, and shallow and ignorant, leading to the experts becoming incapable, and the ruling-class has always depended on manipulating arbitrary and unfounded ideas (or prejudices, most often this is expressed as racism).

What is pursued by the ruling-class is self-interest based on investment (though the economy is not really about markets but rather the economy is an expression of total domination over the clamoring population of wage-slaves, where the worst social strategy for the wage-slaves is to plead for jobs).

That is, an external intent is imposed on an existence which has no explanation within the authoritarian dogmas, which materially support this arbitrary social-hierarchy

The only explanation is based on the arbitrary high-social-value placed on a set of elite people who best support the social-intellectual-hierarchy, and this is done in a context of materialism, both Constantine and Calvin fundamentally worshiped the material-world, and the relation of their control… of the material-world… to social inequality, in this respect the "hidden-knowledge" is to be used for the selfish purposes of the ruling-class, essentially, based on their brutality and their "somewhat hidden" control over society, where "hidden-knowledge" is to be interpreted in this context, to be the knowledge possessed by the experts, whom, in turn, serve the interests of the ruling-class, but because the technical-knowledge is poorly articulated, in fact, the experts themselves do not seem to be able to "place their knowledge in the contexts which make it clear what it is that their constructs are trying to do," so it is knowledge which becomes hidden to the public but science is about: materialism, reduction, and indefinable randomness so as to be consistent with the interests and models regarding both the investor-class and now (2014) nuclear bomb explosions

But technical-knowledge is "constructed to be hidden (or difficult to understand, since it is poorly articulated)" so that it is used to be the basis for a mis-placed reverence by the propagandists for creating an illusion of the great intellectual capabilities of the science and math experts.

That is, the west is a civilization which opposes a true notion about "hidden-knowledge," ie knowledge about existence which lies beyond the idea of materialism (beyond the bogus, science vs. religion (baloney) debate) to the point, where (beyond the limitations of the idea of materialism) mankind has a purpose of using knowledge to extend the context and reach of existence, itself, by controlling the internal processes of existence

That is, at the core of the ignorance of mankind in his civilization is that his knowledge is based on materialism (and reduction to elementary-particles and its relation to indefinable randomness) so that in the math models of material existence, the distinction between quantitatively consistent patterns and quantitatively inconsistent math patterns…., or math processes, or math operations…., is not made,

Where in the descriptive language of the, so called, technical experts, the quantitatively inconsistent math properties stem from the use…, by the so called experts…, of the non-commutative and non-linear local properties for their math-models of measured properties, as well as non-commutative properties of function-space relations,

ie the attempt to diagonalize the harmonic function-space model of containment for material systems has not worked (it can be neither formulated nor calculated ie it cannot be solved)

In such a descriptive context (in math) the practitioners do not distinguish (in a clear manner) between

(1) internal patterns of existence and control,

And

(2) externally motivated patterns of control, ie based on adjustments made in the system in regard to empirically determined properties outside the system, where this type of non-linear based feedback system-model is both vague and always unreliable (ie several levels of feedback are needed to cut-down the failure-rate of such feedback systems),

ie the system's purpose is imposed externally, based on the relation of the position of the system to the external patterns (the position of the system is adjusted to the purpose which is determined in regard to the (unstable) external conditions, [ie the properties of the system are not being described]).

That is, both "the social structures of the hierarchy of society" are arbitrary, and "the math and science models placed within a material world" are arbitrary and unstable,

ie if they really worked then where are the explanations of the observed internal stable patterns of material systems, ie nuclei; atoms; molecules; life-forms; the solar-system etc.

It is in regard to identifying the content of "hidden knowledge" about which the quantitatively inconsistent math constructs should be avoided, in order that the descriptions of higher-dimensions can possess meaningful content, and so as to possess (and to be able to use in practically creative contexts) a precise descriptive language which is based on stable math patterns, which are quantitatively consistent

Ouspensky communicates quite clearly in the first chapter of his "A New Model of the Universe" book 1934, the issues involved in man's relation to "hidden knowledge," ie knowledge about existence which goes beyond the three or four dimensions of the material world

He describes the tales of the three-wishes where the people in the story never know "what they want," and often in these stories about granted-wishes their sequence of wishes leads them back to their original state

Ouspensky makes the statement (p14) "In our times, theories which deny that mankind is capable of gaining 'hidden knowledge'" and these stories about mankind's inadequacies have become only recently very strongly shouted through the propaganda-system by only a few very noisy people.

But, man, in general, still believes in being able to acquire hidden knowledge, but when it shows itself..., {as is provided by the example which B Ehrenreich has given in her book "Living With a Wild God," (talked about below)}, the person is not equipped to deal with the fleeting opportunity.

Ouspensky says that "man is aware of the wall surrounding 'hidden knowledge,' and the unknown, but he also believes that he can get through that wall, and that others have got through that wall, but he cannot imagine what is beyond that wall."

"He does not know what he would like to find there or what it means to possess (hidden) knowledge."

Mankind still cannot articulate what it is that they want.

Do they want, riches and treasures and social domination?

Or

Do they want to have a meaningful relation between their creativity and their knowledge of existence itself?

But now there are math models of existence, which go beyond the idea of materialism, and they maintain the material world as a proper subset, and they are related to the idea that mankind's natural position in existence is to further extend the context of existence itself, eg how to build stable bounded shapes in R(6,3) {contained in R(7,3)} so as to extend the finite spectral set of the containment-set from R(11,1) out to R(12,1) etc

Ouspensky states again (p15) "In this incapacity of man to imagine what exists beyond the wall of the known and the possible lies his chief tragedy, and in this, lies the reason, why so much remains hidden from him and why there are so many questions to which he can never find the answer."

The new ideas being expressed claim that what lies beyond the material-world are many-dimensions whose subspaces are deeply related to stable shapes, which can be used to construct a subspace structure based on stable shapes.

Furthermore,

Note: The new ideas about math and physics solve the problem of the stability of the many-(but-few)-body systems, which exist at all size-scales: from the nucleus to the solar system, and apparently beyond this too.

This brings the discussion into the realm defined by B Ehrenreich, the observed context of being in relation to a "new type of" knowledgeable living-entity, where the sensations transcended the sensations of the material-world.

Ouspensky writes in 1934 that mathematical physics is both too abstract to have any practical value, and none of its math constructs can be solved (though he also a states that the systems are not being sufficiently carefully modeled, "as if this were the problem," but it is not the problem with theoretical mathematical physics (of which general relativity is certainly a part) rather its problem is that it tries to use math structures which are known to "not be quantitatively consistent" such as the math properties of non-commutativity and non-linearity, and the most damaging model... (and now theoretical physics most prevalent physical model).... the model of an indefinably random based, so called, precise descriptive language, which is used to describe the properties of very stable physical systems, where none these very stable physical systems, ie none of these very stable many-(but-few)-body physical systems, has a valid description of its measurable (observed) physical properties based on descriptions which in turn are based-on non-commutativity, non-linearity, and indefinable randomness.

Nnamely the sets of stable spectral-orbital properties of physical systems which exist at all size scales, from nuclei to solar-system, are in need of valid descriptions,

And where it needs to be noted that the physical property of stability, as well as the great prevalence of these stable-systems, together implies that these systems are: linear, metric-invariant, and continuously commutative.... in descriptions which are based on:

"measurable-property-values which are put into local linear, or measurable, relation with the domain set, and which is supposed to exist between the function-values and the domain-set, ie the assumed system-containing set of (locally metric-invariant) coordinates of the domain-set" and this all means that the descriptions of stable systems are solvable and controllable.

That is, the intellectual class of this current social-structure, need to be criticized in the strongest possible way, since "what they are doing is both delusional, and quite obsessively stupid" and these are actions which are strongly connected to the most destructive aspects of the society, ie arbitrary identification of certain people being superior to all the other-people.

That is, the intellectual peer-reviewed bullies of academia are much more dangerous, in regard to causing the destruction of the world, than are the very dangerous arbitrarily motivated people-of-action who define the spying and deeply interfering and quite often very violent actions associated to the processes of the justice system and their militias and diplomats and protectors of freedom (ie hired-guns for the rich).

But

In fact the creation of militias and righteous-right (arbitrary) moralists is done by both the propaganda-system and the justice-system so as to divide the ordinary citizens from their opposition to the investor-class.

The "Right" has its personnel chosen carefully also, where again (as are the intellectual-left) they are chosen to be obedient an obsessively loyal to some arbitrary dogma, which is a dogma and must be considered to be an "absolute law" which is used by their handlers to justify the mayhem, eg protecting women and babies, which these agents, spies, managers, and other justice-system personnel are instructed to inflict onto the population,

These are the same terrorists, which form the militias, which, in turn, terrorized the Bunkerville area of NV, in April, May 2014, around Bundy's ranch, likely also the type of justice-system insider-agents which teamed to blow-up the Olympic games in Atlanta,

where these militia are supposedly going to be "investigated" by the justice-system for threatening violence against the justice-system, but (as is already clear from manager-and-insider actions of the institutional managers in the area) these militia personnel compose the justice-system, where such an "investigation" will be more like a friendly-chat, wherein it is negotiated, which of the few patsies in the crew will be the scapegoats (same type of weird secret scenarios, which are composed of a large list of secret players of which L H Oswald was a part)

Who exactly does this secret state work-for? Are these the same militias as the Roman aristocracy had

Are the lower-classes to always expect a set of organized individual attacks emanating out of the justice system, based on counting words or strings of letters, where language is provided with lethal relationships based on the selfish interests of the ruling-class, that is attacks directed by the aristocracy and their failed statistical models (based on indefinable randomness) which they are now applying to pre-emptive statistics to justify aggressive murder "to protect their selfish interests" in a fantasy-land model used to protect the aristocracy?

While

The intellectual-class is also obsessive and narrow, so as to, essentially, be very uninformed (about the nature of physics and math) as well as unquestioning of the absurd dogmatic basis for their intellectual efforts, where these types of efforts, which are designed to incrementally develop the instruments of greatest interest to the investor-class, are the types of efforts (and the way) in which the intellectual efforts of society are completely dominated by the concerns over these absurd intellectual efforts, where these efforts, which are so narrowly dominated, are much more destructive than the people-of-action, and Looney, terrorists on the "Right,"

But in regard to their extreme destructive affects, on both production and creativity within society, making the society so extremely narrow, and failed in its efforts, since the dogmas are total failures,

but it is the, so called, intellectual-class who continue to revere these failed intellectual efforts,

Where both sides of the psychopathic Looney-bin argue endlessly over absurdities, eg science vs. religion etc etc but it is clear that the murderous, Jesse James style, terrorists working for the justice-system, will have secret groups (within their own larger group of murderous-terrorizing agents) who are intent on instilling murderous fear, which is (going to be) focused-on particular individuals, (for which these Righteous-action types are used by the justice-system, ie to instill fear in society)

It is clear that the voice within the propaganda-system for both "the murderous-spying right" and "the intellectual-social class," whom express intellectual high-value for society, is the expression of arbitrary domination, both through violence and (failed) intellectual-dogmas, so that the subsequent social oppression is expressed in the propaganda-system as the high-value which both sets of Looney-bins are in deep belief about, but "in a word" the main message is that society cultivates the psychopaths, who hold the beliefs which the psychopathic (domineering) investor-class wants them to hold (believe)

Part of the oppressive psychology of the so called, "real world" is the oppressive social context defined by the public being wage-slaves, trapped within the narrow confines of the oligarchy, where the actual framework for the oppression and the oppressive acts are done through the justice-system, and where the confinement is realized by both fear (domestic-terrorism) and propaganda (intellectual-domination), where education is a further expression of the propaganda which best serves the oligarchy.

That is, the militias are motivated to "hate the system" so they want to have dominant relation to the social institutions, thus they are the justice system

The intellectuals want to solve problems but first they compete for the high-positions so they work forever on problems which cannot be solved in the context in which they are being formulated

Everyone wants to manipulate their advantage for their own murderous, destructive, but dominant, intent.

36. Diagrams

An essay about the diagrams of simple geometric-shapes

The figure's numbers are referenced in the essay and where-ever they are mentioned there is a ** in close proximity to the identification of the figure referenced.

The main idea of this new descriptive context is that physical descriptions are, in general, about finding either sufficiently general and also sufficiently precise descriptions based on simple math patterns, or it is about developing patterns (of description) which lead to particular practical creativity, or to new interpretations of observed patterns, or to new directions for new perceptions and new experiences.

The set of diagrams represents a symbolic-map which can be used to help identify a set of analogous higher-dimensional diagrams (or an analogous set of higher-dimensional constructs), where in lower-dimensions these diagrams are consistent with the observed spectral-orbital properties of material patterns, though now the ideas of either materialism…., or existence are changed,

So that in the new mathematical descriptive context, the existence of material systems is to now be contained as a subset within a greater set of higher-dimension, so that the higher-dimensional constructs are analogous to the pictured lower-dimensional properties of shapes given in the diagrams, (analogous) in regard to shapes which can now be both macroscopic geometries and microscopic geometries…., so that shapes and dimensions are constructed in new types of arrangements and in a new interpretive context.

The diagrams provide a succinct outline of the simple math, which is based on stable discrete (hyperbolic) shapes, and higher-dimensional constructs, which is the new basis upon which the descriptions of the observed stability of measurable properties…., for physical systems…, depends.

The diagrams provide a clear picture (or clear analogy, or clear map in which to think about moving into the higher-dimensions) of the context within which these stable discrete shapes (of non-positive constant curvature) are organized, so as to form a many-dimensional context (whose higher-dimensional properties should be thought of as, mostly, being macroscopic) in regard to both component containment, and component interaction.

The many-dimensional containment set, possesses a macroscopic and stable geometric context, which is composed primarily of "discrete hyperbolic shapes," which, in turn, are contained in hyperbolic metric-spaces, (wherein it is true that each dimensional level, except the top dimensional level, also has a discrete hyperbolic shape associated to itself).

The finite set of stable discrete hyperbolic shapes, which model both the different dimensional levels, as well as the different subspaces of the same dimension, where the properties of a stable shape being material or being a metric-space containment set alternate as one ascends into higher dimensional-levels, is the geometric foundation upon which the construct of a finite spectral set depends (a finite spectral set for all existence, contained within a high-dimension containing metric-space, ie an 11-dimensional hyperbolic metric-space is the highest dimensional hyperbolic metric-space).

Each dimensional level (and each subspace of any dimensional level) is associated to a very stable "discrete hyperbolic shape," and each subspace (within the many-dimensional set) is characterized by a size-scale (determined in relation to the finite spectral set), where the size-scale of a dimensional level of a particular subspace is also determined by a set of constant multiplicative factors, which are defined both between dimensional levels and between different subspaces of the same dimension.

The fundamental properties of the high-dimension containing space are determined within an 11-dimensional hyperbolic metric-space.

Hyperbolic space is analogous (or isomorphic) to a general model of space-time defined for various dimensions, eg $R(3,1)=$[space-time], while generally, $R(n,1)$, is a "general space-time."

However, there are also various other "metric-function signature" "types of metric-spaces" of the various dimensions and metric-function signatures, $R(s,t)$, which are involved in the description {where s=space dimension, and t=time dimension, where s must be less than or equal to 11 (it seems, according to the properties of discrete hyperbolic shapes identified by Coxeter), and s+t=n}, most notably the "discrete Euclidean shapes," in $R(s,0)$, which possess properties of continuity (of size), where the property of continuity, in regard to the size of a discrete Euclidean shape, is needed in the interaction process which is continuous.

These diagrams identify a context in which "material" components exist within each dimensional level, where they can identify the contexts of both "free" components (associated to both parabolic and hyperbolic second order partial differential equations in regard to inertial properties), as well perfectly-fitting orbital components (associated to elliptic second order partial differential equations in regard to inertial properties).

These diagrams also (pictorially) can show the basis for "material" interactions, and the relation that a new material system, which is emerging from a material interaction, has to being resonant with values of the "finite spectral set" which is defined for the total containment space.

These diagrams provide a context for the emergence of new stable systems from material interactions.

These diagrams of "'material' component interactions" can be identified at any dimensional level (dimension-2 and above).

Note: Interactions are constrained by:

1. the process itself,

2. dimension,

3. size,

4. subspace,

5. the topology of a metric-space, or the topology associated with a metric-space in a particular

dimensional-level,

and

6. a finite spectral set,

where the basic form for such "material" interactions has an analogous structure (or is "the same") for each different (higher) dimensional level, though there are differences, in regard to the properties of material interaction, between the different dimensional levels, which can be due to dimension, subspace, size, and the relation of an interaction to the geometric properties of the containing metric-space's fiber isometry (or unitary) group.

The diagrams give low-dimension pictures of the very simple, quantitatively consistent, geometric shapes which are stable, ie most notably the discrete hyperbolic shapes, and it is these stable shapes upon which stable mathematical patterns can be described in a context where measuring is reliable, and because the description is geometric this means that the description can be very useful.

The stable shapes are the discrete Euclidean shapes, ie tori (or doughnut shapes, with one-hole, or genus-1), and the discrete hyperbolic shapes which are shapes of linear rows of toral-components, whose genus is at least 2, where there may be folds defined between the toral-components.

The number of holes in a discrete hyperbolic shape is called the shape's genus

2-holes are surrounded by (or caught by) 1-curves, and defined by 2-dimensional discrete hyperbolic shapes

3-holes are surrounded (or caught by) 2-surfaces, and defined by 3-dimensional discrete hyperbolic shapes etc

The descriptions of shapes begin with lattices and the fundamental domains of these lattices from which the discrete Euclidean shapes and discrete hyperbolic shapes are formed by a moding-out process of identifying opposite (congruent, or equal) faces.

**This is shown in figures (1) and (2)

The faces on the fundamental domain (which result from the faces of a hyperbolic shape's rectangular (or "cubical") simplex) form very stable spectral measures on the very stable shapes of the hyperbolic space-forms.

Discrete hyperbolic shapes, or equivalently, hyperbolic space-forms, have open-closed metric-space topological properties, and they may be bounded or unbounded shapes, but all existing hyperbolic space-forms which are 6-dimensional or greater are unbounded shapes, and the dimension of the last known hyperbolic space-form is hyperbolic 10-dimensions (Coxeter).

Orbits can be represented on either discrete Euclidean shapes or discrete hyperbolic shapes since a discrete hyperbolic shape is a linear row of toral-components, thus the orbital properties have similar appearances on the two types of shapes

Subsystems (or sub-metric-spaces) within a metric-space either occupy spectral orbits or they are "free."

**Figure (3) shows orbits of what would be condensed material-components in orbits, or free material, while figure (5) shows orbits on folded discrete hyperbolic shapes, where the folding depends on the structure of Weyl-angles where the context of Weyl-angles is provided in figure (4)

**The material interaction is modeled in figure (9).

Toral-components, which fit in the space between two (interacting) 2-dimensional discrete hyperbolic shapes (or material-components), define an interaction differential 2-form on the geometry of a 3-dimensional Euclidean torus, which is tangent to the interacting 2-dimensional material-components, and which is contained in Euclidean 4-space. This determines the force-field, defined between the interacting material components.

The local vector geometry of the differential 2-form is relatable to the local geometry of the fiber SO(4) Lie group, since the 2-forms in Euclidean 4-space have the same dimension as a local (or tangent) SO(4) matrix-element which is diagonal, thus the geometry of the spatial displacement is determined in SO(4) by it geometric relation to Euclidean 4-space which is given by the 2-forms, so that a local spatial displacement occurs due to a local coordinate transformation which is determined by SO(4) acting on the positions of the vertices [(in relation to the distance, r, defined between the two center-of-mass coordinates of the two material-components) of the original pair of interacting hyperbolic space-forms, which can be averaged so as to appear to be simply a pair of toral-components] in Euclidean 4-space.

If the force is attractive and if the interacting (charged) material (the interaction structure) is contained within either 2-dimensional or 3-dimensional, or 4-dimensional Euclidean space then the force is radial and attractive, and r is made smaller, with each discrete time-interval (see next paragraph). {Note: If the material is of a new type (oscillatory) and contained within Euclidean 4-space then the force-field has a new geometric structure contained in a higher-dimensional Euclidean space.}

Then the same type of process repeats, for time intervals determined by the spin-rotation period of the spin-rotations of opposite metric-space states (about 10^{-18} sec).

In this process the Euclidean torus which forms for each discrete time interval, forms in the context of action-at-a-distance.

Classical partial differential equations are defined within very confining and very rigid sets of both discrete hyperbolic shapes and action-at-a-distance material interaction Euclidean toral components which link the hyperbolic material together, so that the force-field differential 2-form is defined on the torus.

The above interaction for material contained in 3-space results in a 4-dimensional descriptive context, but it can be symbolically represented in 3-space.

The pde has not been able to identify the way in which material-interactions..., which are described by pde..., and which are composed of many-components form into stable systems.

That is, "the pde based description" has failed to be able to describe stable many-(but-few)-body physical systems, but the new description provides a new context in which to understand this most fundamental property of physical systems and how many-(but-few)-body systems can be stable. That is, the new context transcends the descriptive context based on materialism and pde's.

**Forming new stable hyperbolic space-forms from a material interaction. (see figure (10))

Assume an attractive (or repulsive) interaction in 3-space, then the interaction of "free" material components would be similar to a collision of components,

if the material components get very close during the (collision) interaction then, if both the energy of the over-all interaction is within (certain) energy ranges and the closeness allows geometric-based resonances (with the spectral set of the over-all high-dimension containing metric-space) to come into being. Thus, forming a new state of resonance for the interaction geometric-simplex, so that the over-all energy, as well as the resonances, allow a "new" stable "discrete hyperbolic shape" (in the proper dimension of the interaction) to form, so a new hyperbolic space-form rather suddenly emerges.

**There are the properties of orbital structures and angular-momentum which can be understood by using the discrete set of the fiber Lie groups Weyl-angles which, in turn, are related to conjugation classes of the group which, in turn, is related to transforming between the Lie group's set of maximal tori. (see figure (5))

Weyl-transformations between two maximal tori (where a maximal tori has dimension-k) within a Lie group (rank-k compact Lie group) Two intersecting circles of the two maximal tori may be angularly related to one another by Weyl group transformations, where the Weyl group defines the conjugation classes of the maximal tori which "cover" the compact Lie group.

Forming angular changes between toral components of a discrete hyperbolic shape by using Weyl-transformations, which change the angular relations between circles which compose a toral component of a hyperbolic space-form. These Weyl-transformations allow "envelopes of orbital stability".....

for both "free" subsystems (or sub-metric-spaces) which form into condensed-material which follows a stable orbit, in regard to the metric-space shape's geodesic properties, as well as for material-components which can fit naturally into the faces of the orbital-shape's fundamental-domain

..., to be defined.

But rectangles attached at vertices of, say, the integers contained within each coordinate of a rectangular coordinate set, and then moded-out, "without expanding the vertex," shows a model of a discrete hyperbolic shape's toral components, represented as separate tori attached at separate vertices. (This is to emphasize an apparent toral component structure of discrete hyperbolic shapes, which is an important aspect of these discrete shapes, especially, in regard to the interaction toral-shape, which can affect the interacting shape's spectral-geometric properties, so as to, in turn, affect a new shape's ability (or possibility) to resonate.)

** Various types of unbounded 2-dimensional discrete hyperbolic shapes depicting either light, or neutrinos, or neutrinos and electrons, ie an electron-cloud, etc. This is shown in figure (6)

Picturing, the shapes which are used in the partitioning the over-all 11-dimensional hyperbolic metric-space

** Figure (8), titled "Partitioning a many-dimensional containment space" represents two different dimensional levels, where the 2-dimensional level is identified as an un-deformed rectangular lattice, where a deformed lattice shape (in 2-hyperbolic-dimensions) is given in ** figure 4. Whereas in this 2-dimensional figure there are contained representations of 1-dimensional shapes. The other (larger) representation of the 3-dimensional partitioning structure are the un-deformed "cubical," or right-rectangular, 3-dimensional shapes, which contain within itself the stable 2-dimensional discrete hyperbolic shapes, this process of partitioning space can continue up into higher-dimensions, where a deformed 3-dimensional lattice shape can be moded-out to form into a geometric-shape, which would exist in 4-dimensional hyperbolic metric-space.

**Orbital perturbations (are shown in figure (7)), and the subsequent demotion of the importance of a (demoted) pde in physical descriptions

The figure titled "Perturbing material-components on stable shapes:" shows an atomic orbital structure which is a stable geometric structure with electrons in the outer-orbits of concentric toral-components (folded into their stable shape) and the nucleus in the center small orbital shape, where the electron's orbit is mostly held stable by it (the electron) following the geodesic path, which is defined on its toral component, and the electron is also interacting as if in a 2-body interaction with the nucleus, so that this 2-body interaction (most noticeably) perturbs the orbit of the electron, eg perhaps causing the electron to possess an elliptical path, where these orbital deformation may result in the variations in the details of the atom's discrete energy structure, where the stable orbital shapes, perhaps related to various Weyl-angle shapes, are the basis for the atom's stable discrete energy structure.

Chart of the face structure for rectangular simplex geometry

2-rectangular-simplex

1-face

vertices

3-rectangular-simplex

2-faces

1-face

Vertices

Etc

Diagrams brief descriptions:

1 (Figure 1) The lattice structures and fundamental-domains of Euclidean space

2 (Figure 2) The lattice structures and fundamental-domains of a hyperbolic metric-space

3 (Figure 3) orbital properties and containment properties of lower-dimensional shapes within a higher-dimensional shape for discrete hyperbolic shapes

4 (Figure 4) The definition of Weyl-angles on maximal tori in Lie groups

5 (Figure 5) Folding, by Weyl-angles, a discrete hyperbolic shape into a stable orbital structure

6 (Figure 6) Models of unbounded and semi-bounded discrete hyperbolic shapes which mostly model neutrinos but they also model electron-clouds and could be of great significance in modeling a nucleus

7 (Figure 7) Perturbing a stable orbit, which is initially defined by a geodesic path on the orbital-shape, by using a model of a material-interaction which, in turn, defines an elliptic 2nd-order pde

8 (Figure 8) A model for partitioning different dimensional levels by showing fundamental domains of two different dimensions and the (dimensional) types of components which they can contain

9 (Figure 9) The basic two-body interaction, and its average approximation, ie replacing discrete hyperbolic shapes with two discrete Euclidean shapes whose shape is an average shape of the two (or 4) energy-levels of each of the two interacting discrete hyperbolic shapes.

10 (Figure 10) A model of an interaction, its average and a model of an interaction which changes to a new system due to it coming to be in resonance with the finite spectral-set associated to the 11-dimensional hyperbolic metric-space containment space.

Actual Diagrams

Cubical (rectangular) simplexes are related to circle-spaces by means of "equiva-
lence-relation topologies," or equivalently, by a "molding-out" process
On such shapes local geometric measures are based on either measuring rectangular
shapes or (equivalently) by a measuring process based on tangents to the circle, which
is used as a basis for measuring along a circle's curve, eg rdW=dx+dy along a circle.

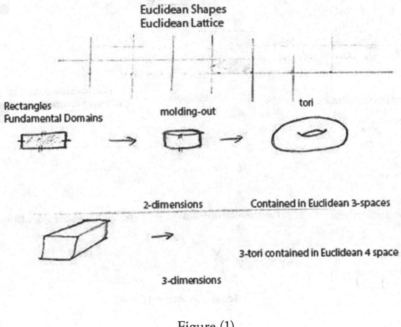

Euclidean Shapes
Euclidean Lattice

Rectangles
Fundamental Domains

molding-out

tori

2-dimensions   Contained in Euclidean 3-spaces

3-tori contained in Euclidean 4 space

3-dimensions

Figure (1)

Hyperbolic shapes
Lattice (in hyperbolic 2-space)

Fundamental domains

molding-out

discrete hyperbolic shapes
(contained in hyperbolic 3-spaces)

Vertices

vertex

Rectangular Simplexes

Fundamental domains

discrete hyperbolic space forms

Edges of Fundamental domain

The vertex is pulled apart

orthogonal parts
of hyperbolae

Figure (2)

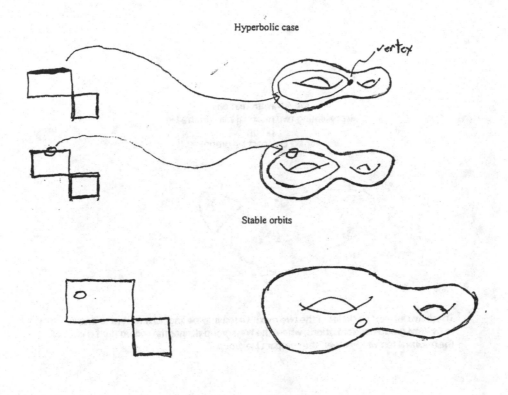

Hyperbolic case

vertex

Stable orbits

"free" subspaces (or "free" subsystrems)

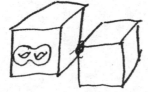

Figure (3)

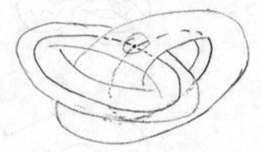

Weyl-trasnfromations
Representing two maximal tori within a Lie
group
(rank-2 compact Lie group)

These two interesting circles of the two maximal tori may be angularly related to one another by Weyl group transformations, where the Weyl group defines the conjugation classes of the maximal tori which "cover" the compact Lie group

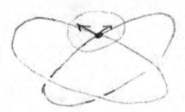

Figure (4)

Envelopes of orbital stability for "free" subsystems (or sub-metric-spaces)

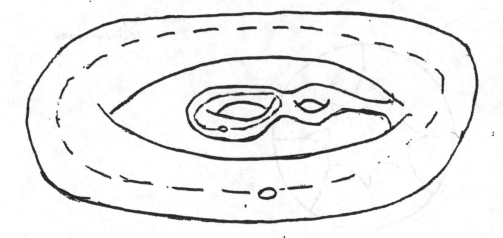

Figure (5)

The folded-shape stable orbital structure for many-(but-few)-body spectral-orbital properties

Various types of unbounded 2-dimensional discrete hyperbolic shapes

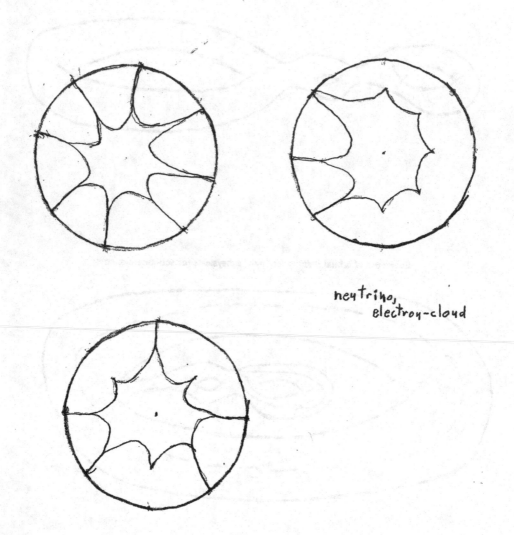

neutrino,
electron-cloud

Figure (6)

Electron-cloud and the shape of either a neutrino or an electromagnetic wave

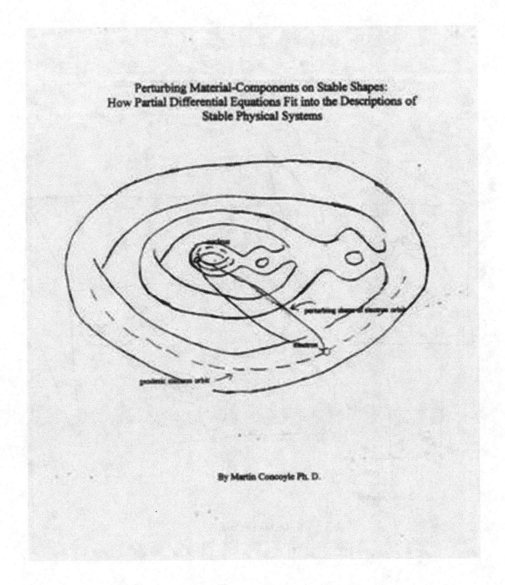

Figure (7)

Perturbing by an interaction construct the geodesic orbits which exist on stable metric-space shapes

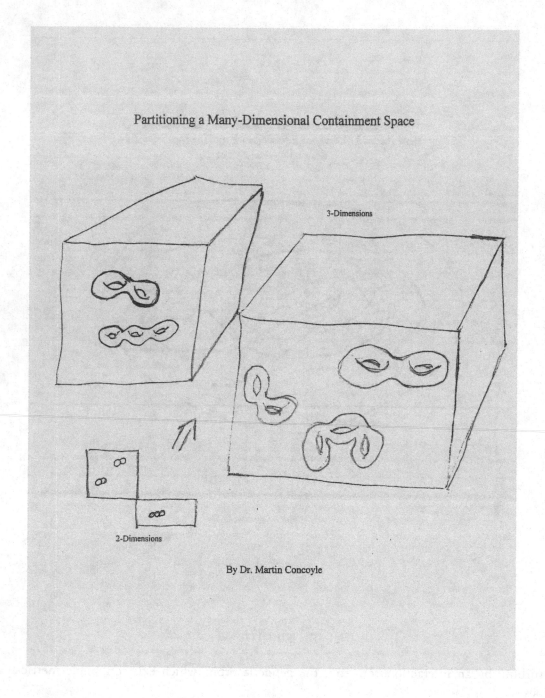

Partitioning a Many-Dimensional Containment Space

3-Dimensions

2-Dimensions

By Dr. Martin Concoyle

Figure (8)

Partitioning space

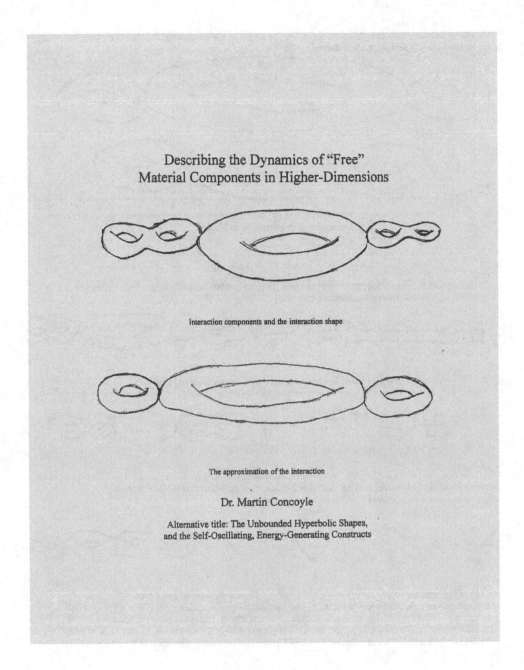

Describing the Dynamics of "Free"
Material Components in Higher-Dimensions

Interaction components and the interaction shape

The approximation of the interaction

Dr. Martin Concoyle

Alternative title: The Unbounded Hyperbolic Shapes,
and the Self-Oscillating, Energy-Generating Constructs

Figure (9)

The shape of interactions between material-components and its average, or approximate, construct

The above interaction for material contained in 3-space results in a 4-dimensional descriptive context, but it can be symbolically represented in 3-space

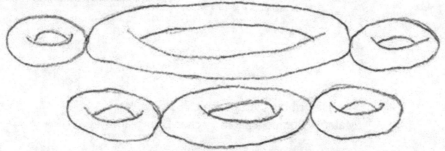

For an attractive interaction in 3-space

It should also be noted that in this new descriptive context eigenfunctions would also be both discrete hyperbolic shapes and discrete Euclidean shapes (tori)

Forming new stable hyperbolic space-forms from a material interaction

Assume an attractive (or repulsive) interaction in 3-space, then the interaction of "free" material components would be similar to a collision of components,

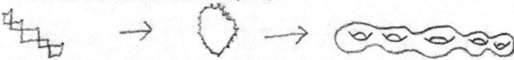

State of resonance   new hyperbolic space-form

if the material components get very close during the (collision) interaction then if the energy of the over-all interaction is within (certain) energy ranges and the closeness allows resonances (with the spectral set of the over-all high-dimension containing metric-space) to begin to form, so that the over-all energy, as well as the resonances, allow a "new" stable "discrete hyperbolic shape" (in the proper dimension of the interaction) to form.

Figure (10)

(lower figures) indicate an interaction where a new system forms due to resonance

References

1. A New Copernican Revolution, Bill G P H Bash and George P Coatimundi, Trafford Publishing, 2004. www.trafford.com/03-1913,

2. The Authority of Material vs. The Spirit, Douglas D Hunter, Trafford Publishing,2006.

www.trafford.com/05-3038

3. Topology and Geometry for Physicists, C. Nash and S. Sen, Academic Press, 1983.

4. The Electromagnetic Field, N Anderson, Plenum Press, 1968.

5. Function Theory, C. L. Siegel,

6. Three-Dimensional Geometry and Topology, W. Thurston, Princeton University, 1997.

7. Gauge Theory and Variational Principles, D Bleeker, Addison, 1981.

8. Geometry II, E B Vinberg, Springer, 1993.

9. Spaces of Constant Curvature, J Wolf, Publish or Perish, 1977.

10. Contemporary College Physics, Jones and Childers, Addison-Wesley, 1993 (High School text).

11. I M Benn and R W Tucker, in, An Introduction to Spinors and Geometry with Applications in Physics, 1987, (Chapter 2)

12. Representations of Compact Lie Groups, T Brocker, T tomDieck, Springer-Verlag, 1985.

13. Dynamical Theories of Brownian Motion, E. Nelson, Princeton University Press, 1967.

14. Quantum Fluctuations, E. Nelson, Princeton University Press, 1985.

15. The Theory and Application of Harmonic Integrals, Cambridge Univ. Press, W V D Hodge, 1941.

16. Electron magnetic moment from gonium spectra, H Dehmelt (Nobel prize winner) et al, Physical

Review D, Vol 34, No. 3, Aug 1, 1986.

17. Newton's Clock, Chaos in the Solar System, I Peterson, W H Freeman and Company, 1993.

18. Quantum Mechanics, J L Powell and B Crasemann, Addison-Wesley Publishing, 1965.

19. The End of Science, J Horgan, Broadway Books, 1996.

20. Riemannian Geometry, L. P. Eisenhart, Princeton University Press, 1925.

21. Reflection Groups and Coxeter Groups, J Humphreys, Cambridge University Press, 1990.

22. Partial Differential Equations, J Rauch, Springer-Verlag, 1991.

23. The Foundation of the General Theory of Relativity, A Einstein, 1916, Annalen der Physik (49).

D Coxeter

Just as Copernicus, Kepler and Galileo provided a quantitative-geometric context for the properties of the solar system, which were then precisely identified by the solutions to (the) solvable differential equations of Newton; Martin Concoyle now provides the stable geometric structures which fit…, both macroscopically and microscopically…, into a many-dimension containment set (hyperbolic 11-dimensional), so that these shapes are the solutions, ie the geometries of the stable spectral-orbital properties, of all the fundamental stable systems which have stable spectra and orbits, and it is the basis for a quantitative system (the spectral set of a measurable existence) which is finite,

and

These ideas are discussed in the following books: (available at math conference, 2013)

1. A New Copernican Revolution (p286), B Bash & P Coatimundi, Trafford, 2004.

2. The Authority of Material vs. The Spirit (p483), D D Hunter, Trafford, 2006.

As well as in the following (new) books from Trafford:

1. The Mathematical Structure of Stable Physical Systems, Martin Concoyle and G. P. Coatimundi, 2013, (p449), Trafford Publishing

2. Partitioning a Many-Dimensional Containment Space, Martin Concoyle, 2013, (p477), Trafford Publishing,

3. Perturbing Material-Components on Stable Shapes:

How Partial Differential Equations Fit into the Descriptions of Stable Physical Systems, Martin Concoyle Ph. D., 2013, (p234), Trafford publishing (Canada)

4. Describing the Dynamics of "Free" Material Components in Higher-Dimensions, Martin Concoyle, 2013, (p478), Trafford Publishing

see: blog-of-martin-concoyle.weebly.com

Copyrights

These new ideas put existence into a new context, a context for both manipulating and adjusting material properties in new ways, but also a context in which life and creativity (practical creativity, ie intentionally adjusting the properties of existence) are not confined to the traditional context of "material existence," and material manipulations, where materialism has traditionally defined the containment of material-existence in either 3-space or within space-time.

Thus, since copyrights are supposed to give the author of the ideas the rights over the relation of the new ideas to creativity [whereas copyrights have traditionally been about the relation that the owners of society have to the new ideas of others, and the culture itself, namely, the right of the owners to steal these ideas for themselves, often by payment to the "wage-slave authors," so as to gain selfish advantages from the new ideas, for they themselves, the owners, in a society where the economics (flow of money, and the definition of social value) serves the power which the owners of society, unjustly, possess within society].

Thus the relation of these new ideas to creativity is (are) as follows:

These ideas cannot be used to make things (material or otherwise) which destroy or harm the earth or other lives.

These new ideas cannot be used to make things for a person's selfish advantage, ie only a 1% or 2% profit in relation to costs and sales (revenues).

These new ideas can only be used to create helpful, non-destructive things, for both the earth and society, eg resources cannot be exploited to make material things whose creation depends on the use of these new ideas, and the things which are made, based on these new ideas, must be done in a social context of selflessness, wherein people are equal creators, and the condition of either wage-slavery, or oppressive intellectual authority, does not exist, but their creations cannot be used in destructive, or selfish, ways.

Last note

This book is an introduction to the simple math patterns used to describe fundamental, stable spectral-orbital physical systems (represented as discrete hyperbolic shapes, ie hyperbolic space-forms), the containment set has many-dimensions, and these dimensions possess macroscopic geometric properties (which are also discrete hyperbolic shapes). Thus, it is a description which transcends the idea of materialism (ie it is higher-dimensional, so that the shapes in the higher-dimensions are not small), and it can also be used to model a life-form as a unified, high-dimension, geometric construct, which generates its own energy, and which has a natural structure for memory, where this construct is made in relation to the main property of the description being, in fact, the spectral-orbital properties of both (1) material systems, and of (2) the metric-spaces, which contain the material systems, as condensed material, where material is simply a lower dimension metric-space, and where both material-components and metric-spaces are in resonance with (or define) the containing space. Partial differential equations (or pde) are defined on both (1) the many metric-spaces of this description and (2) the lower-dimensional material-components which these metric-spaces contain, ie the laws of physics, but the main function of a pde is to act on either the, usually, unimportant free-material components (so as to most often cause non-linear dynamics) or to perturb the orbits of the, quite often condensed, material which has been trapped by (or within) the stable orbits of a very stable hyperbolic metric-space shape.

It could be said that these new ideas about math's new descriptive context are so simple, that some of the main ideas presented in this book may be presented by the handful of diagrams which show these simple shapes, where these diagrams indicate how these simple shapes are formed and folded, or bent, to form the stable shapes, which can carry the stable spectral-orbital properties of the many-(but-few)-body systems…., where these most fundamental-stable-systems now (2014) have no valid quantitative descriptions within the, so called, currently-accepted "laws of physics," (ie the special set of partial differential equations associated to the, so called, physical laws)…. so that the diagrams of these stable geometric shapes are provided at the end of the book.

This new measurable descriptive context is many-dimensional, and thus, it transcends the idea of materialism, but within this new context the 3-dimensional (or 4-dimensional space-time) material-world is a proper subset (in a subspace which has 3-spatial-dimensions),

The, apparent, property of fundamental randomness (in a currently, assumed, absolutely-reducible model of material, and its reducible material-components) is a derived property, but now in a new context in which stable geometric patterns are fundamental,

The property of spherically-symmetric material-interactions is shown to be a special property of material-interactions, which exists (primarily, or only) in 3-spatial-dimensions, of Euclidean space, wherein inertial-properties are to, most naturally, be described,

It is both reductive….,

(to some sets of small material-components, but elementary-particles are most likely about components colliding with higher-dimensional lattice-structures, which are a part of the true geometric context of physical description)

…, and unifying

in its discrete descriptive contexts (relationships) which exist, between both a system's components, and the system's (various) dimensional-levels (where these dimensional-levels are particularly relevant, in regard to understanding both (1) the chemistry and (2) the functional organization of living systems),

But most importantly, this new descriptive language (new context) describes the widely observed properties of stable-physical-systems, which are composed of various dimensional-levels and of various types of components and interaction-constructs, so that this new context provides an explanation about both (1) "how these systems form" and (2) "how they remain stable," wherein, partial differential equations, which model material-interactions, are given a new: context, containment-structure, organization-context, interpretation, and with a new discrete character,

It provides a (relatively easy to follow, in that, the containment set-structure for these different-dimensional stable-geometries are simple dimensional relations) 'map' "up into a higher-dimensional context (or containment set) for existence," wherein some surprising new properties of existence can be modeled, in relation to our own living systems also being modeled as higher-dimensional constructs, and this map can shed-light onto our own higher-dimensional structure, and its relation to both existence, and to the types of experiences into which we may enter (or possess as memory) (or within which we might function), where because any idea about higher-dimensions is difficult to consider, and is relatively easy to hide and ignore these higher-dimensions, especially, if we insist on the idea of materialism.

It turns-out that the problem with materialism is about the issue of quantitative sets, namely, defining quantitative consistency within a precise descriptive context.

Can quantitative sets be formed in an abstract manner, so as to be infinitely dense, and so as to be always relatable to models of measuring (physical) systems, where the models of measuring are mostly based-on models of local measures, which exist within quantitative sets in which "measurable properties (or values) possess an infinite density, ie it is assumed that quantities exist within a continuum," which, it is believed, is needed to be able to model a physical system's set of locally measurable properties, where the properties are assumed to be contained in the abstract context defined by both quantitative-sets, and the idea of materialism?

The short answer is, No!

A physical-and-mathematical structure for "measuring,".... and an associated sets of assumptions about set-containment, in regard to what is measured (what types of measured quantities and the place (space) where the measured quantity exists)...., needs to be built from (and based upon) stable properties.

How to build a quantitative structure which is quantitatively consistent, and which can be related to stable patterns, where it is a set of very stable measured patterns, which are observed for the stable many-(but-few)-body spectral-orbital systems which are so fundamental to our experience?

The five main ideas upon which this (build a quantitatively consistent descriptive construct) can be done

1. Geometrization identifies the set of stable patterns, which, in turn, are associated to the very simple shapes of: lines, rays, line-segments, and circles, which when put into a context so as to identify "continuously" orthogonal sets of shape structures: cubes, cylinders, and "orthogonal" circle-space shapes, so that the stable shapes (associated to geometrization) are: linear, metric-invariant, continuously commutative, and the local property of parallel transport of the natural coordinate directions around (local) closed sets of coordinate-curves is commutative, where a shape's "natural coordinates" are the set of local coordinates which make the metric-function continuously commutative, over the globally defined shape (except possibly at a single-point on the shape).

2. E Noether's symmetries, namely, two main symmetries, (1) invariance of spatial displacements (for a physical system), which implies conservation of momentum, ie this is the context or inertial descriptions and is associated to Euclidean space, and (2) invariance of temporal displacements, which implies conservation of a physical system's energy, ie this is the context of energy, and is

associated with space-time of dimension-(n+1), or equivalently associated with an n-dimensional hyperbolic metric-space. That is, E Noether's symmetries define both a metric-space-type, and the physical property which is associated to that metric-space type. Thus, the discrete Euclidean shapes, ie tori, are continuous structures of shape, and are associated to the descriptions of inertial properties of measuring, while the very stable discrete hyperbolic shapes are spaces of energy, or when the hyperbolic metric-space is associated to a stable discrete hyperbolic shape then these are the very stable energy-shapes.

3. D Coxeter showed that the set of discrete hyperbolic shapes are many-dimensional, up to and including 10-dimensional stable discrete hyperbolic shapes, and that these shapes are discretely separated from one another. Furthermore, these shapes possess holes, ie they are partly identified by their genus (or number of holes in the shape), and each hole is associated to two (usually) distinct geodesic lengths, which define the shape's finite set of spectral properties. [or each hole is associated to a number... equal to "the dimension of the shape"... of (usually) distinct "minimal geometric-measures for the shape's confining faces, in regard to the shape's fundamental domain," which, in turn, define the shape's finite set of (dimensionally dependent) spectral properties.]

4. Thus, in order to make a quantitative set with which one can measure either a stable system's properties, or the inertial properties of a stable system, then one must relate the quantitative set to both (1) the set of stable discrete hyperbolic shapes, which defines the basic set of stable spectral-orbital geometries, which can exist for material systems, and (2) model the continuous nature of the set of discrete Euclidean shapes, so as to be related to a discrete time structure, upon which changes in (inertial) motion will depend, in regard to a component's spatial-displacement properties. [the technical necessity of (2) allows the new descriptive context to be consistent with many of the current properties]
5. So in order to construct a quantitative set based on stable patterns one must "partition" an 11-dimensional hyperbolic metric-space by means of a set of different dimension stable discrete hyperbolic shapes, for the different subspaces of the different dimensional levels, so that each subspace, up to and including hyperbolic-dimension-5, has a maximal discrete hyperbolic shape, where a hyperbolic metric-space already possesses a minimal discrete hyperbolic shape related to a minimal spectral-value (or minimal circumference-length for each hole) of ¼. (where, according to Coxeter, all discrete hyperbolic shapes between dimensions 6 and 10, inclusive, are unbounded shapes). Such a "partition" will identify a set of set-containment trees, depending an both dimension and size, of the shapes, (as well as the subspace structure) which compose the "partition," and through which set-containment is to be defined.

This "partition" of the 11-dimensional hyperbolic metric-space, by a finite set of stable shapes, where this set is defined for each subspace and for each dimensional level, in turn, defines a finite set of discrete hyperbolic shapes, which, in turn, identifies, a finite spectral-set, in relation to both shape-containment and genus of this fixed finite set of shapes.

Note that,

(1) The smaller shapes can exist within larger shapes only if the (smaller) shapes are in resonance with this newly defined finite spectral-set.

(2) Where different "partitions" can allow for different finite spectral-sets, and thus different ways in which to organize existence. But these different "11-dimensional-partition" defined spectral-sets would represent different ways in which to organize the structures of existence. But possessing knowledge of these finite spectral-sets would allow transitions between these spectral-sets by means of resonances, during a discrete transition process.

The quantitative structure needed for measuring a set of physical properties…., in regard to the energy-space, ie the space which defines a stable existence….., is determined by multiplying each value of the finite spectral-set by the integers, and then adding-up all these quantities to form (or to finitely generate) a quantitative set. Where this is a (the) quantitative set upon which all measuring of physically existent properties can be defined, where physical existence is based on an underlying set of stable physical-system shapes, which are used to define the finitely generated quantitative set.

Any operator structures based on (local) measuring which is applied to a context of existence built from this set of stable patterns will be subordinate to this finitely generated quantitative structure which is built upon a finite set of stable patterns.

Changes of spatial position, or more often second-order changes in spatial-position, caused by the spatial geometry of distant material, so as to act on mass (inertia) located at a position in space, ie inertia associated to a material-component's energy-shape and its spatial-position, apparently, are equivalent to the material-component's relation to its always seeking an extrema of (for) its context of action,

ie (action = {energy x time}),

or

what can be thought of as a material-component's relation to non-local toral-components, which define the geometry of the local tangential and radial causes (of) by a force-field (derived from distant material-geometry) acting on an energy-shape.

Without building a quantitative structure upon a stable context (directly related to the description, itself) then one cannot identify stable patterns, within such a quantitative context (structure), where the descriptive construct is both quantitatively consistent, and the quantitatively described patterns are relatable, by measuring, to stable observable patterns which possess stable measurable properties.

In the old viewpoint,

where, the operator structure is imposed on a physical system's (measurable) properties

(ie the solution functions to partial differential equations (or pde), or sets of pde)

and the system's metric-space containment-set (of "natural" metric-space coordinates), where the function-values and the coordinate quantities of a quantitative description are assumed to be quantitative-sets which are built on abstract, arbitrary, indefinably-infinite context (used to describe number-values) which are supposed to represent a continuum of quantitative values, within which operators, or (other) measurable models of a system's quantitative properties, are defined, and act on the system's measurable properties (or solution functions to the system-defining pde's), and which are supposed to relate a system's locally measurable properties to definitive quantitative relations. In this old (current, 2014) construct it is assumed that the quantitative-sets are supposed to exist in a quantitative-containment context, in which it is assumed that all details (of material positions and motions) are measurable, and subsequently describable, so that there always exist sets of operators which define (the associated system's properties) a solution function for (or which identifies) the system's properties, which are then, assumed to be, contained within quantitative containment-sets.

But

When operators are: non-linear, non-commutative, and the descriptive context is indefinably random, then one gets vague descriptions of unstable patterns, which exist within a containment set of quantitative-sets, where the quantitative structures…, ie either the coordinate-systems or the system's properties (or solution function-values)…, are quantitatively inconsistent, and thus, the quantitative-sets are incapable of identifying patterns which are (should be) stable and measurable.

The sets of operators (derived from the, so called, physical laws), applied to either coordinates or to function-spaces, are supposed to identify the "real" context for a physical system's set of identifying measurable properties, but if (when) the operators are either non-linear, or non-commutative, and/or placed in an indefinably random context, then the quantitative sets, ie the coordinates and sets of function-values, are forced to form (or to exist) within quantitatively inconsistent containing-sets, and thus, these quantitatively inconsistent containment-sets cannot carry within their quantitatively inconsistent structures either any stable patterns or any reliably measurable set of properties.

It (the old construct) is a meaningless complicated description, which is better identified as jibber-jabber, and it is a, supposedly, precise description, which is without any meaningful content.

The context in which the, so called, top members of the intellectual-class express ideas about elementary-particles, quantum computing, many-worlds, (gravitational) worm-holes, dark matter etc, etc, etc, all of these patterns are described in an operator context which is either non-linear, or non-commutative, and/or placed in an indefinably random context, and therefore, the quantitative sets, ie the coordinates and sets of function-values, are forced to form (or to exist) within quantitatively inconsistent containing-sets. That is, none of these descriptive contexts can relate patterns (which a describer might imagine, based on their operator structure) to a measurable context, through which there can be

either

stable patterns for systems,

or

control over the patterns of a system.

Thus, these descriptive contexts are all completely absurd, where this is because it is a descriptive construct which are only about quantitative contexts which are all built from operators which are: either non-linear, or non-commutative, and/or indefinably random, and the containment quantitative-sets (function-values and system containment-coordinates) are all quantitatively inconsistent.

That is, it is all a meaningless and absurd descriptive context, in which there are no stable patterns which can exist,

other than irrelevant patterns associated with what might in the imaginations of the people considering the improperly defined operator based descriptive context, which cannot be used to establish actual measurable patterns.

The describers try to impose, from outside, a descriptive context, based on improperly defined sets of operators, (based on improperly defined physical law) which are operator structures, which in the imaginings of the describer, represent a reality, but instead, it is a set of delusions,

which they then try to put, by means of sets of improperly defined operators, within a pattern-less, quantitatively inconsistent context ie they try to impose an imaginary pattern projected onto an operator based descriptive context which is an operator context which cannot be used to establish actual measurable patterns.

This has similar properties to feedback systems, which need feedback since they exist in a non-linear, non-commutative and indefinably random descriptive context, in regard to system models based on improperly defined operators, where in Feedback systems the internal system makes adjustments based on feedback from measurements either made by the component or made by another external system, eg global positioning satellites, where the feedback then signals to the component, where the component's internal properties are then adjusted, eg angular orientation of a set of the component's propellers, where the component is placed into an external pattern, in which internal adjustments are made to adapt the component to the external conditions, and these adjustments must be continually applied.

But,

In the non-feedback descriptive contexts, there does not exist any internal connection to an external context, where an external context can identify a context within which the component can be adjusted, and this is because the descriptive context, based on quantitative inconsistency, does not possess any stable patterns, ie in the feedback system the component, itself, identifies a stable pattern.

However, there is so much mental inertia within the effectively (effective for banker's interests) arrogant absolute dogmatists, who form a set of individual intellectuals, within the intellectual-class, or for any person who has been a product of the propaganda-education system's indoctrination programs (or institutions), where these indoctrinated minds (individual or in the intellectual-class), where the, so called, best-intellects are the highest-paid, ie the real authorities, are considered to be so very well versed in the narrow categories of interest, which have been defined for the intellectual-class by to the bankers, and the well versed intellectual..., in the authoritarian and narrow dogmas associated to these relatively few categories..., supports these narrow dogmas (defined by the bankers) in an authoritarian manner, which, in effect, expresses an absolute truth for society and its intellectual institutions
Eg for bankers physics has come to mean bomb engineering, so all well-paid physicists, basically only know about particle-collisions in a context of chaos.
Furthermore,
The institutions are so heavily invested in the few categories of investment interests, eg slowly developing a few types of instruments, and subsequently, defining narrow (sets of) categorical dogmas, where memorizing these dogmas bequeaths authority to those, so called, best-versed at these institutions, into which the intellectual expressions of the, so called, best-intellects are channeled, to support the knowledge and its related creative projects associated to the investment interests of the bankers.

That this set of criticisms (eg as expressed in the book) this set of correctly critical expressions, of both math and physics, as well as the expression of ideas which correctly adjust and correctly resolve the problems in math and physics, about which the intellectual-class is unable to resolve (or solve), where this is about the inability of the intellectual-class to describe the stable patterns, which are measured for the vast majority of the most fundamental systems which compose our existence
Where this inability to solve problems of physical description is a result of the current (2014) experts to stay within their descriptive context of non-linearity, non-commutativity, and contexts which are indefinably random (eg the elementary-event space is composed of unstable events), so that their quantitative structures are all quantitatively inconsistent.

Nonetheless, ie despite their failures, these experts stay within their dogma, so as to proclaim that the simplest of the stable many-(but-few)-body spectral-orbital systems are too complicated to describe, eg the nucleus the solar-system are stable, but there do not exist any valid explanations (based on the, so called, laws of physics) for this set of stable systems etc, etc, etc

The institutionally arrogant and dogmatic, and, disturbingly, authoritarian personnel of the intellectual-class (intellectuals whom uphold dogmatic beliefs which are as arbitrary as other arbitrary beliefs, such as a belief in racism) have beliefs, and they express ideas and beliefs, which are consistent with the narrowly defined knowledge and creative interests of the investor-class and their highly important creative and knowledgeable institutions

These criticisms, which are both highly critical and very correct expressions, cannot be ideas which the intellectual-class gatekeepers, for the public (and the public's conceptions of truth), can accept, because it is in conflict with the dogmas upon which their high-social standing depends. Thus, the, so called, high-level intellectual cannot read (these new ideas), internalize, and understand the ideas expressed in the book, which are contrary to the dogmas of the intellectual-class, the only form of understanding which is applicable to these so called experts is their realization that these ideas do not fit into their authoritative dogmas, so these new ideas have no value to them, since they are already within the intellectual-class,

Nonetheless, these criticisms (as given in the book) can be seen as being true (or valid), so that the needed corrections to precise description can be adjusted (as these adjustments are provided in this book), and, subsequently, a more correct model of existence considered,
in regard to both experience
    (it is a map into both a new viewpoint [about existence] which is much more interesting than the map of the European-Empire's Columbus)
and
    in regard to a new context for practical creativity.